For a complete listing of titles in the
Antennas and Electromagnetics Analysis Library,
turn to the back of this book.

New Foundations for Applied Electromagnetics

The Spatial Structure of Fields

Said Mikki

Yahia Antar

ARTECH HOUSE

BOSTON | LONDON
artechhouse.com

Library of Congress Cataloging-in-Publication Data
A catalog record for this book is available from the U.S. Library of Congress.

British Library Cataloguing in Publication Data
A catalogue record for this book is available from the British Library.

Cover design by John Gomes

ISBN 13: 978-1-60807-919-3

685 Canton Street
Norwood, MA 02062

10 9 8 7 6 5 4 3 2 1

Contents

Chapter 1

General Outline and Scope of the Book

1.1 MOTIVATION

Facing the bewildering abundance of textbooks and monographs on electromagnetic theory and applications, the reader may wonder why another large book on the subject? In our view, which will be substantiated throughout the progressive devolvement of the work presented here, the field of electromagnetic research has entered during the last few years a turning point. This critical juncture is related to two major changes in our current viewpoint: fundamental pressure emerging internally from within the very discourse of electromagnetics as such, and the enormous demands imposed on us by the industrial and technological evolution of the contemporary world. Those two stimuli may appear for the time being unrelated to each other. Indeed, philosophers and theorists busy themselves with investigations motivated by purely epistemological and sometimes aesthetic considerations, while applied scientists and most engineers are devoted to working on themes close to technology and applications. The separation between these two major thematic orientations has been the predominant mode of "doing science" since at least the end of the nineteenth century. Such bifurcation into pure theory and applications is, in our view, not a very happy chapter of the historical unfolding of scientific thought. In fact, it appears that the present crises of modern science can be traced back to the tension that such lack of communication between theorists and applied scientists has been fostering since the end of World War II and the rise of electrical and electronic engineering as the major force behind technological advance.

Let us first attempt to describe the second motivation for reengaging applied electromagnetic theory at a very fundamental level: The demands of modern society. Every engineer working in the field of antennas and circuits knows that there is an tremendous increase in complexity characterizing every system and almost all working environments. 'Complexity' here refers to the high *density* of signal-processing physical structures within a given volume or surface area. By saying that the environment is also becoming more complex, we mean that there is always something happening close to the device under consideration, an interaction here or an energy coupling there that may radically modify the "normal" or "regular" behaviour of the device. This brings us immediately to the *first* theme referred to above: How can we define what constitutes the "normal" behaviour of an electromagnetic device? How can we define the electromagnetic character of a given physical structure in a way that takes into account the full complexity of the electromagnetic environment? These are the basic questions that this book attempts to investigate and answer. In other words, the double-edged problematic of the present historical conjuncture of electromagnetic research reduces to knowing how to decode increasingly complex operating environments and dense working conditions, *plus* a proper theoretical and conceptual understanding of how to analyze systems embedded into such a milieu in ways going beyond those traditional approaches that have been serving as "proper" or "normal" operational frameworks for electromagnetic research.

This naturally brings us to the first of the two major motivations mentioned above: emerging demands generated by the internal development of applied electromagnetic theory. Indeed, after the standardization of three main full-wave numerical methods (MoM, FDTD, and FEM) and the widespread use of fast and efficient computing machines all around the world, the old and traditional problems that dominated the field began to change rapidly. In the next section, some remarks on the general historical situation will be provided in order to situate this book within the global map of applied electromagnetic research. However, before moving there we mention few important facts about the relation between theory and applications. We believe that it is in the very nature of applied electromagnetics as such to be occupied with the problem of analyzing complex structures with interest in finding new behavior not seen before in nature. The rise of the art and science of metamaterials, for instance, is a striking example illustrating this issue. On the other hand, a fundamental trait of applied electromagnetics is the need to integrate the physical theory of electromagnetic waves with neighboring engineering disciplines like wireless communications and signal processing. While deeper occupation with metamaterials necessarily leads to serious consideration of quantum theory, a neighboring field from physics, it appears that a parallel interest in probing in depth the integration of electromagnetic theory and system theory (communications, signal processing, circuit theory) has not been attempted so far on a large scale. Although admirable attempts do exist, they remain isolated and scattered in the vast literature on the subject. This shortcoming has been further aggravated by the unfortunate massive drift during the last decade toward industrial and very applied research purely at the device level. There is now little patience for long and complex theories or speculative thinking, though exactly such open-minded and free modes of investigation had been the driving force behind innovative developments in the engineering science. With this overemphasis on industrial applications, engineering *science* is now regarded with suspicion while a disproportionately larger recognition is granted to research done on small and short-term problems pertinent to antenna and microwave circuit device development, i.e., commercially-driven design considerations and always with some lab fabrication and measurement inserted into the work whether needed or not.

This last point deserves some additional elaboration. Although it is always interesting to include lab measurements if possible, we believe that this truth has been overexploited during the last decade in a way that deeply affected electromagnetic research all around the world. As is well known, it is very difficult to provide complete fabrication and measurement to support every idea. Therefore, most papers in recent years tend to avoid wide scope formulations and focus instead on small and simple ideas that can be tested in the lab within a reasonable time. Although there is nothing wrong in pursuing such a line of research, it is by no means the only possible one. In fact, electromagnetic theory is one of the very few scientific fields where exact and extremely well-attested laws are known, namely Maxwell's equations. Armed by sound mathematical and conceptual training, a researcher can easily derive far-reaching conclusions within the span of few pages that otherwise may take many years in order to be fully absorbed into mainstream fabrication and measurement methods. The authors believe that this was actually how pioneers in applied electromagnetics and other scientific fields had been already conducting their work. There is an urgent need in the present time to supplement the massive focus on industrial and device research by long and sustained treatments of matter more pertinent to fundamental and even foundational issues. Examples of the latter include energy considerations, causality, applicability of Maxwell's equations to nanostructures, and the complex relations between the microscopic and macroscopic electromagnetic phenomena.

This book attempts to bring into the field an example of this kind of research. However, since it is impossible to address all or even some of these topics with sufficient depth within the span of one book, certain selections were necessary. This was not only to provide a limitation on the size of the final book, but also to propose a specific promising research theme that occupies our investigations to follow, which is the fundamental relation between electromagnetic theory and applications, especially *wireless communications*. Indeed, wireless communications looms large in this book. Every theoretical formulation and each conceptual analysis developed by the authors was in a certain way or another motivated by some mental picture of concrete potential application to the field of wireless communications. This may explain the lack of special chapters addressing propagation in waveguide structures. Although we believe those belong to the major topic of the book – the study of space in applied electromagnetic theory – they were not included in order to provide a focused presentation of the aspects related to wireless communications. Moreover, we believe that the problems arising with the theoretical status of space structures in wireless communications are more challenging than the corresponding case in waveguide problems.

1.2 THE PRESENT STAGE OF ELECTROMAGNETIC SCIENCE

We will not attempt even a brief summary of all what we deem important in applied electromagnetics and classical field theory. Instead, we provide a concise portrait of areas we think present the most essential features of the recent evolution of electromagnetics.

It seems to us that research in electromagnetics can be roughly divided into three main areas:

1. Theoretical (Fundamental) Research
2. Numerical Research
3. Applied Research

There is a marked difference in style among the three domains indicated above. We will start the explication of this threefold classification in reverse. The most popular level, the applied research (level 3), is concerned with the devolvement of working devices for the purpose of functioning within a larger, more complex systems. For example, antenna engineers are concerned with the design, development, and measurements of single antenna elements or arrays in order to efficiently receive and transmit electromagnetic information (or possibly energy) from one location to another. Microwave engineers are concerned with the design and construction of devices that can divide, combine, filter, and amplify various electromagnetic signals propagating in waveguide structures, sometimes connected with antenna fronts. The final goal at this level is the successful operation of a concrete device (e.g., the antenna element itself) for the purpose of establishing a consistent overall functional performance and system integrity. The goal then is to be serviceable to the whole, where the latter is the global system into which the device is embedded.

The second, level, numerical research, had been thriving particularly in the 1980s and 1990s, culminating in the first decade of the 21st century in the wide availability of efficient and reliable computational electromagnetic solvers. The various numerical methods found in use today, most notably the method of moment (MoM), the finite-element method (FEM), and the finite-difference time-domain method (FDTD), went through extensive stages of research and devolvement before becoming robust tools for the analysis and design of antennas and circuits in daily routines at both the academic and industrial levels. Research on the numerical method itself of course continues to attract attention, and even proposals for trying new methods can be found in literature from time to time. However, it appears to us that there is some sort of convergence, at least during the last 10 years, toward the three numerical methods mentioned above (i.e., MoM, FEM, FDTD) as the main choice for commercial solvers, with possible improvement in performance using hybridization and special techniques like the fast multipole methods. It is beyond a doubt that this convergence would never have been possible had computer technologies not evolved to their present level of speed and memory performance. However, the intense efforts pouring into numerical electromagnetic research throughout the last two decades in the 20th century have been, we believe, the main factor behind the impressive level of sophistication found today in commercial EM solvers.

The first area, the one we duped the theoretical research program, is paradoxically both the most fundamental level but at the same time the least explored avenue in the electromagnetic sciences. Before embarking on an explanation of why this has been the case, let us first define what we mean by research at the theoretical level.

The electromagnetic problem, whether being antenna or circuit one, is ultimately governed by a system of coupled partial differential equations discovered by Maxwell in the 1860s and put into their modern vectorial form independently by Hertz and Heaviside in the 1880s and 1890s. This system of equations, Maxwell's equations, contains a complete description of every electromagnetic problem but only if a proper electromagnetic model for the the materials used in the construction of the system under consideration (usually encapsulated in the *constitutive relations* of matter) is at hand. For example, for the purpose of analyzing devices working at the macroscopic level, a set of well-known boundary conditions — delimiting the behavior of the fields at interfaces presenting discontinuities in the material medium under consideration – can be incorporated within the structure of Maxwell's equations in order to derive the main equations to be solved numerically at a subsequent stage. We will come later to a frontier stage where this simple picture is no longer adequate, i.e., the emerging field of nanoscience and nanotechnology. For now, we just add that the macroscopic approach, and the understanding of the delicate and oft-ignored difference between the macroscopic and microscopic descriptions of electromagnetic interactions, can be dated back to the theoretical researches of Lorentz in the 1890s. Once the interaction between the electromagnetic field and matter is well understood, simple linear (or nonlinear) models, such as constitutive constants and dispersion relations, can be employed in conjunction with the vacuum field equations to solve the coupled problem for the unknown fields in both matter and vacuum. The research of the second level, that which is exclusively concerned with numerical methods, made the solution of the coupled equations mentioned above possible and, starting in the last decade, relatively speaking a routine task.

The engineering approach that dominated the second half of the 20th century, accompanying the explosion of new technologies driven by war-related applications, has led to modern network (circuit) theory as the paradigm of all engineering systems. Electromagnetic researchers were quick to seize on the new conceptual breakthrough of the circuit paradigm and proceeded immediately to develop circuit models for all antenna and microwave devices they needed to work with. It is unquestionable that the lack of computing resources at the time motivated the search for simplified circuit models for a realistic description of the devices at hand, and also for guiding the invention and conceptual devolvement of new devices. However, the triumph of the circuit paradigm resulted also in its own self-induced fantasy: a simple circuit model is all that is needed in order to describe a device! For example, the widespread myth that the input impedance of the antenna is the most important performance measure in the near-field zone needed for the direct integration of the antenna element within a larger system. It is true that today an expensive full-wave sophisticated numerical solution of Maxwell's equations is essential in order to obtain accurate data on the input impedance of an antenna or a microwave circuit in agreement with the experimental results, but the philosophical idea remains the same: the input impedance continues to be seen as an equivalent circuit model replacing an otherwise prohibitively complex system. It is then not very surprising to find that recent research on the more fundamental aspects of antennas, the so-called fundamental limitations on radiating structures, in particular the popular field of small antennas, is exclusively concerned with the circuit model through its emphasis on the impedance matching bandwidth as revealed in applications by knowledge of the system's effective quality factor Q.

The upshot of this story is that during the entire devolvement of classical electromagnetic field theory in the 20th century, which was mostly a story in the *applied* side, culminating in the invention of waveguides, antenna communication systems, radar systems, and so forth, the ultimate theoretical bases of classical electromagnetics can be satisfactorily found in the works of Maxwell, Hertz, Heaviside, and Lorentz. This observation is confirmed by a quick glance at the current literature on electromagnetic theory. One cannot help but feel astonished by the huge number of textbooks on electromagnetics that nevertheless are almost always variations on the same theme. It is undeniable that there is a conspicuous redundancy in the academic presentation of the theoretical basis of applied electromagnetics. Even though there is an impressive amount of literature on the solution of numerous *particular* problems, like scattering and propagation in specific environments, there exit only few coherent and comprehensive treatments of the most fundamental in electromagnetic wave phenomena transcending the particular for the sake of the universal, without at the same time departing from concrete reality, the behavior of the electromagnetic field as such, which is our ultimate prize as students of nature.

1.3 FROM STRUCTURES IN SPACE TO SPACE STRUCTURES

A classic textbook definition of applied electromagnetics will probably run like this: Applied electromagnetics is an an engineering science that studies the analysis and design of special structures performing useful functions. Examples of those "useful function" include the all-important wireless communications, cable and optical fiber communications, radar, sensors, imaging, filtering, and energy direction and transfer (through space or cables). More recently, the interesting but often misunderstood topic of metamaterials also entered the picture. But how do we specify those structures? A simple answer will say that those are things like antennas, waveguides, junctions, and arrangements (sometimes very dense) of material objects. At a more abstract level, one may answer the same questions by stating that all those useful structures studied by applied electromagnetics are in fact structures defined *in space*. For instance, an antenna or a junction is a geometrical configuration drawn in space to demarcate the various domains of different electromagnetic materials. This latter definition possesses the virtue of being simpler and more general. Through it we come to think of the engineer's task as simply manipulations of spatial material structures aiming at achieving certain *performance*. Now *this* performance is not measured by purely geometrical details like shape and curvature or distance, but via the most important physical entity in applied electromagnetic: The electromagnetic field itself circulating in and around the spatial structure. The electromagnetic field contains within itself the very essence of what makes a device what it is.

Now, what is the electromagnetic field? *It is a process in space and time.* Assuming time-harmonic excitation, as we will do in the entire book, time can be factored out and what remains are the three spatial variables. In the frequency domain, the electromagnetic field reduces to a process in space. There is then something common between the field and the structure generating it: both are in space. Consequently, both contain structures in space. Now, conventional mathematics provides us with a well-known method for studying the spatial aspects of an object in space like an antenna or a periodic array of cells, namely *geometry*. Indeed, we can analyze and describe the structures of antennas, say arbitrarily-curved systems like a conformal antennas, using differential geometry. But what about the other structures in space associated with the electromagnetic field itself? How can we define and study such structures? And, to start with, how can we know that such structures even exist?

In the following investigations various attempts will be developed in order to propose answers to these fundamental questions. Our approach is not continuous with mainstream theoretical physics and optics. There, emphasis has been always on certain research directions that include quantum field theory, relativistic invariance, and geometrical optics. These are three fundamental viewpoints but — in our opinion — they don't exhaust the subject of what constitutes the spatial structure of the electromagnetic field. Our own approach to the problem is fundamentally that of the engineering science. That is, we are searching for latent spatial structures pertinent to aspects we think useful for applications to energy and wireless communications. Clearly there is already an overlap between the basic sciences and the engineering sciences in areas like information theory and energy, but most of the lines of thought pursued in the following chapters are taken directly from motivations originating in engineering, not fundamental physics, with the exception of Chapter 6.

The major goal of this book is to reexamine the foundations of applied electromagnetics by providing a common abstract substrate unifying the device structure (encoded in its geometry) and its performance (encoded in the behavior of the fields generated by the device). This common substrate is what we call *the spatial structure of the electromagnetic field.* It is a *mathematical* form of the *physical* electromagnetic phenomenon reflecting the *engineering* aspects of the generating device. Since both the device and its fields are somehow "in space," by finding the inner structure of the field and expressing it in mathematical terms there is a hope that this will bring both the fields and their sources into a closer proximity that will help us better understand how to control fields by proper device engineering. To be more specific, attempts will be made to first discover and isolate fundamental mathematical structures in electromagnetic fields, for example the near field, nonlocality or spatial dispersion in the material response, or the system response to external fields. Afterward, attempts will be made to further analyze these structures using topological and geometrical methods. Since topology and geometry are naturally suited to describing spatial structures in general, there is some hope that the two very different structures, those of the fields and the devices, will converge — at some later stage — into a mutually consistent whole.

This book is then an attempt to combine mathematics, physics, and engineering within one consistent discourse, that of the scientific study of the spatial structure of electromagnetic fields in engineering systems. However, we should mention that the grand goal aimed at in the previous passage has not been attained yet. We are here formulating the problem and providing several investigations, each contributing something toward reaching our major goal. How far the book as in its totality has progressed in this direction is something we leave for the future to decide.

1.4 THE RELEVANCE OF SPATIAL STRUCTURES TO THE PRESENT TECHNOLOGICAL WORLD

Moving now from the internal demands of theory back to what we started with, the relation between theory and applications, the natural question is the following: Given the concept of the spatial structure of the electromagnetic field here seen as the common framework for unifying the device and its performance, How does this concept relate to the problems facing the electromagnetic engineer facing today's growing demands on electronics and communication devices?

If one may summarize the most distinguishing characters of the external needs imposed on researchers and design engineers, it is certainly the following:

1. The progressive increase in device density within any given volume or area.
2. The need to operate at multiple spatial scales at the same time.
3. The increase in the electromagnetic complexity of the device's surrounding environment.

Trait 1, the increase in device density, relates to the fact the more people are using communicating devices than any previous time in history. This leads to the unavoidable fact that the standard assumptions of free-space environment, where the transmitter and the receiver are in the far field zones' of each other, are either no longer true or soon will be invalidated. Near-Field Communications (NFC) is one of the emerging fields that deal – among other things – with problems of this kind. Since its inception, traditional applied electromagnetic theory has been relying extensively on the simplifying assumptions of plane-wave illumination, and consequently we believe much of the conventional theories and methods related to antenna arrays and receiver design need to be revisited in light of the fact that near-fields or fields more complicated than simple plane waves are most likely to dominate the device illumination in today's and future world. The near-field theory of Part I and the antenna current Green's functions of Part II are direct responses to this demand.

Trait 2, the need to deal with multiple spatial scale, is more subtle. Its most conspicuous manifestations can be immediately seen in the emerging field of nanoelectromagnetics. There, one deals with two physical objects operating at very different spatial scales: The electromagnetic field (macroscopic object) and the atomic/molecular scattering nanostructures (microscopic object). Much of the confusion in this field in today's nano research stems from the fact that the problematic dialectic between the world of the small and the world of the large is either ignored or improperly treated. The problem, however, is not exclusive to nanoelectromagnetics but seems to be fundamental even in macroscopic structures. One may first mention the interesting research done in metamaterials within the framework of homogenization and effective medium theory. Also, the research areas of subwavelength imaging and periodic structures involve similar problems. In all these cases, it is essential to understand how fine or small-wavelength information around the scattering structure give rise to the observed macroscopic fields. For applications involving multifunction overall plans in which nanostructures, subwavelength interactions, and metamaterials are all combined into one complex system, a new form of electromagnetic theory is needed in which several different spatial scales are dealt with within the same theoretical and conceptual framework. Such a a final theory will not be found in this book, but the search for viable directions and routes that could possibly lead to the construction of a future good theory like that is one of our major motivations for undertaking the effort of writing the investigations to follow.

Finally, we come to Trait 3, the occasional need to embed devices into electromagnetically complex environments. This situation is not totally independent of Traits 1 and 2. Indeed, one may argue that high device density and the presence of multiple spatial scales are what usually make operating environments electromagnetically complex. While this is undoubtedly true, we believe that the issue of complex environments is so fundamental to the degree it merits forming a category by itself within the cluster of challenging demands facing applied electromagnetic theory in today's world. The idea is that complexity may not always be an accidental problem imposed on us because of the development of the surrounding area inside which the system or device of interest happens to operate. Instead, complexity can be a goal in itself. Engineering applications pursuing a known direction of thought may reach a bottleneck after some time, and in this case the only way ahead remaining for developing new applications might be increasing the complexity of the system plus its environments. This appears to be indeed the case in metamaterials research. Although much progress in pertinent electromagnetic theory has been accumulating throughout the last 20 years, we believe that a direct reformulation of theory with emphasis laid from the beginning on the concept of the spatial structure is fundamental. Everywhere in this book, we provide tools and concepts that can be readily exploited to build corresponding complex environments. This is particularly evident, for instance, in Part III, where we see how special "spatially-engineered metamaterials," the nonlocal material, can be used to achieve very unusual electromagnetic behavior.

It is important, however, to call attention to another type of complexity resulting from those electromagnetically-crowded environments so common in today's world, where the main complexity factor may not stem from device density as such, but rather from the presence of undesired nearby targets and scattering objects close to the device under consideration. Here, we are no longer focusing on the problem of near-field communication associated with Trait 1, but instead with the unavoidable presence of nonstandard illumination bombarding the device of interest resulting from scattered fields generated by other objects. In such a scenario, which is becoming more pressing in mobile communications, the naive assumption of plane-wave style illumination within idealized homogeneous and open medium may no longer be satisfactory. Although there has been considerable researches in recent years attempting to address this issue, they appear to be mostly devoted to providing corrections due to mechanisms like the effects of platforms, diffraction, ground planes, etc. On the other hand, the problem of how the device under consideration responds to nonstandard, say near-field illumination, has not been treated in depth so far. This book takes this technological aspect into consideration right from the beginning, with several chapters devoted to detailed proposed solution methods and their implementations.

1.5 THE STRUCTURE OF THE BOOK

All of the investigations to follow revolve around a central theme that will keep recurring as the reader gradually advances in the reading process of the book: The spatial structure of the electromagnetic field. As we explained above, we have selected certain aspects out of many possible others, guided in this choice by the desire to bring into light aspects pertinent to specific applications, here energy and wireless communications. Therefore, the treatments to be found next can be further expanded if other types of applications are taken into account. However, note that the presentations of the first three parts of the book attempt to provide a universal understanding of the subject that is both rigourous, exact, and conceptually clear, without worrying very much about applications at that stage. In the last part of the book, several applications drawing on some – but not all – of the theoretical and conceptual findings of the previous parts will be outlined, and some of these applications, like mutual coupling, will be presented with extensive details.

The overall structure of the book can be mapped into four distinct divisions, referred to as

1. Part I: is devoted to near fields, energy, and the fundamental theory of electromagnetic radiation.
2. Part II: covers the formalisms and methods of the antenna current Green's function.
3. Part III: provides a brief introduction to the emerging field of nonlocal metamaterials.
4. Part IV: develops various applications drawing mainly on Part II and some chapters of Part I, with special focus on electromagnetic mutual coupling in antenna array systems.

Due to the complexity of the various parts of the following investigations, we will provide below detailed overviews of the book's entire argument in order to help the reader in forming a general global grasp of the subject.

1.5.1 Part I: The Theory of Near Fields

Here the authors propose a new theory of electromagnetic radiation, near fields, and energy based on the motivating applications of wireless communications and energy engineering. The fundamental basic idea is that for the sake of such engineering applications, the main need is to control electromagnetic energy *by moving it around from one region to another*. Therefore, we propose to study the inner structure of the field by distinguishing, from the beginning, the propagating part by marking its evolution against a background of a nonpropagating (evanescent) field. In a way that is purely relational, the dynamics of the radiation problem is now understood as how the portion of the field that is moving (propagating part) does in fact flow in space by "kicking" against its unmoving (evanescent) part. Therefore, the dynamics disclosed here is not the conventional time-varying character of the field (the form adopted by Maxwell), but rather the purely relational form between two parts of the field, one moving with respect to the other.

The implementation of this immanent relational theory of dynamics is done gradually, chapter by chapter, until it reaches a climax in Chapter 6, where we propose new differential equations describing the flow of energy around antennas. This very chapter is provisional and will be expanded fully in a future book by one of the authors. For the time being, the preceding Chapter 6 provide comprehensive foundations for the topic more relevant to engineering than fundamental science.

In Chapter 2, we first revisit the existing traditional approach to electromagnetic energy, in particular reactive energy, which we criticize for being insufficient for the program of studying the spatial structure of electromagnetic fields. Here, the presentation summarizes and explains results that will be derived rigorously and in detail throughout the next chapter. The major aim of Chapter 3 is to unpack for the general audience the complex argument at the basis of our rejection of the concept of reactive energy. In order to achieve this, Chapter 3 develops the fine technical details of the conventional spatial theory of near fields to its most possible complete form. Here, the reactive energy is derived in a revealing analytical form and the roots of the troubles previously encountered in its evaluation are clearly spelled out. Based on this result, we provide a critique of reactive energy suggesting the physical incoherence of the concept of reactive energy density. Since energy *densities*, and not total energies, are what are at stake in a the spatial structure of the field, we then conclude that reactive energy cannot be the right spatial structure of the antenna near field.

The alternative new approach to the problem is now formulated in Chapter 4, where we derive the *dynamic* decomposition of the field into propagating and nonpropagating parts. Previously, only a "static" decomposition was used in the engineering community, where a fixed coordinate system was deployed in order effect the separation of the field into sinusoidal and evanescent modes via the Weyl expansion. In order to effectively build our new approach, a rotating local coordinate system is introduced to probe the inner dynamic structure of the field, where the latter is understood as the latent capacity of the field to flow in qualitatively different manners along different directions. The proposed formulation allows us to investigate new forms of energy for the first time. For example, *localized energy* is introduced and derived explicitly using the main conclusions of the dynamic theory of this book. Furthermore, the interactions energy between the propagating and evanescent parts is highlighted in our work and its physical significance and implications briefly discussed. The chapter ends with an explication of the general mechanism of far-field radiation formation in generic antenna systems. We show that at the heart of the process of producing the observable far-field pattern lie simple geometric and filtering operations that can be directly spelled out using the new dynamic theory.

Chapter 5 presents a simplified form of the general theory of Chapter 4. Here, we work with scalar sources and avoid the vectorial complexity of the full electromagnetic problem. This simplification allows us to derive simple analytical expressions for the various new field decomposition, especially the localized energy. Moreover, the results obtained here will form the theoretical basis for new numerical methods developed in Part IV to measure localized energy in general antenna systems.

Numerous applications can makes use of the results of these chapters of which more were developed in Part IV, for example Chapter 17. For completeness, we list some of the current and future related applications:

1. Antenna design.
2. Near-field measurement.
3. Subwavelength imaging.
4. Nanoelectromagnetics.
5. Near-field communications.
6. Electromagnetic mutual coupling.[1]
7. Wireless energy transfer.
8. Near-field radars and sensors.
9. Energy localization, storage, and control.[2]
10. Near-field metamaterials.[3]

1 Cf. Chapter 17.
2 Cf. Chapter 17.
3 Cf. Chapter 13.

1.5.2 Part II: The Antenna Current Green's Function

After the in-depth analysis of the spatial structures of electromagnetic radiation emanating from a single antenna or radiating system, we move now to the fundamental situation demanded by applications when *two* radiating systems interact with each other. The solution to this problem proposed here is what we term *the antenna current Green's function* (ACGF), which is a rigorous generalization of the concept of impulse response in the time analysis of linear systems to the framework of spatial electromagnetics. Part II, in fact, outlines a new program for applied electromagnetics aiming at incorporating into full-wave analysis problems, methods, and techniques borrowed from system theory that have not been widely recognized by the applied electromagnetic community.

Here, we assume that one of two interacting antennas can be replaced by its generated near field while the other behaves more or less like a passive device interacting with this illumination field. The ACGF is the transfer function in space describing how the device generates a radiating current in response to external excitation. The near field generated by the first device and analyzed in Part I is now taken for granted and assumed as an input to the second antenna. Consequently, Part II concentrates on the performance of this second antenna in space.

Chapter 8 provides a careful definition of the problem followed by a proof of the existence of the ACGF for arbitrary antennas. This was accomplished by combining electromagnetic theory, functional analysis, distribution theory, and differential geometry all in one formulation. For simplicity, we consider only perfect electric conductor (PEC) boundary conditions, but the proof ideas can be applied to any other known electromagnetic boundary conditions, although the details are very tedious. (The authors hope to take up this generalization in a future work.) In any case, the method of the existence proof discovered here is of particular interest to the engineer. Indeed, the existence of the ACGF was proved by providing a rigorous and general algorithm allowing the actual construction of the ACGF based on sets of measurements of external fields and induced currents. In other words, while the ACGF is an abstract structure introduced as a tool for clear thinking and alternative computations, the very abstract structure is completely based on observable physical signals. The constructive nature of the existence proof method is directly exploited in Chapter 15 to propose the ACGF as a new and alternative method for computing the response of electromagnetic devices to external near fields. However, the main motivation for the introduction of the ACGF is the conceptual facility of the formalism as a tool for clear thinking about applied electromagnetic problems and as a framework for developing new applications as in Part IV.

Following the foundations of the ACGF laid down in Chapter 8, Chapter 9 supplies in-depth analysis of the complete electromagnetic system comprised of transmitter and coupled receiver. We propose a new look at the overall conceptual structure of such systems reduced to three basic operational modes, which we call Modes A, B, C, explicating the physics and significance of each mode and how it is related to the others. That is, we focus from the beginning on the relational intercourse between various modes in order to attain a deeper understanding of the complete electromagnetic communication system as a whole. The chapter, for example, contains statement and proof of the inverse reciprocity theorem and proposal for designing antennas capable of filtering spatial data according to the illumination field wavelength structures. Here, the inner structure of the near field developed in Part I is directly linked to the spatial filtering behavior of the receiving antenna system. The chapter climax is the new insight obtained on how the geometrical shape of the receiver interacts resonantly (and nonresonantly) with the multiple spatial scales of the illumination fields.

The ACGF formalism, being exact and general, enjoys an enormous range of potential applications that cannot be fully surveyed here. Most of the applications in Part IV are based on the ACGF method, either directly or indirectly. We mention here some of the applications that the authors believe to be important:

1. Antenna design.[4]
2. Near-field communications.
3. Electromagnetic mutual coupling.[5]
4. Wireless energy transfer.
5. Near-field radars and sensors.
6. Direction-or-Arrival (DoA) estimation.
7. MIMO systems.[6]
8. Computational methods (especially response to nonstandard field illumination).[7]
9. Novel statistical methods in electromagnetic analysis.[8]

4 Cf. Chapter 9.
5 Cf. Chapter 16.
6 Cf. Chapter 18.
7 Cf. Chapter 15.
8 Cf. Chapter 14.

1.5.3 Part III: Nonlocal Metamaterials

Part III deals with a third wave of topics different in details from the previous two but still lying within the main theme of the book. Indeed, nonlocal metamaterials are those exhibiting novel *spatial* electromagnetic response, for example spatial dispersion, allowing them to create new electromagnetic phenomena not seen in the conventional type of metamaterials where only temporal dispersion is usually taken into consideration. In this part of the book, the spatial degrees of freedom in the environment are then highlighted right from the start. While we still consider classic temporal dispersion, we show that only with controlled spatial dispersion could novel changes in the electromagnetic field structure be attained.

Chapter 10 first provides a global map of the field, explaining the motivation behind the search for new generations of metamaterials and their potential use in current and future applications. Chapter 11 reviews the established mathematical formalism of spatial electromagnetics, which unfortunately remains not well known in applied electromagnetics. Following that, Chapter 12 develops new theory of nonlocal metamaterials for far-field problems (source-free propagation), we derive the fundamental dispersion engineering equation and show that careful design of the temporal and spatial dispersion profiles can lead to wide range of unusual group velocity profiles, including for instance constant negative group velocity and many others. We then end with Chapter 13, which outlines in a very brief form the near-field theory of nonlocal metamaterials. The chapter is based on deriving the Green's function of a point source in such a nonlocal setting by generalizing the classic Weyl expansion to take into account the new spatial degrees of freedom made available by spatial dispersion. The results show that in the near zone of such a source, there exists backward wave propagation in addition to normal forward waves. Moreover, design equations were derived to describe how a nonlocal medium can be analyzed in order to shape and control the spatial structure of the near field. In the main, some of our conclusions point toward the possibility of completely trapping and localizing energy around antennas embedded into lossless specially designed nonlocal metamaterials.

Applications of nonlocal metamaterials proposed here include

1. Near-field focusing.
2. Energy transfer through controlled near-field shells.
3. Energy localization, storage, and retrieval.

4. Antenna design by spatially engineered nonlocal substrates and/or surrounding.

1.5.4 Part IV: Various Applications

The fourth and final part of the book is concerned with a selection of several applications, most directly based on the theoretical and conceptual results obtained in the previous three parts. The authors were unable to bring into the text some of the more recent applications because of size limitations. The subjects included in this part are very basic and fundamental, but there are also some special topics like the IDM method in Chapter 16 and MIMO systems in Chapter 18.

Chapters 14 and 15 outline how the ACGF formalism can be used in typical and nontypical settings. This includes the use of the ACGF in formulating statistical evaluations in antenna array performance. We also show that the Singularity Expansion Method (SEM) can be used in a new way to describe antenna systems by searching for characteristic poles or resonators buried in the structure of the ACGF itself.

The main bulk of the applications part consists of Chapters 16 and 17, which provide a substantial theory and methods for dealing with the fundamental topic of electromagnetic mutual coupling. This coupling is first defined in a complete and rigorous manner using the ACGF, where a mutual coupling ACGF is introduced. Next, several methods like perturbation techniques and dipole models were outlined to deal with mutual coupling in arbitrary arrays. Chapter 17 integrates the near-field theory of Part I with the mutual coupling framework by proposing new numerical methods to measure localized fields in the antenna-antenna interaction regime.

Finally, the book ends with Chapter 18, which illustrates how the ACGF formalism can be used to electromagnetically derive the channel matrix of a generic MIMO system. Due to limitations in space, only simple cases are given here, while the full analysis will be presented elsewhere. The chapter also contains an introduction to spatial diversity studied thorough a new Green's function, the far-field cross correlation, which, like the ACGF, captures an essential spatial structure in electromagnetic systems.

Part I

The Theory of Electromagnetic Near Fields

Chapter 2

Reactive Energy and the Near Field

2.1 INTRODUCTION

Electromagnetic radiation is both an old and new problem. The basic laws of electromagnetism were laid down by Maxwell in his groundbreaking papers of the 1860s. However, the theory was not totally accepted until Hertz and Heaviside succeeded in reformulating it in a more elegant and economic mathematical form throughout the 1890s. At around the same time, the experimental researches of Hertz confirmed the existence of electromagnetic waves (predicted by Maxwell based on purely theoretical calculations) and the entrepreneurship of Marconi started what succeeded later in constituting a vital strand of modern electrical and electronic technology, the particular form of technification that was destined to dominate the 20th century. Therefore, since the process of utilizing the electromagnetic spectrum for all possible types of applications has been and remains a fundamental dimension of modern applied science, *applied electromagnetics* as such has emerged in the post World War II era as one of the most vibrant and active research sectors in the engineering sciences. In this sense, the study of electromagnetic radiation is as old as the purely scientific curiosity that motivated thinkers like Faraday, Maxwell, Hertz, and Heaviside to contemplate the mere possibility of transferring energy in vacuum; on the other hand, radiation forms a backbone of the modern infrastructure of technology and hence is still a new and fresh topic amenable for further study and development. Indeed, the new requirements imposed by the accelerated growth of human population and the ever-increasing pace of industrialization and technification taking place all over the world, including nations that started recently to catch up with capitalism, make a full mastery of the essentials of classical electromagnetic radiation as relevant and urgent as ever.

In this part of the book, we provide a wide and comprehensive new look at some of the fundamental issues involved in the thinking process of engineers and applied scientists when dealing with radiation problems, particularly antenna devices. The development presented here will be presented again with full details in Chapter 3. While the content there is more technical in nature due to the need to establish a foundational approach based on rigorous and precise mathematical bases, the presentation and reflections given in the this part of the book will stay close to the original conceptual train of thoughts that essentially motivated and stimulated the authors to undergo such investigations.

This investigation will reexamine some fundamental issues related to antenna systems from a specific perspective, that of *electromagnetic energy*. Our choice of this particular theme is motivated by both theoretical and practical factors. From the purely theoretical viewpoint, it seems obvious after a glance at the historical development of field theory (both classical and quantum) that the concept of energy plays a fundamental heuristic role in guiding and organizing the total body of mathematical knowledge accumulated through time concerning the production and propagation of various physical fields, including electromagnetic waves. For example, the Lagrangian formulation of field theory, which is now the dominant theoretical approach to the problem, deals with the field in the form of energy. Indeed, the field variables enter into the Lagrangian functional only in the form of quadratic terms. This is even the case in the classical approach to fields via differential geometry, where one can write down the laws of electromagnetics in a nonvariational form using special tensors where again the field enters through quadratic terms [94]. Therefore, *the energy of the field seems to play a more fundamental role in theory than the fields themselves.*

From the purely practical point of view, almost all thinking of engineers and applied scientists regarding electromagnetic devices centers around energy and energy-related concepts, such as flux, power flow, dissipation, losses, and scattering matrices. The total domination of the energy concept in engineering is probably hard to understand from the theoretical side, but it seems to fit very well with our intuition of what a practical device is supposed to do. In fact, one usually thinks of the antenna, for example, as a system that can send *energy* away, receive *energy* from other sources; a circuit is supposed to process *energies* flowing throughout them, extract information from variation in *energy* levels, redirect pathways of *energy* transfer, and so forth.

Due to all of these considerations, we think that our choice of the theme of electromagnetic energy for this part is well motivated on both the theoretical and applied sides. With this maybe true, it is a little bit surprising how scanty the literature is in applied electromagnetics – especially antenna theory – that has paid directed and focused effort with regard to researching fundamentals of this topic. That does not imply that there have been no contributions for dealing with the problem of defining and computing various antenna-related energies. Indeed, the three major concepts of reactive, evanescent, and stored energies are all taken directly from the existing literature. What we claim here, however, is that there has not been in recent times a comprehensive consideration of the topic from the vantage point of the *conceptual* status of the terms, what they mean and how they can be utilized in practice. The work to follow is an initial contribution to mitigating this shortcoming in the technical literature with particular emphasis on reactive energy.

A few words on the treatment of previous works must be inserted here. In a topic at such a wide general scope, no attempt will be made for complete comprehensive coverage of all the related literature that has dealt with energy in applied electromagnetics. On the other hand, our choices in referring to some previously published works will be guided by our own knowledge and the particular methodology adopted here for attacking the problem. The latter will be admittedly less instrumental than what is usually practiced in mainstream theoretical electromagnetics, where there seems to be much more emphasis on calculation and computation of certain quantities than reflection on general matters. Therefore, we will try to develop the topic from first principles, starting directly from Maxwell's equations and their classical formulations in terms of radiation integrals via suitable Green's functions, series expansion, and so forth. Reference to some of the literature that formed a background to our main line of approach will then be highlighted on the above basis.

The more subtle concepts of evanescent (localized) energy and stored energy will not be treated here. Some of the literature related to them will be referred to but a separate treatment by the authors will be reported elsewhere.

2.2 ELECTROMAGNETIC ENERGY IN ANTENNA THEORY

One of the most important issues in antenna theory is knowing how to deal with energy. In practice, one needs to inject energy coming from the system source to the transmitting antenna in an efficient manner, e.g., avoid reflection and system loss. We need to direct energy at specific location in the far zone and maybe suppress radiation at certain other regions. We perhaps need to couple one antenna to another in the near zone by carefully shaping the form of the energy shell immediately surrounding the antenna's physical body. These are few examples of major concerns in the life of the antenna engineer. It is therefore plain that a fundamental reconsideration of the topic is both enlightening and "useful," especially if we keep in mind that the concept of antenna energy is not well addressed in the physics literature, while in the applied field it is fraught with difficulties.

The magnetic and electric energy densities are given, respectively, by [94], [33], [28]

$$w_e = \frac{1}{4}\varepsilon \mathbf{E}\cdot\mathbf{E}^*, \quad w_h = \frac{1}{4}\mu \mathbf{H}\cdot\mathbf{H}^*. \tag{2.1}$$

These expressions are written here for the time-harmonic case and hence the appearance of $1/4$ instead of $1/2$. In this book, we focus exclusively on such a time dependence.

It is possible, and actually quite natural, to raise the question of why should the *square* of the field amplitude be interpreted as energy. The immediate meaning of energy is taken from mechanics as *the ability to do work* [95]. However, electromagnetic phenomena exhibits mechanical effects only through the Lorentz force law, in which a charged moving particle in electric and magnetic fields experiences certain forms of mechanical force [33]. Based on this force-field connection, a plethora of very convincing evidences has been accumulating, suggesting that the expressions (2.1) *do* represent an authentic form of energy densities. The reader may refer to books like [94], [95], [33], [31] that discuss a such a train of thought. However, it must be kept in mind that the now standard interpretation of electromagnetic theory gives Maxwell's equation the fundamental status of *axioms* or *postulates* of the theory, while Lorentz force law is secondary or derived. This has been in harmony with the major theoretical restructuring of physics by the turn of the last century, especially in the wake of Einstein's special relativity. The latter theory gave the Farady-Maxwellian *field* a foundational status in the sense that the field becomes a physical object as such, and not only an agent manifesting itself through the Lorentz force law. Indeed, the impact of the major classical field theorists Faraday, Maxwell, and Einstein was to shift the focus away from Newtonian mechanics and even sometime (as in Mie's work [94]) attempt to derive classical mechanics from Maxwell's equation. We will not go into this fascinating but neglected history of the topic, but only mention that under this now universally accepted interpretation of Maxwell's theory, the expressions (2.1) should be viewed as *definitions* of energy densities, rather than derived theorems.

This will raise new kinds of problems when dealing with antenna theory. Since radiation occurs in infinite (non-compact) regions of space, there is an interest (as will become clearer later) to evaluate the *total* energy in such regions. In other words, one needs to integrate the expressions (2.1) over *infinite* regions. Unfortunately, it is not difficult to see that any such a computation will lead to absurd results, namely infinite energies or divergent integrals.

This appearance of infinite antenna field energy is *not* related to the famous problem of divergence in quantum field theories. Actually, there seems to be great interest in many of the early papers in demonstrating the divergence of the energy expressions by actually calculating certain special cases. This is not necessary, for this infinite total antenna energy issue is quite natural and expected based on the following very general reasoning.

In the time-harmonic case, all fields assumes the dependence $\exp(-i\omega t)$. The antenna configuration is a radiation problem, i.e., fields in the far zone are expected to be *propagating* waves. Now, the problem of an infinite integral can be divided into the sum of two integrals, one over a *finite* region and the other over the remaining infinite region. It is clear that the second part, that involving the unbounded region, is the source of the infinity trouble. This makes it natural to look for that specific form of the field in the far zone that may cause the divergence of the second integral.

As we have just mentioned, the field in this far-zone region is normally taken as a *propagating wave*. If we look at the problem of propagation in space and time, then the fact that a time-harmonic form $\exp(-i\omega t)$ is assumed makes it necessary that the space dependence takes the form $\exp(i\mathbf{k}\cdot\mathbf{r})$.[1] We focus then on *pure propagating modes*, and by this term we mean fields whose space-time dependence has the special form $\exp(i\mathbf{k}\cdot\mathbf{r}-i\omega t)$. The amplitude squared of this field is $|\exp{(i\mathbf{k}\cdot\mathbf{r}-\omega t)}|^2=1$. Therefore, any integral of the corresponding energy density over infinite regions will produce infinite result.

The above argument will have far-reaching repercussions for the topic of electromagnetic energy in antennas and so it is worthwhile repeating it briefly in a more precise term. Let the radiating antenna be enclosed in a compact region V. The exterior domain or the radiation region D is the complement of V, or $D := V^c = \mathbb{R}^3 - V$. Now, the signature of the antenna problem is the existence of pure propagating waves in at least one infinite subregion in D. In other words, we assume that the exterior region can be decomposed into the sum of two sets D_0 and D_∞ such that $D = D_0 \cup D_\infty$, D_0 is bounded, and D_∞ is unbounded and contains a field spectrum in the form of pure propagating modes $\exp(i\mathbf{k}\cdot\mathbf{r}-i\omega t)$. It follows

1 This can be easily proved from the requirement that, mathematically speaking, a propagating wave must be expressed as a function $f(t-az)$, where a is some real constant (we consider a one-dimensional problem for simplicity). Let the space dependence of the field be $g(z)$. We then need to solve the *functional* equation $\exp(-i\omega t)g(z) = f(t-az)$. It is not difficult to show that one and only one solution is possible, and it is that which makes $g(z)$ the familiar space exponential $\exp(i\beta z)$ for some real constant β.

then that the integral diverges, or we have

$$\int_{D_\infty} d^3r\,|\mathbf{E}(\mathbf{r})|^2 = \infty, \quad \int_{D_\infty} d^3r\,|\mathbf{H}(\mathbf{r})|^2 = \infty. \tag{2.2}$$

Again, this happens because the infinite integral of constant function $|exp(ix)| = 1$ is infinite.

The above argument needs to be worked out more carefully in order to deal with problems involving real analysis. In fact, the entire topic of the antenna energy, with the particular climax surrounding (**??**), centers on what is essentially a *limiting process*. Moreover, any actual computation will inevitably involve interchanging the order of two or more limiting processes, and hence the concrete modes of convergence involved have to be studied carefully. In general, the details will depend on the method adopted for defining and computing the specific type of total energy under consideration. Up to now, there is no comprehensive mathematical method that can be used to study all types of electromagnetic energies. Indeed, it seems that with three major genera of energies, namely reactive, localized, and stored energies, each one will appropriate a special mathematical approach best suited to the purpose at hand.

The general analysis above suggests the following three salient features in antenna electromagnetic energy:

1. The antenna device, being a radiation structure, must involve a noncompact or unbounded (infinite) domains.

2. In unbounded regions, there exists, in at a least a subregion, a sum of the fields in the form of pure propagating modes.

3. The total energy contained in the exterior region is infinite.

Feature 1 motivated the innovative engineering idea of replacing the entire unbounded domain of the antenna by a localized finite region, the input port through which energy is being injected into the system. The proper theoretical analysis of this input port model is the concept of *antenna reactive energy*. Here, the antenna is replaced by an RLC circuit and the reactive energy of this circuit will correspond to the antenna reactive energy, while the resistive loss matches the real radiated power. The discussion of this traditional topic will be given from a new perspective in Section 2.3.

Feature 2 motivates our new approach to the space-time flow of electromagnetic energy using the classic Weyl (plane-wave spectrum) expansion[2] in a away

2 Cf. Chapter 4.

that reflects the physical content of propagation understood literally as "moving energy around in space," to be developed in Chapter 4. This generalizes the concept of antenna energy *beyond* the circuit-model framework of the reactive energy and opens the door for new directions in exploiting the physics of electromagnetic radiation for both theoretical and applied purposes.

Feature 3 will raise the question of what exactly one expects to find physically stored in the near field of the antenna in terms of energy. Indeed, if the naive approach to the antenna energy exemplified by (2.1) points to the (physically) absurd result of infinite capacity to do work, then what actually is the real (necessary finite) ability to perform useful work, or in equivalent terms, what is the concrete energy content stored in the antenna near-field region? Brief and tentative proposals will be given in Section 2.4.

Informally speaking, the response to Feature 1 has emphasized the *engineering* way of thinking. Feature 2 requires more a *mathematical* approach. Finally, the issue raised by Feature 3 focuses on the *physics* of the problem. However, the three aspects of engineering, mathematics, and physics, are all subtly interconnected with each other. Antenna theory requires the three ways of thinking combined together. In the past, it seems that the emphasis on the pure engineering dimensions of the problem was allowed to predominate the discussion. We hope that this book will contribute to bringing the other two dimensions into the picture.

2.3 ON REACTIVE ENERGY

2.3.1 Introduction

For antenna applications, there has been an attempt to deal with the energy problem (??) in a form that is essentially determined by the needs of applications. It was decided that a specific concept of electromagnetic energy, the *reactive energy*, is the one most relevant to practice and hence should be given the greatest focus. One finds sometimes in the literature arguments alluding to the belief that only what can be physically measured at the input port is important. However, the reactive energy concept constructed based on this limited view need not correspond in a comprehensive fashion to everything that an antenna can do. We suggest that more care should be taken in studying energy and that alternative approaches are explored.

In Chapter 3, we will approach the topic through a special detour via both Wilcox and the multipole expansions in order to demonstrate that in the case of the reactive energy no infinite integral is actually involved. This construction can be considered a conclusive statement regarding the computation of reactive energy since the need to deal with unbounded integrals, which are difficult to calculate for general problems using routine codes, is shown to be unnecessary. However, the main motivation for reexamining the reactive energy concept is not mainly computational, but more concerned with the foundational aspect of the concept in addition to a criticism of the prevailing orthodoxy regarding the interpretation of the antenna energy as a circuit energy. The need for a *space-time* approach to energy flow and the various technical proposals in this line of thought are detailed in Chapter 4 and will be developed further elsewhere. In the rest of the present discussion, we first provide a general definition of reactive energy (in conformity with the traditional literature) and then outline how to compute this energy spatial distribution in terms of the far-field radiation pattern. More technical details of this definition can be found in Chapter 3.

2.3.2 Background to the Concept of Reactive Energy

Consider the general radiation problem in Figure 2.1. The complex Poynting theorem states that [33]

$$\nabla \cdot \mathbf{S} = -\frac{1}{2}\mathbf{J}^* \cdot \mathbf{E} + 2i\omega \left(w_h - w_e \right), \tag{2.3}$$

where the complex Poynting vector is given by $\mathbf{S} = (1/2)\,\mathbf{E} \times \mathbf{H}^*$ while the energy densities are defined by (2.1). One can derive by a standard procedure the following result [33], [35]

$$\int_S d\mathbf{s}\frac{1}{2}\left(\mathbf{E} \times \mathbf{H}^*\right) = P_{\text{rad}} - 2i\omega \int_{V_\infty - V} dv \left(w_h - w_e \right), \tag{2.4}$$

where the radiated power is

$$P_{\text{rad}} = \text{Re} \int_S d\mathbf{s}\frac{1}{2}\left(\mathbf{E} \times \mathbf{H}^*\right). \tag{2.5}$$

Relation (2.4) is the theoretical basis for the traditional expression of the antenna input impedance in terms of fields surrounding the radiating structure [33]. For details about how to define the input ports incident wave modes (where transmission line and waveguide theory is used), see [53].

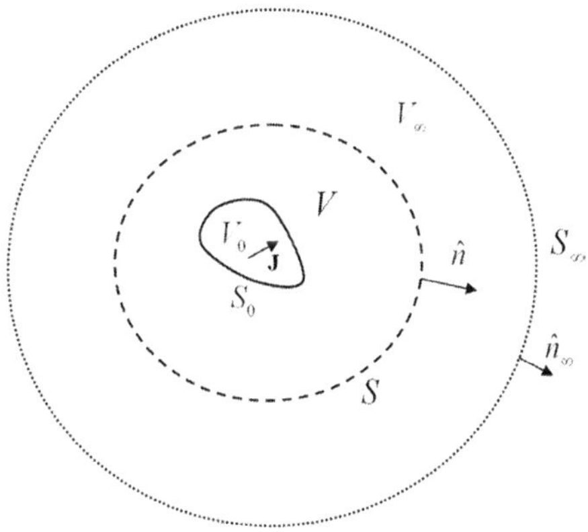

Figure 2.1 General description of antenna system. We assume that an arbitrary electric current $\mathbf{J}(\mathbf{r})$ exists inside a volume V_0 enclosed by the surface S_0. Let the antenna be surrounded by an infinite, isotropic, and homogeneous space with electric permittivity ε and magnetic permeability μ. The antenna current will radiate electromagnetic fields everywhere and we are concerned with the region outside the source volume V_0. We consider two characteristic regions. The first is the region V enclosed by the spherical surface S and this will be the setting for the near fields. The second region V_∞ is the one enclosed by the spherical surface S_∞ taken at infinity and it corresponds to the far fields.

According to the energy balance relation (2.4), one can write the input impedance of the antenna in the form

$$Z_{\text{in}} = R_{\text{in}} + i\,(X_L - X_c)\,, \tag{2.6}$$

where R_{in} will be proportional to the radiation power (2.5), while the inductive and capacitive reactance parts X_L and X_c are proportional to the electric and magnetic *total* energies $\int_{V_\infty - V} dv\, w_e$ and $\int_{V_\infty - V} dv\, w_h$, respectively.

Now, the fundamental technical observation that has provided the theoretical motivation for defining a finite reactive energy is the following

> **The Principle of Finite Energy Difference**. Although the total electric and magnetic energies are (when taken individually) infinite, the *difference* of the electric and magnetic energy densities has a *finite* total integral in the entire exterior (unbounded) domain of the antenna problem.

Stated more precisely, we have

$$\left| \int_{V_\infty - V} dv\, (w_h - w_e) \right| < \infty. \tag{2.7}$$

Recognition of this mathematical fact for various examples abounds in the theoretical antenna literature (e.g., see [51], [49], [53]).

It is important to pay very careful attention to the logic of the discussion so far. The energy balance relation (2.4) is a mathematical *theorem* that can be proved rigorously, together with the Principle of Finite Energy Difference (2.7), directly from Maxwell's equations. *However, the input impedance model in (2.6) is **not** a theorem, but an **engineering model** based on an **interpretation** of a rigorous mathematical theorem.* Moreover, this input impedance model is made possible mainly because of the Principle of Finite Energy Difference. Indeed, the inductive and capacitive reactances X_L and X_c in (2.6), taken separately, *can* have physical meaning as reactive *circuit* energies. However, as *antenna* energies, *no* physical significance can be assigned to either of them because of the divergence problem (2.2). In other words, the individual "*antenna*'s inductive and capacitive energies" are always infinite. This is why no such quantities are ever defined for the antenna *itself*, although they are used all the time in the *circuit models* of the antenna.

It is at this stage that one can see why the input impedance concept is useful. Measurements or computations of the behavior of a general antenna system must conform to the strict law (2.4). Therefore, even though when total electric and magnetic energies – each is taken individually – are infinite, one always measures (or computes) a finite number, i.e, the difference of the total energies. Any RLC circuit can implement this observation since it is only the *total* input impedance of the circuit that is needed in matching applications. An equivalent inductance or capacitance can always be given to describe the input impedance of the actual general antenna system. However, because the relevant physical data depend only on the *difference* between two reactances, *no unique equivalent inductance and capacitance can be given*. For each specific antenna, there is an infinite number of RLC circuits that will model exactly the same antenna, and this fact, which is not really surprising, points to the inherent artificiality of all circuit models even if they can be "useful" sometimes for applications and calculations. Indeed, a model is "good" as long it describes a certain particular aspect of the system under investigation, in this case the total energy balance (2.4) reinterpreted in engineering terms using the circuit model (2.6). However, when more detailed information about the dynamic structure of the near field is needed, the impedance concept need not provide the best complete answer as should be obvious to the reader now.

2.3.3 A Generalized Approach to Reactive Energy

Computations of the reactive energy appear (in the antenna theory literature) to have been motivated mainly by demands for estimating the antenna quality factor Q. This can be seen, for example, in the early work on fundamental antenna limitations [50] and the detailed general analysis of the problem given recently in [53]. As is well known, the antenna matching bandwidth, very roughly, is inversely proportional to Q. Since the latter is directly determined by the reactive energy [51], [49], knowledge of this type of energy is very valuable for characterizing the matching performance of *general* antenna systems.

Theoreticians have paid attention to the particular parameter of the antenna size and how the minimum bounding sphere relates to best achievable bandwidth. This line of attack is very well motivated from both the theoretical and practical side. In the former, there seems to be a genuine effect of the antenna size on the expressions of the total energy since they usually involve series expansions that are sensitive to considerations of this kind. In the latter case, applications typically involve difficult fundamental tradeoffs between various conflicting factors, such as compactness (in terms of size reduction), maximum efficiency (especially matching considerations), and bandwidth performance. Therefore, theory can provide – through careful analysis of the generic antenna configuration used in practice – an insight into how the main performance measures of interest are fundamentally interrelated with each other. Based on this kind of knowledge, certain fundamental limitations – which theoretically cannot be bypassed – are explicitly emphasized in order to avoid wasting recourses on forcing unworkable designs to work. Moreover, theory for its own sake is usually fruitful in opening the door for new ideas for applications utilizing latent physical possibilities that were unheeded hitherto for reasons that have to do with employing limited theoretical tools.

Because of the above motivations, we will outline here a generalized approach to the reactive energy problem in generic antenna systems that aims to connect together as many performance measures as possible. In particular, we will focus our attention on the *interrelation* between the following three fundamental aspects:

1. Antenna size.
2. Antenna reactive energy (e.g., bandwidth, Q).
3. Far-field radiation pattern (e.g., polarization, directivity, beam shaping).

The idea is to derive a general expression that involves the three above factors in a *simultaneous* fashion. In this way, the inherent tradeoff between any *combination* of subfactors, like polarization and bandwidth, directivity and size, bandwidth and size, etc, can be studied according to the need of the application at hand. As we will see in Chapter 3, one of the main motivations for our foundational approach to the topic of the antenna near field is to provide the working community with a canonical machinery allowing for a systematic construction of a wide range of 1) fundamental antenna interrelations between standard performance measures, and 2) fundamental antenna limitations on what is achievable in principle by any design process whatsoever. An approach such as this cannot be implemented using numerical tools (the latter may solve only *special* examples), but instead must rely on a *general* mathematical methodology.[3]

It turns out that the general expression promised above does not contain any infinite integral, and that all terms involved can be evaluated in a closed algebraic form. The antenna size a will manifest itself in the form of an infinite series in powers of $1/a$. The effect of the far field will be taken into account by the amplitudes of the TE and TM modes of the antenna problems (which in turn are well-known functions of the antenna current distribution). The reactive energy will then be formed by a certain series expansion to be given below. In this way, by one and the same expressions the three basic performance measures of size, far field, and reactive energy (input impedance) are organically connected within one whole.

3 Only what is mathematical can deal with the *generic*, since one may define mathematics (informally) as 'the science of the generic *per se*.'

The basic idea of the technical approach is very simple and will be described first in informal fashion before highlighting the conceptual aspects of the construction. For full details of the computations, see Chapter 3. The main idea involves combining the classical Wilcox expansion and the multipole expansion together in order to obtain a relation between the *radial* distribution of the field, which is the essential factor in computing the reactive energy, and the *angular* distribution of energy, which is the way to the far field. Incidentally, the series expansion approach itself will force the antenna size to automatically enter into the picture via the fact that the Wilcox expansion is valid only in the exterior region of the source.

We now turn to a closer examination of the nature of the antenna near fields in the spatial domain. Here, we consider the fields generated by the antenna lying in the intermediate zone, i.e., the interesting case between the far zone $kr \to \infty$ and the static zone $kr \to 0$. We suggest that the natural way to achieve this is the use of the Wilcox expansion. Indeed, since our fields in the volume outside the source region satisfy the homogeneous Helmholtz equation, we can expand the electric and magnetic fields as [47]

$$\mathbf{E}(\mathbf{r}) = \frac{e^{ikr}}{r} \sum_{n=0}^{\infty} \frac{\mathbf{A}_n(\theta, \varphi)}{r^n}, \quad \mathbf{H}(\mathbf{r}) = \frac{e^{ikr}}{r} \sum_{n=0}^{\infty} \frac{\mathbf{B}_n(\theta, \varphi)}{r^n}, \tag{2.8}$$

where $\mathbf{A}_n$ and $\mathbf{B}_n$ are vector angular functions dependent on the far-field radiation pattern of the antenna and $k = \omega\sqrt{\varepsilon\mu}$ is the wavenumber. The far fields are the asymptotic limits of the expansion. That is,

$$\mathbf{E}(\mathbf{r}) \underset{r\to\infty}{\sim} \frac{e^{ikr}}{r}\mathbf{A}_0(\theta, \varphi), \quad \mathbf{H}(\mathbf{r}) \underset{r\to\infty}{\sim} \frac{e^{ikr}}{r}\mathbf{B}_0(\theta, \varphi). \tag{2.9}$$

The reason why this approach is the convenient one can be given in the following manner. We are interested in understanding the structure of the near field of the antenna. In the far zone, this structure is extremely simple; it is simply the zeroth-order term of the Wilcox expansion as singled out by (2.9). Now, as we leave the far zone and descend toward the antenna current distribution, the fields start to get more complicated. Mathematically speaking, *this progressive building up in near fields' complexity corresponds to the addition of more terms into the Wilcox series*. The implication is that more terms (and hence the emerging complexity in the spatial structure) are needed in order to converge to accurate solution of the field as we get closer to the current distribution.

Let us then divide the entire exterior region surrounding the antenna into an infinite number of spherical layers, as shown in Figure 2.1. The outermost layer

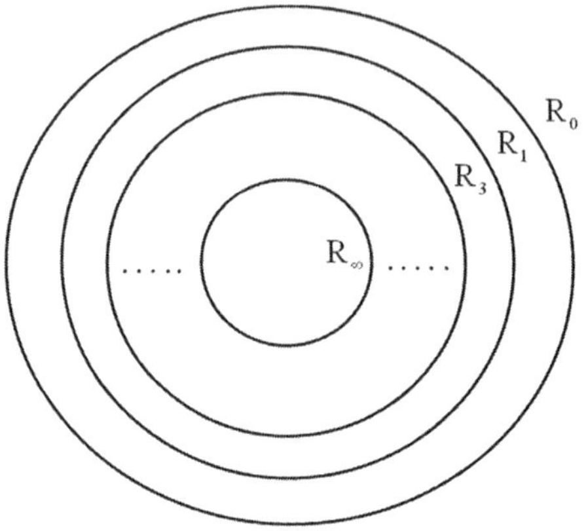

Figure 2.2 General description of antenna near-field spatial structure.

R_0 is identified with the far zone while the innermost layer R_∞ is defined as the minimum sphere totally enclosing the antenna current distribution. In between these two regions, an infinite number of layers exists, each corresponding to a term in the Wilcox expansion. The boundaries between the various regions are not sharply defined, but are to be understood only as indicators in the asymptotic sense.

Since we are interested in the spatial structure of near field, i.e., the variation of the field as we move closer to or farther from the antenna physical body where the current distribution resides, it is natural to average over all the angular information contained in the energy expressions. Therefore, we introduce the *radial energy density function* of the electromagnetic fields by integrating the energy expressions obtained using Wilcox expansion over a full solid angle Ω in order to obtain

$$w_e(r) = \frac{\varepsilon}{4}\sum_{n=0}^{\infty}\frac{\langle \mathbf{A}_n, \mathbf{A}_n\rangle}{r^{2n+2}} + \frac{\varepsilon}{2}\sum_{\substack{n,n'=0\\ n>n'}}^{\infty}\frac{\langle \mathbf{A}_n, \mathbf{A}_{n'}\rangle}{r^{n+n'+2}}, \tag{2.10}$$

$$w_h(r) = \frac{\mu}{4}\sum_{n=0}^{\infty}\frac{\langle \mathbf{B}_n, \mathbf{B}_n\rangle}{r^{2n+2}} + \frac{\mu}{2}\sum_{\substack{n,n'=0\\ n>n'}}^{\infty}\frac{\langle \mathbf{B}_n, \mathbf{B}_{n'}\rangle}{r^{n+n'+2}}, \tag{2.11}$$

where the *mutual interaction* between two angular vector fields $\mathbf{F}$ and $\mathbf{G}$ is defined as

$$\langle \mathbf{F}(\theta,\varphi), \mathbf{G}(\theta,\varphi)\rangle := \int_{4\pi} d\Omega \,\mathrm{Re}\left\{\mathbf{F}(\theta,\varphi)\cdot\mathbf{G}^*(\theta,\varphi)\right\}. \tag{2.12}$$

Equations (2.10) and (2.11) clearly demonstrate the considerable advantage gained by expressing the energy of the antenna fields in terms of Wilcox expansion. The angular functional dependence of the energy density is completely removed by integration over all the solid angles, and we are left afterwards with a power expansion in $1/r$, a result that provides direct intuitive understanding of the structure of the near field since in such a type of series more higher-order terms are needed for accurate evaluation only when we get closer to the antenna body, i.e., for large $1/r$. Moreover, the total energy is then obtained by integrating over the remaining radial variable, which is possible in closed form as can be found in Chapter 3.

We have seen how the Wilcox expansion can be physically interpreted as the mathematical embodiment of a spherical layering of the antenna exterior region understood in a convenient asymptotic sense. The localization of the electromagnetic field within each of the regions appearing in Figure 3.2 suggests that the outermost region R_0, the far zone, corresponds to the simplest field structure possible, while the fields associated with the regions close to the antenna exclusion sphere, R_∞, are considerably more complex.

However, it was pointed out long a time ago that the entire field in the exterior region can be completely determined recursively from the radiation pattern [47]. In Chapter 3 , it is shown that entire exterior region field can be determined (non-recursively) from the far field by a simple construction based on the analysis of the far field into its spherical wavefunctions. Using the classical multipole expansion, we expand the field into TE and TM spherical modes

$$\mathbf{E}(\mathbf{r}) = \sum_{l,m} a_{lm}^{E}\mathbf{E}_{lm}(\mathbf{r}), \mathbf{H}(\mathbf{r}) = \sum_{l,m} a_{lm}^{M}\mathbf{H}_{lm}(\mathbf{r}), \tag{2.13}$$

where a_{lm}^E and a_{lm}^M represent the amplitudes or coefficients of the TE and TM modes, respectively. They can be determined directly if the current distribution on the antenna is known, see [33] for details.

In Chapter 3, the Wilcox series is derived from the multipole expansion, where the exact variation of the angular vector fields $\mathbf{A}_n$ and $\mathbf{B}_n$ were directly determined in terms of the spherical far-field modes of the antenna. In particular, we managed to express all of the interaction integrals appearing in the general expression of the antenna radial energy density (2.10) and (2.11) in the exterior region in closed analytical form involving only the TM_{lm} and TE_{lm} modes excitation amplitudes

$a_M(l,m)$ and $a_E(l,m)$. The results turned out to be intuitive and comprehensible if the entire space of the exterior region is divided into spherical regions understood in the asymptotic sense as shown in Figure 2.2. In this case, the radial energy densities (2.10) and (2.11) are simple power series in $1/r$, where the amplitude of each term is simply the mutual interaction between two regions. From the basic behavior of such expansions, we now see that the closer we approach the exclusion sphere that directly encloses the antenna current distribution, i.e., what we called region R_∞, the more terms we need to include in the energy density series. However, the logic of constructing those higher-order terms clearly shows that only higher-order far-field modes enter into the formation of such increasing powers of $1/r$, confirming the intuitive fact that the complexity of the near field is an expression of richer modal content where more (higher-order) modes are needed in order to describe the intricate details of electromagnetic field spatial variation. As a bonus we also find that the complex behavior of the near field, i.e., that associated with higher-order far-field modes, is localized in the regions closer to the antenna current distribution, so in general the nearer the observation to the limit region R_∞, the more complex becomes the near-field spatial variation.

Finally. it is interesting to note that almost "half" of the interactions giving rise to the amplitudes of the radial energy density series (2.10) and (2.11) are exactly zero— i.e., the interactions between regions R_n and R'_n when $n + n'$ is odd is identically zero. This represents, in our opinion, a significant insight into the nature of antenna near fields in general. For now, *it turns out this observation about the energy exchange processes is fundamentally connected to the Principle of Finite Energy Difference* (2.7).

We next turn to the final step in our generalized approach, constructing the reactive energy directly using the above results in terms of the modal content of the field. Our derivation of the reactive energy will proceed along the following general line of thinking. We have seen that a direct computation of the total energy leads to divergent result (2.2). However, The Principle of Finite Energy Difference (2.7) suggests that there is a *common* term between w_e and w_h, which is the source of the trouble in calculating the total energy of the antenna system. Our approach must eventually prove this principle itself, i.e., show explicitly that (2.7) does hold in general antenna system. In order to achieve this, we first *claim* that there is indeed a common divergent term between the electric and magnetic energy densities and see where this will take us. We write then

$$w_e := w_e^1 + w_{\mathrm{rad}}, \quad w_h := w_h^1 + w_{\mathrm{rad}}. \tag{2.14}$$

Here w_e^1 and w_h^1 are taken *tentatively* as *reactive energy densities* and our goal is to show that they are finite. In Chapter 3, we show that *the total reactive energy is finite*. Moreover, we will see that total reactive energies (expressions (3.69) and (3.70) in Chapter 3) are evaluated completely in analytical form and that *in principle* no computation of infinite numerical integrals is needed for evaluating reactive energy.

2.3.4 The Limits of the Reactive Energy Concept and the Need to Move Beyond

In Chapter 3, an analysis will be presented as a generalized approach to reactive energy (outlined in Section 2.3.3) aiming at understanding the physical meaning of reactive energy. We will not repeat that argument here, which is quite simple from the mathematical point of view, but summarize mainly its conclusion and reflect on its significance for the working scientist and engineer.

It will be found that the terms called "reactive energy density" w_e^1 and w_h^1 in (2.14) are *not* unique. Indeed, there exists an infinite number of equally legitimate reactive energy densities, say w_e^2 and w_h^2, all can replace their corresponding parts in (2.14) and equally reproduce the main results concerning the input impedance of a general antenna system; moreover, this ambiguity is formally captured by the Principle of Finite Energy Difference. We would like now to further emphasize this line of thought by explicitly stating the following proposition

> The individual electric and magnetic reactive energy *densities* are *fundamentally ambiguous* in the sense that there are no *unique* such densities that can reproduce the standard results concerning the input impedance of general antenna systems. Indeed, for each total electric/magnetic energy difference (i.e., input impedance reactance), there exists an infinite number of reactive energy densities, all differing in magnitude, but eventually reproducing the same prediction of the net input antenna reactance.

In practice, since only the physically observable antenna impedance is deemed meaningful (because it is "useful,") this ambiguity was not even noticed. The *conventional* solution was considered *the* solution, although it is actually only one possible solution, probably the most direct one but nevertheless not theoretically special. However, for applications more sophisticated than merely matching the antenna, one needs to know whether the reactive energy previously defined in terms of the input impedance is really the physical energy of the antenna. We now immediately conclude based on the analysis above the following proposition

> In electromagnetic theory, the concept of energy density is fundamental. However, in antenna theory conventional reactive energy density cannot be defined in a unique fashion. Therefore, it is unlikely that the actual physically stored energy in the antenna system is the reactive energy. In other words, reactive energy and stored energies are not identical concepts. In particular, numerical computation of the reactive energy does not give a physical estimate of stored energy.

Engineers should be aware then of the theoretical limitations of some of the widely used concepts, like quality factor and matching bandwidth. The antenna system is a vastly more complex organism in which rich processes of energy conversion, creation, and annihilation seem to be taking place in the near-field zone. This extended picture is described briefly in Chapter 4, where we focus on mainly one concept, *localized* energy, which turns out to be very different from reactive energy.[4] As we mentioned earlier, it seems that each energy concept requires a special methodology. The reactive energy relies heavily on the engineering way of thinking, i.e., circuits and impedances. The localized energy will require the *space-time approach*, first developed by the authors in Chapter 4, where mathematics seems to be providing the main tools. In Section 2.4, however, in which the topic is left in incomplete form, we will speculate that definitive attack on the problem of stored energy may require extending both circuit theory and the time-harmonic localized energy concept in order to deal with time-varying *field* transient phenomena.

2.4 REMARKS ON STORED ENERGY

Now, what then is stored energy? This quantity, regardless of how it will eventually be computed or measured, must not rest on merely the total field, but on some well-defined *sub*-portion of this total field that is somehow both

1. localized in space,
2. recoverable.

By *localization* we understand a nonpropagating field in a suitable sense. The space-time approach in Chapter 4 provides a tentative mathematical approach to the problem of how to define the nonpropagating field in the antenna in a way that is meaningful from both the physical and engineering viewpoint. While the details are technically involved, the progress attained so far suggests that some

4 Although not totally unrelated in certain circumstances.

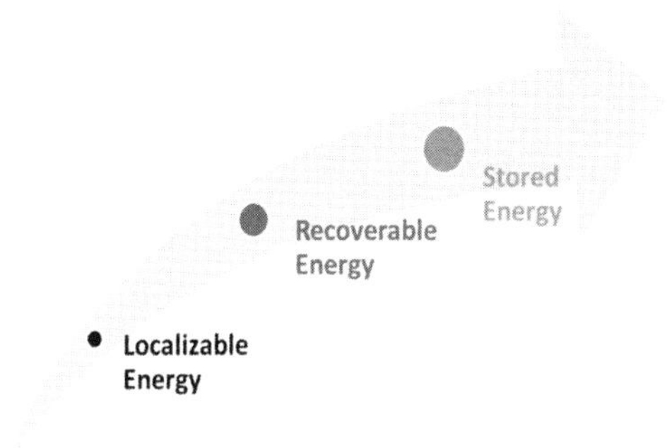

Figure 2.3 The general pattern of working with stored energy in radiating systems. Regardless of the actual final definition to be adopted in measuring/computing the stored energy, it should satisfy the requirement of localizability and recoverability.

version of "radial localization" of energy is more promising than other possible definitions of localization. The study of how energy is moving dynamically in the space surrounding the antenna can be put into a more fundamental level by deriving a set of differential equations describing the flow as outlined in Chapter 6.

By *recoverable* we require that the physically stored energy be real in the sense of being capable of either measurement or reutilization in a real actual world setting. Stored energy should be, at least in principle, extractable from the antenna system. This idea is advanced informally in Chapter 4 and still remain undeveloped from experimental viewpoint. The basic proposal here is to construct a working setup in which a radiating antenna system is suddenly switched off. It is expected that the proper localized energy existing in the local vicinity of the source will undergo two possibilities: either part of it will convert into radiating field and escape into the far zone, or/and part of the localized energy will couple into the feed and emerge in the antenna circuit part. By measuring or computing both energies, estimations of the physically realizable stored energy in the antenna can be found. It is suggested in Chapter 4 that there are no reasons to expect that reactive energy, localized energy, and stored energy are numerically equal. To our knowledge, this experiment has not been carried out yet.

2.5 BEYOND REACTIVE ENERGY

It is very important at this juncture to set the stage for a more decisive contribution to the problematic issue of how *electromagnetic energy* and the *near field* are related to each other. This will first require coming back into the topic of what is electromagnetic energy *as such* and how it is important in applications.

It is standard in classical electromagnetic theory to work with a definition of energy densities given by the following expressions

$$w_e(\mathbf{r}) = \frac{1}{4\varepsilon}|\mathbf{E}(\mathbf{r})|^2, \quad w_h(\mathbf{r}) = \frac{1}{4\mu}|\mathbf{H}(\mathbf{r})|^2. \tag{2.15}$$

Here, frequency-domain (time-harmonic) conditions are assumed. The medium is described by the isotropic constitutive relations ε and μ. For simplicity, we don't treat anisotropic and/or spatially dispersive media. In fact, most of the discussion to follow will even refer to energy processes in free space.

Based on the expressions (2.15), we can compute the *total* electromagnetic energies inside any region $\mathcal{D}$ in space by the integrals

$$W_e(\mathcal{D}) = \int_{\mathcal{D}} d^3r\, w_e(\mathbf{r}), \quad W_h(\mathcal{D}) = \int_{\mathcal{D}} d^3r\, w_h(\mathbf{r}). \tag{2.16}$$

Mathematically speaking, classical electromagnetic theory provides us only with knowledge encapsulated by (2.16), i.e., that a certain continuous map exists between closed subsets of the antenna exterior region (say in free space) and the real line $\mathbb{R}$. Let the exterior region be V_{ext}.[5]

Unfortunately, the two energy maps described above can seldom be of much practical value in applied electromagnetics. Indeed, although the computation of the total energy appears to be adequate for the tasks and aims usually prevailing in a discussion of classical electromagnetic theory in the physics literature, the situation appears to be different in the case of the applied community. Explicating the nature of the difference between theoretical and applied electromagnetics is one of the main goals of the present discussion.

In applied settings, the main object of investigation is a *device* working within an electromagnetic environment. The device, by definition, is not merely a physical structure, but possesses also a *function* to be performed. This juxtaposition of both

5 This map associates with each region $\mathcal{D} \subseteq V_{\text{ext}}$ real numbers W_e and W_h measuring the total electric and magnetic energies, respectively, enclosed in the interior of $\mathcal{D}$; in symbols, the maps $W_e : \mathcal{P}(V_{\text{ext}}) \to \mathbb{R}$ and $W_h : \mathcal{P}(V_{\text{ext}}) \to \mathbb{R}$.

structure and function in one object is the signature of the engineering sciences and how it is usually differentiated from the physical sciences. Taking into account such a well-known demarcation between the two fields, the natural question is now the following: What is the main function to be performed by electromagnetic devices? Our most general and simple answer will the following: *Moving electromagnetic energy from one location to another.*

This function can be used to explain the operations of all known electromagnetic devices. Radiation, filtering, coupling, interaction, localization, and so forth, all involve in some way or another a question about how some energy quantum is being transferred. In a more precise manner, we speak about how the energy contents of the total system are being differentiated, divided, manipulated, stored, and so forth. *Energy motion* is then the key guiding concept to be taken into account throughout the current work.

It is at this point that we find the classical approach in theoretical electromagnetics to energy not fully suitable for our needs. The total energies given by (2.16) don't describe in detail how the energy content is actually flowing inside the region. The Poynting theorem only gives a net flow along a surface inclosing $\mathcal{D}$, which is numerically equal to the time rate of the *total* energy change. However, the total energy itself need not be actually composed fully of propagating modes. This is particularly true in the near field. As is well known, close to the device the total field tend to be "static," or nonpropagating. Application of the Poynting theorem in this region usually do not provide detailed knowledge of the dynamic structure of the field. In contrast, we find that in most of the computations involving the Poynting vector one typically works in the far zone, as in optics for example.

Besides the problem of the near field, there is also the important issue of analyzing the performance of radiating systems in the entire exterior region. In this case, the region V_{ext} becomes infinite. Computations of the total energy using the expressions (2.16) give divergent values. There has been a long history of research in applied electromagnetic theory aiming at extracting some useful *finite* numbers from the otherwise infinite integrals (2.16). Such approaches leads to topics involving reactive energy, quality factors, input impedance matching bandwidth, and so forth. However, they also point to a more fundamental problem, that involving the nature of *stored energy* as such. Stored energy can be defined in various ways. It is not very clear at the present time what is the definitive approach to the subject. No exhaustive survey of the topic is therefore necessary here. Instead, some general conceptual considerations will be mentioned and their connection with existing applied electromagnetic research will be pointed out.

Technical discussions of the topic of stored energy tend to be mired with tedious computational details that may unfortunately obscure the essentials, which are nevertheless easy to grasp. For this reason, we will develop here a simple model, basically an example, where the main ingredients of the topic are illustrated. Suppose that the field existing in the exterior regions of some antenna is given by the expression

$$\mathbf{E}(\mathbf{r}) = A_e e^{-\alpha r} + A_p e^{ikr}, \tag{2.17}$$

where both α and k are real while the amplitudes A_e and A_p are complex. In other words, the total field consists of two parts, an evanescent mode $A_e \exp(\alpha r)$ and a propagating wave $A_p \exp(-kr)$. This decomposition is permissible since both modes satisfy the Helmholtz equation in free space and consequently a superposition of two solutions is another solution satisfying Maxwell's equations.

Next, suppose we are concerned with studying the nature of the energy content inside a region $\mathcal{D}$. Traditional electromagnetic theory knows only the total field, that given by the sum of both the propagating and evanescent modes in (2.17). That is, we can compute the numbers $W_e(\mathcal{D})$ and $W_h(\mathcal{D})$, and those involve the total field levels. However, from the perspective of the antenna or device engineer, it is plain that the energy associated with the evanescent modes is somehow localized or fixed in the space region around the origin. The reason is that the decaying exponential $A_p \exp(-\alpha r)$ has insignificant values away from $r = 0$. On the other hand, the energy content of the pure propagating wave $A_p \exp(ikr)$ is constant and is given by $|A_p|^2$. *This implies that the energy associated with the propagating part spreads equally in the entire region while the part corresponding to the evanescent wave is concentrated in a smaller subregion.*

If we let $\mathcal{D}$ approach the entire exterior domain where the expression (2.17) is valid, then we obtain the following by simple calculations

$$W_e(\mathcal{D}) = \int_{\mathcal{D}} d^3r \left(A_e e^{-2\alpha r} + 2\Re(A_e A_p^*) e^{-\alpha r} + |A_p|^2 \right). \tag{2.18}$$

Although the first and second integrals in (2.18) *do* converge in infinite regions, the third term, i.e., the constant $|A_p|^2$ *causes the integral to diverge*. In other words,

$$\lim_{\mathcal{D} \to V_{\text{ext}}} W_e(\mathcal{D}) = \infty. \tag{2.19}$$

What is the physical meaning of such a straightforward calculation? We believe the most important lesson to be learned from the simple example above is that *energy localization, and related to it somehow energy storage, is not connected*

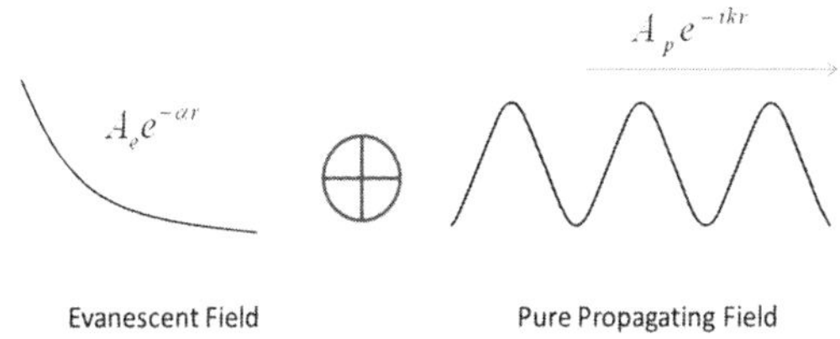

Figure 2.4 Simple example of an electromagnetic field divided into two modes, one is pure propagating and the other is evanescent.

with the total field, but to a part of the total field. This direct conclusion was implicitly taken for granted in the traditional approach to reactive energy, where a term proportional to radiation density (in the far field) was subtracted from the total energy. The remaining quantity (after subtraction) turned out to be convergent and is sometimes interpreted to be the stored energy of the antenna system.

Regardless of the details, what is relevant here is that *applied* electromagnetic theory requires a more subtle treatment of the energy concept than what has been traditionally achieved in theoretical electromagnetics. This increase in subtlety is dictated by the demands of applications, in particular to the fact that understanding electromagnetic *devices* involves delicate matters pertinent to the problem of energy motion. The latter necessitates in turns differentiating subcomponents in the total field that are "moving" and delimitating them in contrast to a background of "unmoving" or "static" fields.

Now what then is stored energy? This quantity, regardless of how it will eventually be computed or measured, must involve not the total field, but some well-defined subpart of this field that is somehow both 1) localized in space and 2) recoverable. By *localization* we understand a nonpropagating field in a suitable sense. By *recoverable* we require that the physically stored energy be real in the sense of being capable of either measurement or reutilization in actual world setting. Stored energy should be, at least in principle, extractable from the antenna system.

Let us go back now to the simple example (2.17). It is very clear that the pure propagating part $A_p \exp{(ikr)}$ violates both of the defining conditions of stored energy proposed above. This wave is the extreme opposite of anything localized: it actually spreads everywhere. Worse still, being a pure wave, it is always escaping to infinity, and hence cannot be recaptured anywhere else. On the other hand, the evanescent part $A_e \exp{(-\alpha r)}$ satisfies a condition of localizability. However, it is not immediately clear whether the entire energy content of this mode can always be recovered. The topic of stored energy gets its peculiar difficulty from exactly such considerations.

2.6 RELEVANCE OF A FUNDAMENTAL UNDERSTANDING OF ELECTROMAGNETIC ENERGY TO OTHER MAINSTREAM RESEARCH DIRECTIONS

The issues of main considerations in the topic of electromagnetic energy includes the following:

Renewable Energy. There has recently been a great interest in managing energy resources not only in electromagnetics, but also in other physical domains like the mechanical, chemical, and biological. In the case of applied electromagnetics, there is certainly the question of what constitutes energy utilization as such. It should be clear from the above passages that even in a very basic device such as radiating antenna, the ultimate physical status of the energy concept is by no means clear. This motivates, we think, the need to invest more in research targeting the fundamentals of the subject matter in a way that is general and rigorous. The benefits from this line of investigations are manifolds. First, we expect that new results regarding the definition of energy will open the door for fresh ideas regarding just how to efficiently handle the available energy content of arbitrary electromagnetic systems, leading to improvements in design, efficiency, multifunctionality, etc. Second, knowing more about the nature of electromagnetic energy may permit deeper understanding of how other physical processes (mechanical, chemical, biological) can be related in a fruitful manner to the electromagnetic case. This will lead to further progress in multidisciplinary areas like nanotechnology, biotechnology, health systems, where the interaction between varieties of natural fields becomes indispensable. Third, our planet is now fully penetrated by carbon-biotic materials such as silicon and electromagnetic systems are contributing radiation almost everywhere. There has not been much research on the fundamental nature of this type of radiation coming from manmade devices and how it affects the biosphere and the environment at large. Asking general questions about energy in electromagnetic systems is therefore not irrelevant to current concerns with global environment aspects.

Nanotechnology. The issue of energy in nanotechnology is quite complex due to the inherently multidisciplinary nature of the topic. Indeed, thermodynamic considerations plays an essential role in nano-transport problems where the connection with electromagnetic radiation and interaction is often dropped in order to simplify the analysis. In pure radiation problems, for example nano-antennas, the effect of spatial scale is also not taken explicitly into consideration. Indeed, due to the smallness of the spatial extent of the nano-system, subwavelength components (evanescent modes) may become physically very prominent, bringing into the picture new phenomena typically not addressed in classical macroscopic electromagnetics. We think that a clarification of the energy concept acquires an acute importance in this field of research because of the unexplored richness of the *electromagnetic* nano-environment and the need also to understand the interaction between multiple spatial scales (e.g., a nano-system performs its hardcore function at the nano-scale, but real-world measurement occurs at a large or macroscopic scale, so one must have a theory for how energy is computed in the two scales and their interfaces). Finally, the topic of the *near field* becomes fundamental in nano-systems since this is usually the setting of the device (each nano-antenna or nano-circuit is always in close interaction with other nano-object nearby. Nano-devices are always in the near-field zone of each other and their environment). The connection between the near field and electromagnetic = energy has been noticed by many and was discussed briefly above. It represents, in our opinion, one of the most promising directions in which a reconsideration of the energy concept in applied electromagnetics can bring valuable and positive contribution to existing research fields.

Metamaterials. A metamaterial is defined as an effective artificial medium exhibiting an electromagnetic performance not typically seen in natural materials. The main issue here is the homogenization of a large number of basic unit cells ("atoms") in order to produce a macroscopic behavior that is interesting for applications. In our opinion, energy considerations are crucial here. Indeed, one of the most basic features exploited in artificial materials in order to achieve new electromagnetic behavior is a sort of dispersion engineering. Since most metamaterials are highly dispersive, and given the well-known fact that the definition of energy is problematic when there is strong dispersion, it immediately follows that a fundamental investigation of energy in this direction of research is very important and timely. On the other hand, it is necessary to obtain a better understanding of stored energy for the purpose of researching the potential of using metamaterials for energy harvesting, storage, localization, processing, etc. The conventional definitions of energy in classical electromagnetics are not sufficient enough for this regard since they treat mainly the total field. Based on what we saw previously, it is highly probable that developing a theory of electromagnetic energy in dispersive media may involve dealing not with total field, but with dynamically relevant sub-components, for example proper nonpropagating waves.

Fundamental antenna limitations. Deign directives and insights. This is by large the most well researched field in which energy is usually related to a mainstream topic. Energy here enters in the guise of "reactive energy" within the context of computing the Q factor of radiating systems. The main objective of this calculation is usually estimating the input impedance matching bandwidth. The area of small antennas has dominated this research direction where considerable progress was attained. It appears, however, that detailed examination of the reactive energy can be made not only for small antennas (which in any case have simple well-known field structure, that of infinitesimal dipoles), but also for arbitrary size antennas. It is of interest to develop fundamental antenna limitations for antenna types other than the small antennas, and to bring into focus not only impedance matching considerations to applications, but also the device size and the far field into interplay in a unified and systematic manner. To summarize, understanding the spatial structure of the reactive energy may help bringing out new insights into design issues not restricted to matching small antennas.

2.7 CONCLUSION

The most important conceptual lesson we aim to achieve from writing on the topic of energy is the following. Applications necessarily involve not the total field, but the *part* of the field that is propagating. In terms of energy, that means the *portion* of energy that is moving. For example, stored energy is some *part* of the total energy that is both localizable and recoverable. In this sense, we look at stored energy as something similar to available energy in thermodynamics. As is well known, not all the heat supplied to an engine can be converted into real work, but only a part of it called available energy. The situation in electromagnetic theory is somehow analogous but has not received the attention it deserves from either theoreticians or practitioners. In conclusion, we believe that for *devices* focus must shift away from the total energy into some reasonable definition of "nonpropagating" energy, localized energy, and finally stored energy.

We have not discussed the technical literature on the topic. Some of the few papers dealing with the subject approach the problem from very different perspectives, for example fundamental antenna limitations, time domain, and near fields. It is expected that nanoelectromagnetics, plasmonics, and metamaterials will provide a strong stimulus for new research in the foundations of applied electromagnetics.

The chapter has provided a comprehensive but compact view on the topic of electromagnetic energy, with special emphasis on the widely used reactive energy in antenna practice and research. The concept of energy was discussed from a general perspective before analyzing in detail the conceptual structure of a new approach to the topic developed recently by the authors. The main goal of the analysis was to highlight the essential key points in our argument and how it fits into existing themes in the community that we believe needs more careful examination. It was shown that the new results derived by the authors can be used within certain contexts to generalize the topic of fundamental antenna limitations by including the far field, antenna size, and input impedance in a unified framework. In particular, it was suggested that fundamental antenna limitations need not be all about small antennas, and that arbitrary shaped and sized devices can be perfectly investigated using our formulation. We have shown subsequently that there are limitations to the physical validity of reactive energy when interpreted as stored energy. Indeed, certain formal considerations demonstrated the physical ambiguity of reactive energy densities when attempts are made to extend their range of physical significance beyond the input impedance, for example, such as studying the structure of the near field and energy localization and control in the vicinity of the source. Finally, brief remarks about stored energy were given and some clues about its nature and how to measure or compute it were briefly discussed.

Chapter 3

The Spatial Theory of Electromagnetic Fields

3.1 INTRODUCTION

3.1.1 Motivations for the Search for a Theory of Antenna Near Fields

Antenna practice has been dominated since its inception by the research of Hertz by pragmatic considerations, such as how to generate and receive electromagnetic waves with the best possible efficiency, how to design and build large and complex systems, including arrays, circuits to feed these arrays, and the natural extension toward a more sophisticated signal processing done on site. However, we believe that the other aspects of the field, such as the purely theoretical, nonpragmatic study of antennas for the sake of knowledge-for-itself, is in a state altogether different. We believe that to date the available literature on antennas still appears in need of a sustained, comprehensive, and rigorous treatment for the topic of *near* fields, a treatment that takes into account the peculiar nature of the electromagnetic behavior at this zone.

Near fields are important because they are operationally complex and structurally rich. Away from the antenna, in the far zone, things become predictable; the fields take a simple form and approach plane waves. There is not much to know about the behavior of the antenna aside from the radiation pattern. However, in the near zone, the field form cannot be anticipated in advance like the corresponding case in the far zone. Instead, we have to live with a generally very complicated field pattern that may vary considerably in qualitative form from one point to another. In such situations, it is meaningless to search for an answer to the question: What is the near field everywhere? since one has at least to specify what kinds of structures he is looking for. In light of being totally ignorant about the particular source excitation of the antenna, the best one can do is to rely on general theorems derived from Maxwell's equations, most prominently the dyadic Green's function theorems. But even this is not enough. It is required, in order to develop a significant, nontrivial theory of near fields, to look for further structures separated off from this Green's functions of the antenna. We propose in this work (Chapter 4) the idea of propagating and nonpropagating fields as the remarkable features in the electromagnetic fields of relevance to understanding how antennas work.

The common literature on antenna theory does not seem to offer a systematic treatment of the near field in a *general* way, i.e., when the type and excitation of the antenna are not known *a priori*. In this case, one has to resort to the highest possible abstract level of theory in order to formulate propositions general enough to include all antennas of interest. The only level in theory where this can be done, of course, is the mathematical one. Since this represents the innermost core of the structure of antennas, one can postulate valid conclusions that may describe the majority of current or potential applications. In this context, engineering practice is viewed methodologically as being commensurate with physical theory as such, with the difference that the main object of study in the former, antennas, is an artificially created system, not a natural object per se.

Antenna theory has focused for a long time on the problems of analysis and design of radiating elements suitable for a wide variety of scientific and engineering applications. The demand for a reliable tool helping to guide the design process led to the invention and devolvement of several numerical tools, such as method of moments, finite element method, and finite difference time-domain method, and so on, which can efficiently solve Maxwell's equations for almost any geometry, and corresponding to a wide range of important materials. While this development is important for antenna engineering practice, the numerical approach, obviously, does not shed light on the deep structure of the antenna system in general. The reason for this is that numerical tools accept a given geometry and generate a set of numerical data corresponding to certain electromagnetic properties of interest related to that particular problem at hand. The results, being firstly numerical, and secondly related only to a *particular* problem, cannot lead to significant insights on *general* questions, such as the nature of electromagnetic radiation or the inner structure of the antenna near field. An insight like this, however, can be gained by reverting to some traditional methods in the literature, most conveniently expansion theorems for quantities that proved to be of interest in electromagnetic theory, and then applying similar tools creatively to the antenna problem in order to gain a knowledge as much general as possible.

The engineering community is generally interested in this kind of research for several reasons. First, the antenna system is an engineering system *par excellence*; it is not a natural object, but an artificial entity created by humans to satisfy certain pragmatic needs. As such, the theoretical task of studying the general behavior of antennas, especially the structural aspects of the system, falls, in our opinion, into the lot of engineering science, not physics proper. Second, the working engineer can make use of several general results obtained within the theoretical program of the study of antenna systems as proposed in this book, and pioneered previously by many [50, 51, 49, 56, 52, 53]. These general results can give useful information about the fundamental limitation on certain measures, such as quality factor, bandwidth, cross-polarization, and gain, and so on. It is exactly the generality of such theoretical derivations that makes them extremely useful in practice. Third, more knowledge about fields and antennas is always a positive contribution even if it does not lead to practical results at the immediate level. Indeed, future researchers, with fertile imagination, may manage to convert some of the mathematical results obtained through a theoretical program of research into a valuable design and devolvement criterion.

3.1.2 Philosophy of the New Theoretical Program

This chapter, as can be discerned from its title, is the laying down for a new foundation upon which the edifice of the theoretical structure of antenna near fields may be constructed by researchers in the future. The guiding thread in our investigation is the search for a suitable theoretical level at which general information about the antenna can be summarized in the most economical form. By observing the pattern of the evolution of the physical and mathematical sciences throughout the last one hundred years, it appears to us that the most fundamental level at which a scientific investigation can be enacted is the *topological* one.

There are two levels of structures that may be exploited in formulating a general theory of the antenna near field. The first is the topological layer. It consists of systematic abstraction from the real number system used in representing electromagnetic phenomena within the framework of Maxwell's equations. The second level is the physical layer. This consists of any input arising from observing the way physical quantities actually behave when represented by quantities selected from the topological layer. For example, we mention here the invariance properties of the electromagnetic fields, which are expressions of the *continuous* symmetry of the underlying topological base structure when equipped with additional structures like differentiability and geometry.[1]

The crux of our theoretical program is to start from the physical layer and build our way back to the highest abstract level, that of the topological layer. This task is accomplished in the following manner. First, the rotational invariance of scalar sources in electromagnetic theory are observed.[2] We exploit this symmetry in the interesting case where there are several point sources coexisting next to each other. The rotational symmetry of each source is broken by choosing a particular orientation of the coordinate system along which we attempt a mathematical description of the electromagnetic near field.[3] We therefore move from perfect (rotational) symmetry to superposition of a multiplicity of broken symmetries, all coexisting and hence resonating with each other. This picture will be described in detail in Chapter 4 from an engineering point of view while the fundamental level will be treated elsewhere. In the more advanced stage, we will derive a system of nonlinear ordinary differential equations describing the actual dynamic structure of the field. We then search for the *singularities* of this system (critical points in the phase space of the problem). It is the topological structure of these singularities that is, in our opinion, the broadest abstract level in the scientific description of antennas. Since singularities are eminent points in the phase space structuring *potential* solutions of the problem starting from given initial conditions, a topological description of the distribution of such singularities is naturally the best general way so far to encode all the relevant information of a given antenna. This encoding provides a deeper understanding of how the current distribution may vary in order to effect certain structural changes in the near fields, or how sensitive the whole antenna system field is to a continuous change in some parameters of interest, and so on. Once this topological understanding of the antenna problem is attained, we can work our way again toward the physical layer, and then back to the topological layer in light of new questions, problems, and proposals. *It is this immanent resonant interrelation that communicates back and forth between the topological and physical layers that should be taken as the ultimate object of any scientific theory.* A physical theory sensitive to both the fundamental structural principles and the operational behavior

1 The rotational and translational symmetry of the electromagnetic fields are consequences of modeling the physical phenomena using the topology of Euclidean spacetime. However, it is *only* through *observations* that we come to know that phenomena appearing in the real world are indeed well-behaved with respect to certain general symmetry principles.

2 Notice again that this is an expression of structures at the physical layer.

3 As will appear later, this 'mathematical description' is chosen to be a decomposition of the total field into propagating and nonpropagating parts.

of radiating electromagnetic systems must address this dual mode of theoretical description.

3.1.3 Overview of the Chapter

At the most general level, this chapter, will study the antenna near-field structure in the *spatial* domain, while the main emphasis of Chapter 4 will be the analysis this time conducted in the *spectral* domain. The spatial domain analysis will be performed via the Wilcox expansion while the spectral approach will be pursued using the Weyl expansion. The relation between the two approaches will be addressed in the final stages of Chapter 4.

In Section 3.2, we formulate the antenna system problem at the general level related to the near-field theory to be developed in the following sections. We don't consider at this stage additional specifications such as dispersion, losses, anisotropicity, since these are not essential factors in the near-field description to be developed in chapter 3 using the Wilcox expansions and in Chapter 4 using the Weyl expansion. Our goal will be to set the antenna problem in terms of power and energy flow in order to satisfy the demands of the subsequent sections, particularly our treatment of reactive energy in Section 3.6.

In Section 3.3, we start our conceptualization of the near field by providing a physical interpretation of the Wilcox expansion of the radiation field in the antenna exterior region. Here, the spatial structure is defined as a layering of this region into spherical regions understood in the asymptotic sense such that each region corresponds to a term in the Wilcox expansion. In Section 3.4, we support this description by showing how to construct the electromagnetic field in all these regions starting from the far-field radiation pattern and in a direct, nonrecursive fashion. This will provide a complete and exact mathematical description for the near field of the class of antennas that are compatible with a given radiation pattern and also can be fit inside the innermost region defined in the spatial configuration introduced in Section 3.3. We then use these results to study the phenomenon of electromagnetic interaction between all the spherical regions comprising the antenna field in the exterior region. Section 3.5 provides a complete set of expressions for the self and mutual interactions, quantifying then the details of the energy exchange processes occurring between various spatial regions in the antenna surrounding domain. Of particular interest, we prove that the mutual interaction between "halfs" of these regions is exactly zero.

In Section 3.6, we reexamine the traditional concept of reactive energy. The main contribution here resides in utilizing the Wilcox expansion of the exterior electromagnetic fields in order to compute the reactive energy in a complete analytical form. As it turns out, no infinite numerical integral is needed in principle for computing the antenna reactive energy and hence the quality factor. We also show that the reason why the reactive energy is finite has its roots in the general theorem proved in Section 3.5, which states that the energy exchange between some regions in the exterior domain is exactly zero. The application of this theorem will show that a term in the energy density series cancels out what would otherwise give rise to logarithmic divergence in the total reactive energy.

We then provide a demonstration of the inherent ambiguity in the traditional definition of the reactive energy when the field distribution in the near zone is examined more carefully. The existence of such an ambiguity renders the concept of reactive energy, designed originally for the study of the RLC circuit model of the antenna input impedance, of limited value in describing the antenna as a *field* oscillator, rather than being the corresponding circuit concept. Finally, to prepare for the transition to Chapter 4, we compute the total energy in a spherical shell around the antenna and express it as a power series in $1/r$. This analysis of the near-field shell reveals the maximum information that can be discerned about the near-field structure in the spatial domain from the far-field perspective.

3.2 GENERAL CONSIDERATION FOR ENERGETICS AND POWER FLOW IN ANTENNA SYSTEMS

The purpose of this section is to carefully review the general knowledge we can infer from Maxwell's equations regarding the energy and power dynamics surrounding arbitrary antenna systems. The radiation problem is very complicated. At this preliminary stage, we need to examine how much information can be deduced from the mathematical formalism of electromagnetic theory concerning radiation problems in a way that does not fall under the restrictions of particular antenna geometries and/ord excitations. Given the complexity of the problem thus described, we need to critically reflect on what has been already achieved so far in antenna theory, particularly as developed by the electrical engineering community.

Consider the general radiation problem in Figure 3.1. We assume that an arbitrary electric current $\mathbf{J}(\mathbf{r})$ exists inside a volume V_0 enclosed by the surface S_0. Let the antenna be surrounded by an infinite, isotropic, and homogeneous space with electric permittivity ε and magnetic permeability μ. The antenna current will

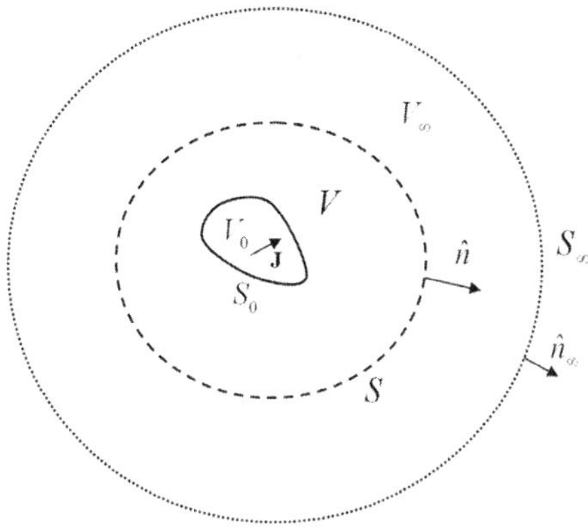

Figure 3.1 General description of antenna system.

radiate electromagnetic fields everywhere and we are concerned with the region outside the source volume V_0. We consider two characteristic regions. The first is the region V enclosed by the spherical surface S and this will be the setting for the near fields. The second region V_∞ is that enclosed by the spherical surface S_∞ taken at infinity and it corresponds to the far fields. The complex Poynting theorem states that [33]

$$\nabla \cdot \mathbf{S} = -\frac{1}{2}\mathbf{J}^* \cdot \mathbf{E} + 2i\omega\left(w_h - w_e\right), \tag{3.1}$$

where the complex Poynting vector is defined as $\mathbf{S} = (1/2)\,\mathbf{E}\times\mathbf{H}^*$ and the magnetic and electric energy densities are given, respectively, by

$$w_e = \frac{1}{4}\varepsilon\mathbf{E}\cdot\mathbf{E}^*, \quad w_h = \frac{1}{4}\mu\mathbf{H}\cdot\mathbf{H}^*. \tag{3.2}$$

Let us integrate (3.1) throughout the volume V (near-field region). We find

$$\begin{aligned}\int_S d\mathbf{s}\tfrac{1}{2}\left(\mathbf{E}\times\mathbf{H}^*\right) = \int_{V_0} dv\left(-\tfrac{1}{2}\mathbf{J}^*\cdot\mathbf{E}\right)\\ +2i\omega\int_V dv\left(w_m - w_e\right).\end{aligned} \tag{3.3}$$

The divergence theorem was employed in writing the LHS while the integral of the first term on the RHS was restricted to the volume V_0 because the source current is vanishing outside this region. The imaginary part of this equation yields

$$\begin{aligned}\operatorname{Im} \int_S d\mathbf{s}\tfrac{1}{2}\left(\mathbf{E}\times\mathbf{H}^*\right) = \operatorname{Im} \int_{V_0} dv \left(-\tfrac{1}{2}\mathbf{J}^*\cdot\mathbf{E}\right) \\ +2\omega \int_V dv \left(w_h - w_e\right).\end{aligned} \tag{3.4}$$

The real part leads to

$$\operatorname{Re} \int_S d\mathbf{s}\frac{1}{2}\left(\mathbf{E}\times\mathbf{H}^*\right) = \operatorname{Re} \int_{V_0} dv \left(-\frac{1}{2}\mathbf{J}^*\cdot\mathbf{E}\right). \tag{3.5}$$

This equation stipulates that the real time-averaged power, which is conventionally defined as the real part of the complex Poynting vector, is given in terms of the work done by the source on the field right at the antenna current. Moreover, since this work is evaluated only over the volume V_0, while the surface S is chosen at arbitrary distance, we can see then that the net time-averaged energy flux generated by the antenna is the same throughout any closed surface as long as it does enclose the source region V_0.[4]

We need to eliminate the source-field interaction (work) term appearing in (3.3) in order to focus entirely on the fields. To do this, consider the spherical surface S_∞ at infinity. Applying the complex Poynting theorem there and noticing that the far-field expressions give *real* power flow, we conclude from (3.4) that

$$\operatorname{Im} \int_{V_0} dv \left(-\frac{1}{2}\mathbf{J}^*\cdot\mathbf{E}\right) = -2\omega \int_{V_\infty} dv \left(w_h - w_e\right). \tag{3.6}$$

Substituting (3.6) into the near-field energy balance (3.4), we find

$$\operatorname{Im} \int_S d\mathbf{s}\frac{1}{2}\left(\mathbf{E}\times\mathbf{H}^*\right) = -2\omega \int_{V_\infty - V} dv \left(w_h - w_e\right). \tag{3.7}$$

This equation suggests that the imaginary part of the complex Poynting vector, when evaluated in the near-field region, is dependent on the *difference* between the electric and magnetic energy in the region enclosed between the observation surface S and the surface at infinity S_∞, i.e., the *total* energy difference outside the observation volume V. In other words, we now know that the energy difference $W_h - W_e$ is a convergent quantity because the LHS of (3.7) is finite.[5] Since this condition is going

4 That is, the surface need not be spherical. However, in order to facilitate actual calculations in later parts of this chapter, we restrict ourselves to spherical surfaces.

5 We remind the reader that all source singularities are assumed to be inside the volume V_0.

to play an important role later, we stress it again as

$$\left| \int_{V_\infty - V} dv \left(w_h - w_e \right) \right| < \infty. \tag{3.8}$$

Combining equations (3.5) and (3.7), we reach

$$\int_S d\mathbf{s} \frac{1}{2} \left(\mathbf{E} \times \mathbf{H}^* \right) = P_{\text{rad}} - 2i\omega \int_{V_\infty - V} dv \left(w_h - w_e \right), \tag{3.9}$$

where the radiated power is defined as

$$P_{\text{rad}} = \text{Re} \int_S d\mathbf{s} \frac{1}{2} \left(\mathbf{E} \times \mathbf{H}^* \right). \tag{3.10}$$

We need to be careful about the interpretation of equation (3.9). Strictly speaking, what this result tells us is only the following. Form an observation sphere S at an arbitrary distance in the near-field zone. As long as this sphere encloses the source region V_0, then the real part of the power flux, the surface integral of the complex Poynting vector, will give the net real power flow through S, while the imaginary part is the total difference between the electric and magnetic energies in the infinite region outside the observation volume V. We repeat: the condition (3.8) is satisfied and this energy difference is finite. Relation (3.9) is the theoretical basis for the traditional expression of the antenna input impedance in terms of fields surrounding the radiating structure [33], [53].

3.3 THE STRUCTURE OF THE ANTENNA NEAR FIELD IN THE SPATIAL DOMAIN

We now turn to a closer examination of the nature of the antenna near fields in the spatial domain, while the spectral approach is deferred to Chapter 4. Here, we consider the fields generated by the antenna that is lying in the intermediate zone, i.e., the interesting case between the far zone $kr \to \infty$ and the static zone $kr \to 0$. The objective is not to obtain a list of numbers describing the numerical spatial variation of the fields away from the antenna, a task well achieved with present day computer packages. Instead, we aim to attain a conceptual insight on the nature of the near field by mapping out its inner structure in details. We suggest that the natural way to achieve this is the use of the Wilcox expansion [47]. Indeed, since our fields in the volume outside the source region satisfy the homogeneous Helmholtz equation, we can expand the electric and magnetic fields as [47]

$$\mathbf{E}(\mathbf{r}) = \frac{e^{ikr}}{r}\sum_{n=0}^{\infty}\frac{\mathbf{A}_n(\theta,\varphi)}{r^n}, \quad \mathbf{H}(\mathbf{r}) = \frac{e^{ikr}}{r}\sum_{n=0}^{\infty}\frac{\mathbf{B}_n(\theta,\varphi)}{r^n}, \tag{3.11}$$

where $\mathbf{A}_n$ and $\mathbf{B}_n$ are vector angular functions dependent on the far-field radiation pattern of the antenna and $k = \omega\sqrt{\varepsilon\mu}$ is the wavenumber. The far fields are the asymptotic limits of the expansion. That is,

$$\mathbf{E}(\mathbf{r}) \underset{r\to\infty}{\sim} \frac{e^{ikr}}{r}\mathbf{A}_0(\theta,\varphi), \quad \mathbf{H}(\mathbf{r}) \underset{r\to\infty}{\sim} \frac{e^{ikr}}{r}\mathbf{B}_0(\theta,\varphi). \tag{3.12}$$

The reason why this approach is the convenient one can be given in the following manner. We are interested in understanding the structure of the near field of the antenna. In the far zone, this structure is extremely simple; it is simply the zeroth-order term of the Wilcox expansion as singled out in (3.12). Now, as we leave the far zone and descend toward the antenna current distribution, the fields start to get more complicated. Mathematically speaking, this corresponds to the addition of more terms into the Wilcox series. The implication is that more terms (and hence the emerging complexity in the spatial structure) are needed in order to converge to accurate solution of the field as we get closer and closer to the current distribution. Let us then divide the entire exterior region surrounding the antenna into an infinite number of spherical layers as shown in Figure 3.2. The outermost layer R_0 is identified with the far zone while the innermost layer R_∞ is defined as the minimum sphere totally enclosing the antenna current distribution.[6] In between these two regions, an infinite number of layers exists, each corresponding to a term in the Wilcox expansion as we now explain. The boundaries between the various regions are not sharply defined, but taken only as indicators in the asymptotic sense to be described momentarily.[7] The outermost region R_0 corresponds to the far zone. The value of, say, the electric field there is $\mathbf{A}_0 \exp(ikr)/r$. As we start to descend toward the antenna, we enter into the next region R_1, where the mathematical expression of the far field given in (3.12) is no longer valid

6 Strictly speaking, there is no reason why R_∞ should be the *minimum* sphere. Any sphere with larger size satisfying the mentioned condition will do in theory.

7 To be precise, by definition only region R_∞ possesses a clear-cut boundary (the minimum sphere enclosing the source distribution. This sphere encloses the region V_0 in Figure 3.1. The field between V_0 and R_∞ belongs to the interior region and is not included in our present treatment, which is concerned exclusively with the study of the exterior region in general antenna systems.

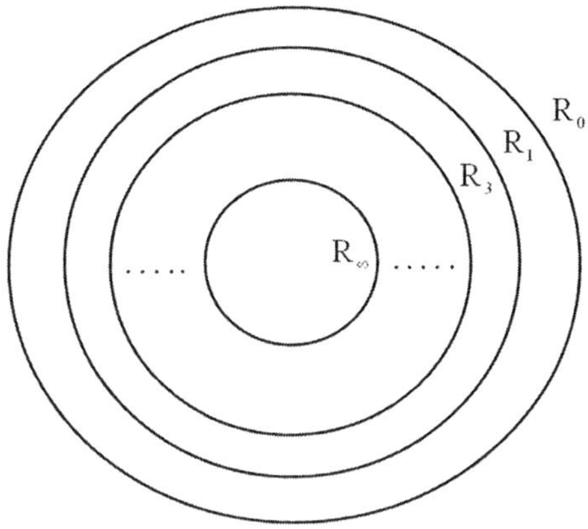

Figure 3.2 General description of antenna near-field spatial structure.

and has to be augmented by the next term in the Wilcox expansion. Indeed, we find that for $\mathbf{r} \in \mathrm{R}_1$, the electric field takes (approximately, asymptotically) the form $\mathbf{A}_0 \exp{(ikr)}/r + \mathbf{A}_1 \exp{(ikr)}/r^2$. Subtracting the two fields from each other, we obtain the difference $\mathbf{A}_1 \exp{(ikr)}/r^2$. Therefore, it appears to us very natural to interpret the region R_1 as the "seat" of a field in the form $\mathbf{A}_1 \exp{(ikr)}/r^2$. Similarly, the nth region R_n is associated (in the asymptotic sense just sketched) with the field form $\mathbf{A}_n \exp{(ikr)}/r^{n+1}$. We immediately mention that this *individual* form of the field does *not* satisfy Maxwell's equations. The nth field form given above is a mathematical depiction of the effect of getting closer to the antenna on the total (Maxwellian) field structure; it represents the contribution added by the layer under consideration when passed through by the observer while descending from the far zone to the antenna current distribution. By dividing the exterior region in this way, we become able to mentally visualize progressively the various contributions to the total near field expression *as they are mapped out spatially*.[8]

It is important here to mention that, as is proved in Chapter 4, localized *and* nonlocalized energies exist in *each* layer in turn; that is, each region R_n contains *both* propagating and nonpropagating energies, which amounts to the observation

8 It is for this reason that we refrain from rigorously defining the near field as all the terms in the Wilcox expansion with $n \geq 1$ as is the habit with some writers. The reason is that such a field is not Maxwellain.

that in each region part of the field remains there, while the remaining part of the field moves to the next larger layer.[9] What concerns us here is not this more sophisticated spectral analysis of the field associated with each layer, but the simple mapping out of the antenna near fields into such a rough spatial distribution of concentric layers understood in the asymptotic sense.

To be sure, this spatial picture, illuminating as it is, will remain a mere definition unless it is corroborated by some interesting consequences. This actually turns out to be the case. As pointed out in the previous paragraph, it is possible to show that certain theorems about the physical behavior of each layer can be proved. Better still, it is possible to investigate the issue of the mutual *electromagnetic interaction* between different regions appearing in Figure 3.2. It turns out that a general theorem (to be proved in Section 3.5) can be established, which shows that exactly "half" of these layers don't electromagnetically interact with each other. In order to understand the meaning of this remark, we need first to define precisely what is expressed in the term 'interaction.' Let us use the Wilcox expansion (4.85) to evaluate the electric and magnetic energies appearing in (3.2). Since the series expansion under consideration is absolutely convergent [47], and the conjugate of an absolutely convergent series is still absolutely convergent, the two expansions of $\mathbf{E}$ and $\mathbf{E}^*$ can be freely multiplied and the resulting terms can be arranged as we please. The result is

$$w_e = \frac{\varepsilon}{4}\mathbf{E}\cdot\mathbf{E}^* = \frac{\varepsilon}{4}\sum_{n=0}^{\infty}\sum_{n'=0}^{\infty}\frac{\mathbf{A}_n\cdot\mathbf{A}_{n'}^*}{r^{n+n'+2}}, \tag{3.13}$$

$$w_h = \frac{\mu}{4}\mathbf{H}\cdot\mathbf{H}^* = \frac{\mu}{4}\sum_{n=0}^{\infty}\sum_{n'=0}^{\infty}\frac{\mathbf{B}_n\cdot\mathbf{B}_{n'}^*}{r^{n+n'+2}}. \tag{3.14}$$

We rearrange the terms of these two series to produce the following illuminating form

$$w_e(\mathbf{r}) = \frac{\varepsilon}{4}\sum_{n=0}^{\infty}\frac{\mathbf{A}_n\cdot\mathbf{A}_n^*}{r^{2n+2}} + \frac{\varepsilon}{2}\sum_{\substack{n,n'=0\\ n>n'}}^{\infty}\frac{\mathrm{Re}\left\{\mathbf{A}_n\cdot\mathbf{A}_{n'}^*\right\}}{r^{n+n'+2}}, \tag{3.15}$$

$$w_h(\mathbf{r}) = \frac{\mu}{4}\sum_{n=0}^{\infty}\frac{\mathbf{B}_n\cdot\mathbf{B}_n^*}{r^{2n+2}} + \frac{\mu}{2}\sum_{\substack{n,n'=0\\ n>n'}}^{\infty}\frac{\mathrm{Re}\left\{\mathbf{B}_n\cdot\mathbf{B}_{n'}^*\right\}}{r^{n+n'+2}}. \tag{3.16}$$

9 The process is still even more complicated because of the interaction (energy exchange) between the propagating and nonpropagating parts. See Chapter 4 for analysis and conclusions.

In writing equations (3.15) and (3.16), we made use of the reciprocity in which the energy transfer from layer n to layer n' is equal to the corresponding one from layer n' to layer n. The first sums on the RHS of (3.15) and (3.16) represent the *self* interaction of the nth layer with itself. Those are the self interaction of the far field, the so-called radiation density, and the self-interactions of all remaining (inner) regions R_n with $n \geq 1$. The second sum in both equations represents the interaction between *different* layers. Notice that those interactions can be grouped into two categories, the interaction of the far field (0th layer in the Wilcox expansion) with all other layers, and the remaining mutual interactions between different layers before the far-field zone (again R_n with $n \geq 1$).

Now because we are interested in the spatial structure of near field; that is, the variation of the field as we move closer to or farther from the antenna physical body where the current distribution resides, it is natural to average over all the angular information contained in the energy expressions (3.15) and (3.16). That is, we introduce the *radial energy density function* of the electromagnetic fields by integrating (3.15) and (3.16) over the entire solid angle Ω in order to obtain

$$w_e(r) = \frac{\varepsilon}{4} \sum_{n=0}^{\infty} \frac{\langle \mathbf{A}_n, \mathbf{A}_n \rangle}{r^{2n+2}} + \frac{\varepsilon}{2} \sum_{\substack{n,n'=0 \\ n>n'}}^{\infty} \frac{\langle \mathbf{A}_n, \mathbf{A}_{n'} \rangle}{r^{n+n'+2}}, \tag{3.17}$$

$$w_h(r) = \frac{\mu}{4} \sum_{n=0}^{\infty} \frac{\langle \mathbf{B}_n, \mathbf{B}_n \rangle}{r^{2n+2}} + \frac{\mu}{2} \sum_{\substack{n,n'=0 \\ n>n'}}^{\infty} \frac{\langle \mathbf{B}_n, \mathbf{B}_{n'} \rangle}{r^{n+n'+2}}, \tag{3.18}$$

where the mutual interaction between two angular vector fields $\mathbf{F}$ and $\mathbf{G}$ is defined as[10]

$$\langle \mathbf{F}(\theta, \varphi), \mathbf{G}(\theta, \varphi) \rangle := \int_{4\pi} d\Omega \, \mathrm{Re}\{\mathbf{F}(\theta, \varphi) \cdot \mathbf{G}^*(\theta, \varphi)\}. \tag{3.19}$$

In deriving (3.17) and (3.18), we made use of the fact that the energy series is uniformly convergent in θ and φ in order to interchange the order of integration and summation.[11]

Equations (3.17) and (3.18) clearly demonstrate the considerable advantage gained by expressing the energy of the antenna fields in terms of Wilcox expansion. The angular functional dependence of the energy density is completely removed by

10 For example, in terms of this notation, the principle of reciprocity used in deriving (3.15) and (3.16) can now be expressed economically in the form $\langle \mathbf{A}_n, \mathbf{A}_{n'} \rangle = \langle \mathbf{A}_{n'}, \mathbf{A}_n \rangle$.

11 See the appendix in Section 3.8.1.

integration over all the solid angles, and we are left afterwards with a power expansion in $1/r$, a result that provides direct intuitive understanding of the structure of the near field since in such a type of series more higher-order terms are needed for accurate evaluation only when we get closer to the antenna body, i.e., for large $1/r$. Moreover, the total energy is then obtained by integrating over the remaining radial variable, which is possible in closed form as we will see later in Section 3.6.2.

A particularly interesting observation, however, is that almost "half" of the mutual interaction terms appearing in (3.17) and (3.18) are exactly zero. Indeed, we will prove later that if the integer $n + n'$ is odd, then the interactions are identically zero, i.e., $\langle \mathbf{A}_n, \mathbf{A}_{n'} \rangle = \langle \mathbf{B}_n, \mathbf{B}_{n'} \rangle = 0$ for $n + n' = 2k + 1$ and k is integer. This represents, in our opinion, a significant insight on the nature of antenna near fields in general. In order to prove this theorem and deduce other results, we need to express the angular vector fields $\mathbf{A}_n(\theta, \varphi)$ and $\mathbf{B}_n(\theta, \varphi)$ in terms of the antenna spherical TE and TM modes. This we accomplish next by deriving the Wilcox expansion from the multipole expansion.

3.4 DIRECT CONSTRUCTION OF THE ANTENNA NEAR-FIELD STARTING FROM A GIVEN FAR-FIELD RADIATION PATTERN

3.4.1 Introduction

We have seen how the Wilcox expansion can be physically interpreted as the mathematical embodiment of a spherical layering of the antenna exterior region understood in a convenient asymptotic sense. The localization of the electromagnetic field within each of the regions appearing in Figure 3.2 suggests that the outermost region R_0, the far zone, corresponds to the simplest field structure possible, while the fields associated with the regions close to the antenna exclusion sphere, R_∞, are considerably more complex. However, as was pointed out long ago, the entire field in the exterior region can be completely determined recursively from the radiation pattern [47]. In this section we further develop this idea by showing that the entire region field can be determined from the far field directly, i.e., nonrecursively, by a simple construction based on the analysis of the far field into its spherical wavefunctions. In other words, we show that a modal analysis of the radiation pattern, a process that is computationally robust and straightforward, can lead to complete knowledge of the exterior domain near field, in an analytical form, as it is increasing in complexity while progressing from the far zone to the near zone. This description is meaningful because it has been expressed in terms of physical radiation modes. The derivation will help to appreciate the general nature of the near field spatial structure that was given in Section 3.3 by gaining some insight into the mechanism of electromagnetic coupling between the various spatial regions defined in Figure 3.2, a task we address in detail in Section 3.5.

3.4.2 Mathematical Description of the Far-Field Radiation Pattern and the Concomitant Near-Field

Our point of departure is the far-field expressions (3.12), where we observe that because $\mathbf{A}_0(\theta, \varphi)$ and $\mathbf{B}_0(\theta, \varphi)$ are well-behaved angular vector fields tangential to the 2-sphere, it is possible to expand their functional variations in terms of infinite sum of vector spherical harmonics [89], [33]. That is, we write

$$\mathbf{E}(\mathbf{r}) \underset{r\to\infty}{\sim} \eta \frac{e^{ikr}}{kr} \sum_{l=0}^{\infty} \sum_{m=-l}^{l} (-1)^{l+1} \left[a_E(l,m)\,\mathbf{X}_{lm} - a_M(l,m)\,\hat{r} \times \mathbf{X}_{lm}\right], \tag{3.20}$$

$$\mathbf{H}(\mathbf{r}) \underset{r\to\infty}{\sim} \frac{e^{ikr}}{kr} \sum_{l=0}^{\infty} \sum_{m=-l}^{l} (-1)^{l+1} \left[a_M(l,m)\,\mathbf{X}_{lm} + a_E(l,m)\,\hat{r} \times \mathbf{X}_{lm}\right], \tag{3.21}$$

the series being absolutely-uniformly convergent [69], [45]. Here, $\eta = \sqrt{\mu/\varepsilon}$ is the wave impedance. $a_E(l,m)$ and $a_M(l,m)$ stand for the coefficients of the expansion TE_{lm} and TM_{lm} modes, respectively.[12] The definition of these modes will be given in a moment. The vector spherical harmonics are defined as $\mathbf{X}_{lm} = \left(1\big/\sqrt{l(l+1)}\right) \mathbf{L} Y_{lm}(\theta, \varphi)$, where $\mathbf{L} = -i\,\mathbf{r} \times \nabla$ is the angular momentum operator; Y_{lm} is the spherical harmonics of degree l and order m defined as

$$Y_{lm}(\theta, \varphi) = \sqrt{\frac{(2l+1)(l-m)!}{4\pi (l+m)!}} P_l^m(\cos\theta)\, e^{im\varphi}, \tag{3.22}$$

where P_l^m stands for the associated Legendre function.

Since the asymptotic expansion of the spherical vector wavefunctions is exact,[13] the electromagnetic fields throughout the *entire* exterior region of the antenna problem can be expanded as a series of complete set of of vector multipoles

12 These coefficients can also be determined from the antenna current distribution, i.e., the source point of view. For derivations and discussion, see [33].

13 That is, *exact* because of the expansion of the spherical Hankel function given in (3.28).

in the form [33]

$$\mathbf{E}(\mathbf{r}) = \eta \sum_{l=0}^{\infty} \sum_{m=-l}^{l} \Big[a_E(l,m)\, h_l^{(1)}(kr)\, \mathbf{X}_{lm} + \tfrac{i}{k} a_M(l,m)\, \nabla \times h_l^{(1)}(kr)\, \mathbf{X}_{lm} \Big], \tag{3.23}$$

$$\mathbf{H}(\mathbf{r}) = \sum_{l=0}^{\infty} \sum_{m=-l}^{l} \Big[a_M(l,m)\, h_l^{(1)}(kr)\, \mathbf{X}_{lm} - \tfrac{i}{k} a_E(l,m)\, \nabla \times h_l^{(1)}(kr)\, \mathbf{X}_{lm} \Big], \tag{3.24}$$

which is absolutely and uniformly convergent. The spherical Hankel function of the first kind $h_l^{(1)}(kr)$ is used to model the radial dependence of the outgoing wave in antenna systems. In this formulation, we define the TE and TM modes as follows

$$\mathrm{TE}_{lm}\ \text{mode} := \begin{cases} \mathbf{r} \cdot \mathbf{H}_{lm}^{\mathrm{TE}} \\ = a_E(l,m)\, \frac{l(l+1)}{k} h_l^{(1)}(kr)\, Y_{lm}(\theta,\varphi), \\ \mathbf{r} \cdot \mathbf{E}_{lm}^{\mathrm{TE}} = 0, \end{cases} \tag{3.25}$$

$$\mathrm{TM}_{lm}\ \text{mode} := \begin{cases} \mathbf{r} \cdot \mathbf{E}_{lm}^{\mathrm{TE}} \\ = a_M(l,m)\, \frac{l(l+1)}{k} h_l^{(1)}(kr)\, Y_{lm}(\theta,\varphi), \\ \mathbf{r} \cdot \mathbf{H}_{lm}^{\mathrm{TE}} = 0. \end{cases} \tag{3.26}$$

Strictly speaking, the adjective 'transverse' in the labels TE and TM is meaningless for the far field because there both the electric and magnetic fields have zero radial components. However, the terminology is still mathematically pertinent because the two linearly independent angular vector fields $\mathbf{X}_{lm}$ and $\hat{r} \times \mathbf{X}_{lm}$ form a complete set of basis functions for the space of tangential vector fields on the 2-sphere. For this reason, and only for this, we still may frequently use phrases like 'far field TE and TM modes.' In conclusion we find that the far-field radiation pattern (3.20) and (3.21) determines *exactly* the electromagnetic fields *everywhere* in the antenna *exterior* region. This observation was corroborated by deriving a recursive set of relations constructing the entire Wilcox expansion starting only from the far field [47]. In the remaining part of this section, we provide an alternative nonrecursive derivation of the same result in terms of the far-field spherical TE and TM modes. The upshot of this argument is the unique determinability of the antenna near field in the various spherical regions appearing in Figure 3.2 by a specified far field taken as the starting point of the engineering analysis of general radiating structures.[14]

14 We stress, however, that in general the interior-domain field cannot be recovered from the far field pattern. The field close to the source contains mostly evanescent modes and these cannot in general be recovered from the far field. For extended discussion, see [35].

3.4.3 Derivation of the Exterior Domain Near-Field from the Far-Field Radiation Pattern

The second terms on the RHS of (3.23) and (3.24) can be simplified with the help of the following relation[15]

$$\begin{aligned}\nabla \times h_l^{(1)}(kr)\,\mathbf{X}_{lm} &= \hat{r} i \frac{\sqrt{l(l+1)}}{r} h_l^{(1)}(kr)\,Y_{lm}(\theta,\varphi) \\ &+ \frac{1}{r}\frac{\partial}{\partial r}\left[r h_l^{(1)}(kr)\right]\hat{r}\times\mathbf{X}_{lm}(\theta,\varphi).\end{aligned} \tag{3.27}$$

We expand the outgoing spherical Hankel function $h_l^{(1)}(kr)$ in a power series of $1/r$ using the following well-known series [89],[88]

$$h_l^{(1)}(kr) = \frac{e^{ikr}}{r}\sum_{n=0}^{l}\frac{b_n^l}{r^n}, \tag{3.28}$$

where

$$b_n^l = (-i)^{l+1}\frac{i^n}{n!2^n k^{n+1}}\frac{(l+n)!}{(l-n)!}. \tag{3.29}$$

That is, in contrast to the situation with cylindrical wavefunctions, the spherical Hankel function can be expanded only in *finite* number of powers of $1/r$, the highest power coinciding with the order of the Hankel function l. Substituting (3.28) into (3.27), we obtain after some manipulations

$$\begin{aligned}\nabla \times h_l^{(1)}\mathbf{X}_{lm} &= i\sqrt{l(l+1)}\frac{e^{ikr}}{r}\sum_{n=0}^{l}\frac{b_n^l}{r^{n+1}}\hat{r}Y_{lm} \\ &-\frac{e^{ikr}}{r}\sum_{n=0}^{l}\frac{nb_n^l}{r^{n+1}}\hat{r}\times\mathbf{X}_{lm} + \frac{e^{ikr}}{r}\sum_{n=0}^{l}\frac{ikb_n^l}{r^n}\hat{r}\times\mathbf{X}_{lm}.\end{aligned} \tag{3.30}$$

By relabeling the indices in the summations appearing on the RHS of (3.30) involving powers $1/r^{n+2}$, the following is obtained

$$\begin{aligned}\nabla \times h_l^{(1)}(kr)\,\mathbf{X}_{lm} &= i\sqrt{l(l+1)}\frac{e^{ikr}}{r}\sum_{n=1}^{l+1}\frac{b_{n-1}^l}{r^n}\hat{r}Y_{lm} \\ &-\frac{e^{ikr}}{r}\sum_{n=1}^{l+1}\frac{(n-1)b_{n-1}^l}{r^n}\hat{r}\times\mathbf{X}_{lm} + \frac{e^{ikr}}{r}\sum_{n=0}^{l} ik\frac{b_n^l}{r^n}\hat{r}\times\mathbf{X}_{lm}.\end{aligned} \tag{3.31}$$

15 Equation (3.27) can be readily derived from the definition of the operator $\mathbf{L} = -i\mathbf{r}\times\nabla$ above and the expansion $\nabla = \hat{\mathbf{r}}(\hat{\mathbf{r}}\cdot\nabla) - \hat{\mathbf{r}}\times\hat{\mathbf{r}}\times\nabla$, and by making use of the relation $L^2 Y_{lm} = l(l+1)\,Y_{lm}$.

Now it will be convenient to write this expression in the following succinct form

$$\nabla \times h_l^{(1)} \mathbf{X}_{lm} = \frac{e^{ikr}}{r} \sum_{n=0}^{l+1} \frac{c_n^l \hat{r} Y_{lm} + d_n^l \hat{r} \times \mathbf{X}_{lm}}{r^n}, \tag{3.32}$$

where

$$c_n^l = \begin{cases} 0, & n = 0, \\ i\sqrt{l\,(l+1)} b_{n-1}^l, & 1 \leq n \leq l+1. \end{cases} \tag{3.33}$$

and

$$d_n^l = \begin{cases} ikb_0^l, & n = 0, \\ ikb_n^l - (n-1)\, b_{n-1}^l, & 1 \leq n \leq l, \\ -lb_l^l, & n = l+1. \end{cases} \tag{3.34}$$

Using (3.32), the expansions (3.23) and (3.24) can be rewritten as

$$\begin{aligned} \mathbf{E}(\mathbf{r}) = \eta \sum_{l=0}^{\infty} \sum_{m=-l}^{l} & \left[a_E\,(l,m) \frac{e^{ikr}}{r} \sum_{n=0}^{l+1} \frac{b_n^l \mathbf{X}_{lm}}{r^n} \right. \\ & \left. + \frac{i}{k} a_M\,(l,m) \frac{e^{ikr}}{r} \sum_{n=0}^{l+1} \frac{c_n^l \hat{r} Y_{lm} + d_n^l \hat{r} \times \mathbf{X}_{lm}}{r^n} \right], \end{aligned} \tag{3.35}$$

$$\begin{aligned} \mathbf{H}(\mathbf{r}) = \sum_{l=0}^{\infty} \sum_{m=-l}^{l} & \left[a_M\,(l,m) \frac{e^{ikr}}{r} \sum_{n=0}^{l+1} \frac{b_n^l \mathbf{X}_{lm}}{r^n} \right. \\ & \left. - \frac{i}{k} a_E\,(l,m) \frac{e^{ikr}}{r} \sum_{n=0}^{l+1} \frac{c_n^l \hat{r} Y_{lm} + d_n^l \hat{r} \times \mathbf{X}_{lm}}{r^n} \right]. \end{aligned} \tag{3.36}$$

Assuming that the electromagnetic field in the antenna exterior region is well-behaved, it can be shown that the infinite double series in (3.35) and (3.36) involving the l- and n- sums are absolutely convergent. Subsequently, these sums are invariant to any permutation (rearrangement) of terms [67]. Now let us consider the first series on the RHS of (3.36). We can easily see that each power r^{-n} will arise from contributions coming from all the multipoles of degree $l \geq n$. That is, we rearrange as

$$\begin{aligned} & \sum_{l=0}^{\infty} \sum_{m=-l}^{l} a_M\,(l,m) \frac{e^{ikr}}{r} \sum_{n=0}^{l} \frac{b_n^l}{r^n} \mathbf{X}_{lm} \\ & \qquad = \frac{e^{ikr}}{r} \sum_{n=0}^{\infty} \frac{1}{r^n} \sum_{l=n}^{\infty} \sum_{m=-l}^{l} a_M\,(l,m)\, b_n^l \mathbf{X}_{lm}. \end{aligned} \tag{3.37}$$

The situation is different with the second series on the RHS of (3.36). In this case, contributions to the 0th and 1st powers of $1/r$ originate from the *same* multipole, that of degree $l = 0$. Afterwards, all higher power of $1/r$, i.e., terms with $n \geq 2$, will receive contributions from multipoles of the $(n-1)$th degree, but yet with *different* weighting coefficients. We unpack this observation by writing

$$\begin{aligned}
&\sum_{l=0}^{\infty}\sum_{m=-l}^{l}\tfrac{i}{k}a_E(l,m)\tfrac{e^{ikr}}{r}\sum_{n=0}^{l+1}\tfrac{\left(c_n^l\,\hat{r}Y_{lm}+d_n^l\,\hat{r}\times\mathbf{X}_{lm}\right)}{r^n}\\
&=-\tfrac{e^{ikr}}{ikr}\left[\sum_{l=0}^{\infty}\sum_{m=-l}^{l}a_E(l,m)\left(c_0^l\,\hat{r}Y_{lm}+d_0^l\,\hat{r}\times\mathbf{X}_{lm}\right)\right.\\
&\left.+\sum_{n=1}^{\infty}\tfrac{1}{r^n}\sum_{l=n-1}^{\infty}\sum_{m=-l}^{l}a_E(l,m)\left(c_n^l\,\hat{r}Y_{lm}+d_n^l\,\hat{r}\times\mathbf{X}_{lm}\right)\right].
\end{aligned}\tag{3.38}$$

That is, from (3.37) and (3.38), (3.36) takes the form

$$\mathbf{H}(\mathbf{r}) = \frac{e^{ikr}}{r}\sum_{n=0}^{\infty}\frac{\mathbf{B}_n(\theta,\varphi)}{r^n},\tag{3.39}$$

where

$$\begin{aligned}
\mathbf{B}_0(\theta,\varphi) = \sum_{l=0}^{\infty}\sum_{m=-l}^{l}\tfrac{(-i)^{l+1}}{k}&\left[a_M(l,m)\,\mathbf{X}_{lm}\right.\\
&\left.+a_E(l,m)\,\hat{r}\times\mathbf{X}_{lm}\right],
\end{aligned}\tag{3.40}$$

$$\begin{aligned}
&\mathbf{B}_n(\theta,\varphi) = \sum_{l=n}^{\infty}\sum_{m=-l}^{l}a_M(l,m)\,b_n^l\mathbf{X}_{lm}\\
&-\sum_{l=n-1}^{\infty}\sum_{m=-l}^{l}\tfrac{ia_E(l,m)}{k}\left(c_n^l\,\hat{r}Y_{lm}+d_n^l\,\hat{r}\times\mathbf{X}_{lm}\right),\quad n\geq 1.
\end{aligned}\tag{3.41}$$

By exactly the same procedure, we derive from (3.35) the following result

$$\mathbf{E}(\mathbf{r}) = \frac{e^{ikr}}{r}\sum_{n=0}^{\infty}\frac{\mathbf{A}_n(\theta,\varphi)}{r^n},\tag{3.42}$$

where

$$\begin{aligned}
\mathbf{A}_0(\theta,\varphi) = \eta\sum_{l=0}^{\infty}\sum_{m=-l}^{l}\tfrac{(-i)^{l+1}}{k}&\left[a_E(l,m)\,\mathbf{X}_{lm}\right.\\
&\left.-a_M(l,m)\,\hat{r}\times\mathbf{X}_{lm}\right],
\end{aligned}\tag{3.43}$$

$$\begin{aligned}
&\mathbf{A}_n(\theta,\varphi) = \eta\sum_{l=n}^{\infty}\sum_{m=-l}^{l}a_E(l,m)\,b_n^l\mathbf{X}_{lm}\\
&+\eta\sum_{l=n-1}^{\infty}\sum_{m=-l}^{l}\tfrac{ia_M(l,m)}{k}\left(c_n^l\,\hat{r}Y_{lm}+d_n^l\,\hat{r}\times\mathbf{X}_{lm}\right),\quad n\geq 1.
\end{aligned}\tag{3.44}$$

Therefore, the Wilcox series is derived from the multipole expansion and the exact variation of the angular vector fields $\mathbf{A}_n$ and $\mathbf{B}_n$ are directly determined in terms of the spherical far-field modes of the antenna. We notice that these two nth vector fields take the form of an infinite series of spherical harmonics of degrees $l \geq n$, i.e., the form of the *tail* of the infinite series appearing in the far field expression (3.20) and (3.21). The coefficients, however, of the same modes appearing in the latter series are now modified by the simple n-dependence of c_n^l and d_n^l as given in (3.33) and (3.34). Conversely, the contribution of each l-multipole to the respective terms in the Wilcox expansion is determined by the weights c_n^l and d_n^l, which are varying with l. There is no dependence on m in this derivation of the Wilcox terms in terms of the electromagnetic field multipoles.

3.4.4 General Remarks

As can be seen from the direct relations (3.43), (3.44), (3.40), and (3.41), the antenna near field in the various regions R_n defined in Figure 3.2 is developable in a series of higher-order TE and TM modes, those modes being uniquely determined by the content of the far-field radiation pattern. Some observations on this derivation are worthy of mention. We start by noticing that the expressions of the far field (3.43) and (3.40), the initial stage of the analysis, are not homogeneous with the expressions of the inner regions (3.44) and (3.41). This can be attributed to mixing between two adjacent regions. Indeed, in the scalar problem only modes of order $l \geq n$ contribute to the content of the region R_n. However, due to the effect of radial differentiation in the second term of the RHS of (3.27), the aforementioned mixing between two adjacent regions emerges to the scene, manifesting itself in the appearance of contributions from modes with order $n - 1$ in the region R_n. This, however, always comes from the *dual* polarization. For example, in the magnetic field, the TM_{lm} modes with $l \geq n$ contribute to the field localized in region R_n, while the contribution of the TE_{lm} modes comes from order $l \geq n - 1$. The dual statement holds for the electric field. As will be seen in Section 3.5, this will lead to similar conclusion for electromagnetic interactions between the various regions.

We also bring to the reader's attention the fact that the derivation presented in this section does *not* imply that the radiation pattern determines the antenna itself, if by the antenna we understand the current distribution *in*side the innermost region R_∞. There is an infinite number of current distributions that can produce the same exterior-domain field (and consequently the far-field pattern).[16] Our results indicate, however, that the entire field in the *exterior* region, i.e., outside the region R_∞, is determined exactly and nonrecursively by the far field. We believe that the advantage of this observation is considerable for the engineering study of electromagnetic radiation. Antenna designers usually specify the goals of their devices in terms of radiation pattern characteristics like sidelobe level, directivity, cross polarization, null location, and son. It appears from our analysis that an exact analytical relation between the near field and these design goals do exist in the form derived above. Since the engineer can still choose any type of antenna that fits within the enclosing region R_∞, the results of this chapter should be viewed as a kind of canonical machinery for generating fundamental relations between the far-field performance and the lower bound formed by the field behavior in the entire exterior region compatible with any antenna current distribution that can be enclosed inside R_∞. For example, relations (3.69) and (3.70) provide the exact analytical form for the reactive energy in the exterior region. This then forms a *lower bound* on the actual reactive energy for a *specific* antenna, because the field inside R_∞ will only add to the reactive energy calculated for the exterior region. To summarize this important point, our conclusions in this work apply only to a *class*[17] of antennas compatible with a given radiation pattern, not to a particular antenna current distribution. This, we repeat, is a natural theoretical framework for the *engineering* analysis of antenna fundamental performance measures.[18]

16 We repeat that there *is* one-to-one correspondence between the exterior-domain field and the far field; however, there is *no* one-to-one correspondence between the exterior-domain field and the antenna source distribution.

17 Potentially infinite in size.

18 The extensively-researched area of fundamental limitations of electrically small antennas is a special case in this general study. We don't presuppose any restriction on the size of the innermost region R_∞, which is required only to enclose the entire antenna in order for the various series expansions used in the present work to converge nicely. Strictly speaking, electrically small antennas are more challenging for the *impedance matching* problem than the *field* point of view. The field structure of an electrically small antenna approaches the field of an infinitesimal dipole and hence does not motivate the more sophisticated treatment developed here, particularly the spectral approach of Chapter 4.

3.5 A PHENOMENOLOGICAL EXAMINATION OF THE SPATIAL DISTRIBUTION OF ELECTROMAGNETIC ENERGY IN THE ANTENNA EXTERIOR REGION

3.5.1 Introduction

In this section, we utilize the results obtained in Section 3.4 in order to evaluate and analyze the energy content of the antenna near field in the spatial domain. We continue to work within the overall picture sketched in Section 3.3 in which the antenna exterior domain was divided into spherical regions understood in the asymptotic sense (Figure 3.2), and the total energy viewed as the sum of self and mutual interactions among these regions. Indeed, we will treat now in detail the various types of interactions giving rise to the radial energy density function in the form introduced in (3.17) and (3.18). The calculation will make use of the following standard orthogonality relations

$$\begin{aligned}
&\int_{4\pi} d\Omega\, \mathbf{X}_{lm} \cdot \mathbf{X}^*_{l'm'} = \delta_{ll'}\delta_{mm'}, \\
&\int_{4\pi} d\Omega\, \mathbf{X}_{lm} \cdot (\hat{r} \times \mathbf{X}^*_{l'm'}) = 0, \\
&\int_{4\pi} d\Omega\, (\hat{r} \times \mathbf{X}_{lm}) \cdot (\hat{r} \times \mathbf{X}^*_{l'm'}) = \delta_{ll'}\delta_{mm'}, \\
&\hat{r} \cdot (\hat{r} \times \mathbf{X}_{lm}) = \hat{r} \cdot \mathbf{X}_{lm} = 0,
\end{aligned} \tag{3.45}$$

where δ_{lm} stands for the Kronecker delta function.

3.5.2 Self-Interaction of the Outermost Region (Far Zone, Radiation Density)

The first type of terms is the self-interaction of the fields in region R_0, i.e., the far zone. These are due to the terms involving $\langle \mathbf{A}_0, \mathbf{A}_0 \rangle$ and $\langle \mathbf{B}_0, \mathbf{B}_0 \rangle$ for the electric and magnetic fields, respectively. From (3.19), (3.43), (3.40), and (3.45), we readily obtain the familiar expressions

$$\langle \mathbf{A}_0, \mathbf{A}_0 \rangle = \frac{\eta^2}{k^2} \sum_{l=0}^{\infty} \sum_{m=-l}^{l} \Big[|a_E(l,m)|^2 + |a_M(l,m)|^2 \Big], \tag{3.46}$$

$$\langle \mathbf{B}_0, \mathbf{B}_0 \rangle = \frac{1}{k^2} \sum_{l=0}^{\infty} \sum_{m=-l}^{l} \Big[|a_M(l,m)|^2 + |a_E(l,m)|^2 \Big]. \tag{3.47}$$

That is, *all* TE_{lm} and TM_{lm} modes contribute to the self-interaction of the far field. As we will see immediately, the picture is different for the self-interactions of the inner regions.

3.5.3 Self-Interactions of the Inner Regions

From (3.19), (3.44), and (3.45), we obtain

$$\begin{aligned} \langle \mathbf{A}_n, \mathbf{A}_n \rangle &= \eta^2 \sum_{l=n}^{\infty} \sum_{m=-l}^{l} \left| a_E(l,m)\, b_n^l \right|^2 \\ &+ \frac{\eta^2}{k^2} \sum_{l=n-1}^{\infty} \sum_{m=-l}^{l} \left| a_M(l,m) \right|^2 \left(\left| c_n^l \right|^2 + \left| d_n^l \right|^2 \right), \quad n \geq 1, \end{aligned} \tag{3.48}$$

Similarly, from (3.19), (3.41), and (3.45) we find

$$\begin{aligned} \langle \mathbf{B}_n, \mathbf{B}_n \rangle &= \sum_{l=n}^{\infty} \sum_{m=-l}^{l} \left| a_M(l,m)\, b_n^l \right|^2 \\ &+ \frac{1}{k^2} \sum_{l=n-1}^{\infty} \sum_{m=-l}^{l} \left| a_E(l,m) \right|^2 \left(\left| c_n^l \right|^2 + \left| d_n^l \right|^2 \right), \quad n \geq 1. \end{aligned} \tag{3.49}$$

Therefore, in contrast to the case with the radiation density, the 0th region, the self-interaction of the nth inner region ($n > 0$) consists of two types: the contribution of TE_{lm} modes to the electric energy density, which involves only modes with $l \geq n$; and the contribution of the TM_{lm} modes to the same energy density, which comes this time from modes with order $l \geq n-1$. The dual situation holds for the magnetic energy density. This qualitative splitting of the modal contribution to the energy density into two distinct types is ultimately due to the *vectorial* structure of Maxwell's equations.[19]

3.5.4 Mutual Interaction Between the Outermost Region and The Inner Regions

We turn now to the mutual interactions between two different regions, i.e., to an examination of the second sums in the RHS of (3.17) and (3.18). We first evaluate here the interaction between the far field and an inner region with index n. From

19 Cf. Section 3.4.4.

(3.19), (3.43), (3.44), and (3.45), we compute

$$\begin{aligned}\langle \mathbf{A}_0, \mathbf{A}_n \rangle &= \tfrac{\eta^2}{k} \sum_{l=n}^{\infty} \sum_{m=-l}^{l} g_n^1 (l,m) \left| a_E (l,m) \right|^2 \\ &+ \tfrac{\eta^2}{k^2} \sum_{l=n-1}^{\infty} \sum_{m=-l}^{l} g_n^2 (l,m) \left| a_M (l,m) \right|^2 , \quad n \geq 1. \end{aligned} \tag{3.50}$$

From (3.19), (3.40), (3.41), and (3.45), we also reach to

$$\begin{aligned}\langle \mathbf{B}_0, \mathbf{B}_n \rangle &= \tfrac{1}{k} \sum_{l=n}^{\infty} \sum_{m=-l}^{l} g_n^1 (l,m) \left| a_M (l,m) \right|^2 \\ &+ \tfrac{1}{k^2} \sum_{l=n-1}^{\infty} \sum_{m=-l}^{l} g_n^2 (l,m) \left| a_E (l,m) \right|^2 , \quad n \geq 1. \end{aligned} \tag{3.51}$$

From (3.29), we calculate

$$g_n^1 (l,m) := \mathrm{Re} \left\{ (-i)^{l+1} b_n^{l*} \right\} = \begin{cases} 0, \ n \text{ odd}, \\ \frac{(-1)^{3n/2}}{n! 2^n k^{n+1}} \frac{(l+n)!}{(l-n)!}, \quad n \text{ even}. \end{cases} \tag{3.52}$$

Similarly, we use (3.34) to calculate

$$\begin{aligned} g_n^2 (l,m) &:= \mathrm{Re} \left\{ (-i)^{l+1} i d_n^{l*} \right\} \\ &= \begin{cases} k g_n^1 (l,m) - (n-1) \, g_n^3 (l,m) , \ 1 \leq n \leq l, \\ -l g_{l+1}^3 (l,m) , \ \ n = l+1. \end{cases} \end{aligned} \tag{3.53}$$

Here, we define

$$g_n^3 (l,m) := \begin{cases} 0, \ n \text{ odd}, \\ \frac{(-1)^{3n/2-1}}{(n-1)! 2^{n-1} k^n} \frac{(l+n-1)!}{(l-n-1)!}, \quad n \text{ even}. \end{cases} \tag{3.54}$$

Therefore, it follows that *the interaction between the far field zone and any inner region* R_n*, with odd index* n *is exactly zero*. This surprising result means that "half" of the mutual interactions between the regions comprising the core of the antenna near field on one side, and the far field on the other side, is exactly zero. Moreover, the non-zero interactions, i.e., when n is even, are evaluated exactly in simple analytical form. We also notice that this nonzero interaction with the nth region R_n involves only TM_{lm} and TE_{lm} modes with $l \geq n$ and $l \geq n-1$. The appearance of terms with $l = n-1$ is again due to the polarization structure of the radiation field.[20]

20 Cf. Section 3.4.4.

3.5.5 Mutual Interaction Between Different Inner Regions

We continue the examination of the mutual interactions appearing in the second term of the RHS of (3.17) and (3.18), but this time we focus on mutual interactions of only inner regions, i.e., interaction between region R_n and $\mathrm{R}_{n'}$ where both $n \geq 1$ and $n' \geq 1$. From (3.19), (3.44), and (3.45), we arrive at

$$\begin{aligned}\langle \mathbf{A}_n, \mathbf{A}_{n'} \rangle = \eta^2 \sum_{l=\vartheta_n^{n'}}^{\infty} \sum_{m=-l}^{l} g_{n,n'}^4 (l,m) \left| a_E (l,m) \right|^2 \\ + \frac{\eta^2}{k^2} \sum_{l=\vartheta_n^{m}}^{\infty} \sum_{m=-l}^{l} g_{n,n'}^5 (l,m) \left| a_M (l,m) \right|^2 \\ + \frac{\eta^2}{k^2} \sum_{l=\vartheta_{n-1}^{n'-1}}^{\infty} \sum_{m=-l}^{l} g_{n,n'}^6 (l,m) \left| a_M (l,m) \right|^2, \quad n, n' \geq 1.\end{aligned} \tag{3.55}$$

Similarly, from (3.19), (3.41), and (3.45), we reach to

$$\begin{aligned}\langle \mathbf{B}_n, \mathbf{B}_{n'} \rangle = \sum_{l=\vartheta_n^{n'}}^{\infty} \sum_{m=-l}^{l} g_{n,n'}^4 (l,m) \left| a_M (l,m) \right|^2 \\ + \frac{1}{k^2} \sum_{l=\vartheta_{n-1}^{n'-1}}^{\infty} \sum_{m=-l}^{l} g_{n,n'}^5 (l,m) \left| a_E (l,m) \right|^2 \\ + \frac{1}{k^2} \sum_{l=\vartheta_{n-1}^{n'-1}}^{\infty} \sum_{m=-l}^{l} g_{n,n'}^6 (l,m) \left| a_E (l,m) \right|^2, \quad n, n' \geq 1.\end{aligned} \tag{3.56}$$

Here we define $\vartheta_n^m := \max(n,m)$. Finally, formulas for $g_{n,n'}^4$, $g_{n,n'}^5$, and $g_{n,n'}^6$ are derived in Appendix 3.8.2.

Now, it is easy to see that if $n + n'$ is even (odd), then $n - 1 + n' - 1$ is also even (odd). Therefore, we conclude from the above and Appendix 3.8.2 that *the mutual interaction between two inner regions* R_n *and* $\mathrm{R}_{n'}$ *is exactly zero if* $n + n'$ *is odd.* For the case when the interaction is not zero, the result is evaluated in simple analytical form. This nonzero term involves only TM_{lm} and TE_{lm} modes with $l \geq \max(n, n')$ and $l \geq \max(n-1, n-1')$. Therefore, there exist modes satisfying $\min(n, n') \leq l < \max(n, n')$ and $\min(n-1, n'-1) \leq l < \max(n-1, n-1')$ that do not contribute to the electromagnetic interaction between regions R_n and $\mathrm{R}_{n'}$. The appearance of terms with $l = n - 1$ is again a consequence of coupling through different modal polarization in the electromagnetic field under consideration.[21]

21 Cf. Section 3.4.4.

3.5.6 Summary and Conclusion

In this Section, we managed to express all of the interaction integrals appearing in the general expression of the antenna radial energy density (3.17) and (3.18) in the exterior region in closed analytical form involving only the TM_{lm} and TE_{lm} modes excitation amplitudes $a_M(l,m)$ and $a_E(l,m)$. The results turned out to be intuitive and comprehensible if the entire space of the exterior region is divided into spherical regions understood in the asymptotic sense as shown in Figure 3.2. In this case, the radial energy densities (3.17) and (3.18) are simple power series in $1/r$, where the amplitude of each term is simply the mutual interaction between two regions. From the basic behavior of such expansions, we now see that the closer we approach the exclusion sphere that directly encloses the antenna current distribution, i.e., what we called region R_∞, the more terms we need to include in the energy density series. However, the logic of constructing those higher-order terms clearly shows that only higher-order far-field modes enter into the formation of such increasing powers of $1/r$, confirming the intuitive fact that the complexity of the near field is an expression of richer modal content where more (higher-order) modes are needed in order to describe the intricate details of electromagnetic field spatial variation. As a bonus we also find that the complex behavior of the near field, i.e., that associated with higher-order far-field modes, is localized in the regions closer to the antenna current distribution, so in general the nearer the observation to the limit region R_∞, the more complex becomes the near-field spatial variation.

Finally. it is interesting to note that almost "half" of the interactions giving rise to the amplitudes of the radial energy density series (3.17) and (3.18) are exactly zero— i.e., the interactions between regions R_n and R'_n when $n + n'$ is odd. There is no immediate *a priori* reason why this should be the case or even obvious, the logic of the verification presented here being after all essentially computational. We believe that further theoretical research is needed to shed light on this conclusion from the conceptual point of view, not merely the computational one.

3.6 THE CONCEPT OF REACTIVE ENERGY: THE CIRCUIT POINT OF VIEW OF ANTENNA SYSTEMS

3.6.1 Introduction

In the common literature on antennas, relation (3.9) has been taken as an indication that the so-called 'reactive field' is responsible of the imaginary part of the complex

Poynting vector. Since it is this term that enters into the imaginary part of the input impedance of the antenna system, and since from *circuit* theory we usually associate the energy *stored* in the circuit with the imaginary part of the impedance, a trend developed in regarding the convergent integral (3.7) as an expression of the energy "stored" in the antenna's surrounding fields, and even sometimes call it 'evanescent field.' Hence, there is a confusion resulting from the uncritical use of the formula: reactive energy = stored energy = evanescent energy. However, there is nothing in (3.9) that speaks about such a profound conclusion! The equation, read at its face value, is an energy balance derived based on certain convenient definitions of *time*-averaged energy and power densities. The fact that the integral of the energy difference appears as the imaginary part of the complex Poynting vector is quite accidental and relates to the contingent utilization of time-harmonic excitation condition. However, the concepts of stored and evanescent field are, first of all, *spatial* concepts, and, secondly, are thematically broad; rightly put, these concepts are fundamental to the *field* point of view of general antenna systems. The conclusion that the stored energy is the sole contributor to the reactive part of the input impedance of the antenna system is an exaggeration of the circuit model that was originally advanced to study the antenna through its input port. The field structure of the antenna is richer and more involved than the limited 'terminal-like' point of view implied by circuit theory. The concept of reactance is isomorphic to neither stored nor evanescent energy.

In this section, we will first carefully construct the conventional reactive energy and show that its natural definition emerges only after the use of the Wilcox expansion in writing the radiated electromagnetic fields. In particular, we show that the general theorem we proved above about the null result of the interaction between the far field and inner layers with odd index is one of the main reasons why a finite reactive energy throughout the entire exterior region is possible. Moreover, we show that such a reactive energy can be evaluated directly in closed form and that no numerical infinite integral is involved in its computation. We then end this section by demonstrating the existence of certain ambiguity in the achieved definition of the reactive energy when attempts to extend its use beyond the circuit model of the antenna system are made.

3.6.2 Construction of the Reactive Energy Densities

We will call any energy density calculated with the point of view of those quantities appearing in the imaginary part of (3.9) *reactive* densities.[22] When someone tries to calculate the total electromagnetic energies in the region $V_\infty - V$, the result is divergent integrals. In general, we have

$$\int_{V_\infty - V} dv \left(w_h + w_e\right) = \infty. \tag{3.57}$$

However, condition (3.8) clearly suggests that there is a *common* term between w_e and w_h which is the source of the trouble in calculating the total energy of the antenna system. We postulate then that

$$w_e := w_e^1 + w_{\text{rad}}, \quad w_h := w_h^1 + w_{\text{rad}}. \tag{3.58}$$

Here w_e^1 and w_h^1 are taken as *reactive energy densities* and we hope to prove that they are finite. The common term w_{rad} is divergent in the sense

$$\int_{V_\infty - V} dv w_{\text{rad}} = \infty. \tag{3.59}$$

Therefore, it is obvious that $w_h - w_e = w_h^1 - w_e^1$, and therefore we conclude from (3.8) that

$$\left| \int_{V_\infty - V} dv \left(w_m^1 - w_e^1\right) \right| < \infty. \tag{3.60}$$

Next, we observe that the asymptotic analysis of the complex Poynting theorem allows us to predict that the energy difference $w_h - w_e$ approaches zero in the far-field zone. This is consistent with (3.58) only if we assume that

$$w_h\left(r\right) \underset{r\to\infty}{\sim} w_{\text{rad}}\left(r\right), \quad w_e\left(r\right) \underset{r\to\infty}{\sim} w_{\text{rad}}\left(r\right). \tag{3.61}$$

That is, in the asymptotic limit $r \to \infty$, the postulated quantities $w_{h,e}^1$ can be neglected in comparison with w_{rad}. In other words, the common term w_{rad} is easily identified as the radiation density *at the far-field zone*.[23] It is well-known that the

22 The question of the reactive *field* is usually ignored in literature for the reason for having difficulty treating the cross or interaction terms [51].

23 As will be seen shortly, it is meaningless to speak of a radiation density in the near-field zone.

integral of this density is not convergent and hence our assumption in (3.59) is confirmed. Moreover, this choice for the common term in (3.58) has the merit of making the energy difference, the imaginary part of (3.9), "devoid of radiation," and hence the common belief in the indistinguishability between the reactive energy and the stored energy. As we will show later, this conclusion cannot be correct, at least not in terms of field concepts.

The final step consists in showing that the total energy is finite. Writing the appropriate sum with the help of (3.58), we find

$$\begin{aligned} W_h^1 + W_e^1 &:= \int_{V_\infty - V} dv \left(w_h^1 + w_e^1\right) \\ &= \lim_{r' \to \infty} \int_{V(r') - V} dv \left[w_h(r) + w_e(r) - 2w_{\mathrm{rad}}\right]. \end{aligned} \tag{3.62}$$

To prove that this integral is finite, we make use of the Wilcox expansion of vectorial wavefunctions. First, we notice that the far-field radiation patterns are related to each other by

$$\mathbf{B}_0(\theta,\varphi) = (1/\eta)\hat{r} \times \mathbf{A}_0(\theta,\varphi), \tag{3.63}$$

This relation is the origin of the equality of the radiation density of the electric and magnetic types when evaluated in the far-field zone. That is, we have

$$w_{\mathrm{rad}}(r) = (\varepsilon/4)(\mathbf{A}_0 \cdot \mathbf{A}_0^*)/r^2 = (\mu/4)(\mathbf{B}_0 \cdot \mathbf{B}_0^*)/r^2. \tag{3.64}$$

Employing the Wilcox expansion (2.8) in the expressions of the energy densities (3.2) obtained above, it is found that

$$\begin{aligned} w_e(r) &= w_{\mathrm{rad}}(r) + \frac{\varepsilon}{2}\frac{\langle \mathbf{A}_0, \mathbf{A}_1\rangle}{r^3} \\ &+ \frac{\varepsilon}{4}\sum_{n=1}^{\infty}\frac{\langle \mathbf{A}_n, \mathbf{A}_n\rangle}{r^{2n+2}} + \frac{\varepsilon}{2}\sum_{\substack{n,n'=1 \\ n>n'}}^{\infty}\frac{\langle \mathbf{A}_n, \mathbf{A}_{n'}\rangle}{r^{n+n'+2}}, \end{aligned} \tag{3.65}$$

$$\begin{aligned} w_h(r) &= w_{\mathrm{rad}}(r) + \frac{\mu}{2}\frac{\langle \mathbf{B}_0, \mathbf{B}_1\rangle}{r^3} \\ &+ \frac{\mu}{4}\sum_{n=1}^{\infty}\frac{\langle \mathbf{B}_n, \mathbf{B}_n\rangle}{r^{2n+2}} + \frac{\mu}{2}\sum_{\substack{n,n'=1 \\ n>n'}}^{\infty}\frac{\langle \mathbf{B}_n, \mathbf{B}_{n'}\rangle}{r^{n+n'+2}}. \end{aligned} \tag{3.66}$$

By carefully examining the radial behavior of the total energies, we notice that the divergence of their volume integral over the exterior region arises from two types of terms:

1. The first type is that associated with the radiation density w_{rad}, which takes a functional form like $\langle \mathbf{A}_0, \mathbf{A}_0\rangle/r^2$ and $\langle \mathbf{B}_0, \mathbf{B}_0\rangle/r^2$. The volume integral of such terms will give rise to *linearly* divergent energy.

2. The second type is that associated with functional forms like $\langle \mathbf{A}_0, \mathbf{A}_1 \rangle / r^3$ and $\langle \mathbf{B}_0, \mathbf{B}_1 \rangle / r^3$. The volume integral of these terms will result in energy contribution that is *logarithmically* divergent.

However, we make use of the fact proved in Section 3.5.4 stating that the interactions $\langle \mathbf{A}_0, \mathbf{A}_1 \rangle$ and $\langle \mathbf{B}_0, \mathbf{B}_1 \rangle$ are identically zero. Therefore, only singularities of the first type will contribute to the total energy. Making use of the equality (3.64) and the definitions (3.58), those remaining singularities can be eliminated and we are then justified in reaching the following series expansions for the *reactive* radial energy densities

$$w_e^1(r) = \frac{\varepsilon}{4} \sum_{n=1}^{\infty} \frac{\langle \mathbf{A}_n, \mathbf{A}_n \rangle}{r^{2n+2}} + \frac{\varepsilon}{2} \sum_{\substack{n,n'=1 \\ n>n'}}^{\infty} \frac{\langle \mathbf{A}_n, \mathbf{A}_{n'} \rangle}{r^{n+n'+2}}, \tag{3.67}$$

$$w_h^1(r) = \frac{\mu}{4} \sum_{n=1}^{\infty} \frac{\langle \mathbf{B}_n, \mathbf{B}_n \rangle}{r^{2n+2}} + \frac{\mu}{2} \sum_{\substack{n,n'=1 \\ n>n'}}^{\infty} \frac{\langle \mathbf{B}_n, \mathbf{B}_{n'} \rangle}{r^{n+n'+2}}. \tag{3.68}$$

For the purpose of demonstration, let us take a hypothetical spherical surface that encloses the source region V_0. Denote by a the radius of smallest such a sphere, i.e., $\mathrm{R}_\infty = \{(r, \theta, \varphi) : r \leq a\}$. The evaluation of the total reactive energy proceeds then in the following way. The expansions (3.67) and (3.68) are uniformly convergent in r and therefore we can interchange the order of summation and integration in (3.62). After integrating the resulting series term by term, we finally arrive at the following results

$$W_e^1 = \sum_{n=1}^{\infty} \frac{(\varepsilon/4) \langle \mathbf{A}_n, \mathbf{A}_n \rangle}{(2n-1)\, a^{2n-1}} + \sum_{\substack{n,n'=1 \\ n>n'}}^{\infty} \frac{(\varepsilon/2) \langle \mathbf{A}_n, \mathbf{A}_{n'} \rangle}{(n+n'-1)\, a^{n+n'-1}}, \tag{3.69}$$

$$W_h^1 = \sum_{n=1}^{\infty} \frac{(\mu/4) \langle \mathbf{B}_n, \mathbf{B}_n \rangle}{(2n-1)\, a^{2n-1}} + \sum_{\substack{n,n'=1 \\ n>n'}}^{\infty} \frac{(\mu/2) \langle \mathbf{B}_n, \mathbf{B}_{n'} \rangle}{(n+n'-1)\, a^{n+n'-1}}. \tag{3.70}$$

Therefore, the total reactive energy is finite. It follows then that the definitions postulated above for the reactive energy densities w_h^1 and w_e^1 are consistent. Moreover, from the results of Section 3.5, we now see that total reactive energies (3.69) and

(3.70) are evaluated completely in analytical form and that *in principle* no computation of infinite numerical integrals is needed here.[24]

We stress here that the contribution of the expressions (3.69) and (3.70) is not merely having at hand a means to calculate the reactive energy of the antenna. The main insight here is the fact that the same formulas contain information about the mutual dependence of 1) the quality factor Q (through the reactive energy), 2) the size of the antenna (through the dependence on a), and 3) the far-field radiation pattern (through the interaction terms and the results of Section 3.5). The derivation above points to the *relational* structure of the antenna from the engineering point of view in the sense that the quantitative and qualitative interrelations of performance measures like directivity and polarization (far field), matching bandwidth (the quality factor), and the physical size become all united within one look. The being of the antenna is *not* understood by computing few numbers, but rather by the interconnection of all measures within an integral whole. The relational structure of the antenna system will be further developed with increasing sophistication in Chapter 4.

3.6.3 Remarks and General Discussion

The expressions (3.69) and (3.70) form the central result of the generalized approach to the reactive energy problem in general antenna systems. Using standard definitions of quality factor and matching bandwidth, one can use (3.69) and (3.70) to compute these important performance measures. However, our main intention at the moment is to illustrate the kind of knowledge now made available to antenna engineers by working with such very general results. As we promised earlier, in (3.69) and (3.70) we see the three important factors of antenna size, radiation pattern, and input impedance coming into mutual interaction with each other.

First, the **radiation pattern** is hidden into the interactions terms defined in (3.19) and reappearing in the numerators of all the terms in (3.69) and (3.70). The far field is determined by the modal content of the field (2.13), i.e., by the weights of each TE and TM modes and those in turns determine the interactions involving $\mathbf{A}_n$ and $\mathbf{B}_n$ for all orders n using the closed-form analytical expressions derived in Chapter 3. Every desired form in the radiation pattern, for example, polarization, directivity, beam shape, can be translated into the actual numerical values of the modes coefficients a^E and a^M, hence influencing the reactive energy as dictated by the laws (3.69) and (3.70).

24 Special cases of (3.69) and (3.70) appeared throughout literature. For example, see [51, 49, 52, 53].

Second, the **size of the antenna** enters the picture through the minimum radius a in (3.69) and (3.70). Its manifestation in the general expression of the reactive energy takes the form of an infinite sequence of terms proportional to negative powers of a. Physically speaking, each successive term models with increasing accuracy the behavior of the near field from the antenna length-scale perspective. Therefore, the effect of the antenna size can be studied systematically using this expression by detailed investigation of how each higher-order term contribute *purturbatively* to the behavior of the energy.

Third, the reactive energies themselves W_e^1 and W_h^1 can be used to study the **input impedance** as mentioned above. Therefore, the tradeoffs between radiation pattern measures on one hand, and the input impedance on the other hand, can be studied carefully for each successive term in (3.69) and (3.70), i.e., for each antenna size order. This *perturbation approach* needs to be carried out systematically for a wide variety of examples, not necessarily only small antennas. Indeed, our main motivation is to present the fundamental interrelations and limitation in general antenna systems, in which compact and minute devices are one chapter of the problem among others. *To the best of our knowledge, a systematic investigation of fundamental antenna limitations in this comprehensive and systematic fashion has not been carried out yet, but will most likely require intensive group effort comprising several research styles and recourses.*

3.6.4 Additional Remarks

The previous generalized derivation of the reactive energy of the antenna systems presents some important observations that are properly worth considering in some details.

First, we note that in Section 2.3.2, when we spoke freely about reactive energy, we have not at the same time made allusion to the apparently identical concept of reactive energy *density*. A careful reading of our work will show that the concept of 'density' was proposed only in the stipulation (3.58). Indeed, as will be explained in more details later, 1) *there is no unique reactive energy density for the antenna problem*, while we already know from Section 2.3.2 that 2) the individual electric and magnetic reactive energies (not densities) are infinite and 3) by The Principle of Finite Energy Difference, the difference of the total electric and magnetic energies is finite, and hence can be physically real when interpreted as the reactance of the input impedance of the antenna system. Now, in literature we find that the identification of the radiation density w_{rad} as the origin of the infinity trouble encountered while computing the total reactive energy, and hence the proposal of subtracting it, both go back to the work [51], or possibly earlier, and the idea has been intensively elaborated since then in various antenna theory circles. Our point here is that this idea was developed in our work on a more general basis and that the expression (3.58) is simply a *stipulation* or a starting point which was proved subsequently to be fruitful. In other words, *reactive energy densities are not "derived" in the strict sense of the word, but assumed or stipulated as part of the thinking process of the engineer regarding how antennas work.* This point will be further elaborated shortly.

We come to the second observation. In writing (3.58), a critic may state that The Principle of Finite Energy Difference was presupposed instead of being proved (as we claimed later). This is not true. The two new terms introduced by (3.58) as a matter of implicitly defined quantities, mainly the reactive energy densities w_e^1 and w_h^1, were proved in the sequel to have the properties expected in such quantities. The expression (3.58) can best be thought of formally as a "guess" that proved right afterwards. The upshot of this point is that, contrary to what seem to be prevailing nowadays, *there is no unique, unequivocal derivation of reactive energy densities.* Maxwell's equations don't contain in themselves a fundamental physical quantity under this label of reactive energy. The concept of reactive energy is an engineering idea invented to deal with a specific practical situation, modeling the antenna system in terms of a fictitious circuit model called the input impedance. Certain measuring procedures were correlated with the real and imaginary parts of this impedance and a universally accepted circuit model approach to the design and analysis of general antenna systems has been developed very successfully. This, however, does not imply that a tool that is very widespread in practice such as the input impedance and the network parameters need to exhaust the radiation problem.

3.6.5 Ambiguity of the Concept of Reactive Field Energy

It is often argued in literature that the procedure outlined here is a "derivation" of the energy 'stored' in antenna systems. Unfortunately, this matter is questionable. The confusion arises from the bold interpretation of the term w_{rad} as a radiation energy density *everywhere*. This cannot be true for the following reason. When we write $w_{\text{rad}} = (\varepsilon/4)\mathbf{E}_{\text{rad}} \cdot \mathbf{E}^*_{\text{rad}} = (\mu/4)\mathbf{H}_{\text{rad}} \cdot \mathbf{H}^*_{\text{rad}}$, the resulting quantity is a function of the radial distance r. However, the expression loses its meaning when the observation is at the *near*-field zone. Indeed, if one applies the complex Poynting theorem there, we still get the same value of the net real power flow, but the *whole* field expression must now be taken into account, not just the far-field terms. Such *field* terms, whose amplitudes squared were used to calculate w_{rad}, simply don't satisfy Maxwell's equations in the near-field zone. For this reason, it is incoherent to state something like "since energy is summable quantity, then we can split the total energy into radiation density and non-radiation density" as we already did in (3.58). These two equations are *definitions* for the quantities w_h^1 and w_e^1, not derivations of them by a physical argument.[25]

To make this argument transparent, let us imagine the following scenario. Scientist X has already gone through all the steps of the previous procedure and ended up with mathematically sound definitions for the quantities w_h^1 and w_e^1, which he termed 'reactive energy densities.' Now, another person, say Scientist Y, is trying to solve the same problem. However, for some reason he does not hit directly on the term $w_{\rm rad}$ found by Scientist X, but instead considers the positive term Υ appearing in the equation

$$w_{\rm rad} = \alpha + \Upsilon, \tag{3.71}$$

where we assume

$$\int_{V_\infty - V} dv\Upsilon = \infty \tag{3.72}$$

and

$$\left| \int_{V_\infty - V} dv\alpha \right| < \infty. \tag{3.73}$$

That is, the divergent density $w_{\rm rad}$ is composed of two terms, one convergent and the other divergent. We further require that

$$w_{\rm rad}(r) = \alpha(r) + \Upsilon(r) \underset{r\to\infty}{\sim} \Upsilon(r). \tag{3.74}$$

That is, the asymptotic behavior of $w_{\rm rad}$ is dominated by the term Υ. The equations of the total energy density now become

$$w_e = w_e^1 + w_{\rm rad} = \left(w_e^1 + \alpha\right) + \Upsilon = w_e^2 + \Upsilon \tag{3.75}$$

and

$$w_h = w_h^1 + w_{\rm rad} = \left(w_h^1 + \alpha\right) + \Upsilon = w_h^2 + \Upsilon, \tag{3.76}$$

where

$$w_e^2 = w_e^1 + \alpha, \quad w_h^2 = w_h^1 + \alpha. \tag{3.77}$$

25 One has always to remember that the concept of energy in electromagnetism is not straightforward. All energy relations must be viewed as rigorous mathematical propositions derived from the calculus of Maxwell's equations, and *afterwards* interpreted as energies and power in the usual mechanical sense.

Now, it is easily seen that the conditions required for the "derivation" of w_h^1 and w_e^1 are already satisfied for the new quantities w_h^2 and w_e^2. That is, we have

$$w_{h,e}\left(r\right)=w_{h,e}^{1}\left(r\right)+\alpha\left(r\right)+\Upsilon\left(r\right)\underset{r\to\infty}{\sim}\Upsilon\left(r\right)\underset{r\to\infty}{\sim}w_{\mathrm{rad}}\left(r\right), \tag{3.78}$$

which states that the large argument approximation of $\Upsilon(r)$ coincides with the radiation density $w_{\mathrm{rad}}\left(r\right)$ at the far-field zone. Furthermore, it is obvious that

$$\begin{aligned}\int_{V_\infty-V}dv\left(w_h-w_e\right)&=\int_{V_\infty-V}dv\left(w_h^1-w_e^1\right)\\&=\int_{V_\infty-V}dv\left(w_h^2-w_e^2\right),\end{aligned} \tag{3.79}$$

and hence is convergent. Also,

$$\begin{aligned}&\int_{V_\infty-V}dv\left(w_h^2+w_e^2\right)\\&\quad=\int_{V_\infty-V}dv\left[\left(w_h-w_e\right)-2w_{rad}\right]-2\int_{V_\infty-V}dv\alpha\end{aligned} \tag{3.80}$$

and hence is also convergent. Therefore, the quantities w_h^2 and w_e^2 will be identified by Scientist Y as legitimate 'stored' energy in his quest for calculating the reactive energy density of the antenna. This clearly shows that the reactive energy calculated this way cannot be a legitimate physical quantity *in the sense that it is not unique*. In our opinion, the procedure of computing the reactive energy is artificial since it is tailored to fit an artificial requirement, the engineering circuit description of the antenna port impedance. Subtracting the radiation energy from the total energy is not a unique recipe for removing infinities. As should be clear by now, nobody seems to have thought that maybe the subtracted term w_{rad} itself contains a non-divergent term that is part of a physically genuine stored energy density defined through a non-circuit approach, i.e., field formalism *per se*.[26]

3.6.6 Critical Reexamination of the Near-Field Shell

We turn now to a qualitative and quantitative analysis of the magnitude of the ambiguity in the identification of the stored energy with the reactive energy. Let a be the minimum size of the hypothetical sphere enclosing the source region V_0. Denote by b the radial distance $b > a$ at which the term w_{rad} dominates asymptotically the reactive energy densities w_h^1 and w_e^1. It is the contribution of w_{rad} to the energy density lying in the interval $a < r < b$ which is ambiguous in the sense that it can be arbitrarily decomposed into the sum of two positive functions $\alpha(r)+\Upsilon(r)$

26 In Chapter 4, we show explicitly that this is indeed the case.

in the indicated interval. However, if the total contribution of the decomposable energy density within this interval is small compared with the overall contributions of the higher-order terms, then the ambiguity in the definition of the reactive energy densities does not lead to serious problems in practice. The evaluation of all the integrals with respect to r gives an expression in the form[27]

$$W_e^1 + W_h^1 = \left(\tfrac{\varepsilon}{4}\left\langle \mathbf{A}_0, \mathbf{A}_0\right\rangle + \tfrac{\mu}{4}\left\langle \mathbf{B}_0, \mathbf{B}_0\right\rangle\right)(b-a) + \sum_{n=1}^{\infty}\sum_{n'=1}^{\infty} \frac{\varepsilon\langle \mathbf{A}_n, \mathbf{A}_{n'}\rangle + \mu\langle \mathbf{B}_n, \mathbf{B}_n\rangle}{4(n+n'-1)}\left(\frac{1}{a^{n+n'-1}} - \frac{1}{b^{n+n'-1}}\right). \tag{3.81}$$

The integration with respect to the solid angle yields quantities with the same order of magnitude. Therefore, we focus in our qualitative examination on the radial dependance. It is clear that when a becomes very small, i.e., $a \ll 1$, the higher-order terms dominate the sum and the contribution of the lowest-order term can be safely neglected, with all its ambiguities. On the other hand, when a approaches the antenna operating wavelength and beyond, the higher-order terms rapidly decay and the lowest-order term dominates the contribution to the total energy in the interval $a < r < b$. Since it is in this very interval that we find the ambiguity in defining the reactive energy, we conclude that the reactive energy as defined in circuit theory cannot correspond to a physically meaningful definition of 'stored' *field* energy, and that the results calculated in literature as a fundamental limit to antenna Q are incoherent when the electrical size of the exclusion volume approaches unity and beyond.

One more point that needs to be examined in the above argument relates to the choice of b. Of course, b cannot be fixed arbitrarily because it is related to the behavior of the higher-order terms, i.e., b is the radius of the radiation sphere, the sphere through which most of the field is converted into radiation field.[28] Therefore, in our argument above a reaches the critical value of unit wavelength but cannot increase significantly because it is bounded from above by b, which is not freely varying like a. The upshot of the argument is that the vagueness in the precise value of b is simply the vagueness in any asymptotic expansion in general where accuracy is closely tied to the physical conditions of the particular situation under consideration. In this situation, the one corresponding to computing the reactive

27 In writing (3.81), we explicitly dropped the zero terms involving $\langle \mathbf{A}_0, \mathbf{A}_1\rangle$ and $\langle \mathbf{B}_0, \mathbf{B}_1\rangle$ in order simplify the notation.

28 Radiation field does not mean here propagating wave, but fields that contribute to the *real* part of the complex Poynting vector. Strictly speaking, the propagating field is close to the radiation field but not exactly the same because the nonpropagating field contributes to the far field. See also Chapter 4.

energy as defined above, the value of the reactive field energy $W_h^1 + W_e^1$ becomes very small with increasing a for the obvious reason that reactive energy is mostly localized in the near field close to the antenna. However, it is not clear at what precise value b one should switch from near field into radiation field. Indeed, it is exactly in this way that the entire argument of this part of the chapter was motivated: The circuit approach to antennas cannot give a coherent picture of genuine field problems. All what the common approach requires is that at a distance "large enough," the energy density converges (asymptotically) to the radiation density. However, while the total energy density is approaching this promised limit, the reactive energy is rapidly decaying in magnitude, and in such a case any ambiguity or error in the definition of the separation of the two densities (which, again, we believe to be non-physical) may produce a very large error, or at least render the results of the Q factor not so meaningful.[29]

3.7 CONCLUSION

In this chapter, we started the formulation of a comprehensive theoretical program for the analysis of the antenna electromagnetic field in general, and without restriction to a particular or specific configuration in the source regions. The study in Chapter 3 dealt with the analysis conducted in the spatial domain, that is, by mapping out the various spatial regions in the antenna exterior domain and explicating their electromagnetic behavior. We studied the phenomena of energy transfer between these regions and derived exact expressions for all types of such an energy exchange in closed analytical form in terms of the antenna TE and TM modes. The formulation shows that this detailed description can be obtained nonrecursively merely from knowledge of the antenna far-field radiation pattern. The resulting construction shows explicitly the contribution of each mode in the various spatial regions of the exterior domain, and also the coupling between different polarizations.

29 One can even reach this conclusion without any evaluation of total energy. The energy density itself is assumed to be a physically meaningful quantity. At around $a = 1$, all the radial factors in the terms appearing in (3.65) and (3.66) become roughly comparable in magnitude (assuming normalization to wavelength, i.e., $a = 1$ is taken here to be the intermediate-field zone boundary). However, the lowest-order term has an ambiguity in its definition that can be varied freely up to its full positive level. Thus, there seems to be a serious problem beginning in the intermediate-field zone. Even for larger a, since the overall reactive energy density becomes very small, slight changes in the value of the contribution of the radiation density resulting from the aforementioned ambiguity render, in our opinion, the Q factors curves reported in literature of limited physical relevance as indicators of the size of the actually stored field.

Of special interest is the discovery that the mutual interaction between regions with odd sum of indices is exactly zero, regardless to the antenna under study. Such a general result appears to be the reason why the infinite integral of the radial energy density giving rise to the antenna reactive energy is finite. The final parts of the chapter reexamined the concept of reactive energy when extended to study the field structure of the antenna. We showed how ambiguities in the definition of this circuit quantity render it of limited use in antenna near-field theory proper (matching considerations put aside). This prepares for the transition to Chapter 4, which is concerned with the analysis of the antenna near field in the spectral domain.

On the side of antenna practice, we believe that the proposed theory will play a role in future advanced research and devolvement of antenna systems. Indeed, this chapter has provided a formalism suitable for the visualization of the important spatial regions surrounding the antenna and the details of energy exchange processes taking place there. It has been found during the long history of electromagnetic theory and practice that the best intuitive but also rigorous way for understanding the operation and performance of actual devices and systems is the energy point of view. For this reason, the theory proposed did not stop at the field formalism, but also went ahead to investigate how this formalism can be used to provide general concrete results concerning the pathways of energy transfer between various regions in the antenna surrounding domain of interest. For example, we mention the interaction theorems developed in this chapter, which provide a quantitative measure of the field modal content passing from one spatial region to another. As we emphasized repeatedly before, this proved to be a natural way in understanding better the reactive energy, the quantity of fundamental importance in the determining the behavior of the antenna input impedance. Furthermore, the specification of all these descriptions in terms of the antenna physical TE and TM modes is continuous with established traditions in the electromagnetic community in which basic well-understood solutions of Maxwell's equations are used to determine and understand the complex behavior of the most general field. We believe that the generality of the formalism developed here will help future researchers to investigate special cases arising from particular applications within their range of interest for the community-at-large.

The far-field perspective can provide a different kind of valuable information for the antenna engineer. Here, one starts with a specification of a class of antennas compatible with a given far field radiation pattern, and then proceeds in constructing the near field of all antennas belonging to this class, in both the spatial and spectral domain, in order to relate far field performance measures, such as directivity, polarization, null formation, etc, to near field characteristics, such as input impedance and antenna size. A set of fundamental relations, understood in this sense, can be generated using our formalism for any set of objectives of interest found in a particular application, and hence guide the design process by deciding what kind of inherent conflicts and tradeoffs exist between various antagonistic measures. In this way, one can avoid cumbersome efforts to enforce a certain design goal that cannot be achieved in principle with any configuration whatsoever because it happens to violate one of the fundamental limitations mentioned above.

3.8 APPENDICES AND SUPPLEMENTARY MATERIALS

3.8.1 On the Uniform Convergence of the Energy Series using Wilcox Expansion

From [47], we know that the single series converges both absolutely and uniformly in all its variables. We prove that the energy (double) series is uniformly convergent in the following way. First, convert the double sum into a single sum by introducing a map $(n, n') \to l$. From a basic theorem in real analysis, the multiplication of two absolutely convergent series can be rearranged without changing its value. This guarantee that our new single series will give the same value regardless to the map $l = l(n, n')$. Finally, we apply the Cauchy criterion of uniform convergence [67] to deduce that the energy series, i.e., the original double sum, is uniformly convergent in all its variables.

3.8.2 Computation of the Functions $g^4_{n,n'}(l,m)$, $g^5_{n,n'}(l,m)$, and $g^6_{n,n'}(l,m)$

From (3.29), we calculate

$$g^4_{n,n'}(l,m) := \mathrm{Re}\left\{b^l_n b^{l*}_{n'}\right\} = \begin{cases} 0, & n+n' \text{ odd}, \\ (-1)^{(n+3n')/2} A_1(n,n';k), & n+n' \text{ even}, \end{cases} \tag{3.82}$$

where

$$A_1(n,n';k) = \frac{(l+n)!\,(l+n')!}{\left(n!2^n k^{n+1}\right)\left(n'!2^{n'} k^{n'+1}\right)(l-n)!\,(l-n')!}. \tag{3.83}$$

From (3.33), we also compute

$$\begin{aligned} g^5_{n,n'}(l,m) &:= \text{Re}\left\{c^l_n c^{l*}_{n'}\right\} \\ &= l(l+1)\, g^4_{n-1,n'-1}(l,m), \quad 1 \le n, n' \le l+1. \end{aligned} \tag{3.84}$$

From (3.34) we find

$$g^6_{n,n'}(l,m) := \text{Re}\left\{d^l_n d^{l*}_{n'}\right\}$$
$$= \begin{cases} (n-1)(n'-1)\,\text{Re}\left\{b^l_{n-1} b^{l*}_{n-1}\right\} \\ \quad +k^2 \text{Re}\left\{b^l_n b^{l*}_{n'}\right\} \\ \quad -k(n'-1)\,\text{Re}\left\{b^l_n i b^{l*}_{n'-1}\right\} \\ \quad +k(n-1)\,\text{Re}\left\{b^{l*}_{n'} i b^l_{n-1}\right\}, \ 1 \le n \le l, \\ l^2 \text{Re}\left\{b^l_l b^{l*}_l\right\}, \ n = l+1. \end{cases} \tag{3.85}$$

From (3.29), we compute

$$\text{Re}\left\{b^l_n i b^{l*}_{n'-1}\right\} = \begin{cases} 0, \ n+n' \text{ odd}, \\ (-1)^{(n+3n')/2-1} \\ \quad \times A_2(n,n';k), \ n+n' \text{ even}. \end{cases} \tag{3.86}$$

Similarly, we have

$$\text{Re}\left\{b^l_n i b^{l*}_{n'-1}\right\} = \begin{cases} 0, \ n+n' \text{ odd}, \\ (-1)^{(n'+3n)/2-1} \\ \quad \times A_2(n',n;k), \ n+n' \text{ even}. \end{cases} \tag{3.87}$$

Here we define

$$\begin{aligned} A_2(n,n';k) := &\frac{(l+n)!}{(n!2^n k^{n+1})(l-n)!} \\ &\times \frac{(l+n'-1)!}{(n'-1)!2^{n'-1} k^{n'} (l-n'+1)!}. \end{aligned} \tag{3.88}$$

We have used in obtaining (3.16) and (3.17), and also all similar calculations in Section 3.5, the manipulation $(i^n)^* = (i^*)^n = (-i)^n = i^n(-1)^n$.

3.8.3 On Rearrangement of the Y_{lm}/r^n Terms in the Multipole Expansion

We prove here that the series (3.35) and (3.36) are absolutely convergent and uniformly convergent in all their variables. The absolute convergence is our main goal since it provides the necessary and sufficient condition for arbitrary rearrangement of terms in infinite series. It is important to notice that the classical absolute and uniform convergence of both the multipole expansion and the Wilcox expansion does *not* necessarily imply that the series comprised of terms in the form $Y_{lm}(\theta,\varphi)/r^n$ for solutions of the Helmholtz equation are also absolutely and uniformly convergent. They strongly suggest that this is the case, but the conclusion has to be established in a rigorous fashion, which we endeavor to achieve in this appendix.

Due to limitation on space, we focus on the scalar problem. The generalization to the vector case is tedious but can be accomplished using the scalar case developed here in addition to a suitable Huygens-like integral formulation.[30]

Consider a scalar field $\psi(\mathbf{r})$ satisfying the scalar Helmholtz equation

$$\left(\nabla^2 + k^2\right)\psi(\mathbf{r}) = 0. \tag{3.89}$$

Our basic prototype of the multipole expansion is the series [89],[33]

$$\psi(\mathbf{r}) = \sum_{l=0}^{\infty}\sum_{m=-l}^{l} a_m^l h_l^{(1)}(kr)\, Y_{lm}(\theta,\varphi), \tag{3.90}$$

where a_m^l stands for the expansion coefficients. Using the series (3.28), (12.49) can be rewritten as

$$\psi(\mathbf{r}) = \sum_{l=0}^{\infty}\sum_{m=-l}^{l}\sum_{n=0}^{l} a_m^l b_n^l \frac{e^{ikr}}{r^{n+1}} Y_{lm}(\theta,\varphi). \tag{3.91}$$

Our goal is to show that the rearranged series

$$\frac{e^{ikr}}{r}\sum_{n=0}^{\infty}\frac{A_n}{r^n}, \tag{3.92}$$

where

$$A_n = \sum_{l=n}^{\infty}\sum_{m=-l}^{l} a_m^l b_n^l Y_{lm}(\theta,\varphi), \tag{3.93}$$

converges and is equal to (12.49). First, notice that the two inner sums in (3.130) are finite and hence can be freely interchanged. Therefore, we focus our study on the interaction between the l-sum and the n-sum.

30 For example, see [47].

The solution of (3.89) can be expressed in the form

$$\psi(\mathbf{r}) = \int_V d^3r' j(\mathbf{r}') \frac{e^{ik|\mathbf{r}-\mathbf{r}'|}}{|\mathbf{r}-\mathbf{r}'|}, \tag{3.94}$$

where the source function $j(\mathbf{r})$ is assumed to be continuous on a compact support V. Let the positions of the observation and source points be given in spherical coordinates as $\mathbf{r} = \mathbf{r}(r, \theta, \varphi)$ and $\mathbf{r}' = \mathbf{r}'(r', \theta', \varphi')$, respectively. Let the angle between the two vectors $\mathbf{r}$ and $\mathbf{r}'$ be γ. We have then

$$u := \cos\gamma = \cos\theta\cos\theta' + \sin\theta\sin\theta'\cos(\varphi - \varphi') \tag{3.95}$$

and

$$|\mathbf{r}-\mathbf{r}'| = r\sqrt{1 - 2(r'/r)\cos\gamma + (r'/r)^2}. \tag{3.96}$$

From potential theory, we know that the following expansion holds true [69],[89]

$$\frac{1}{|\mathbf{r}-\mathbf{r}'|} = \sum_{n=0}^{\infty} (r'/r)^n P_n(u), \tag{3.97}$$

where $P_n(u)$ is the Legendre polynomials. Furthermore, this series is absolutely convergent and uniformly convergent in all its variables [69].

Let us turn now the remaining factor in the scalar Green's function, $\exp(ik|\mathbf{r}-\mathbf{r}'|)$. We first consider instead the form

$$\exp(ik|\mathbf{r}-\mathbf{r}'|) = \exp(ikr)\exp\{ikr[(1/r)|\mathbf{r}-\mathbf{r}'| - 1]\} \tag{3.98}$$

and then expand the square root (3.96) in Taylor series. We obtain

$$(1/r)|\mathbf{r}-\mathbf{r}'| - 1 = \sum_{n=1}^{\infty} \alpha_n \left[-2(r'/r)u + (r'/r)^2\right]^n, \tag{3.99}$$

which is valid for $\left|-2(r'/r)\cos\gamma + (r'/r)^2\right| \leq 1$, a condition that is automatically satisfied in the exterior region. The coefficients α_n are given by $\alpha_n = (-1)^2 (2n)! \big/ (1-2n)(2n!)^2 (4^n)$. It is important to notice that the constant term in (3.99) dropped out, which will insure that the final series of the scalar Green's function takes the form of the Wilcox expansion.

Next, by substituting (3.99) into the exponential function, expanding the latter in its Taylor series, we arrive at

$$\begin{aligned}&\exp\left\{ikr\left[(1/r)\left|\mathbf{r}-\mathbf{r}'\right|-1\right]\right\}\\&=\sum_{l=0}^{\infty}\tfrac{1}{l!}\left(ikr\sum_{n=1}^{\infty}\alpha_n\left[-2\left(r'/r\right)u+\left(r'/r\right)^2\right]^n\right)^l\end{aligned}\tag{3.100}$$

In order to proceed, we need to expand each n-term in this series and then collect the resulting $l-$ and $n-$ terms, a process that involves freedom of rearrangement. To justify this, we need to show that the series (3.100) is absolutely convergent. This we accomplish most easily in the following way.

We replace all negative signs appearing in the square root function (3.99) with positive sign. Also, we replace each trigonometric function appearing there with its absolute value. That is, we define

$$\begin{aligned}u' :=\ &\left|\cos\theta\right|\left|\cos\theta'\right|\\&+\left|\sin\theta\right|\left|\sin\theta'\right|\left(\left|\cos\varphi\right|\left|\cos\varphi'\right|+\left|\sin\varphi\right|\left|\sin\varphi'\right|\right)\end{aligned}\tag{3.101}$$

We consider then the *new* series

$$\begin{aligned}&\sqrt{1+2\left(r'/r\right)u'+\left(r'/r\right)^2}-1\\&\qquad=\sum_{n=1}^{\infty}\alpha_n\left[2\left(r'/r\right)u'+\left(r'/r\right)^2\right]^n,\end{aligned}\tag{3.102}$$

which is convergent for $\left|2\left(r'/r\right)u'+\left(r'/r\right)^2\right|\leq 1$, again a condition satisfied in the exterior regions because all the trigonometric functions are bounded. It is well-known that this series is absolutely convergent for $\left|2\left(r'/r\right)u'+\left(r'/r\right)^2\right|<1$. Since the terms in brackets are already positive, we find then that

$$\sum_{n=1}^{\infty}\left|\alpha_n\right|\left[2\left(r'/r\right)u'+\left(r'/r\right)^2\right]^n<\infty.\tag{3.103}$$

Finally, since the Taylor series expansion of the exponential function is always convergent, we conclude that

$$\begin{aligned}&\exp\left(kr\sum_{n=1}^{\infty}\left|\alpha_n\right|\left[2\left(r'/r\right)u'+\left(r'/r\right)^2\right]^n\right)\\&=\sum_{l=0}^{\infty}\tfrac{1}{l!}\left(kr\sum_{n=1}^{\infty}\left|\alpha_n\right|\left[2\left(r'/r\right)u'+\left(r'/r\right)^2\right]^n\right)^l<\infty\end{aligned}\tag{3.104}$$

Since all the terms appearing in this series are positive and dominate the series (3.100), we conclude that the expansion (3.100) is absolutely convergent. To be more precise, let us isolate the "atomic" terms of our final expansion. By 'atomic terms' we mean the ultimate indivisible terms comprising the expansion of the Green's function obtained by multiplying (3.97) with (3.100). This general term takes the form

$$\begin{aligned}(r'/r)^n \cos^{l_1}\theta \cos^{l_2}\theta' \sin^{l_3}\theta \sin^{l_4}\theta' \\ \times \cos^{l_5}\varphi \cos^{l_6}\varphi' \sin^{l_7}\varphi \sin^{l_8}\varphi',\end{aligned} \tag{3.105}$$

where all the exponents are integers. The result (3.103) then tells us that the infinite series comprised of the atomic terms (3.105) is absolutely convergent in the exterior region. From the triangular identity, any grouping of the atomic terms results in a series that is also absolutely convergent. This fact will be used throughout the remaining part of this appendix.

Knowing that all the series expansions involved are absolutely convergent, we can now multiply (3.97) with (3.100) and collect all terms involving like powers of $1/r$. The coefficients of such powers are all continuous functions of u, which is in the range $-1 \leq u \leq 1$. Since the Legendre functions form a complete set of basis functions for the space of continuous functions on the interval $[-1,1]$ [69], [89], we conclude that the Green's functions can be written in the form

$$\frac{e^{ik|\mathbf{r}-\mathbf{r}'|}}{|\mathbf{r}-\mathbf{r}'|} = \sum_{n=0}^{\infty}\sum_{l=0}^{\infty} \frac{A_n^l(r')}{r^n} P_l(u), \tag{3.106}$$

where the functional dependence of the coefficient A_n^l on r' takes the simple form of r' raised to some integer power whose exact form does not concern us here. From the addition theorem of the Legendre function we have [89]

$$P_l(\cos\gamma) = \frac{4\pi}{2l+1} \sum_{m=-l}^{l} Y_{lm}^*(\theta',\varphi')\, Y_{lm}(\theta,\varphi). \tag{3.107}$$

Substituting this into (3.106), we find

$$\frac{e^{ik|\mathbf{r}-\mathbf{r}'|}}{|\mathbf{r}-\mathbf{r}'|} = \sum_{n=0}^{\infty}\sum_{l=0}^{\infty}\sum_{m=-l}^{l} \frac{4\pi A_n^l(r')\, Y_{lm}^*(\theta',\varphi')}{2l+1} \frac{Y_{lm}(\theta,\varphi)}{r^n}. \tag{3.108}$$

It is not difficult to show that this series is uniformly convergent in all of its variables. For the angular dependence, this follows readily from the boundedness of the trigonometric functions and the Weierstrass-M test. For the radial dependence, this can be shown by rearranging the relevant series of the atomic terms (3.105) in the form of a power series in the radial variable under consideration, an operation permissible since we already know that the original series is absolutely convergent. The convergence of a power series at one point implies that it is also uniformly convergent in the exterior region [67]. Now knowing that (3.106) is uniformly convergent, we can substitute it into (3.94) and interchange the order of integration and summation. The result is

$$\psi(\mathbf{r}) = \sum_{n=0}^{\infty}\sum_{l=0}^{\infty}\sum_{m=-l}^{l} \xi_n^l \frac{Y_{lm}(\theta, \varphi)}{r^n}, \tag{3.109}$$

where the expansion coefficients are given by

$$\xi_n^l = \frac{4\pi}{2l+1}\int_V d^3r' j(\mathbf{r}') A_n^l(r') Y_{lm}^*(\theta', \varphi'). \tag{3.110}$$

The last step of our proof is to show that the series given by (3.109) is absolutely (and also uniformly) convergent. To accomplish this, we make use of the fact that the source function is continuous on a compact support V. From classical analysis, it follows that $j(\mathbf{r}')$ achieves its bound in V, that is, there exists a positive real number M such that

$$|j(\mathbf{r}')| \leq M \ , \mathbf{r}' \in V. \tag{3.111}$$

Consider the series of positive terms

$$\sum_{n=0}^{\infty}\sum_{l=0}^{\infty}\sum_{m=-l}^{l} |\xi_n^l| \frac{|Y_{lm}(\theta, \varphi)|}{r^n} \tag{3.112}$$

From (3.110) and (3.111), we find

$$|\xi_n^l| \leq \frac{4\pi M}{2l+1}\int_V d^3r' \left|A_n^l(r') Y_{lm}^*(\theta', \varphi')\right|. \tag{3.113}$$

Now, multiply this inequality by $|Y_{lm}(\theta, \varphi)|/r^n$ and sum over all indices. From the absolute and uniform convergence of the series of atomic terms (3.105), we find that

$$\int_V d^3r' \sum_{n=0}^{\infty}\sum_{l=0}^{\infty}\sum_{m=-l}^{l} \frac{4\pi M \left|A_n^l(r') Y_{lm}^*(\theta', \varphi') Y_{lm}(\theta, \varphi)\right|}{(2l+1)\, r^n} \tag{3.114}$$

is finite, where the uniform convergence was needed to move the integral outside the sums and hence relate (3.114) to the RHS of (3.113). From (3.112), (3.113), (3.114), we conclude that (3.109) is absolutely convergent. The proof that it is also uniformly convergent in all its variables also follows from the same argument, mainly from using (3.111) and the Weierstrass-M test.

Consider the limit

$$\lim_{l\to\infty}\left|b_n^l\right| = \lim_{l\to\infty}\frac{1}{n!2^n k^n}\frac{(l+n)!}{(l-n)!} = \frac{1}{n!2^n k^n}. \tag{3.115}$$

Since this limit exists, the sequence b_n^l is bounded. That is, there exists a positive number M_n, generally dependent on n, while the following is true

$$\exists\ N \geq 0\ \text{ such that}\ \forall\ l \geq N,\ \left|b_n^l\right| \leq M_n. \tag{3.116}$$

We can write then

$$\left|a_m^l b_n^l Y_{lm}(\theta,\varphi)\right| \leq \left|a_m^l Y_{lm}(\theta,\varphi)\right| M_n, \tag{3.117}$$

which is valid for sufficiently large l. Now, by definition, the series comprising the far field, which is evaluated by taking the asymptotic limit $kr \to \infty$ of the expansion (12.49) and using (3.28), is convergent. Being a Laplace expansion of a well-behaved field [69],[31] it is also absolutely convergent [45]. Therefore, we have

$$M_n \sum_{l=n}^{\infty}\sum_{m=-l}^{l}\left|a_m^l Y_{lm}(\theta,\varphi)\right| < \infty. \tag{3.118}$$

It follows then from (3.135) and the comparison test that the family of series

$$\sum_{l=n}^{\infty}\sum_{m=-l}^{l}\left|a_m^l b_n^l Y_{lm}(\theta,\varphi)\right| < \infty \tag{3.119}$$

is convergent for all $n \geq 0$. Now consider the series of positive terms

$$\sum_{n=0}^{\infty}\frac{D_n}{n!2^n k^n r^n}, \tag{3.120}$$

31 For example, it is enough here to assume that the field in the exterior region has continuous first-order derivatives.

where

$$D_n = \sum_{l=n}^{\infty} \sum_{m=-l}^{l} \left| a_m^l Y_{lm}(\theta, \varphi) \right| \frac{(l+n)!}{(l-n)!}. \tag{3.121}$$

The series (3.120) is the one needed in deciding the absolute convergence of (3.130). We apply the ratio test to (3.120) and investigate the limit

$$\lim_{n\to\infty} \frac{D_{n+1}}{D_n} \frac{1}{2nkr}. \tag{3.122}$$

First, we notice that

$$\lim_{n\to\infty} \frac{1}{2nkr} = 0. \tag{3.123}$$

Next, we consider the behavior of the ratio D_{n+1}/D_n. The general term D_n has the form of the *tail* of a convergent series. Therefore,

$$\lim_{n\to\infty} D_n = \lim_{n\to\infty} \sum_{l=n}^{\infty} \sum_{m=-l}^{l} \left| a_m^l Y_{lm}(\theta, \varphi) \right| \frac{(l+n)!}{(l-n)!} = 0. \tag{3.124}$$

This implies that the ratio D_{n+1}/D_n is bounded at infinity. For suppose that it approaches infinity. This amounts to saying that for any chosen number $M > 0$, it is possible to find $N > 0$ such that $D_{n+1}/D_n > M$ for $n \geq N$. Therefore, the sequence D_n is increasing at infinity (to see this, choose $M > 1$). But this contradicts (3.124). Therefore, the ratio D_{n+1}/D_n is bounded and we can write that there exists $0 < M < \infty$ such that for sufficiently large n, we have

$$0 \leq \frac{D_{n+1}}{D_n} \leq M. \tag{3.125}$$

Now choose any small positive number ε. The result (3.123) implies that we can always find an integer $N > 0$ such that

$$n \geq N \;\Rightarrow\; 0 \leq \frac{1}{2nkr} \leq \frac{\varepsilon}{M}. \tag{3.126}$$

Using (3.125), we conclude

$$0 \leq \frac{1}{2nkr} \frac{D_{n+1}}{D_n} \leq \varepsilon. \tag{3.127}$$

Therefore, we find

$$\lim_{n\to\infty}\frac{D_{n+1}}{D_n}\frac{1}{2nkr}=0. \tag{3.128}$$

It follows from the ratio test that the series (3.120) is convergent. Therefore, the series (3.131) is absolutely convergent and hence can be rearranged in any way leading always to the same value. One possible rearrangement can reproduce (3.130), which is the result needed.

3.8.4 On Rearrangement of the $1/r^n$ Terms in the Multipole Expansion

Our basic prototype of the multipole expansion is the series of the scalar field [89], [33]

$$\psi(\mathbf{r})=\sum_{l=0}^{\infty}\sum_{m=-l}^{l}a_m^l h_l^{(1)}(kr)\,Y_{lm}(\theta,\varphi), \tag{3.129}$$

where a_m^l stands for the expansion coefficients. By using the series (3.28), (12.49) can be rewritten as

$$\psi(\mathbf{r})=\sum_{l=0}^{\infty}\sum_{m=-l}^{l}\sum_{n=0}^{l}a_m^l b_n^l\frac{e^{ikr}}{r^{n+1}}Y_{lm}(\theta,\varphi). \tag{3.130}$$

Our goal is to show that the rearranged series

$$\frac{e^{ikr}}{r}\sum_{n=0}^{\infty}\frac{1}{r^n}\sum_{l=n}^{\infty}\sum_{m=-l}^{l}a_m^l b_n^l Y_{lm}(\theta,\varphi) \tag{3.131}$$

converges and is equal to (12.49). Probably the easiest way to achieve this is by noticing the special form of (12.49), which consists formally of an infinite series where each summand itself is expanded in a finite number of terms, each being a power of $1/r$. From complex analysis, the following theorem exists [93] (The Weierstrass Double Series Theorem). Let

$$w(z)=w_1(z)+w_2(z)+w_3(z)+\cdots \tag{3.132}$$

be an infinite series of functions of the complex variable z, all analytic (holomorphic) in domain T and let the series converges uniformly in T. Then the sum $w(z)$ is analytic in any closed region $D\subseteq T$. Furthermore, if each function $w_n(z)$ is expanded in a power series around some point z_0

$$w_n(z) = \sum_{m=0}^{\infty} c_m^n (z - z_0)^m, \tag{3.133}$$

and if

$$w(z) = \sum_{n=0}^{\infty} c_n (z - z_0)^n, \tag{3.134}$$

then each series $\sum_{m=0}^{\infty} c_m^n$ converges and we have

$$c_n = \sum_{m=0}^{\infty} c_m^n. \tag{3.135}$$

This theorem can be directly applied to (3.130) since it has the same form for $z = 1/r$ and $z_0 = 0$. Since the series (3.129) can be originally derived by expanding the Green's function in a power series of $1/r$ [69], the expansion is valid in the complex domain as such. In this case, the exterior region of the antenna $|r| > r_0 > 0$ corresponds to the analytically continued complex domain $|z| < 1/r_0$. It remains to show that (3.129) converges uniformly in r. This can be easily done by majorizing (3.129) by the series

$$\sum_{l=0}^{\infty} \sum_{m=-l}^{l} M_m^l = \sum_{l=0}^{\infty} \sum_{m=-l}^{l} \left|a_m^l\right| \left|h_l^{(1)}(kr_0)\right| \left|Y_{lm}(\theta, \varphi)\right| \tag{3.136}$$

which is possible because $\left|h_l^{(1)}(kr)\right| \leq \left|h_l^{(1)}(kr_0)\right|$ for $r > r_0$. We know that the Laplace series (multipole expansion) (3.129) is absolutely convergent [69] and hence $\sum_{l,m} M(l, m)$ converges. It follows then from the classical Weierstrass-M test that the series (3.129) is uniformly convergent in r.

Chapter 4

The Spectral Theory of Electromagnetic Fields

4.1 INTRODUCTION

The results of the Chapter 3 have provided us with an insight into the structure of what we particularly called 'the near-field shell' in the spatial domain. This concept has been important particularly in connection with the computation of the reactive energy of the antenna system, the quantity needed in the estimation of the quality factor and hence the input impedance bandwidth. We have shown, however, that since the concept of reactive energy is mainly a circuit concept, it is incapable of describing adequately the more troublesome concept of stored field energy. In this chapter, we propose a new look into the structure of the near fields by examining the *evanescent* part of the electromagnetic radiation in the vicinity of the antenna. The mathematical treatment will be fundamentally based on the Weyl expansion [94], and hence this will be essentially a spectral method. Such an approach, in our opinion, is convenient from both the mathematical and physical viewpoints. For the former, the availability of the general form of the radiated field via the dyadic Green's function theorem allows the applicability of the Weyl expansion to Fourier-analyze any field form into its spectral components. From the physical point of view, we notice that in practice the the focus is mainly on 'moving energy around' from that location to another. Therefore, it appears to us natural to look for a general mathematical description of the antenna near fields in terms of, speaking informally, 'parts that do not move' (nonpropagating field), and 'parts that do move' (propagating field). As we will see shortly, the Weyl expansion is well suited to exactly this; it combines both the mathematical and physical perspectives in one step. Such a field decomposition into two parts can therefore be seen as a logical step toward a fundamental insight into the nature of the electromagnetic near field.

Because of the complexity involved in the argument presented in this chapter, we review here the basic ideas and motivations behind each section. In Section 4.2, we provide a more sophisticated analysis of the near field that goes beyond the customary (circuit) view of reactive fields and energies. To start with, we recruit the Weyl expansion in expanding the scalar Green's function into propagating and nonpropagating (evanescent) parts. By substituting this expansion into the dyadic Green's function theorem, an expansion of the total fields into propagating and nonpropagating parts becomes feasible. We then break the rotational symmetry by introducing two coordinate system, once is fixed (the global frame), while the other can rotate freely with respect to the fixed frame (the local frame). We then systematically develop the mathematical machinery that allows us to describe the decomposition of the electric field into the two modes above along the local frame. It turns out that an additional rotation of the local frame around its z-axis does not change the decomposition into total propagating and nonpropagating parts along this axis. This crucial observation, which can be proved in a straightforward manner, is utilized to introduce the concept of radial streamlines. This concept is a description of how the electromagnetic fields split into propagating and nonpropagating modes along radial streamlines, like the situation in hydrodynamics, but defined here only in terms of fields. The concept of radial streamlines will appear with the progress of our study to be the most important structure of the antenna near field from an engineering point of view. We also show that the propagating and nonpropagating parts both satisfy Maxwell's equations individually. This important observation will be needed later in building the energy interpretation. The section ends with a general flow chart illustrating how the spectral composition of the electric field is constructed. This is indeed the essence of the formation of the antenna near field, which we associate here with the nonpropagating part.

In Section 4.3, we further study the near-field streamlines by systematically investigating the energy associated with our previous field decomposition. The fact that the propagating and nonpropagating parts are Maxwellian fields is exploited to generalize the Poynting theorem to accommodate for the three different contributions to the total energy, the self energy of the propagating field, the self energy of the nonpropagating field, and the interaction energy between the two fields, which may be positive or negative, while the first two self energies are always positive. We then investigate various types of near field energies. It appears that two important classes of energies can be singled out for further consideration: localization energy and stored energy. We notice that the latter may not be within the reach of the time-harmonic theory we develop in this work, but provide expressions to compute the former energy type. One conclusion here is that the radial streamline nonpropagating energy is convergent in the antenna exterior region, another positive evidence of its importance.

In Section 4.4, we investigate the near-field structure from the far-field point of view, i.e., using the Wilcox expansion. To achieve this, a generalization of the Weyl expansion is needed, which we derive and then use to devise a hybrid Wilcox-Weyl expansion. The advantage of the hybrid expansion is this. While the recursive structure of the Wilcox expansion, and the direct construction outlined in Chapter 3, allow us to obtain all the terms in the series by starting from a given far-field radiation pattern, the generalized Weyl expansion permits a spectral analysis of each term into propagating and nonpropagating streamlines. We notice that only radial streamlines are possible here, which can be interpreted as a strong relation between the the far field and the near field of antennas that was not suspected previously. A more thorough study of this last observation will be conducted in a later publication.

Finally, in Section 4.5 we go back to the analysis of the antenna from the source point of view where we provide a very general explication of the way in which the far field of antennas is produced starting from a given current distribution. The theory explains naturally why some antennas like linear wires and patch antennas possess broadside radiation patterns. It turns out that the whole process of the far field formation can be described in terms of geometrical transformations and spatial filtering, two easy-to-understand processes.

4.2 SPECTRAL ANALYSIS OF ANTENNA NEAR FIELDS: THE CONCEPT OF RADIAL STREAMLINES

4.2.1 Spectral Decomposition Using the Weyl Expansion

We start by assuming that the current distribution of an arbitrary antenna is given by a continuous electric current volume density $\mathbf{J}(\mathbf{r})$ defined on a compact support (finite and bounded volume) V. Let the antenna be surrounded by an infinite, isotropic, and homogeneous space with electric permittivity ε and magnetic permeability μ. The electric field radiated by this current distribution is given by the dyadic Green's function theorem [35]

$$\mathbf{E}(\mathbf{r}) = i\omega\mu \int_V d^3r' \, \bar{\mathbf{G}}(\mathbf{r}, \mathbf{r}') \cdot \mathbf{J}(\mathbf{r}'), \tag{4.1}$$

where the dyadic Green's function is given by

$$\bar{\mathbf{G}}\left(\mathbf{r},\mathbf{r}'\right)=\left[I+\frac{\nabla\nabla}{k^2}\right]g\left(\mathbf{r},\mathbf{r}'\right), \tag{4.2}$$

while the scalar Green's function is defined as

$$g\left(\mathbf{r},\mathbf{r}'\right)=\frac{e^{ik\left|\mathbf{r}-\mathbf{r}'\right|}}{4\pi\left|\mathbf{r}-\mathbf{r}'\right|}. \tag{4.3}$$

Therefore, the electromagnetic fields radiated by the antenna[1] can be totally determined by knowledge of the dyadic Green's function and the current distribution on the antenna. We would like to further decompose the former into two parts, one pure propagating and the other evanescent. This task can be accomplished by using the Weyl expansion [94],[35]

$$\frac{e^{ikr}}{r}=\frac{ik}{2\pi}\int_{-\infty}^{\infty}\int_{-\infty}^{\infty}dpdq\frac{1}{m}e^{ik(px+qy+m|z|)}, \tag{4.4}$$

where[2]

$$m(p,q)=\begin{cases}\sqrt{1-p^2-q^2} & ,p^2+q^2\leq 1\\ i\sqrt{p^2+q^2-1} & ,p^2+q^2>1\end{cases}. \tag{4.5}$$

Our mathematical devolvement has been constrained to the condition of time-harmonic excitation, i.e., all time variations take the form $\exp(-i\omega t)$. From the basic definition of waves [91], we know that wave propagation occurs only if the mathematical solution of the problem can be expressed in the form $\Psi(r-ct)$, where c is a constant and Ψ is some function.[3] Since the time variation and the spatial variation are separable, it is not difficult to see that the only spatial variation that can lead to a total spatio-temporal solution that conforms to the expression of a propagating wave mentioned above is the exponential form $\exp(imr)$, where m is a *real* constant. The part of the field that cannot be put in this form is taken simply as the *nonpropagating* portion of the total field.[4] Indeed, the Weyl expansion shows that the total scalar Green's function can be divided into the sum of two parts, one

1 The magnetic field can be easily obtained from Maxwell's equations.

2 Throughout this chapter, the explicit dependance of m on p and q will be suppressed for simplicity.

3 Here, a one-dimensional problem is assumed for simplicity.

4 This convention supplies the incentive for our whole treatment of the concept of energies localized and stored in the antenna fields as presented in this chapter.

as pure propagating waves and the other as evanescent, hence nonpropagating part. Explicitly, we write

$$g(\mathbf{r},\mathbf{r}') = g_{\text{ev}}(\mathbf{r},\mathbf{r}') + g_{\text{pr}}(\mathbf{r},\mathbf{r}'), \tag{4.6}$$

where the propagating and nonpropagating (evanescent) parts are given, respectively, by the expressions

$$\begin{aligned} g_{\text{ev}}(\mathbf{r},\mathbf{r}') = \tfrac{ik}{8\pi^2}\textstyle\int_{p^2+q^2>1} dpdq \tfrac{1}{m} e^{ik\left[p(x-x')+q(y-y')\right]} \\ \times e^{im\left|z-z'\right|}, \end{aligned} \tag{4.7}$$

$$\begin{aligned} g_{\text{pr}}(\mathbf{r},\mathbf{r}') = \tfrac{ik}{8\pi^2}\textstyle\int_{p^2+q^2<1} dpdq \tfrac{1}{m} e^{ik\left[p(x-x')+q(y-y')\right]} \\ \times e^{im\left|z-z'\right|}. \end{aligned} \tag{4.8}$$

The Weyl expansion can be significantly simplified by transforming the double integrals into cylindrical coordinates and then making use of the integral representation of the Bessel function [88]. The final results are[5]

$$g_{\text{ev}}(\mathbf{r},\mathbf{r}') = \frac{k}{4\pi}\int_0^{\infty} du J_0\left(k\rho_s\sqrt{1+u^2}\right) e^{-k\left|z-z'\right|u}, \tag{4.9}$$

$$g_{\text{pr}}(\mathbf{r},\mathbf{r}') = \frac{ik}{4\pi}\int_0^{1} du J_0\left(k\rho_s\sqrt{1+u^2}\right) e^{ik\left|z-z'\right|u}, \tag{4.10}$$

where $\rho_s = \sqrt{(x-x')^2+(y-y')^2}$. A routine but important observation is that the integral (4.9), which gives the total evanescent part of the electric field, is both uniformly and absolutely convergent for $z \neq z'$.[6]

By substituting the Weyl identity (4.4) into (4.1) and using (4.3), we obtain easily the following expansion for the dyadic Green's function[7]

$$\begin{aligned} \bar{\mathbf{G}}(\mathbf{r},\mathbf{r}') = \tfrac{ik}{8\pi^2}\textstyle\int_{-\infty}^{\infty}\int_{-\infty}^{\infty} dpdq \tfrac{\bar{\mathbf{I}}k^2-\mathbf{KK}}{k^2 m} \\ \times e^{ik\left[p(x-x')+q(y-y')+m\left|z-z'\right|\right]}, \end{aligned} \tag{4.11}$$

5 The details of similar transformation will be given explicitly in Section 4.2.6.

6 See Appendix 4.7.1.

7 First, we bring the differentiation operators into the integral (see Appendix 4.7.2 for justification). Next, the vector identities $\nabla \exp(\mathbf{A}\cdot\mathbf{r}) = \mathbf{A}\exp(\mathbf{A}\cdot\mathbf{r})$ and $\nabla\cdot\mathbf{B}\exp(\mathbf{A}\cdot\mathbf{r}) = \mathbf{A}\cdot\mathbf{B}\exp(\mathbf{A}\cdot\mathbf{r})$ are used.

where the spectral variable (wavevector) is given by

$$\mathbf{K} = \hat{x}kp + \hat{y}kq + \hat{z}\text{sgn}\left(z - z'\right) km. \tag{4.12}$$

Here, sgn stands for the signum function.[8] Throughout this chapter, we will be concerned only with the *exterior* region of the antenna, i.e., we don't investigate the fields within the source region. For this reason, the singular part that should appear explicitly in the Fourier expansion of the dyadic Green's function (4.11) in the form of a delta function was dropped.

The dyadic Green's function can be decomposed into two parts, evanescent and propagating, and the corresponding expressions are given by[9]

$$\begin{aligned}\bar{\mathbf{G}}_{\text{ev}}\left(\mathbf{r},\mathbf{r}'\right) = \tfrac{ik}{8\pi^2}\int_{p^2+q^2>1} dpdq \tfrac{\bar{\mathbf{I}}k^2-\mathbf{KK}}{k^2m} \\ \times e^{ik\left[p\left(x-x'\right)+q\left(y-y'\right)+m\left|z-z'\right|\right]},\end{aligned} \tag{4.13}$$

$$\begin{aligned}\bar{\mathbf{G}}_{\text{pr}}\left(\mathbf{r},\mathbf{r}'\right) = \tfrac{ik}{8\pi^2}\int_{p^2+q^2<1} dpdq \tfrac{\bar{\mathbf{I}}k^2-\mathbf{KK}}{k^2m} \\ \times e^{ik\left[p\left(x-x'\right)+q\left(y-y'\right)+m\left|z-z'\right|\right]}.\end{aligned} \tag{4.14}$$

Substituting the spectral expansion of the dyadic Green's function as given by (4.11) into (4.1), we obtain after interchanging the order of integration

$$\mathbf{E}\left(\mathbf{r}\right) = \frac{-\omega k\mu}{8\pi^2}\int_{-\infty}^{\infty} dpdq \frac{\bar{\mathbf{I}}k^2 - \mathbf{KK}}{k^2m} \cdot \tilde{\mathbf{J}}\left(\mathbf{k}\right) e^{i\mathbf{K}\cdot\mathbf{r}}, \tag{4.15}$$

where $\tilde{\mathbf{J}}\left(\mathbf{K}\right)$ is the spatial Fourier transform of the source distribution

$$\tilde{\mathbf{J}}\left(\mathbf{K}\right) = \int_V d^3r'\, \mathbf{J}\left(\mathbf{r}'\right) e^{-i\mathbf{K}\cdot\mathbf{r}'}. \tag{4.16}$$

The expansion (4.15) is valid only in the region $z > L$ and $z < -L$, i.e., the region exterior to the infinite slab $-L \leq z \leq L$. The reason is that in the integral representation of the dyadic Green's function (4.11), the integration contour actually does not vary smoothly on the source variables $\mathbf{r}'$. However, for the region $|z| > L$, it is possible to justify this exchange of order.[10]

8 The signum function is defined as

$$\text{sgn}\left(z\right) = \begin{cases} z, & z \geq 0 \\ -z, & z < 0 \end{cases}$$

9 For the purpose of numerical evaluation, the reader must observe that the expressions of the dyadic Green's function decomposition (4.13) and (4.14) contain more than two basic integrals because of the dependence of $\mathbf{K}$ on p and q as indicated by (16.17).

10 See Appendix 4.7.3.

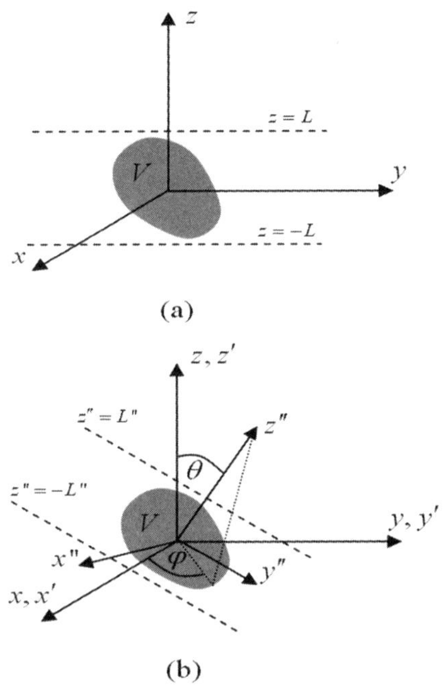

Figure 4.1 The geometrical description of the antenna source distribution (shaded volume V) suitable for the application of Weyl expansion. (a) Global observation coordinate system. The spectral representation of the radiated field given by (4.15) is valid only in the region $|z| > L$. (b) Global and local coordinate system. Here, for any orientation of the local frame described by θ and ϕ, L'' will be greater than the maximum dimension of the source region V in that direction.

4.2.2 Concept of Propagation in the Antenna Near-Field Zone

As can be seen from equation (4.13) for the antenna fields expressed in terms of evanescent modes, the expansion itself depends on the choice of the coordinate system while the total field does not. Actually, there are two types of coordinates to be taken into account here, those needed for the mathematical description of the antenna current distribution $\mathbf{J}(\mathbf{r}')$, i.e., the point $\mathbf{r}'$, and those associated with the observation point $\mathbf{r}$. In Figure 4.1(a), we show only the observation frame since the source frame is absorbed into the dummy variables of the integral defining the Fourier transform of the antenna current distribution (4.16). In the Weyl expansion as originally given in (4.4), the orientation of the observation frame of reference is unspecified. This is simply the mathematical expression of the fact that scalar electromagnetic sources possess rotational symmetry, i.e., the field generated by a point source located at the origin depends only on the distance of the observation point from the origin. At a deeper level, we may take this symmetry condition as an integral trait of the underlying spacetime structure upon which the electromagnetic field is defined.[11] What is relevant to our present discussion, which is concerned with the nature of the antenna near field, is that the observation frame of reference can be rotated in an arbitrary manner around a fixed origin. Let us start then by fixing the choice for the origin of the source frame x', y', and z'. Next, we define a *global* frame of reference and label its axis by x, y, and z. Without loss of generality, we assume that the source frame is coincident with the global frame. We then introduce another coordinate system with the same origin of the both the global and source frames and label its coordinates by x'', y'', and z''. This last frame will act as our *local* observation frame. It can be orientated in an arbitrary manner as is evident from the freedom of choice of the coordinate system in the Weyl expansion (4.4). We allow the z''-axis of our local observation frame to be directed at an arbitrary direction specified by the spherical angles θ and φ, i.e., the z''-axis will coincide with the unit vector $\hat{r}$ in terms of the global frame. The situation is geometrically

11 This observation can be further formalized in the following way. The field concept is defined at the most primitive level as a function on space and time. Now what is called space and time is described mathematically as a *manifold*, which is a precise way of saying that space and time are topological spaces that admit differentiable coordinate charts (frames of references). We find then that *the electromagnetic fields are functions defined on manifolds*. The manifold itself may possess certain symmetry properties, which in the case of our Euclidean space are a rotational and translational symmetry. Although only the rotational symmetry is evident in the form of Weyl expansion given by (4.4), the reader should bear in mind that the translational invariance of the radiated fields has been already used implicitly in moving from (4.4) to expressions such as (4.13) and (4.14), where the source is located at $\mathbf{r}'$ instead of the origin.

described in Figure 4.1(b). There, the Weyl expansion will be written in terms of the local frame x'', y'', and z'' with region of validity given by $|z| > L''$, where $L'' = L''(\theta, \varphi)$ is chosen such that it will be greater than the maximum size of the antenna in the direction specified by θ and φ. It can be seen then that our radiated electric fields written in terms of the global frame but spectrally expanded using the (rotating) local frame are given[12]

$$\begin{aligned}\mathbf{E}(\mathbf{r}) = \tfrac{-\omega k\mu}{8\pi^2} \int_{-\infty}^{\infty}\int_{-\infty}^{\infty} dpdq \tfrac{\bar{\mathbf{I}}k^2 - \mathbf{K}''\mathbf{K}''}{k^2 m} \cdot \int_V d^3r' \, \mathbf{J}(\mathbf{r}') \\ \times e^{ik\left[px'' + qy'' + \mathrm{sgn}\left(z'' - L''\right) mz''\right]} \\ \times e^{ik\left[-px'_s - qy'_s - \mathrm{sgn}\left(z' - L''\right) mz'_s\right]},\end{aligned} \tag{4.17}$$

where the new spectral vector is given by

$$\mathbf{K}'' = \hat{x}'' kp + \hat{y}'' kq + \hat{z}'' \mathrm{sgn}\left(z'' - L''\right) km. \tag{4.18}$$

The cartesian coordinates $\mathbf{r}'_s = \langle x'_s, y'_s, z'_s \rangle$ in (4.17) represent the source coordinates $\mathbf{r}' = \langle x', y', z' \rangle$ after being transformed into the language of the new frame $\mathbf{r}'' = \langle x'', y'', z'' \rangle$.[13] In terms of this notation, (4.17) is rewritten in the more compact form

$$\begin{aligned}\mathbf{E}(\mathbf{r}) = \tfrac{-\omega k\mu}{8\pi^2} \int_{-\infty}^{\infty} dpdq \tfrac{\bar{\mathbf{I}}k^2 - \mathbf{K}''\mathbf{K}''}{k^2 m} \\ \times \cdot \int_V d^3r' \, \mathbf{J}(\mathbf{r}') \, e^{-i\mathbf{K}\cdot\mathbf{r}'_s} e^{i\mathbf{K}\cdot\mathbf{r}''}.\end{aligned} \tag{4.19}$$

To proceed further, we need to write down the local frame coordinates explicitly in terms of the global frame. To do this, the following rotation matrix is employed[14]

$$\bar{\mathbf{R}}(\theta, \varphi) := \begin{pmatrix} R_{11} & R_{12} & R_{13} \\ R_{21} & R_{22} & R_{23} \\ R_{31} & R_{32} & R_{33} \end{pmatrix}, \tag{4.20}$$

where the elements are given by

$$\begin{aligned}R_{11} &= \sin^2\varphi + \cos^2\varphi\cos\theta, \\ R_{12} &= -\sin\varphi\cos\varphi\left(1 - \cos\theta\right), \\ R_{13} &= -\cos\varphi\sin\theta, \; R_{21} = -\sin\varphi\cos\varphi\left(1 - \cos\theta\right), \\ R_{22} &= \cos^2\varphi + \sin^2\varphi\cos\theta, \; R_{23} = -\sin\varphi\sin\theta, \\ R_{31} &= \cos\varphi\sin\theta, \; R_{32} = \sin\varphi\sin\theta, \; R_{33} = \cos\theta.\end{aligned} \tag{4.21}$$

12 That is, we expand the dyadic Green's function (**??**) in terms of the local frame and then substitute the result into (4.1).

13 These are required only in the argument of the dyadic Green's function.

14 See Appendix 4.7.4 for the derivation of the matrix elements (4.21).

In terms of this matrix, we can express the local frame coordinates in terms of the global frame's using the following relations

$$\mathbf{r}'' = \bar{\mathbf{R}}(\theta,\varphi)\cdot\mathbf{r},\ \ \mathbf{r}'_s = \bar{\mathbf{R}}(\theta,\varphi)\cdot\mathbf{r}'. \tag{4.22}$$

It should be immediately stated that this rotation matrix will also rotate the $x''y''$-plane around the z''-axis with some angle. We can further control this additional rotation by multiplying (4.20) by the following matrix

$$\bar{\mathbf{R}}_\alpha := \begin{pmatrix} \cos\alpha & -\sin\alpha & 0 \\ \sin\alpha & \cos\alpha & 0 \\ 0 & 0 & 1 \end{pmatrix}, \tag{4.23}$$

where α here represents some angle through which we rotate the $x''y''$-plane around the z''-axis. However, as will be shown in Section 4.2.4, a remarkable characteristic of the field decomposition based on Weyl expansion is that it does not depend on the angle α if we restrict our attention to the *total* propagating part and the *total* evanescent part of the electromagnetic field radiated by the antenna.

From (4.18) and (5.13), it is found that $\mathbf{K}'' = \bar{\mathbf{R}}^T\cdot\mathbf{K}$ and therefore $\mathbf{K}''\mathbf{K}'' = \left(\bar{\mathbf{R}}^T\cdot\mathbf{K}\right)\left(\mathbf{K}\cdot\bar{\mathbf{R}}\right) := \bar{\mathbf{R}}^T\cdot\mathbf{K}\mathbf{K}\cdot\bar{\mathbf{R}}$, where T denotes matrix transpose operation. Moreover, it is easy to show that $\mathbf{K}\cdot\left(\bar{\mathbf{R}}\cdot\mathbf{r}\right) = \left(\bar{\mathbf{R}}^T\cdot\mathbf{K}\right)\cdot\mathbf{r}$. Using these two relation, (4.19) can be put in the form

$$\begin{aligned}\mathbf{E}(\mathbf{r}) = \tfrac{-\omega k\mu}{8\pi^2}\int_{-\infty}^{\infty}\int_{-\infty}^{\infty} dpdq\,\tfrac{\bar{\mathbf{I}}k^2-\bar{\mathbf{R}}^T\cdot\mathbf{K}\mathbf{K}\cdot\bar{\mathbf{R}}}{k^2 m} \\ \times\cdot\int_V d^3r'\mathbf{J}(\mathbf{r}')\,e^{-i\left(\bar{\mathbf{R}}^T\cdot\mathbf{K}\right)\cdot\mathbf{r}'}e^{i\mathbf{K}\cdot\left(\bar{\mathbf{R}}\cdot\mathbf{r}\right)}.\end{aligned} \tag{4.24}$$

Therefore, from the definition of the spatial Fourier transform of the antenna current as given by (16.16), (4.24) can be reduced into the form

$$\begin{aligned}\mathbf{E}(\mathbf{r}) = \tfrac{-\omega k\mu}{8\pi^2}\int_{-\infty}^{\infty}\int_{-\infty}^{\infty} dpdq\,\tfrac{\bar{\mathbf{I}}k^2-\bar{\mathbf{R}}^T\cdot\mathbf{K}\mathbf{K}\cdot\bar{\mathbf{R}}}{k^2 m} \\ \times\cdot\tilde{\mathbf{J}}\left(\bar{\mathbf{R}}^T\cdot\mathbf{K}\right)e^{i\mathbf{K}\cdot\left(\bar{\mathbf{R}}\cdot\mathbf{r}\right)}.\end{aligned} \tag{4.25}$$

Separating this integral into nonpropagating (evanescent) and propagating parts, we obtain, respectively,

$$\begin{aligned}\mathbf{E}_{\text{ev}}(\mathbf{r};\hat{u}) = \tfrac{-\omega k\mu}{8\pi^2}\int_{p^2+q^2>1} dpdq\,\tfrac{\bar{\mathbf{I}}k^2-\bar{\mathbf{R}}^T(\hat{u})\cdot\mathbf{K}\mathbf{K}\cdot\bar{\mathbf{R}}(\hat{u})}{k^2 m} \\ \times\cdot\tilde{\mathbf{J}}\left[\bar{\mathbf{R}}^T(\hat{u})\cdot\mathbf{K}\right]e^{i\mathbf{K}\cdot\left[\bar{\mathbf{R}}(\hat{u})\cdot\mathbf{r}\right]},\end{aligned} \tag{4.26}$$

$$\begin{aligned}\mathbf{E}_{\mathrm{pr}}(\mathbf{r};\hat{u}) = \tfrac{-\omega k\mu}{8\pi^2}\int_{p^2+q^2<1} dpdq\tfrac{\bar{\mathbf{I}}k^2-\bar{\mathbf{R}}^T(\hat{u})\cdot\mathbf{KK}\cdot\bar{\mathbf{R}}(\hat{u})}{k^2 m} \\ \times\cdot\tilde{\mathbf{J}}\left[\bar{\mathbf{R}}^T(\hat{u})\cdot\mathbf{K}\right]e^{i\mathbf{K}\cdot\left[\bar{\mathbf{R}}(\hat{u})\cdot\mathbf{r}\right]}.\end{aligned} \tag{4.27}$$

We will refer to the expansions (4.26) and (4.27) as the *general decomposition theorem* of the antenna fields. They express the decomposition of the field at location **r** into total evanescent and propagating parts measured along the direction specified by the unit vector $\hat{u} = \hat{x}\sin\theta\cos\varphi + \hat{y}\sin\theta\sin\varphi + \hat{z}\cos\theta$, i.e., when the z''-axis of the local observation frame is oriented along the direction of $\hat{u}$. Moreover, since it can be proved that this decomposition is independent of an arbitrary rotation of the local frame around $\hat{u}$ (see Section 4.2.4), it follows that the quantities appearing in (4.26) and (4.27) are unique. However, it must be noticed that the expansions (4.26) and (4.27) are valid only in an *exterior* region, for example $|z''| > L$, where here L is taken as the maximum dimension of the antenna current distribution. Using the explicit form of the rotation matrix (4.20) given in (4.21), we find that the general decomposition theorem is valid in the region exterior to the infinite slab enclosed between the two planes

$$\begin{aligned}\left[\sin^2\varphi + \cos^2\varphi\cos\theta\right]x - \left[\sin\varphi\cos\varphi\,(1-\cos\theta)\right]y \\ -\cos\varphi\sin\theta z = \pm L.\end{aligned} \tag{4.28}$$

This region will be refereed to here as the *antenna horizon*, meaning the horizontal range inside which the simple expressions in (4.26) and 4.27) are *not* valid.[15] We immediately notice that the antenna horizon is changing in orientation with every angles θ and φ. This will restrict the usefulness of the expansions (4.26) and (4.27) in many problems in field theory as we will see later. However, a particularly attractive field structure, the radial streamline concept, will not suffer from this restriction. Toward this form we now turn.

4.2.3 The Concept of Antenna Near-Field Radial Streamlines

We focus our attention on the description of the radiated field surrounding the antenna physical body using spherical coordinates. In particular, notice that by inserting $\mathbf{r} = \hat{x}r\sin\theta\cos\varphi + \hat{y}r\sin\theta\sin\varphi + \hat{z}\cos\theta$ into (5.13), and using the form of the rotation matrix given by (4.20) and (4.21), one can easily calculate

15 For an example of calculations made *in*side the antenna horizon, see Appendix 4.7.6.

$\bar{\mathbf{R}}(\theta,\varphi)\cdot\mathbf{r} = \langle 0,0,r\rangle$.[16] Therefore, the expansion (4.25) becomes

$$\begin{aligned}\mathbf{E}(\mathbf{r}) = \tfrac{-\omega k\mu}{8\pi^2}\int_{-\infty}^{\infty}\int_{-\infty}^{\infty} dpdq\tfrac{\bar{\mathbf{I}}k^2-\bar{\mathbf{R}}^T\cdot\mathbf{KK}\cdot\bar{\mathbf{R}}}{k^2m}\cdot\tilde{\mathbf{J}}\left(\bar{\mathbf{R}}^T\cdot\mathbf{K}\right)\\ \times e^{i\mathrm{sgn}(r-L)kmr},\end{aligned} \tag{4.29}$$

where $L := \max_{\theta,\varphi} L''(\theta,\varphi)$. Since the observation is of the field propagating or nonpropagating *away* from the antenna, we are always on the branch $r > L$. Furthermore, by dividing the expansion (4.29) into propagating and nonpropagating parts, it is finally obtained

$$\begin{aligned}\mathbf{E}_{\mathrm{ev}}(\mathbf{r}) = \tfrac{-\omega k\mu}{8\pi^2}\int_{p^2+q^2>1} dpdq\bar{\mathbf{R}}^T(\theta,\varphi)\cdot\bar{\mathbf{\Omega}}(p,q)\cdot\bar{\mathbf{R}}(\theta,\varphi)\\ \times\cdot\tilde{\mathbf{J}}\left[\bar{\mathbf{R}}^T(\theta,\varphi)\cdot\mathbf{K}\right]e^{-kr\sqrt{p^2+q^2-1}},\end{aligned} \tag{4.30}$$

$$\begin{aligned}\mathbf{E}_{\mathrm{pr}}(\mathbf{r}) = \tfrac{-\omega k\mu}{8\pi^2}\int_{p^2+q^2<1} dpdq\bar{\mathbf{R}}^T(\theta,\varphi)\cdot\bar{\mathbf{\Omega}}(p,q)\cdot\bar{\mathbf{R}}(\theta,\varphi)\\ \times\cdot\tilde{\mathbf{J}}\left[\bar{\mathbf{R}}^T(\theta,\varphi)\cdot\mathbf{K}\right]e^{ikr\sqrt{1-p^2+q^2}},\end{aligned} \tag{4.31}$$

where we have introduced the *spectral polarization dyad* defined as[17]

$$\bar{\mathbf{\Omega}}(p,q) := \frac{\bar{\mathbf{I}}k^2 - \mathbf{KK}}{k^2m}. \tag{4.32}$$

We notice that in this way the general decomposition theorems (4.26) and (4.27) are always satisfied since for each direction specified by θ and φ, the slab enclosed between the two planes given by (4.28) will also rotate such that the observation point is always in the exterior region. This desirable fact is behind the great utility of the radial streamline concept (to be defined momentarily) in the antenna theory we are proposing in this work.

The expansions (4.30) and (4.31) can be interpreted as the decomposition of the electromagnetic fields into propagating and nonpropagating waves in the *radial* directions described by the spherical angles θ and φ. That is, we do *not* obtain a plane wave spectrum in this formulation, but instead, what we prefer to call *radial streamlines emanating from the origin of the coordinate system* (conveniently chosen at the center of the actual radiating structure). The physical meaning of 'streamlines' here is analogous to the situation encountered in hydrodynamics,

16 This computation can be considered as an alternative derivation of the rotation matrix compared with the one presented in Appendix 4.7.4.

17 For a discussion of the physical meaning of this dyad, and hence a justification of the proposed name, see Section 4.5.

where material particles move in trajectories embedded within continuous fluids. In the case considered here, streamlines have the mathematical form $\Psi(r - ct)$ for a propagating mode with constant phase speed c, and hence are defined completely in terms of fields. As explained earlier, it is only such solutions that represent a genuine propagating mode; the remaining part, the evanescent mode in the electromagnetic problem, represents clearly the nonpropagating part of the radiated field. The concept of 'electromagnetic field streamlines' developed above is a logical deduction from a peculiarity in the Weyl expansion, namely the symmetry breaking of the rotational invariance of the scalar Green's function, a mathematical trait we propose to elevate to the level of a genuine physical process at the heart of the dynamics of the antenna near fields.[18] It is this form of radial streamlines that appears to the authors to be the most natural representation of the inner structure of the antenna near fields since it is viewed from the perspective of the far fields, which in turn is most conventionally expressed in terms of spherical coordinates. Since antenna engineers almost always describe the antenna in the far-field zone (among other measures like the input impedance), and since such mathematical description necessitates a choice of a spherical coordinate system, we take our global frame introduced in the previous parts to coincide with the spherical coordinate system employed by engineers in the characterization of antennas. Therefore, our near field picture, although it starts from a given current distribution in the antenna region, still partially reflects the perspective of the far field. In Section 4.4, we will develop the near field picture completely from the far field perspective by employing the Wilcox expansion.

4.2.4 Independence of the Spectral Expansion to Arbitrary Rotation Around the Main Axis of Propagation/Nonpropagation

We now turn to the issue of the effect of rotation around the main axis chosen to perform the spectral expansion. As we have already seen, the major idea behind the near-field theory is the interpretation of the rotational invariance of the scalar Green's function in terms of its Weyl expansion. It turned out that with respect to a given antenna current distribution, as long as one is concerned with the exterior region, the observation frame of reference can be arbitrarily chosen in order to enact a Weyl expansion with respect to this frame. It is our opinion that such a freedom of choice is not an arbitrary consequence of the mathematical identity per se, but rather the deeper expression of the being of electromagnetic radiation as such. Indeed,

18 The generalized concept of non-radial streamlines will be developed by the authors in separate publications. For example, see Chapter 4.

the very essence of how antenna works is the scientific explication of a definite mechanism through which the near field genetically gives rise to the far field; in other words, the genesis of electromagnetic radiation out from the near field shell. Although the full analysis of this problem will be addressed in future publications by the authors, we have introduced so far the concept of radial streamlines to describe the conversion mechanism above mentioned in precise terms. It was found that we can orient the z-axis of the observation frame along the unit radial vector $\hat{r}$ of the global frame in order to obtain a decomposition of the total fields propagating and nonpropagating away from the antenna origin along the direction of $\hat{r}$.[19] It remains to see how our spectral expansion is affected by a rotation of the *local* frame xy-plane around the radial direction axis. More precisely, the problem is stated in the following manner. Consider a point in space described by the position vector $\mathbf{r}$ in the language of the global frame of observation. Assume further that the expansion of the electric field into propagating and nonpropagating modes along the direction of the z-axis of this frame was achieved, with values $\mathbf{E}_{\rm ev}(\mathbf{r})$ and $\mathbf{E}_{\rm pr}(\mathbf{r})$ giving the evanescent and propagating parts, respectively. Now, keeping the the direction of the z-axis fixed, we merely rotate the xy-plane by an angle α around the z-axis. The electric field is now expanded into evanescent and propagating modes again along the same z-axis, and the results are $\mathbf{E'}_{\rm ev}(\mathbf{r})$ and $\mathbf{E'}_{\rm pr}(\mathbf{r})$, respectively. The question we now investigate is the relation between these two sets of fields.

To accomplish this, let us start from the original expansion (4.24) but replace $\bar{\mathbf{R}}(\theta, \varphi)$ with a rotation around the z-axis through an angle α, which can be used to obtain the transformed spatial and spectral variables through the equations $\mathbf{r}' = \bar{\mathbf{R}}_\alpha \cdot \mathbf{r}$ and $\mathbf{K}' = \bar{\mathbf{R}}_\alpha^T \cdot \mathbf{K}$, where $\bar{\mathbf{R}}_\alpha$ is given by (4.23). By direct calculation, we obtain $\mathbf{K} \cdot \mathbf{r} = k\,(p\cos\alpha + \sin\alpha)\,x + k\,(-p\sin\alpha + q\cos\alpha)\,y + \operatorname{sgn}(z - L)\,kmz$ and $\mathbf{K}' = \hat{x}k\,(p\cos\alpha + q\sin\alpha) + \hat{y}k\,(-p\sin\alpha + q\cos\alpha) + \hat{z}\operatorname{sgn}(z' - L')\,km$. These results suggest introducing the substitutions $p' = p\cos\alpha + q\sin\alpha$ and $q' = -p\sin\alpha + q\cos\alpha$, which are effectively a rotation of the pq-plane by and angle $-\alpha$ around the origin. Being a rotation, the Jacobian of this transformation is one, i.e., $J\left(\bar{\mathbf{R}}_\alpha\right) = 1$, where $J(\cdot)$ denotes the Jacobian of the transformation matrix applied to its argument. Also, it is evident that $m' = \sqrt{1 - p'^2 + q'^2} = \sqrt{1 - p^2 + q^2} = m$. Moreover, this implies that the two regions $p^2 + q^2 < 1$ and $p^2 + q^2 > 1$ transform into the regions $p'^2 + q'^2 < 1$ and $p'^2 + q'^2 > 1$, respectively.

19 Cf. equations (4.30) and (4.31).

After dividing (4.24) into evanescent and propagating part, then rotating the pq-plane and changing the spectral variables from p and q to p' and q', we find

$$\begin{aligned}\mathbf{E'}_{\mathrm{ev}}(\mathbf{r}) = \tfrac{-\omega k\mu}{8\pi^2}\int_{p'^2+q'^2>1} dp'dq' \tfrac{\bar{\mathbf{I}}k^2-\mathbf{K}(p',q')\mathbf{K}(p',q')}{k^2 m'} \\ \times J\left(\bar{\mathbf{R}}_\alpha\right)\cdot\tilde{\mathbf{J}}\left[\mathbf{K}\left(p',q'\right)\right]e^{i\mathbf{K}(p',q')\cdot(\mathbf{r})},\end{aligned} \tag{4.33}$$

$$\begin{aligned}\mathbf{E'}_{\mathrm{pr}}(\mathbf{r}) = \tfrac{-\omega k\mu}{8\pi^2}\int_{p'^2+q'^2<1} dp'dq' \tfrac{\bar{\mathbf{I}}k^2-\mathbf{K}(p',q')\mathbf{K}(p',q')}{k^2 m'} \\ \times J\left(\bar{\mathbf{R}}_\alpha\right)\cdot\tilde{\mathbf{J}}\left[\mathbf{K}\left(p',q'\right)\right]e^{i\mathbf{K}(p',q')\cdot(\mathbf{r})}.\end{aligned} \tag{4.34}$$

Applying the results of the paragraph preceding the two equations above, we conclude that

$$\mathbf{E}_{\mathrm{ev}}(\mathbf{r}) = \mathbf{E'}_{\mathrm{ev}}(\mathbf{r}), \quad \mathbf{E}_{\mathrm{pr}}(\mathbf{r}) = \mathbf{E'}_{\mathrm{pr}}(\mathbf{r}). \tag{4.35}$$

Therefore, the *total* evanescent and *total* propagating parts of the antenna radiated fields are invariant to rotation around the z-axis of the observation frame. *This result is true only when we are interested in field decomposition into regions in the spectral* pq-*plane that do not change through rotation.* For example, if we are interested in studying part of the radiated field such that it contains the modes propagating along the z-direction, but with spectral content in the pq-plane inside, say, a square, then since not every rotation is a symmetry operation for a square, we conclude that the quantity of interest above *does* vary with rotation of the observation frame around the z-axis for this special case. In this chapter, however, our interest will focus on the total propagating and nonpropagating parts since these are the quantities that help rationalize the overall behavior of antennas in general. However, it should be kept in mind that for more general and sophisticated understanding of near-field interactions, it is better to retain a general region in the pq-plane as the basis for a broad spectral analysis of the electromagnetic fields (see Figure 4.2).

4.2.5 The Propagating and Nonpropagating Parts are Maxwellian

Our formalism concerning the expansion of the electromagnetic field produced by a given antenna current distribution into propagating and evanescent modes is still that directly reflecting the physics of the phenomena under consideration, which is the laws dictated by Maxwell's equations. We will show now that both the propagating and nonpropagating parts obeys individually Maxwell's equations.

The frequency-domain Maxwell's equations in source-free homogeneous space described by electric permittivity ε and magnetic permeability μ are given

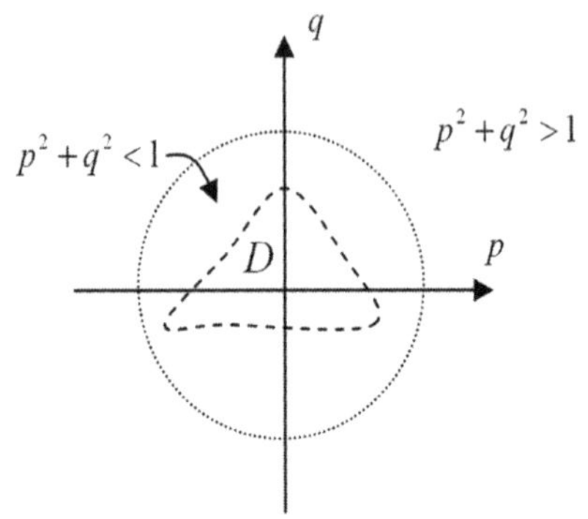

Figure 4.2 Regions in the spectral pq-plane in terms of which the decomposition of the electromagnetic field into propagating and nonpropagating modes is conducted. The circle $p^2+q^2 = 1$ marks the boundary between the so-called invisible region $p^2+q^2 > 1$ and the visible region $p^2+q^2 < 1$ (a circular disk). In general, the mathematical description of the field can be accomplished with any region in the spectral plane, not necessary the total regions inside and outside the circle $p^2+q^2 = 1$. In particular, we show an arbitrary region D located inside the propagating modes disk $p^2+q^2 < 1$. In general, D need not be a proper subset of the region $p^2+q^2 < 1$, but may include arbitrary portions of both this disk and its complement in the plane.

by

$$\begin{array}{c}\nabla\times\mathbf{E} = i\omega\mu\mathbf{H}, \quad \nabla\times\mathbf{H} = -i\omega\varepsilon\mathbf{E} \\ \nabla\cdot\mathbf{E} = 0, \, \nabla\cdot\mathbf{H} = 0.\end{array} \tag{4.36}$$

The first curl equation in (4.36) can be used to compute the magnetic field if the electric field is known. We assume that the latter can be expressed by the general decomposition theorem as stated in (4.26) and (4.27). Noticing the vector identity $\nabla\times(\psi\mathbf{A}) = \nabla\psi\times\mathbf{A}+\psi\nabla\cdot\mathbf{A}$ and the relation $\nabla\exp(\mathbf{A}\cdot\mathbf{r}) = \mathbf{A}\exp(\mathbf{A}\cdot\mathbf{r})$, which are true in particular for constant vector $\mathbf{A}$ and a scalar field $\psi(\mathbf{r})$, we easily obtain

$$\begin{array}{r}\mathbf{H}(\mathbf{r}) = \frac{ik}{8\pi^2}\int_{-\infty}^{\infty}\int_{-\infty}^{\infty} dpdq\frac{\bar{\mathbf{I}}k^2-\bar{\mathbf{R}}^T\cdot\mathbf{K}\mathbf{K}\cdot\bar{\mathbf{R}}}{k^2m}\cdot\tilde{\mathbf{J}}\left(\bar{\mathbf{R}}^T\cdot\mathbf{K}\right) \\ \times\bar{\mathbf{R}}^T\cdot\mathbf{K}e^{i\mathbf{K}\cdot\left(\bar{\mathbf{R}}\cdot\mathbf{r}\right)},\end{array} \tag{4.37}$$

where the curl operator was brought inside the spectral integral. Next, from the dyadic identity $\mathbf{ab}\cdot\mathbf{c} = \mathbf{a}(\mathbf{b}\cdot\mathbf{c})$, we write

$$\bar{\mathbf{R}}^T\cdot\mathbf{K}\mathbf{K}\cdot\bar{\mathbf{R}}\cdot\tilde{\mathbf{J}}\left(\bar{\mathbf{R}}^T\cdot\mathbf{K}\right) = \bar{\mathbf{R}}^T\cdot\mathbf{K}\left[\left(\mathbf{K}\cdot\bar{\mathbf{R}}\right)\cdot\tilde{\mathbf{J}}\left(\bar{\mathbf{R}}^T\cdot\mathbf{K}\right)\right]. \tag{4.38}$$

This allows us to conclude that

$$\bar{\mathbf{R}}^T \cdot \mathbf{KK} \cdot \bar{\mathbf{R}} \cdot \tilde{\mathbf{J}} \left(\bar{\mathbf{R}}^T \cdot \mathbf{K}\right) \times \bar{\mathbf{R}}^T \cdot \mathbf{K} = 0. \tag{4.39}$$

Therefore, after separating the integral into propagating and evanescent parts, (4.37) becomes

$$\begin{aligned}\mathbf{H}_{\text{ev}}(\mathbf{r};\hat{u}) = \tfrac{ik}{8\pi^2}\int_{p^2+q^2>1} dpdq \tfrac{1}{m}\tilde{\mathbf{J}}\left[\bar{\mathbf{R}}^T(\hat{u})\cdot\mathbf{K}\right] \\ \times\bar{\mathbf{R}}^T(\hat{u})\cdot\mathbf{K}e^{i\mathbf{K}\cdot\left(\bar{\mathbf{R}}\cdot\mathbf{r}\right)},\end{aligned} \tag{4.40}$$

$$\begin{aligned}\mathbf{H}_{\text{pr}}(\mathbf{r};\hat{u}) = \tfrac{ik}{8\pi^2}\int_{p^2+q^2<1} dpdq \tfrac{1}{m}\tilde{\mathbf{J}}\left[\bar{\mathbf{R}}^T(\hat{u})\cdot\mathbf{K}\right] \\ \times\bar{\mathbf{R}}^T(\hat{u})\cdot\mathbf{K}e^{i\mathbf{K}\cdot\left(\bar{\mathbf{R}}\cdot\mathbf{r}\right)}.\end{aligned} \tag{4.41}$$

The radial streamline magnetic fields corresponding to those given for the electric field in (4.30) and (4.31) are

$$\begin{aligned}\mathbf{H}_{\text{ev}}(\mathbf{r}) = \tfrac{ik}{8\pi^2}\int_{p^2+q^2>1} dpdq \tfrac{1}{m}\tilde{\mathbf{J}}\left[\bar{\mathbf{R}}^T(\theta,\varphi)\cdot\mathbf{K}\right] \\ \times\bar{\mathbf{R}}^T(\theta,\varphi)\cdot\mathbf{K}e^{-kr\sqrt{p^2+q^2-1}},\end{aligned} \tag{4.42}$$

$$\begin{aligned}\mathbf{H}_{\text{pr}}(\mathbf{r}) = \tfrac{ik}{8\pi^2}\int_{p^2+q^2<1} dpdq \tfrac{1}{m}\tilde{\mathbf{J}}\left[\bar{\mathbf{R}}^T(\theta,\varphi)\cdot\mathbf{K}\right] \\ \times\bar{\mathbf{R}}^T(\theta,\varphi)\cdot\mathbf{K}e^{ikr\sqrt{1-p^2+q^2}}.\end{aligned} \tag{4.43}$$

It is evident from the original (4.37) that the evanescent (propagating) magnetic field is found by applying the curl operator to the evanescent (propagating) part of the electric field. That is,

$$\mathbf{H}_{\text{ev}} = (1/i\omega\mu)\,\nabla\times\mathbf{E}_{\text{ev}},\ \mathbf{H}_{\text{pr}} = (1/i\omega\mu)\,\nabla\times\mathbf{E}_{\text{pr}}. \tag{4.44}$$

Moreover, the divergence of the evanescent and propagating parts of both the electric and magnetic fields is identically zero. To see this, take the divergence of (4.26), interchange the order of integration and differentiation, and apply the identity $\nabla\cdot\mathbf{B}\exp(\mathbf{A}\cdot\mathbf{r}) = \mathbf{A}\cdot\mathbf{B}\exp(\mathbf{A}\cdot\mathbf{r})$. It follows that

$$\begin{aligned}\nabla\cdot\mathbf{E}_{\text{ev}}(\mathbf{r};\hat{u}) = \tfrac{-\omega k\mu}{8\pi^2}\int_{p^2+q^2>1} dpdq\tfrac{\bar{\mathbf{I}}k^2-\bar{\mathbf{R}}^T(\hat{u})\cdot\mathbf{KK}\cdot\bar{\mathbf{R}}(\hat{u})}{k^2m} \\ \times\cdot\tilde{\mathbf{J}}\left[\bar{\mathbf{R}}^T(\hat{u})\cdot\mathbf{K}\right]\cdot\left[\bar{\mathbf{R}}^T(\hat{u})\cdot\mathbf{K}\right]e^{i\mathbf{K}\cdot\left(\bar{\mathbf{R}}\cdot\mathbf{r}\right)}.\end{aligned} \tag{4.45}$$

We calculate by $\mathbf{ab}\cdot\mathbf{c} = \mathbf{a}(\mathbf{b}\cdot\mathbf{c})$ and obtain

$$\begin{aligned}&\left\{\bar{\mathbf{R}}^T\cdot\mathbf{KK}\cdot\bar{\mathbf{R}}\cdot\tilde{\mathbf{J}}\left[\bar{\mathbf{R}}^T\cdot\mathbf{K}\right]\right\}\cdot\left(\bar{\mathbf{R}}^T\cdot\mathbf{K}\right) \\ &= \left(\bar{\mathbf{R}}^T\cdot\mathbf{K}\right)\cdot\left(\bar{\mathbf{R}}^T\cdot\mathbf{K}\right)\left\{\left(\mathbf{K}\cdot\bar{\mathbf{R}}\right)\cdot\tilde{\mathbf{J}}\left[\bar{\mathbf{R}}^T\cdot\mathbf{K}\right]\right\}.\end{aligned} \tag{4.46}$$

However, since the rotation matrix is orthogonal, i.e., $\bar{\mathbf{R}}^T \cdot \bar{\mathbf{R}} = \bar{\mathbf{I}}$, we have $\left(\bar{\mathbf{R}}^T \cdot \mathbf{K}\right) \cdot \left(\bar{\mathbf{R}}^T \cdot \mathbf{K}\right) = k^2$ and (11.23) becomes

$$\begin{aligned} \left\{\bar{\mathbf{R}}^T \cdot \mathbf{K}\mathbf{K} \cdot \bar{\mathbf{R}} \cdot \tilde{\mathbf{J}}\left[\bar{\mathbf{R}}^T \cdot \mathbf{K}\right]\right\} \cdot (\bar{\mathbf{R}}^T \cdot \mathbf{K}) \\ = k^2 \tilde{\mathbf{J}}\left[\bar{\mathbf{R}}^T \cdot \mathbf{K}\right] \cdot \left(\bar{\mathbf{R}}^T \cdot \mathbf{K}\right). \end{aligned} \tag{4.47}$$

Substituting this result into (11.19), we find that $\nabla \cdot \mathbf{E}_{\text{ev}}(\mathbf{r}; \hat{u}) = 0$. A similar procedure can now be applied to all other field parts and the divergence is also zero. We conclude from this together with (4.44) that

$$\begin{aligned} \nabla \times \mathbf{E}_{\text{ev}} = i\omega\mu\mathbf{H}_{\text{ev}}, \ \nabla \times \mathbf{H}_{\text{ev}} = -i\omega\varepsilon\mathbf{E}_{\text{ev}} \\ \nabla \cdot \mathbf{E}_{\text{ev}} = 0, \ \nabla \cdot \mathbf{H}_{\text{ev}} = 0. \end{aligned} \tag{4.48}$$

$$\begin{aligned} \nabla \times \mathbf{E}_{\text{pr}} = i\omega\mu\mathbf{H}_{\text{pr}}, \ \nabla \times \mathbf{H}_{\text{pr}} = -i\omega\varepsilon\mathbf{E}_{\text{pr}} \\ \nabla \cdot \mathbf{E}_{\text{pr}} = 0, \ \nabla \cdot \mathbf{H}_{\text{pr}} = 0. \end{aligned} \tag{4.49}$$

These are the main results of this section. They show that each field part satisfies *individually* Maxwell's equations. In other words, whatever is the direction of decomposition, the resultant fields are always Maxwellian. For the case when the observation point lies within the antenna horizon, it is still possible to apply the same procedure of this section but to the most general expressions given by (4.114) and (4.115). It follows again the the propagating and nonpropagating parts are still Maxwellian.

4.2.6 Summary and Interpretation

By now we know that our expansion of the electromagnetic field into propagating and nonpropagating modes along a changing direction is well justified by the result of Section 4.2.4, namely that such an expansion along a given direction is independent of an arbitrary rotation of the local observation frame around this direction. This important conclusion significantly simplifies the analysis of the antenna near fields. The reason is that the full rotation group requires *three* independent parameters in order to specify an arbitrary 3D orientation of the rotated observation frame. Instead, our formulation depends only on *two* independent parameters, namely θ and φ, which are the same parameters used to characterize the degrees of freedom of the antenna far field. This step then indicates an intimate connection between the antenna near and far fields, which is, relatively speaking, not quite obvious.

However, our knowledge of the structure of the near field, as can be discerned from the expansions (4.30) and (4.31), is enhanced by the record of the exact manner, as we progress away from the antenna along the radial direction $\hat{r}$, in which the evanescent field, i.e., the nonpropagating part, is being continually converted into propagating modes. As we reach the far-field zone, most of the field contents reduce to propagating modes, although the evanescent part still contributes asymptotically to the far field. For each direction θ and φ, the functional form of the integrands in (4.30) and (4.31) will be different, indicating the 'how' of the conversion mechanism we are concerned with.

Since close to the antenna most of the near field content is nonpropagating, we focus now our attention on the evanescent mode expansion of the electric magnetic field as given by (4.30).[20] Let us introduce the cylindrical variables v and α such that $p = v\cos\alpha$ and $q = v\sin\alpha$. Therefore in the region $p^2 + q^2 > 1$,

$$\mathbf{K}(v,\alpha) = \hat{x}kv\cos\alpha + \hat{y}kv\sin\alpha + \hat{z}ik\sqrt{v^2-1}. \tag{4.50}$$

The integral (4.30) then becomes

$$\begin{aligned}\mathbf{E}_{\text{ev}}(\mathbf{r}) = \tfrac{-\omega k\mu}{8\pi^2}\textstyle\int_1^\infty v dv \int_0^{2\pi} d\alpha\, \bar{\mathbf{F}}(\theta,\varphi,v,\alpha) \\ \times\cdot\tilde{\mathbf{J}}\left[\bar{\mathbf{R}}^T(\theta,\varphi)\cdot\mathbf{K}(v,\alpha)\right]e^{-k\sqrt{v^2-1}r},\end{aligned} \tag{4.51}$$

where

$$\begin{aligned}&\bar{\mathbf{F}}(\theta,\varphi,v,\alpha) = \\ &\tfrac{\bar{\mathbf{I}}k^2-\bar{\mathbf{R}}^T(\theta,\varphi)\cdot\mathbf{K}(v,\alpha)\mathbf{K}(v,\alpha)\cdot\bar{\mathbf{R}}(\theta,\varphi)}{ik^2\sqrt{v^2-1}}.\end{aligned} \tag{4.52}$$

Next, perform another substitution $u = \sqrt{v^2-1}$. Since $du = v/\sqrt{v^2-1}dv$, it follows that the integral (4.51) reduces to

$$\mathbf{E}_{\text{ev}}(\mathbf{r}) = \frac{-\omega k\mu}{8\pi^2}\int_0^\infty du\,\mathbf{G}(\theta,\varphi,u)\,e^{-kur}, \tag{4.53}$$

where

$$\begin{aligned}\mathbf{G}(\theta,\varphi,u) = \textstyle\int_0^{2\pi} d\alpha\,\bar{\mathbf{F}}\left(\theta,\varphi,\sqrt{1+u^2},\alpha\right) \\ \times\cdot\tilde{\mathbf{J}}\left[\bar{\mathbf{R}}^T(\theta,\varphi)\cdot\mathbf{K}\left(\sqrt{1+u^2},\alpha\right)\right].\end{aligned} \tag{4.54}$$

Therefore, for a fixed radial direction θ and φ, the functional form of the evanescent part of the field along this direction takes the expression of a Laplace transform

20 The subsequent formulation in this section can be also developed for the evanescent part of the magnetic field (4.42).

in which the radial position r plays the role of frequency. This fact is interesting, and suggests that certain economy in the representation of the field decomposition along the radial direction has been already achieved by the expansions (4.30) and (4.31). To appreciate better this point, we notice that since $\bar{\mathbf{R}}(\theta, \varphi)$ is a rotation matrix, it satisfies $\bar{\mathbf{R}}^{-1} = \bar{\mathbf{R}}^{T}$. In light of this, the change in the integrands of (4.30) and (4.31) with the orientation of the decomposition axis given by θ and φ can be viewed as, firstly, a rotation of the spatial Fourier transform of the current by the *inverse* rotation originally applied to the local observation frame, and, secondly, as applying a *similarity transformation* to transform the spectral polarization dyad $\bar{\mathbf{\Omega}}(p, q)$ to $\bar{\mathbf{R}}^{-1}(\theta, \varphi) \cdot \bar{\mathbf{\Omega}}(p, q) \cdot \bar{\mathbf{R}}(\theta, \varphi)$, that is, the spectral matrix $\bar{\mathbf{\Omega}}(p, q)$ is undergoing a similarity transformation under the transformation $\bar{\mathbf{R}}^{-1}$, the inverse rotation. *These results indicate that there is a simple geometrical transformation at the core of the change of the spectral content of the electromagnetic fields,*[21] *which enacts the decomposition of the electromagnetic fields into nonpropagating and propagating modes.* These transformations are simple to understand and easy to visualize. We summarize the entire process in the following manner

1. Calculate the spatial Fourier transform of the antenna current distribution in a the global observation frame.
2. Rotate this Fourier transform by the inverse rotation $\bar{\mathbf{R}}^{-1}$.
3. Transform the spectral polarization dyad by the similarity transformation generated by the inverse rotation $\bar{\mathbf{R}}^{-1}$.
4. Multiply the rotated Fourier transform by the transformed spectral polarization dyad. Convert the result from cartesian coordinates p and q to cylindrical coordinates v and α and evaluate the angular (finite) integral with respect to α. That is, *average out* the angular variations α.
5. Transform as $v = \sqrt{1 + u^2}$ and compute the Laplace transform of the remanning function of u. This will give the functional dependence of the antenna evanescent field on the radial position where r will play the role of frequency in the Laplace transform.

The overall process is summarized in the flowchart of Figure 4.3. The significance of this picture is that it provides us with a detailed explication of the actual route to the far field. Indeed, since the radiation observed away from the antenna emerges from the concrete way in which the nonpropagating part is being transformed into propagating modes that escape to the far field zone, it follows that all of

21 The functional form of the integrands of (4.30) and (4.31)

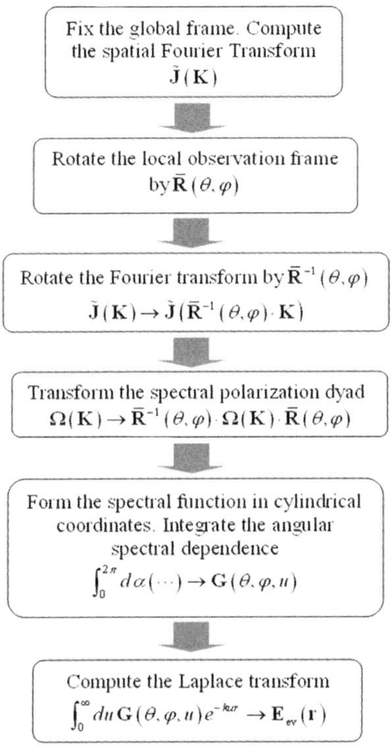

Figure 4.3 The process of forming the near field for general antenna system. The flowchart describes the details in which the mechanism of conversion from evanescent mode to propagating mode unfolds. This is delimited by the variation of the nonpropagating part along the radial direction θ and φ, with distance r. The flowchart indicates that the changes in the spectral functions can be understood in terms of simple geometrical transformations applied to basic antenna quantities like the spatial Fourier transform of the antenna current distribution and the spectral polarization tensor of the dyadic free space Green's function.

the radiation characteristics of antennas, like the formation of single beams, multiple beams and nulls, polarization, etc, can be traced back into the particular functional form of the spectral function appearing in the Laplace transform expression (4.53). Moreover, we now see that the generators of the variation of this key functional form are basically geometrical transformation associated with the rotation matrix $\bar{\mathbf{R}}\,(\theta,\varphi)$ through which we orient the local observation frame of reference. In Section 4.5, the theoretical narrative of the far field formation started here will be further illuminated.

4.3 THE CONCEPT OF LOCALIZED AND STORED ENERGIES IN THE ANTENNA ELECTROMAGNETIC FIELD

4.3.1 Introduction

Armed with the concrete but general results of the previous chapter, we now turn our attention to a systematic investigation of the phenomena usually associated with the energy stored in the antenna surrounding field. We have already encountered the term 'energy' in our general investigation of the antenna circuit model in Chapter 3, where an effective reactive energy was defined in conjunction with the circuit interpretation of the complex Poynting theorem. We have seen that this concept is not adequate when attempts to extend it beyond the confines of the circuit approach are made, pointing to the need to develop a deeper general understanding of antenna near fields before turning to an examination of various candidates for a physically meaningful definition of stored energy. In this section, we employ the understanding of the near-field structure attained in terms of the Weyl expansion of the free space Green's function in order to build a solid foundation for the phenomenon of energy localization in general antenna systems. The upshot of this argument will be our proposal that there is a subtle distinction between localization energy and stored energy. The former is within the reach of the time-harmonic theory developed here, while the latter may require in general an extension to transient phenomena.

4.3.2 Generalization of the Complex Poynting Theorem

Since we know at this stage how to decompose a given electromagnetic field into propagating and nonpropagating parts, the natural next step is to examine the power flow in a closed region. Our investigation will lead to a form of the Poynting theorem that is more general than the customary one (where the latter results from treating only the total fields).

Start by expanding both the electric and magnetic fields as

$$\mathbf{E}(\mathbf{r}) = \mathbf{E}_{\mathrm{ev}}(\mathbf{r}) + \mathbf{E}_{\mathrm{pr}}(\mathbf{r}), \ \mathbf{H}(\mathbf{r}) = \mathbf{H}_{\mathrm{ev}}(\mathbf{r}) + \mathbf{H}_{\mathrm{pr}}(\mathbf{r}). \tag{4.55}$$

The complex Poynting vector is given by [33]

$$\mathbf{S}(\mathbf{r}) = \frac{1}{2}\mathbf{E}(\mathbf{r}) \times \mathbf{H}^*(\mathbf{r}). \tag{4.56}$$

Substituting (15.6) into (4.56), we find

$$\begin{aligned}\mathbf{S}(\mathbf{r}) = {} & \tfrac{1}{2}\mathbf{E}_{\mathrm{ev}} \times \mathbf{H}^*_{\mathrm{ev}} + \tfrac{1}{2}\mathbf{E}_{\mathrm{pr}} \times \mathbf{H}^*_{\mathrm{pr}} \\ & + \tfrac{1}{2}\mathbf{E}_{\mathrm{ev}} \times \mathbf{H}^*_{\mathrm{pr}} + \tfrac{1}{2}\mathbf{E}_{\mathrm{pr}} \times \mathbf{H}^*_{\mathrm{ev}}.\end{aligned} \tag{4.57}$$

Since it has been proved in Section 4.2.5 that each of the propagating and nonpropagating part of the electromagnetic field is Maxwellian, it follows immediately that the first and the second terms of the RHS of (4.57) can be identified with complex Poynting vectors

$$\mathbf{S}_{\mathrm{ev}}(\mathbf{r}) = \frac{1}{2}\mathbf{E}_{\mathrm{ev}}(\mathbf{r}) \times \mathbf{H}^*_{\mathrm{ev}}(\mathbf{r}), \tag{4.58}$$

$$\mathbf{S}_{\mathrm{pr}}(\mathbf{r}) = \frac{1}{2}\mathbf{E}_{\mathrm{pr}}(\mathbf{r}) \times \mathbf{H}^*_{\mathrm{pr}}(\mathbf{r}). \tag{4.59}$$

From the complex Poynting theorem [33] applied to a source-free region, we also find

$$\nabla \cdot \mathbf{S}_{\mathrm{ev}}(\mathbf{r}) = -2i\omega\left(w^e_{\mathrm{ev}} - w^h_{\mathrm{ev}}\right), \tag{4.60}$$

$$\nabla \cdot \mathbf{S}_{\mathrm{pr}}(\mathbf{r}) = -2i\omega\left(w^e_{\mathrm{pr}} - w^h_{\mathrm{pr}}\right), \tag{4.61}$$

with electric and magnetic energy densities defined as

$$w^e_{\mathrm{ev}}(\mathbf{r}) = \frac{\varepsilon}{4}\mathbf{E}_{\mathrm{ev}} \cdot \mathbf{E}^*_{\mathrm{ev}}, \ w^h_{\mathrm{ev}}(\mathbf{r}) = \frac{\mu}{4}\mathbf{H}_{\mathrm{ev}} \cdot \mathbf{H}^*_{\mathrm{ev}}, \tag{4.62}$$

$$w^e_{\mathrm{pr}}(\mathbf{r}) = \frac{\varepsilon}{4}\mathbf{E}_{\mathrm{pr}} \cdot \mathbf{E}^*_{\mathrm{pr}}, \ w^h_{\mathrm{pr}}(\mathbf{r}) = \frac{\mu}{4}\mathbf{H}_{\mathrm{pr}} \cdot \mathbf{H}^*_{\mathrm{pr}}. \tag{4.63}$$

It remains to deal with the cross terms (third and fourth term) appearing on the RHS of (4.57). To achieve this, we need to derive additional Poynting-like theorems.

Take the dot product of the first curl equation in (4.48) with $\mathbf{H}^*_{\mathrm{pr}}$. The result is

$$\mathbf{H}^*_{\mathrm{pr}} \cdot \nabla \times \mathbf{E}_{\mathrm{ev}} = i\omega\mu\mathbf{H}^*_{\mathrm{pr}} \cdot \mathbf{H}_{\mathrm{ev}}. \tag{4.64}$$

Next, take the dot product of the complex conjugate of the second curl equation in (4.49) with $\mathbf{E}_{\mathrm{ev}}$. The result is

$$\mathbf{E}_{\mathrm{ev}} \cdot \nabla \times \mathbf{H}_{\mathrm{pr}}^{*} = i\omega\varepsilon \mathbf{E}_{\mathrm{ev}} \cdot \mathbf{E}_{\mathrm{pr}}^{*}. \tag{4.65}$$

Subtracting (4.65) and (4.64), we obtain

$$\begin{aligned} &\mathbf{H}_{\mathrm{pr}}^{*} \cdot \nabla \times \mathbf{E}_{\mathrm{ev}} - \mathbf{E}_{\mathrm{ev}} \cdot \nabla \times \mathbf{H}_{\mathrm{pr}}^{*} \\ &\qquad = -i\omega \left(\varepsilon \mathbf{E}_{\mathrm{pr}} \cdot \mathbf{E}_{\mathrm{ev}}^{*} - \mu \mathbf{H}_{\mathrm{pr}}^{*} \cdot \mathbf{H}_{\mathrm{ev}}\right). \end{aligned} \tag{4.66}$$

Using the vector identity $\nabla \cdot (\mathbf{A} \times \mathbf{B}) = \mathbf{B} \cdot (\nabla \times \mathbf{A}) - \mathbf{A} \cdot (\nabla \times \mathbf{B})$, equations (4.66) finally becomes

$$\nabla \cdot \left(\mathbf{E}_{\mathrm{ev}} \times \mathbf{H}_{\mathrm{pr}}^{*}\right) = -i\omega \left(\varepsilon \mathbf{E}_{\mathrm{ev}} \cdot \mathbf{E}_{\mathrm{pr}}^{*} - \mu \mathbf{H}_{\mathrm{ev}} \cdot \mathbf{H}_{\mathrm{pr}}^{*}\right). \tag{4.67}$$

By exactly the same procedure, the following dual equation can also be derived

$$\nabla \cdot \left(\mathbf{E}_{\mathrm{pr}} \times \mathbf{H}_{\mathrm{ev}}^{*}\right) = -i\omega \left(\varepsilon \mathbf{E}_{\mathrm{pr}} \cdot \mathbf{E}_{\mathrm{ev}}^{*} - \mu \mathbf{H}_{\mathrm{pr}} \cdot \mathbf{H}_{\mathrm{ev}}^{*}\right). \tag{4.68}$$

Adding (4.67) and (4.68), the following result is obtained

$$\nabla \cdot \mathbf{S}_{\mathrm{int}} = -2i\omega \left(w_{\mathrm{int}}^{e} - w_{\mathrm{int}}^{h}\right), \tag{4.69}$$

where we defined the *complex interaction Poynting vector* by

$$\mathbf{S}_{\mathrm{int}} := \frac{1}{2}\left(\mathbf{E}_{\mathrm{ev}} \times \mathbf{H}_{\mathrm{pr}}^{*} + \mathbf{E}_{\mathrm{pr}} \times \mathbf{H}_{\mathrm{ev}}^{*}\right), \tag{4.70}$$

and the *time-averaged interaction electric and magnetic energy densities* by

$$w_{\mathrm{int}}^{e} := \frac{\varepsilon}{2}\mathrm{Re}\left\{\mathbf{E}_{\mathrm{pr}} \cdot \mathbf{E}_{\mathrm{ev}}^{*}\right\}, \tag{4.71}$$

$$w_{\mathrm{int}}^{h} := \frac{\mu}{2}\mathrm{Re}\left\{\mathbf{H}_{\mathrm{pr}} \cdot \mathbf{H}_{\mathrm{ev}}^{*}\right\}, \tag{4.72}$$

respectively. It is immediate that

$$w^{e} = w_{\mathrm{pr}}^{e} + w_{\mathrm{ev}}^{e} + w_{\mathrm{int}}^{e}, \tag{4.73}$$

$$w^{h} = w_{\mathrm{pr}}^{h} + w_{\mathrm{ev}}^{h} + w_{\mathrm{int}}^{h}, \tag{4.74}$$

$$\mathbf{S} = \mathbf{S}_{\text{ev}} + \mathbf{S}_{\text{pr}} + \mathbf{S}_{\text{int}}. \tag{4.75}$$

The justification for calling the quantities appearing in (4.71) and (4.72) *energy densities* is the following. Maxwell's equations for the evanescent and propagating parts, namely (4.48) and (4.49), can be rewritten in the original time-dependent form. By repeating the procedure that led to equation (4.69) but now in the time domain, it is possible to derive the following continuity equation[22]

$$\nabla \cdot \bar{\mathcal{S}}_{\text{int}} + \frac{\partial}{\partial t}\left(u_{\text{int}}^{e} + u_{\text{int}}^{h}\right) = 0. \tag{4.76}$$

Here, we match the *time-dependent* 'interaction' Poynting vector

$$\bar{\mathcal{S}}_{\text{int}} = \bar{\mathcal{E}}_{\text{pr}} \times \bar{\mathcal{H}}_{\text{ev}} + \bar{\mathcal{E}}_{\text{ev}} \times \bar{\mathcal{H}}_{\text{pr}} \tag{4.77}$$

with the *time-dependent* electric and magnetic energy densities

$$u_{\text{int}}^{e} = \varepsilon \bar{\mathcal{E}}_{\text{pr}} \cdot \bar{\mathcal{E}}_{\text{ev}}, \quad u_{\text{int}}^{h} = \mu \bar{\mathcal{H}}_{\text{pr}} \cdot \bar{\mathcal{H}}_{\text{ev}}, \tag{4.78}$$

where $\bar{\mathcal{E}}$ and $\bar{\mathcal{H}}$ stand for the time-dependent (real) fields. We follow in this treatment the convention of electromagnetic theory in interpreting the quantities (4.78) as energy densities. It is easy now to verify that the expressions (4.71) and (4.72) give the time-average of the corresponding densities appearing in (4.78). Moreover, it follows that the time-average of the instantaneous Poynting vector (4.77) is given by $\text{Re}\left\{\mathbf{S}_{\text{int}}\right\}$.

Therefore, the complex Poynting theorem can be generalized in the following manner. In each source-free space region, the total power flow outside the volume can be separated into three parts, $\mathbf{S}_{\text{ev}}$, $\mathbf{S}_{\text{pr}}$, and $\mathbf{S}_{\text{int}}$. Each term individually is interpreted as a Poynting vector for the corresponding field. The evidence for this interpretation is the fact that a continuity-type equation Poynting theorem can be proved for each individual Poynting vector with the appropriate corresponding energy density.[23]

22 See Appendix 4.7.5 for the derivation of (4.76).

23 For example, consider the energy theorem (4.76). This results states the following. Inside any source-free region of space, the amount of the interaction power flowing outside the surface enclosing the region is equal to negative the time rate decrease of the interaction energy located inside the surface. This interaction energy itself can be either positive or negative, but its "quantity," is always conserved as stated by (4.69) or (4.76).

4.3.3 The Multifarious Aspects of the Energy Flux in the Near Field

According to the fundamental expansion given in the general decomposition theorem of (4.26) and (4.27), at each spatial location $\mathbf{r}$, the field can be split into total nonpropagating and propagating parts along a direction given by the unit vector $\hat{u}$.[24] Most generally, this indicates that if the near field *stored* energy is to be associated with that portion of the total electromagnetic field that is *not* propagating, then it follows immediately that the definition of stored energy in this way cannot be unique. The reason, obviously, is that along different directions $\hat{u}$, the evanescent part will have different expansions, giving rise to different total energies. Summarizing this mathematically, we find that the energy of the evanescent part of the fields is given by

$$W_{\rm ev}^{e}\left(\hat{u}\right) = \frac{\varepsilon}{4}\int_{V_{\rm ext}} d^3r \left|\mathbf{E}_{\rm ev}\left[\mathbf{r};\hat{u}\left(\mathbf{r}\right)\right]\right|^2, \tag{4.79}$$

where $V_{\rm ext}$ denotes a volume exterior to the antenna (and possibly the power supply). In writing down this expression, we made the assumption that the directions along which the general decomposition theorem (4.26) is applied form a vector field $\hat{u} = \hat{u}\left(\mathbf{r}\right)$.

The first problem we encounter with the expression (4.79) is that it need not converge if the volume $V_{\rm ext}$ is infinite. This can be most easily seen when the vector field $\hat{u}(\mathbf{r})$ is taken as the constant vector $\hat{u}_0$. That is, we fix the observation frame for all points in space, separate the evanescent part, and integrate the amplitude square of this quantity throughout all space points exterior to the antenna current distribution. It is readily seen that since the field decays exponentially only in one direction (away from the antenna current along $\hat{u}_0$), then the resulting expression will diverge along the *perpendicular* directions. The divergence of the total evanescent energy in this special case is discussed mathematically in Appendix 4.7.6. There, we proved that the total evanescent energy will diverge unless certain volumes around the antenna are excluded. Carrying the analysis in spherical coordinates, we discover that the exterior region can be divided into four regions as shown in Figure 4.5, in which the total energy converges only in the upper and lower regions.

24 Although the particular mathematical expression given in (4.26) and (4.27) are not valid if the point at which this decomposition is considered lies within the antenna horizon, the separation into propagating and nonpropagating remains correct in principle but the appropriate expression is more complicated.

4.3.4 The Concept of Localized Energy in the Electromagnetic Field

We now define the *localized energy as the energy that is not propagating along certain directions of space.* Notice that the term 'localized energy' is 1) not necessarily isomorphic to 'stored energy' and 2) is dependent on certain vector field $\hat{u} = \hat{u}(\mathbf{r})$. The first observation will be discussed in detail later.[25] The second observation is related to the fundamental insight gained from the freedom of choosing the observation frame in the Weyl expansion. It seems then that the mathematical description of the wave structure of the electromagnetic field radiated by an antenna cannot be attained without reference to a particular local observation frame. We have now learned that only the orientation of the z-axis of this local frame is necessary, reducing the additional degrees of freedom needed in explicating the wave structure of the near field into two parameters, e.g., the spherical angles θ and φ. This insight can be generalized by extending it to the energy concept. 'Localization' here literally means to restrict or confine something into a limited volume. The electromagnetic near field possesses a rich and complex structure in the sense that it represents a latent potential of localization into various forms depending on the local observation frame chosen to enact the mathematical description of the problem. It is clear then that the localized energy will be a function of such directions and hence inherently not unique.[26] The overall picture boils down to this: to localize or confine the electromagnetic energy around the antenna, you first separate the nonpropagating field along the directions in which the potential localization is to be actualized, and then the amplitude square of this field is taken as a measure of the energy density of the localized field in question. By integrating the resulting energy density along the volume of interest, the total localized energy is obtained. The uncritical approach to the energy of the antenna fields confuses the stored energy with the localized energy, and then postulates – without justification – that this energy must be independent of the observation frame.

One may hope that although the energy density of the evanescent part is not unique, the total energy, i.e., the volume integral of the density, may turn out to be unique. Unfortunately, this is not true in general, as can be seen from the results of Appendix 4.7.6. The total *convergent* evanescent energy in a give volume depends

25 Cf. Section 4.3.7.

26 The reader should compare this with the definition of quantities like potential and kinetic energies in mechanics. These quantities will vary according to the frame of reference chosen for the problem. This does not invalidate the physical aspect of these energies since relative to any coordinate system, the total energy must remain fixed in a (conservative) closed system. Similarly, relative to any local observation frame, the sum of the total propagating and nonpropagating fields yields the same actually observed electromagnetic field.

in general on the orientation of the decomposition axis $\hat{u}$. The 'near-field pattern'[27] is the quantity of interest that antenna engineers may consider in studying the local field structure. Such new measure describes the localization of electromagnetic energy around the antenna in a way that formally resembles the concept of directivity in the far field. Moreover, based on the general mathematical expression of the near-field pattern (4.121), it is possible to search for antenna current distributions $\mathbf{J}(\mathbf{r})$ with particular orientations of $\hat{u}$ in which the obtained evanescent energy density is invariant. In other words, concepts like omnidirectionality, which is a far-field concept, can be analogously invented and applied to the analysis of the antenna near field. Due to the obvious complexity of the near-field energy expression (4.121), one expects that a richer symmetry pattern may develop with no straightforward connection with the physical geometry of the antenna body. It is because the far-field perspective involves an integration operation that the rich sub-wavelength effects of the antenna spatial current distribution on the generated field tend to be smoothed out when viewed from the perspective of the antenna radiation pattern. In the more careful approach of this chapter, the crucial information of the antenna near zone corresponds to the *short*-wavelength components, i.e., the spectral components $p^2 + q^2 > 1$, which are responsible of giving the field its intricate terrain of fine details. These components dominate the field as we approach the antenna current distribution and may be taken as the main object of physical interest at this localized level.

4.3.5 The Radial Evanescent Field Energy in the Near-Field Shell

We now reexamine the concept of the near-field shell at a greater depth. The idea was introduced in Chapter 3 in the context of the reactive energy, i.e., the energy associated with the circuit model of the antenna input impedance. As it has been concluded there, this circuit concept was not devised based on the field vantage point, but mainly to fit the circuit perspective related to the input impedance expressed in terms of the antenna fields as explicated by the complex Poynting theorem. We now have the refined model of the radial evanescent field developed in Section 4.2.3. *We define the localized energy in the near-field spherical shell as the self energy of the nonpropagating modes along the radial streamlines enclosed in the region* $a < r < b$. The total local energy then is the limit of the previous expression when $b \to \infty$.

27 Cf. equation (4.121) in Appendix 4.7.6.

To derive an expression for the localized electric[28] radial energy defined this way, substitute (4.30) to (4.79) with the identification $\hat{u} = \hat{r}$. It is obtained[29]

$$\begin{aligned} W_{\mathrm{ev}}^{e,\,\mathrm{rd}} &= \frac{\omega^2 k^2 \mu^2 \varepsilon}{256\pi^4} \int_{V_{\mathrm{ext}}} d^3 r \\ &\times \int_{p^2+q^2>1} dp dq \int_{p'^2+q'^2>1} dp' dq' \\ &\times \bar{\mathbf{R}}^T(\theta,\varphi) \cdot \bar{\boldsymbol{\Omega}}(p,q) \cdot \bar{\mathbf{R}}(\theta,\varphi) \\ &\times \bar{\mathbf{R}}^T(\theta,\varphi) \cdot \bar{\boldsymbol{\Omega}}^*(p',q') \cdot \bar{\mathbf{R}}(\theta,\varphi) \\ &\times \cdot \tilde{\mathbf{J}}\left[\bar{\mathbf{R}}^T(\theta,\varphi) \cdot \mathbf{K}\right] \cdot \tilde{\mathbf{J}}^*\left[\bar{\mathbf{R}}^T(\theta,\varphi) \cdot \mathbf{K}\right] \\ &\times e^{-kr\left(\sqrt{q^2+p^2-1}+\sqrt{q'^2+p'^2-1}\right)}. \end{aligned} \tag{4.80}$$

By converting the space integral in (4.80) into spherical coordinates, and using identity (4.120) to evaluate the radial integral in the region $a < r < b$, we end up with the following expression

$$\begin{aligned} W_{\mathrm{ev}}^{e,\,\mathrm{rd}}(a \le r \le b) &= \frac{\omega^2 k^2 \mu^2 \varepsilon}{256\pi^4} \int_0^{2\pi} \int_0^{\pi} d\theta d\varphi \sin\theta \\ &\times \int_{p^2+q^2>1} dp dq \int_{p'^2+q'^2>1} dp' dq' \\ &\times \bar{\mathbf{R}}^T(\theta,\varphi) \cdot \bar{\boldsymbol{\Omega}}(p,q) \cdot \bar{\mathbf{R}}(\theta,\varphi) \\ &\times \cdot \bar{\mathbf{R}}^T(\theta,\varphi) \cdot \bar{\boldsymbol{\Omega}}^*(p',q') \cdot \bar{\mathbf{R}}(\theta,\varphi) \\ &\times \cdot \tilde{\mathbf{J}}\left[\bar{\mathbf{R}}^T(\theta,\varphi) \cdot \mathbf{K}\right] \cdot \tilde{\mathbf{J}}^*\left[\bar{\mathbf{R}}^T(\theta,\varphi) \cdot \mathbf{K}'\right] \\ \times \Bigg\{ &\frac{e^{ik(m+m')b}}{ik(m+m')}\left[b^2 - \frac{2b}{ik(m+m')} - \frac{2}{k^2(m+m')^2}\right] \\ - &\frac{e^{ik(m+m')a}}{ik(m+m')}\left[a^2 - \frac{2a}{ik(m+m')} - \frac{2}{k^2(m+m')^2}\right]\Bigg\}, \end{aligned} \tag{4.81}$$

where $m = i\sqrt{q^2 + p^2 - 1}$ and $m' = i\sqrt{q'^2 + p'^2 - 1}$. It appears to the authors that the radial evanescent mode expansion is the simplest type of near-field decomposition, and one of the most natural way to mathematically describe the near field of antennas in general, especially from the engineering point of view.

4.3.6 Electromagnetic Interactions Between Propagating and Nonpropagating Fields

We turn our attention now to a closer examination of the interaction electromagnetic field energy in the near-field shell of a general antenna system. The electric field

28 For reasons of economy, throughout this section we give only the expressions of the electric energy. The magnetic energy is obtained in the same way.

29 Throughout this chapter, the conversion of the multiplication of two integrals into a double integral, interchange of order of integration, and similar operations are all justified by the results of the appendices concerning the convergence of the Weyl expansion.

will again be decomposed into propagating and evanescent parts as $\mathbf{E}(\mathbf{r}) = \mathbf{E}_{\text{ev}}(\mathbf{r}) + \mathbf{E}_{\text{pr}}(\mathbf{r})$. The energy density becomes then

$$w^e = \frac{\varepsilon}{4}\left|\mathbf{E}_{\text{ev}}(\mathbf{r})\right|^2 + \frac{\varepsilon}{4}\left|\mathbf{E}_{\text{pr}}(\mathbf{r})\right|^2 + \frac{\varepsilon}{2}\text{Re}\left\{\mathbf{E}^*_{\text{ev}}(\mathbf{r}) \cdot \mathbf{E}_{\text{pr}}(\mathbf{r})\right\}. \tag{4.82}$$

The first term is identified with the *self* energy density of the evanescent field, the second with the *self* energy of the pure propagating part. The third term is a new event in the near field shell: it represents a measure of *interaction* between the propagating and nonpropagating parts of the antenna electromagnetic fields. While it is relatively easy to interpret the first two terms as energies, the third term, that which we duped the interaction link between the first two types of fields, presents some problems. We first notice that contrary to the two self energies, it can be either positive or negative. Hence, this term cannot be understood as a representative of an entity standing alone by itself like the self energy, but, instead, it must be viewed as a *relative* energy, a relational component in the description of the total energy of the electromagnetic system. To understand better this point, we imagine that the two positive energies standing for the self-interaction of both the propagating and nonpropagating parts subsist individually as physically existing energies associated with the corresponding field in the way usually depicted in Maxwell's theory. The third term, however, is a mutual interaction that relates the two self energies to each other such that the total energy will be either be *larger* than the sum of the two self-subsisting energies (positive interaction term) or *smaller* than this sum (negative interaction term.) In other words, although we imagine the self energy density to be a reflection of an actually existing physical entity, i.e., the corresponding field, the two fields nevertheless exists in a state of mutual interdependence on each other in a way that affects the actual total energy of the system.

Consider now the total energy in the near field shell. This will be given by the volume integral of the terms of (4.82). In particular, we have for the interaction term the following *total interaction energy*

$$\begin{aligned} W^{\text{e,rd}}_{\text{int}} = \tfrac{\omega^2 k^2 \mu^2 \varepsilon}{256\pi^4}\text{Re}\Big\{ &\textstyle\int_{V_{\text{ext}}} d^3r \int_{p^2+q^2<1} dpdq \\ &\times \textstyle\int_{p'^2+q'^2>1} dp'dq' \bar{\mathbf{R}}^T(\theta,\varphi)\cdot\bar{\boldsymbol{\Omega}}(p,q)\cdot\bar{\mathbf{R}}(\theta,\varphi) \\ &\times \bar{\mathbf{R}}^T(\theta,\varphi)\cdot\bar{\boldsymbol{\Omega}}^*(p',q')\cdot\bar{\mathbf{R}}(\theta,\varphi) \\ &\times\cdot\tilde{\mathbf{J}}\left[\bar{\mathbf{R}}^T(\theta,\varphi)\cdot\mathbf{K}\right]\cdot\tilde{\mathbf{J}}^*\left[\bar{\mathbf{R}}^T(\theta,\varphi)\cdot\mathbf{K}'\right] \\ &\times e^{ikr\left(\sqrt{1-p^2+q^2}+i\sqrt{p'^2+q'^2-1}\right)}\Big\}. \end{aligned} \tag{4.83}$$

For a particular spherical shell, expressions corresponding to (4.81) and (**??**) can be easily obtained. Again, the total interaction energy (4.83) may be negative. Notice that from the Weyl expansion, most of the field very close to the antenna current distribution is evanescent. On the other hand, most of the field in the far-field zone is propagating. It turns out that the interaction density is very small in those two limiting cases. Therefore, most of the contribution to the total interaction energy in (4.83) comes from the intermediate-field zone, i.e., the crucial zone in any theory striving to describe the formation of the antenna radiated fields.

It is the opinion of the present authors that the existence of the interaction term in (4.82) is not an accidental or side phenomenon, but instead lies at the heart of the genesis of electromagnetic radiation out of the near-field shell. The theoretical treatment we have been developing so far is based on the fact that the antenna near field consists of streamlines along which the field "flows" not in a metaphorical sense, but in the mathematically precise manner through which the evanescent mode is being converted to a propagating modes, and vice versa. The two modes transform into each other according to the direction of the streamlines under consideration. *This indicates that effectively there is an energy exchange between the propagating and nonpropagating parts within the near-field shell.* Expression (4.83) is simply an evaluation of the *net* interaction energy transfer in the case of radial streamlines. Since this quantity is a single number, it only represents the overall average of an otherwise extremely complex process. A detailed theory analyzing the exact interaction mechanism is beyond the scope of this book and will be addressed elsewhere.

4.3.7 The Concept of Stored Energy

There exists a long history of investigations in the antenna theory literature concerning the topic of 'stored energy' in radiating systems, both for concrete particular antennas and general electromagnetic systems.[30] The quality factor Q is the most widely cited quantity of interest in the characterization of antennas. As we have already seen in Chapter 3, all such calculations of Q are essentially those related to an equivalent RLC circuit model for the antenna input impedance. In such a simple case, the stored energy can be immediately understood as the energy stored in the inductor and capacitor appearing in the circuit representation. In the case of resonance, both are equal so one type of energy is usually required. Mathematically speaking, underlying the RLC circuit there is a second-order ordinary differential equation that is formally identical to the governing equation of a harmonic oscillator with damping term. It is well-known that a mechanical analogy exists for the

30 For a comprehensive view on the topic of antenna reactive energy and the associated quantities like quality factor and input impedance, see [53].

electrical circuit model in which the mechanical kinetic and potential energies will correspond to the magnetic and electric energies. The stored mechanical energy can be shown to be the sum of the two mechanical energies mentioned above, while the friction term will then correspond to the resistive loss in the oscillator [95]. Now, when attempting to extend this basic understanding beyond the circuit model toward the antenna as a *field* oscillator, we immediately face the difficult task of identifying what stands for the stored energy in the field problem.

The first observation we make is that the concept of Q is well-defined and clearly understood in the context of harmonic oscillators, which are mainly physical systems governed by *ordinary* differential equations. The antenna problem, on the other hand, is most generally governed by *partial* differential equations. This implies that the number of degrees of freedom in the field problem is infinitely larger than the number of degrees of freedom in the circuit case. While it is enough to characterize the circuit problem by only measuring or computing the input impedance as seen when looking into the antenna terminals, the field oscillator problem requires generally the determination of the spatio-temporal variation of six field components throughout the entire domain of interest. In order to bring this enormous complexity into the simple level of second-order oscillatory systems, we need to search for ordinary differential equations that summarily encapsulate the most relevant parameters of interest. We will not attempt such an approach here, but instead endeavor to clarify the general requirements for such a study.

We start from the following quote by Feynman made as preparation for his introduction of the concept of quality factor [95]:

> Now, when an oscillator is very efficient ... the stored energy is very high—we can get a large stored energy from a relatively small force. The force does a great deal of work in getting the oscillator going, but then to keep it steady, all it has to do is to fight the friction. The oscillator can have a great deal of energy if the friction is very low, and even though it is oscillating strongly, not much energy is being lost. The efficiency of an oscillator can be measured by how much energy is stored, compared with how much work the force does per oscillation.

The 'efficiency' of the oscillator is what Feynman will immediately identify as the conventional quality factor. Although his discussion focused mainly on mechanical and electric (circuit) oscillator, i.e., simple systems that can be described accurately enough by second-order ordinary differential equations, we notice that the above quote is a fine elucidation of the general phenomenon of stored energy in oscillatory systems. To see this, let us jump directly to our main object of study, the antenna as a field oscillator. Here, we are working in the time-harmonic regime, which means that the problem *is* an oscillatory one. Moreover, we can identify mechanical friction with radiation loss, or the power of the radiation escaping into the far-field zone. In such a case, the antenna system can be viewed as an oscillator driven by external force, which is simply the power supplied to the antenna through its input terminal, such that a constant amount of energy per cycle is being injected in order to keep the oscillator "running." Now this oscillator, our antenna, will generate a near-field shell, i.e., a localized field surrounding the source, which will persist in existence as long as the antenna is "running," an operation that we can insure by continuing to supply the input terminal with steady power. The oscillator function, as is well-known, is inverted: in antenna systems the radiation loss is the main object of interest that has to be maximized, while the stored energy (whatever that be) has to be minimized. The *stored* energy in the field oscillator problem represents then an inevitable side effect of the system: a nonpropagating field has to exist in the near field. We say *non*propagating because anything that is propagating is associated automatically with the oscillator loss; what we are left with belongs only to the energy stored in the fields and which averages to zero in the long run.

The next step then is to find a means to calculate this stored energy. In the harmonic oscillator problem, this is an extremely easy task. However, in our case, in which we are not in possession of such a simple second-order differential equations governing the problem, one has to resort to indirect method. We suggest that the quantitative determination of the antenna stored energy must revert back to the basic definition of energy as such. *We define the energy stored in the antenna surrounding fields as the latent capacity to perform work when the power supply of the system is switched off.* To understand the motivation behind this definition, let us make another comparison with the time evolution of damped oscillators. Transient phenomena can be viewed as a discharge of initial energy stored in the system.[31] When the antenna power supply is on, the radiation loss is completely compensated for by the power removed by the antenna terminals from the source generator, *while the antenna stored energy remains the same.* Now, when the power supply is switched off, the radiation loss can no longer by accounted for by the energy flux through the antenna port. The question here is about what happens to the stored energy. In order to answer this question, we need to be more specific about the description of the problem. It will be assumed that a load is immediately connected arcos the antenna input terminals after switching off the generator. The new problem is still governed by Maxwell's equations and hence can be solved under the appropriate initial and boundary conditions. It is expected

31 "By a *transient* is meant a solution of the differential equation when there is no force present, but when the system is not simply at rest." [95].

that a complicated process will occur, in which part of the stored energy will be converted to electromagnetic radiation, while another portion will be absorbed by the load. We define then the actual stored energy as the total amount of radiated power *and* the power supplied to the load *after* switching off the source generator. In this case, the answer to the question about the *quantity* of the stored energy can in principle be answered.

Based on this formulation of the problem, we find that our near field theory can *not* definitely answer the quantitative question concerning the amount of energy stored in the near field since it is essentially a time-harmonic theory. A transient solution of the problem is possible but very complicated. However, our derivations have demonstrated a phenomenon that is closely connected with the current problem. This is the energy *exchange* between the evanescent and propagating modes. As could be seen from equation (4.83), the two parts of the electromagnetic field interact with each other. Moreover, by examining the field expression of the interaction energy *density*, we discover that this 'function over space' extends in a localized fashion in a way similar to the localization of the self evanescent field energy. This strongly suggests that the interaction energy density is part of the "non-moving" field energy, and hence should be included with the self evanescent field energy as one of the main constituents of the total energy stored in the antenna surrounding fields. Unfortunately, such a proposal faces the difficulty that this total sum of the two energies may very well turn out to be negative, in which its physical interpretation becomes problematic. One way out of this difficulty is to put things in their appropriate level: the time-harmonic theory is incapable of giving the fine details of the temporal evolution of the system; instead, it only gives averaged steady state quantities. The interaction between the propagating and nonpropagating field, however, is a genuine electromagnetic process and is an expression of the essence of the antenna as a device that helps converting a nonpropagating energy into a propagating one. In this sense, the interaction energy term predicted by the time-harmonic theory measures the net average energy exchange process that occurs between propagating and nonpropagating modes *while the antenna is running*, i.e., supplied by steady power through its input terminals. The existence of this time-averaged harmonic interaction indicates the possibility of energy conversion between the two modes *in general*. When the generator is switched off, *another* energy conversion process (the transient process) will take place, which might not be related in a simple manner to the steady-state quantity.[32]

4.3.8 Dependence of the Radial Localized Energy on the Choice of the Origin

In this section, we investigate the effect of changing the location of the origin of the local observation frame used to compute the radial localized energy in antenna systems. In (4.80), we presented the expression of such an energy in terms of a local coordinate system with an origin fixed in advance. If the location of this origin is shifted to the position $\mathbf{r}_0$, then it follows from (16.16) that the only effect will be to multiply the spatial Fourier transform of the antenna current distribution by $\exp\left(i\mathbf{K}\cdot\mathbf{r}_0\right)$. Therefore, the new total radial localized energy will become

$$\begin{aligned} W_{\mathrm{ev}}^{e,\mathrm{rd}}\left(\mathbf{r}_0\right) = & \tfrac{\omega^2 k^2 \mu^2 \varepsilon}{256\pi^4} \int_{V_{\mathrm{ext}}} d^3 r \\ & \times \int_{p^2+q^2>1} dp dq \int_{p'^2+q'^2>1} dp' dq' \\ & \times \bar{\mathbf{R}}^T\left(\theta,\varphi\right)\cdot\bar{\boldsymbol{\Omega}}\left(p,q\right)\cdot\bar{\mathbf{R}}\left(\theta,\varphi\right) \\ & \times \bar{\mathbf{R}}^T\left(\theta,\varphi\right)\cdot\bar{\boldsymbol{\Omega}}^*\left(p',q'\right)\cdot\bar{\mathbf{R}}\left(\theta,\varphi\right) \\ & \times \cdot\tilde{\mathbf{J}}\left[\bar{\mathbf{R}}^T\left(\theta,\varphi\right)\cdot\mathbf{K}\right]\cdot\tilde{\mathbf{J}}^*\left[\bar{\mathbf{R}}^T\left(\theta,\varphi\right)\cdot\mathbf{K}'\right] \\ & \times e^{i\left(\mathbf{K}-\mathbf{K}'^*\right)\cdot\mathbf{r}_0} e^{-kr\left(\sqrt{q^2+p^2-1}+\sqrt{q'^2+p'^2-1}\right)}. \end{aligned} \tag{4.84}$$

It is obvious that in general $W_{\mathrm{ev}}^{\mathrm{rad}}\left(\mathbf{r}_0\right) \neq W_{\mathrm{ev}}^{\mathrm{rad}}\left(0\right)$, that is, the new localized energy corresponding to the shifted origin with respect to the antenna is not unique. This nonuniqueness, however, has nothing alarming or even peculiar about it. It is a logical consequence from the Weyl expansion. To see this, consider Figure 4.4 where we show the old origin O, the new origin located at $\mathbf{r}_0$, and an arbitrary observation point $\mathbf{r}$ outside the antenna current region. With respect to the frame O, the actually computed field at the location $\mathbf{r}$ is the evanescent part along the unit vector $\hat{u}_1 = \mathbf{r}/r$. On the other hand, for the computation of the contribution at the very same point but with respect to the frame at $\mathbf{r}_0$, the field added there is the evanescent part along the direction of the unit vector $\hat{u}_2 = \left(\mathbf{r}-\mathbf{r}_0\right)/\left|\mathbf{r}-\mathbf{r}_0\right|$. Clearly then the two localized energies cannot be exactly the same in general.

The reader is invited to reflect on this conclusion in order to remove any potential misunderstanding. If two different coordinate systems are used to describe

32 The reader may observe that the situation in circuit theory is extremely simple compared with the field problem. There, the transient question of the circuit can be answered by parameters from the time-harmonic theory itself. For example, in an RLC circuit, the Q factor is a simple function of the capacitance, inductance, and resistance, all are basic parameters appearing throughout the steady state *and* the transient equations. It is not obvious that such a simple parallelism will remain the case in the transient field problem.

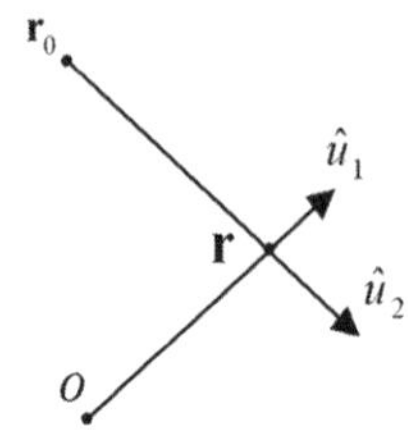

Figure 4.4 Geometric illustration for the process of forming the radial localized energy with respect to different origins.

the radial energy localized around the same origin, i.e., an origin with the same relative position compared to the antenna, then the two results will be exactly the same. The situation illustrated in Figure 4.4 does not refer to two coordinate systems per se, but to two different choices of the origin of the radial directions utilized in computing the localized energy of the antenna under consideration. There is no known law of physics necessitating that the localized energy has to be the same regardless to the observation frame. The very term 'localization' is a purely spatial concept, which must make use of a particular frame of reference in order to draw mathematically specific conclusion. In our particular example, by changing the *relative* position of the origin with respect to the antenna, what is meant by the expression "*radial* localization" has also to undergo certain change. Equation (4.84) gives the exact quantitative modification of this meaning.[33]

4.4 THE NEAR-FIELD RADIAL STREAMLINES FROM THE FAR FIELD POINT OF VIEW

4.4.1 Introduction

In this section, we synthesize the knowledge that has been achieved in Chapter 3, concerning the near field in the spatial domain, and Section 4.2, which focused

33 An example illustrating this relativity can be found in the area of rigid-body dynamics. There, the fundamental equations of motion involve the moment of inertia around certain axes of rotation. It is a well-known fact that this moment of inertial, which plays a role similar to mass in translational motion, does depend on the choice of the axis of rotation, and varies even if the new axis is parallel to the original one.

mainly on the concept of radial streamlines developed from the spectral domain perspective. The main mathematical device utilized in probing the spatial structure of the near field was the Wilcox expansion

$$\mathbf{E}(\mathbf{r}) = \frac{e^{ikr}}{r}\sum_{n=0}^{\infty}\frac{\mathbf{A}_n(\theta,\varphi)}{r^n}, \quad \mathbf{H}(\mathbf{r}) = \frac{e^{ikr}}{r}\sum_{n=0}^{\infty}\frac{\mathbf{B}_n(\theta,\varphi)}{r^n}, \tag{4.85}$$

On the other hand, the Weyl expansion (17.23) represented the major mathematical tool used to analyze the near field into its constituting spectral components. There is, however, a deeper way to look into the problem. The view of the antenna presented in Chapter 3 is essentially an *exterior* region description. Indeed, inside the sphere $r = a$, which encloses the antenna physical body, there is an infinite number of current distributions that can be compatible with the Wilcox expansion in the exterior region. Put differently, we are actually describing the antenna system from the far-field point of view. Indeed, as was already shown by Wilcox [47], it is possible to recursively compute all the higher-order terms in the expansion (4.85) starting from a given far field. Now, the approach presented in Section 4.2 is different essentially for the opposite reason. There, the mathematical description of the problem starts from an actual antenna current distribution using the dyadic Green's function as shown in (16.12). This means that even when inquiring about the fields radiated outside some sphere enclosing the antenna body, the fields themselves are determined uniquely by the current distribution. It is for this reason that the analysis following Section 4.2 is inevitably more difficult than Chapter 3.

Our purpose in the present section is to reach for a kind of compromise between the two approaches. From the engineering point of view, the Wilcox series approach is more convenient since it relates directly to familiar antenna measures like far field and minimum Q. On the other hand, as we have already demonstrated in detail, the reactive energy concept is inadequate when extensions beyond the antenna circuit models are attempted. The Weyl expansion supplied us with a much deeper understanding of the near-field structure by decomposing electromagnetic radiation into propagating and nonpropagating parts. What is required is an approach that directly combines the Wilcox series with the deeper perspective of the Weyl expansion. This we proceed now to achieve in the present section. We first generalize the classical Weyl expansion to handle the special form appearing in the Wilcox series. This allows us then to derive new Wilcox-Weyl expansion, a hybrid series that combines the best of the two approaches. The final result is a sequence of higher-order terms explicating how the radial streamlines split into propagating and nonpropagating modes as we progressively approach the antenna physical body, all computed starting from a given far-field pattern,

4.4.2 Generalization of the Weyl Expansion

We start by observing the following from the product rule

$$\frac{\partial}{\partial r}\frac{e^{ikr}}{r^n}=ik\frac{e^{ikr}}{r^n}-n\frac{e^{ikr}}{r^{n+1}}, \tag{4.86}$$

which is valid for $n \geq 1$. We will be interested in deriving a spectral representation for e^{ikr}/r^{n+1} since it is precisely this factor that appears in the Wilcox expansion (4.85). From (4.86) write

$$\frac{e^{ikr}}{r^{n+1}}=\frac{1}{n}\left(ik\frac{e^{ikr}}{r^n}-\frac{\partial}{\partial r}\frac{e^{ikr}}{r^n}\right). \tag{4.87}$$

The Weyl expansion (17.23) written in spherical coordinates reduces to

$$\frac{e^{ikr}}{r}=\frac{ik}{2\pi}\int_{-\infty}^{\infty}\int_{-\infty}^{\infty}dpdq\frac{1}{m}e^{i\mathbf{K}\cdot\mathbf{r}}, \tag{4.88}$$

where

$$\hat{\mathbf{K}}=\hat{x}p+\hat{y}q+\hat{z}\operatorname{sgn}(\cos\theta)\,m, \tag{4.89}$$

$$\hat{r}=\hat{x}\cos\varphi\sin\theta+\hat{y}\sin\varphi\sin\theta+\hat{z}\cos\theta. \tag{4.90}$$

By bringing the differentiation inside the integral, it is possible to achieve

$$\frac{\partial}{\partial r}\frac{e^{ikr}}{r}=\frac{(ik)^2}{2\pi}\int_{-\infty}^{\infty}\int_{-\infty}^{\infty}dpdq\frac{\hat{r}\cdot\hat{\mathbf{K}}}{m}e^{i\mathbf{K}\cdot\mathbf{r}}. \tag{4.91}$$

Substituting (4.88) and (4.91) into (4.87), it is found that

$$\frac{e^{ikr}}{r^2}=\frac{(ik)^2}{2\pi}\int_{-\infty}^{\infty}\int_{-\infty}^{\infty}dpdq\frac{1}{m}\left(1-\hat{r}\cdot\hat{\mathbf{K}}\right)e^{i\mathbf{K}\cdot\mathbf{r}}. \tag{4.92}$$

Iterating, the following general expansion is attained

$$\frac{e^{ikr}}{r^3}=\frac{1}{2}\frac{(ik)^3}{2\pi}\int_{-\infty}^{\infty}\int_{-\infty}^{\infty}dpdq\frac{1}{m}\left(1-\hat{r}\cdot\hat{\mathbf{K}}\right)^2e^{i\mathbf{K}\cdot\mathbf{r}}. \tag{4.93}$$

Observing the repeated pattern, we arrive at the *generalized Weyl expansion*[34]

$$\frac{e^{ikr}}{r^{n+1}} = \frac{1}{n!}\frac{(ik)^n}{2\pi}\int_{-\infty}^{\infty}\int_{-\infty}^{\infty} dpdq\frac{1}{m}\left(1 - \hat{r}\cdot\hat{\mathbf{K}}\right)^n e^{i\mathbf{K}\cdot\mathbf{r}}. \tag{4.94}$$

In reaching into this result, the differentiation and integration were freely interchanged. The justification for this is very close to the argument in Appendix 4.7.2 and will not be repeated here. On a different notice, the singularity $\theta = \pi/2$ (i.e., $z = 0$) is avoided in this derivation because our main interest is in the antenna exterior region.

4.4.3 The Hybrid Wilcox-Weyl Expansion

We now substitute the generalized Weyl expansion (4.94) into the wilcox expansion (4.85) to obtain

$$\begin{aligned}\mathbf{E}(\mathbf{r}) = \sum_{n=0}^{\infty}\int_{-\infty}^{\infty}\int_{-\infty}^{\infty} dpdq\tfrac{1}{n!}\tfrac{(ik)^n}{2\pi m}\mathbf{A}_n(\theta,\varphi)\\ \times\left[1 - \hat{r}(\theta,\varphi)\cdot\hat{\mathbf{K}}(p,q)\right]^n e^{i\mathbf{K}(p,q)\cdot\hat{r}(\theta,\varphi)r},\end{aligned} \tag{4.95}$$

$$\begin{aligned}\mathbf{H}(\mathbf{r}) = \sum_{n=0}^{\infty}\int_{-\infty}^{\infty}\int_{-\infty}^{\infty} dpdq\tfrac{1}{n!}\tfrac{(ik)^n}{2\pi m}\mathbf{B}_n(\theta,\varphi)e^{i\mathbf{K}\cdot\mathbf{r}}\\ \times\left[1 - \hat{r}(\theta,\varphi)\cdot\hat{\mathbf{K}}(p,q)\right]^n e^{i\mathbf{K}(p,q)\cdot\hat{r}(\theta,\varphi)r}.\end{aligned} \tag{4.96}$$

By separating the spectral integral into propagating and evanescent parts, we finally arrive at our main results

$$\mathbf{E}_{\text{ev}}(\mathbf{r}) = \sum_{n=0}^{\infty}\Xi_n^e(\mathbf{r}), \tag{4.97}$$

$$\mathbf{H}_{\text{ev}}(\mathbf{r}) = \sum_{n=0}^{\infty}\Xi_n^e(\mathbf{r}), \tag{4.98}$$

where

$$\begin{aligned}\Xi_n^e(\mathbf{r}) = \int_{p^2+q^2>1} dpdq\tfrac{1}{n!}\tfrac{(ik)^n}{2\pi m}\mathbf{A}_n(\theta,\varphi)\\ \times\left[1 - \hat{r}(\theta,\varphi)\cdot\hat{\mathbf{K}}(p,q)\right]^n e^{i\mathbf{K}(p,q)\cdot\hat{r}(\theta,\varphi)r},\end{aligned} \tag{4.99}$$

34 This result can be rigorously proved by applying the principle of mathematical induction.

$$\begin{aligned}\Xi_n^h(\mathbf{r}) = \int_{p^2+q^2>1} & dpdq \frac{1}{n!}\frac{(ik)^n}{2\pi m}\mathbf{B}_n(\theta,\varphi) \\ & \times \left[1 - \hat{r}(\theta,\varphi)\cdot\hat{\mathbf{K}}(p,q)\right]^n e^{i\mathbf{K}(p,q)\cdot\hat{r}(\theta,\varphi)r}.\end{aligned} \tag{4.100}$$

Also, we have

$$\mathbf{E}_{\text{pr}}(\mathbf{r}) = \sum_{n=0}^{\infty} \mathrm{P}_n^e(\mathbf{r}), \tag{4.101}$$

$$\mathbf{H}_{\text{pr}}(\mathbf{r}) = \sum_{n=0}^{\infty} \mathrm{P}_n^e(\mathbf{r}), \tag{4.102}$$

where

$$\begin{aligned}\mathrm{P}_n^e(\mathbf{r}) = \int_{p^2+q^2<1} & dpdq \frac{1}{n!}\frac{(ik)^n}{2\pi m}\mathbf{A}_n(\theta,\varphi) \\ & \times \left[1 - \hat{r}(\theta,\varphi)\cdot\hat{\mathbf{K}}(p,q)\right]^n e^{i\mathbf{K}(p,q)\cdot\hat{r}(\theta,\varphi)r},\end{aligned} \tag{4.103}$$

$$\begin{aligned}\mathrm{P}_n^h(\mathbf{r}) = \int_{p^2+q^2<1} & dpdq \frac{1}{n!}\frac{(ik)^n}{2\pi m}\mathbf{B}_n(\theta,\varphi) \\ & \times \left[1 - \hat{r}(\theta,\varphi)\cdot\hat{\mathbf{K}}(p,q)\right]^n e^{i\mathbf{K}(p,q)\cdot\hat{r}(\theta,\varphi)r}.\end{aligned} \tag{4.104}$$

The expansion electric and magnetic functions (4.99) and (4.100) can be interpreted in the following manner. The factor $i\mathbf{K}(p,q)\cdot\hat{r}(\theta,\varphi)$ appearing in $\exp\left[i\mathbf{K}(p,q)\cdot\hat{r}(\theta,\varphi)r\right]$ has an attenuating part $-mr\left|\cos\theta\right| = -r\sqrt{p^2+q^2-1}\left|\cos\theta\right|$. Therefore, the field described here consists of evanescent modes along the radial direction specified by the spherical angles θ and φ. Similarly, the expansion electric and magnetic functions (4.103) and (4.104) are *pure* propagating modes along the same radial direction. Thus, we have achieved a mathematical description similar to the radial streamline in Section 4.2.3, mainly equations (4.30) and (4.31).

In the new expansion, the rich information encompassing the near-field *spectral* structure are given by the functions $\left(i^n k^2/n!2\pi m\right)\mathbf{A}_n(\theta,\varphi)\left[1-\hat{r}(\theta,\varphi)\cdot\hat{\mathbf{K}}(p,q)\right]^n$ and $\left(i^n k^2/n!2\pi m\right)\mathbf{B}_n(\theta,\varphi)\left[1-\hat{r}(\theta,\varphi)\cdot\hat{\mathbf{K}}(p,q)\right]^n$ for the electric and magnetic fields, respectively. We immediately notice that this spectral function consists of direct multiplication of two easily identified contributions, the first is the Wilcox-type expansion given by the angular functions $\mathbf{A}_n$ and $\mathbf{B}_n$, and the second is a *common* Weyl-type spectral factor given by $\left(i^n k^2/n!2\pi m\right)\left[1-\hat{r}(\theta,\varphi)\cdot\hat{\mathbf{K}}(p,q)\right]^n$. This latter is function of both the spectral variable p and q, *and* the spherical angles θ and φ.

We can now understand the structure of the antenna near field from the point of view of the far field in the following manner. Start from a given far field pattern for a class of antennas of interest. Strictly speaking, an infinite number of actually realized antennas can be built such that they all agree on the supposed far field. Mathematically, this is equivalent to stating that the hybrid Wilcox-Weyl expansions above are valid only in the exterior region $r > a$. We then proceed by computing (recursively as in [47] or directly as in Chapter 3 all the vectorial angular functions $\mathbf{A}_n$ and $\mathbf{B}_n$ starting from the radiation pattern. With respect to this basic step, a radial streamline spectral description of the near-field structure can be be constructed by just multiplying the obtained angular vector field $\mathbf{A}_n$ and $\mathbf{B}_n$ by $\left(i^n k^2/n!2\pi m\right)\left[1 - \hat{r}\left(\theta,\varphi\right)\cdot\hat{\mathbf{K}}\left(p,q\right)\right]^n$. This will generate the dependence of the spectral content of the near field on the radial streamline orientation specified by θ and φ. The actual spatial dependence of the propagating and nonpropagating fields can be recovered by integrating the result of multiplying the above obtained spectrum with the radial streamline functions $\exp\left[i\mathbf{K}\left(p,q\right)\cdot\hat{r}\left(\theta,\varphi\right)r\right]$ over the regions $p^2 + q^2 < 1$ and $p^2 + q^2 > 1$, respectively.

A striking feature in this picture is its simplicity. For arbitrary antennas, it seems that the *spectral* effect of including higher-order terms in the hybrid Wilcox-Weyl expansion is nothing but multiplication by higher-order polynomials of p, q, and m,[35] with coefficients directly determined *universally* by the direction cosines of the radial vector along which a near-field streamline is considered. On the other hand, *antenna-specific* details of the radial streamline description seem to be supplied directly by the angular vector fields $\mathbf{A}_n$ and $\mathbf{B}_n$, which are functions of the (far-field) radiation pattern.

It appears then that the expansions (4.97), (4.101),(4.98), (4.102), provide further information about the antenna, namely the importance of size. Indeed, the smaller the sphere $r = a$ (inside where the antenna is located), the more terms in those expansions are needed in order to converge to accurate values of the electromagnetic fields. Taking into consideration that the angular vector fields $\mathbf{A}_n$ and $\mathbf{B}_n$ are functions of the far-field radiation pattern, we can see now how the hybrid Wilcox-Weyl expansion actually relates many parameters of interest in a unified whole picture: the far-field radiation pattern, the near-field structure as given by the radial streamlines, the size of the antenna, and the minimum Q (for matching bandwidth consideration). It is for these reasons that the authors believe the results of this chapter to be of direct interest to the antenna engineering community. More

35 This is intuitively clear since, as we have found in Chapter 3, higher-order terms in the Wilcox-type expansion correspond to more complex near-field radial structure as we descend from the far zone toward the source region, which in turns necessities the need to include significant short-wavelength components (i.e., large p and q components).

extensive analysis of specific antenna types within the lines sketched above will be considered elsewhere.

4.4.4 General Remarks

We end this section with few remarks on the Wilcox-Weyl expansion. Notice first that the reactive energy, as defined in Chapter 3, is the form of the total energy expressed through the Wilcox series with the $1/r^2$ term excluded. It is very clear from the results of this section that this reactive energy includes both nonpropagating *and* propagating modes. This may provide an insight into the explanations and analysis normally attached to the relationship between reactive energy, localized energy, and stored energy.[36]

The second remark is about the nature of the new streamline here. Notice that although we ended up in the hybrid Wilcox-Weyl expansion with a radial streamline picture of the near field, there is still a marked difference between this particular streamline and those introduced in Section 4.2.3 from the source point of view. The difference is that the nonpropagating fields in (4.99) and (4.100) are damped sinusoidal functions while those appearing, for example, in (4.30), are pure evanescent modes.

This is related to a deeper difference between the two approaches of Section 4.2.3 and the present one. In using the Wilcox expansion for the mathematical description of the antenna electromagnetic fields, we are asserting a far-field point of view and hence our obtained near-field insight is already biased. This appears behind the fact that the generalized Weyl integral (4.94), when separated into the two regions inside and outside the circle $p^2 + q^2 = 1$, will *not* give a decomposition into propagating and nonpropagating modes in general. The reason is that there exists in the integrand spatial variables, mainly the spherical angles θ and φ. Only when these two angles are fixed can we interpret the resulting quantity as propagating and nonpropagating modes with respect to the remaining spatial variable, namely r. *It follows then that from the far-field point of view, the only possible meaningful decomposition of the near field into propagating and nonpropagating parts is the radial streamline picture.*

36 Cf. Section 4.3 Chapter 3.

4.5 THE MECHANISM OF FAR FIELD FORMATION

We are now in a position to put together the theory developed throughout this chapter into a more concrete presentation by employing it to explain the structural formation of the far field radiation. This we aim to achieve by relying on the insight into the spectral composition of the near field provided by the Weyl expansion. In the remaining parts of this section, our focus will be on applying the source point of view developed in Section 4.2. The theory of Section 4.4, i.e., the far-field point of view, will be taken up in separate work.

Let us assume that the current distribution on the antenna physical body was obtained by a numerical solution of Maxwell's equations, ideally using an accurate, preferably higher-order, method of moment.[37] We will now explicate the details of how the far-field pattern is created starting from this information.

We focus on the electric field. Since the far-field pattern is a function of the angular variables θ and φ, the most natural choice of the appropriate mathematical tool for studying this problem is the concept of radial streamlines as developed in Section 4.2.3. A glance at equations (4.30) and (4.31) shows that the quantity pertinent to the antenna current distribution is the spatial Fourier transform of this current $\tilde{\mathbf{J}}(\mathbf{K})$ as defined in (16.16). Now, to start with, we choose a global cartesian frame of reference xyz. Relative to this frame we fix the spherical angles θ and φ used in the description of both the far-field pattern and the radial streamline picture of the near field. The global frame is chosen such that the z-axis points in the direction of the broadside radiation. For example, if we are analyzing a linear wire antenna or a planner patch, the global frame is chosen such that the z-axis is perpendicular to the wire in the former case and to the plane containing the patch in the latter case. Although we don't prove this here, it can be shown that under these condition the Fourier transform of the current distribution in the previous two special cases, as a function of the spectral variables p and q, has its maximum value around the origin of the pq-plane as shown in Figure 4.2. Since the majority of the contribution to the far field comes from the propagating modes appearing in (4.31), the rest being attenuated exponentially as shown in (4.30), we can picture the antenna operation as a two-dimensional low-pass spatial filer in the following manner. All spectral components within the unit circle $p^2 + q^2 = 1$ (the visible domain) will pass to the far field, while components outside this region will be

37 It is evident that the problem formulated this way is not exact. However, since the integral operator of the problem is bounded, the approximate finite dimensional matrix representation of this operator will approach the correct exact solution in the limit when $N \to \infty$.

filtered out. Let us call this filter *the visible domain filter*.[38] Now, the fact that when the global frame is chosen such that its z-axis is oriented in the direction along which the spatial Fourier transform of the current distribution $\tilde{\mathbf{J}}(\mathbf{K})$, as a function of p and q, will have most of its values concentrated around the region $p = q = 0$ immediately explains why some antennas, such as linear wires and planner patches, have broadside radiation pattern to begin with.

We unpack this point by first noticing how the near field splits into propagating and nonpropagating streamlines. The mechanism here, as derived in (4.30) and (4.31), is *purely geometrical*. To see this, let us call the region around which $\tilde{\mathbf{J}}(\mathbf{K})$ is maximum $D(p,q)$; e.g., in the case of planner patch this region will be centered around $p = q = 0$. What happens is that for varying spherical angles θ and φ, we have to rotate the spatial Fourier transform $\tilde{\mathbf{J}}(\mathbf{K})$ by the matrix $\bar{\mathbf{R}}^T(\theta,\varphi)$. This will translates into the introduction of new *nonlinear* transformation of p and q as given by $\mathbf{K}' = \bar{\mathbf{R}}^T(\theta,\varphi)\cdot\mathbf{K}$.[39] The region $D(p,q)$ is now transformed into $D(p',q')$. Since we are viewing the antenna operation in producing the far field pattern as a global two-dimensional spatial filter, we must transform back into the language of the global frame. The newly transformed region $D(p',q')$ will be written in the old language as $D'(p,q)$. Therefore, varying the observation angles θ and φ is effectively equivalent to a nonlinear stretching of the original domain $D(p,q)$ given by

$$D(p,q) \xrightarrow{\mathbf{K}'=\bar{\mathbf{R}}^T(\theta,\varphi)\cdot\mathbf{K}} D'(p,q). \tag{4.105}$$

This implies that a re-shaping of the domain $D(p,q)$ is the main cause for the formation of the far-field pattern. Indeed, by relocating points within the pq-plane, the effect of the visible domain filter will generate the far-field pattern.

However, there is also a universal part of the filtering process that does not depend on the antenna current distribution. This is the spectral polarization dyad $\bar{\boldsymbol{\Omega}}(p,q)$ defined by (4.32). The multiplication of this dyad with m, i.e., the spectral quantity $m\bar{\boldsymbol{\Omega}}(p,q)$, is the outcome of the fact that the electromagnetic field has polarization, or that the problem is vector in nature.[40] It is common to all radiation processes.

We now see that the overall effect of varying the observation angles can be summarized in the tertiary process

1. Rotate the spatial Fourier transform by $\bar{\mathbf{R}}^T(\theta,\varphi)$.

38 Similar construction of this filter exists in optics.

39 This transformation is nonlinear because m depends nonlinearly on p and q via the relation $m = \sqrt{1 - p^2 + q^2}$.

40 Cf. Section 4.2.3.

2. Multiply (filter) the rotated Fourier transform by the spectral polarization dyad $\bar{\boldsymbol{\Omega}}(p, q)$ after applying to the latter a similarity transformation.
3. Filter the result by the visible domain filter of the antenna.

This process fully explicates the formation of the far-field pattern of any antenna from the source point of view. As it can be seen, our theoretical narrative utilizes only two types of easy-to-understand operations: 1) geometrical transformations (rotation, stretching, similarity transformation), and 2) spatial filtering (spectral polarization filtering, visible domain filtering).

4.6 CONCLUSION

This chapter provided a broad outline for the understanding of the electromagnetic near fields of general antenna systems in the spectral domain. The concept of streamlines was introduced using the Weyl expansion in order to picture the field dynamically as a process of continuous decomposition into propagating and nonpropagating streamlines viewed here from the source point of view. We then used the new insight to reexamine the topic of the antenna energy, suggesting that there are multiple possible views of what best characterizes the near-field structure from the energy point of view. The concept of the near-field radial streamlines was then developed but this time from the far-field point of view by deriving a hybrid Wilcox-Weyl expansion to mathematically describe the splitting of the near field into radial propagating and nonpropagating streamlines constructed recursively or directly from a given far field radiation pattern. The source point of view was finally used to provide an explanation for why and how antennas produce far-field radiation patterns.

It seems from the overall consideration of this work that there exists a deep connection between the near and far fields different from what is seen in the first look. Indeed, the results of Section 4.2.4 suggested that only *two* degrees of freedom are needed to describe the splitting of the electromagnetic field into propagating and nonpropagating parts, which supplied the theoretical motivation to investigate the radial streamline structure of the near field. Furthermore, the results of Section 4.4 showed that the only near-field decomposition into propagating and nonpropagating modes possible from the far-field point of view is the radial streamline picture introduced previously from the source point of view. This shows that there exists an intimate relation between the far and near field structures, and we suggest that further research in this direction is needed in order to understand the deep implications of this connection for electromagnetic radiation in general.

On the side of antenna practice, we believe that the proposed theory will play a role in future advanced research and devolvement of antenna systems. Indeed, Chapter 3 provided a formalism suitable for the visualization of the important spatial regions surrounding the antenna and the details of energy exchange processes taking place there. It has been found during the long history of electromagnetic theory and practice that the best intuitive but also rigorous way for understanding the operation and performance of actual devices and systems is the energy point of view. For this reason, the theory proposed did not stop at the field formalism, but also went ahead to investigate how this formalism can be used to provide general concrete results concerning the pathways of energy transfer between various regions in the antenna surrounding domain of interest. For example, we mention the interaction theorems developed in Chapter 3, which provide a quantitative measure of the field modal content passing from one spatial region to another. As we emphasized repeatedly before, this proved to be a natural way in understanding better the reactive energy, the quantity of fundamental importance in the determining the behavior of the antenna input impedance. Furthermore, the specification of all these descriptions in terms of the antenna physical TE and TM modes is continuous with the established tradition in the electromagnetic community in which basic well-understood solutions of Maxwell's equations are used to determine and understand the complex behavior of the most general field. We believe that the generality of the formalism developed here will help future researchers to investigate special cases arising from particular applications within their range of interest to the community.

The more fundamental treatment presented in this chapter aimed at providing foundations for the analysis of Chapters 2 and 3. The strategy we followed here was the classical Fourier analysis of mathematical physics and engineering in which complex arbitrary field forms are developed in a series of well-behaved basic solution, i.e., the sinusoidal or harmonic functions. This not only provide a solid grounding for the results obtained in the direct study conducted in the spatial domain, but also opens the door for new windows that may be needed in characterizing the field structure in emerging advanced applications and experimental setups. The spectral theory, which decomposes the fields into evanescent and propagating modes together with a fundamental understanding of their mutual interrelation, can be related to the ongoing research in nanooptics, imaging, and other areas relevant to nanostructures and artificial materials. Indeed, the crux of this new devolvement is the manipulation of the intricate way in which the electromagnetic fields interact with subwavelength (nano) objects. Mathematically and physically, the resonance of such subwavelength structures occurs upon interaction with evanescent modes, because the latter correspond to the high-wavenumber k-components. Therefore, the analysis in this chapter regarding the fine details of the process in which the total field is being continually split into propagating and evanescent modes appears as a natural approach for studying the interaction of a nanoantenna or any radiating structure with complex surrounding environments. What is even more interesting is to see how such a kind of applications (interaction with complex environments) can be studied by the same mathematical formalism used to understand how the far field of any antenna (in free space) is formed, as suggested particularly in Section 4.5. The advantage of having one coherent formalism that can deal with a wide variety of both theoretical and applied issues is one of main incentives that stimulated us in carrying out this program of antenna near-field theory research.

On the more conventional side, the design and devolvement of antennas radiating in free space, we have tried to illuminate the near-field structure from both the source point of view and the far field perspective at the same time. Both views are important in the actual design process. For the source point of view, our analysis in Chapter 4, especially Section 4.2.3, relates in a fundamental way the exact variation in the antenna current distribution to the details of how the near field converts continually from evanescent to propagating modes. This can help antenna engineers in devising clues about how to modify the antenna current distribution in order to meet some desirable design or performance goals. The advantage gained from such an outcome is reducing the dependence on educated guess, random trial and error, and expensive optimization tasks, by providing a solid base for carrying the antenna devolvement process in a systematic fashion.

The far-field perspective, which was developed in Chapter 3 and continued in Section 4.5, could provide a different kind of valuable information for the antenna engineer. Here, one starts with a specification of a class of antennas compatible with a given far field radiation pattern, and then proceeds in constructing the near field of all antennas belonging to this class, in both the spatial and spectral domain, in order to relate far field performance measures, such as directivity, polarization, null formation, etc, to near field characteristics, such as input impedance and antenna size. A set of fundamental relations, understood in this sense, can be generated using our formalism for any set of objectives of interest found in a particular application, and hence guide the design process by deciding what kind of inherent conflicts and tradeoffs exist between various antagonistic measures. In this way, one can avoid cumbersome efforts to enforce a certain design goal that cannot be achieved in principle with any configuration whatsoever because it happens to violate one of the fundamental limitations mentioned above.

4.7 APPENDICES AND SUPPLEMENTARY MATERIALS

4.7.1 Absolute and Uniform Convergence of the Weyl Expansion

We prove this observation by using the integral representation (5.11). First, notice that from the definition of the Bessel function, $\left|u^2 J_0\left(\rho\sqrt{1+u^2}\right)e^{-k|z|u}\right| \leq \left|u^2 e^{-k|z|u}\right|$. Next, by L'Hopital rule, we have $\lim_{u\to\infty}\left|u^2 e^{-k|z|u}\right| = 0$ for $z \neq 0$. We conclude then that $\lim_{u\to\infty}\left|u^2 J_0\left(\rho\sqrt{1+u^2}\right)e^{-k|z|u}\right| = 0$ for $z \neq 0$. This allows as to write $\left|J_0\left(\rho\sqrt{1+u^2}\right)e^{-k|z|u}\right| < \frac{1}{u^2}$ for sufficiently large u, say $u \geq u_0$. Notice that this is valid for any $\rho \geq 0$ and for any $|z| \geq z_0 > 0$, which is the case here because we are working in the *exterior* region of the antenna system. We now apply the Weierstrass-M [90] test for uniform convergence. Specifically, identify $M(u) = \frac{1}{u^2}$ and notice that $\int_{u_0}^{\infty} M(u)\, du < \infty$. It follows then that the integral is absolutely convergent and uniformly convergent in all its variables.

4.7.2 Interchange of Integration and Differentiation in Weyl Expansion

Here we interchange the order of integration and differentiation. To prove this, we make use of the following theorem [90]: If $f(x,\alpha)$ is continuous and has continuous partial derivatives with respect to α for $x \geq a$ and $\alpha_1 \leq \alpha \leq \alpha_2$, and if $\int_a^\infty \frac{\partial}{\partial\alpha} f(x,\alpha)\, dx$ converges uniformly in the interval $\alpha_1 \leq \alpha \leq \alpha_2$, and if a dose not depend on α, then

$$\frac{\partial}{\partial\alpha}\int_a^\infty f(x,\alpha)\, dx = \int_a^\infty \frac{\partial}{\partial\alpha} f(x,\alpha)\, dx.$$

We now consider the derivative of the Weyl expansion (5.11) with respect to x, y, and z. The last case gives

$$\begin{aligned}&\textstyle\int_0^\infty du \frac{\partial}{\partial z} J_0\left(k\rho\sqrt{1+u^2}\right)e^{-k|z|u}\\ &\quad = -\operatorname{sgn}(z)\, k \textstyle\int_0^\infty du\, u J_0\left(k\rho\sqrt{1+u^2}\right)e^{-k|z|u}.\end{aligned}$$

We notice that $\left|uJ_0\left(k\rho\sqrt{1+u^2}\right)e^{-k|z|u}\right| \leq \left|ue^{-k|z|u}\right|$. Moreover, it can be easily shown that $\lim_{u\to\infty} u^2 ue^{-k|z|u} = 0$ which implies $\left|uJ_0\left(k\rho\sqrt{1+u^2}\right)e^{-k|z|u}\right| \leq \left|ue^{-k|z|u}\right| < M(u) = \frac{1}{u^2}$ for sufficiently large u. Therefore, $\int_0^\infty du \frac{\partial}{\partial z}$ is uniformly convergent. Also, the integrand is continuous. All these requirement are valid for $\rho \geq 0$ and $z \neq 0$. We conclude then by the theorem stated above that $\int_0^\infty du \frac{\partial}{\partial z} = \frac{\partial}{\partial z}\int_0^\infty du$.

We now consider the derivatives with respect to x (the case with respect to y is essentially the same). It is possible to write

$$\begin{aligned}&\int_0^\infty du \frac{\partial}{\partial x} J_0\left(k\rho\sqrt{1+u^2}\right) e^{-k|z|u}\\&\quad = k\cos\varphi \int_0^\infty du\, \sqrt{1+u^2} J_1\left(k\rho\sqrt{1+u^2}\right) e^{-k|z|u},\end{aligned}$$

where the recurrence relation of the derivative of the bessel function was used. Again, from the properties of bessel functions that, $|J_1(x)| < 1$ for all positive real x, so we can write $\left|\sqrt{1+u^2} J_1\left(k\rho\sqrt{1+u^2}\right) e^{-k|z|u}\right| < \sqrt{1+u^2} e^{-k|z|u}$. From L'Hopital rule, we compute $\lim_{u\to\infty} u^2\sqrt{1+u^2} e^{-k|z|u} = 0$. It follows that $\left|\sqrt{1+u^2} J_1\left(k\rho\sqrt{1+u^2}\right) e^{-k|z|u}\right| < \sqrt{1+u^2} e^{-k|z|u} < M(u) = \frac{1}{u^2}$ for sufficiently large u and the Weierstrass-M test guarantee that the integral of the derivative is absolutely and uniformly convergent [90]. From the theorem stated earlier on the exchange of the derivative and integral operators, it follows that $\frac{\partial}{\partial x}\int_0^\infty du = \int_0^\infty du \frac{\partial}{\partial x}$.

4.7.3 Exchange of Order of Integrations in the Radiated Field Formula Via the Spectral Representation of the Dyadic Green's Function

We can exchange the order of integrations by using the following theorem from real analysis [90]: If $f(x,\alpha)$ is continuous for $x \geq a$, and $\alpha_1 \leq \alpha \leq \alpha_2$, and if $\int_a^\infty f(x,\alpha)\, dx$ is uniformly convergent for $\alpha_1 \leq \alpha \leq \alpha_2$, we conclude that $\int_{\alpha_1}^{\alpha_2}\int_a^\infty f(x,\alpha)\, dx d\alpha = \int_a^\infty \int_{\alpha_1}^{\alpha_2} f(x,\alpha)\, d\alpha dx$. Now, we already proved that the Weyl expansion converges uniformly. In addition, since the antenna current distribution is confined to a *finite* region it immediately follows by repeated application of the theorem above that we can bring the integration with respect to the source elements inside the spectral integral.

4.7.4 Derivation of the Rotation Matrix

We know that the matrix describing 3D rotation by an angle θ around an axis described by the unit vector $\hat{u}$ is given by

$$\begin{pmatrix} u_x^2 + e_x c & u_x u_y d - u_z s & u_x u_z d + u_y s \\ u_x u_y d + u_z s & u_y^2 + e_y c & u_y u_z d - u_x s \\ u_x u_z d - u_y s & u_y u_z d + u_x s & u_z^2 + e_z c \end{pmatrix}$$

with $c = \cos\theta$, $s = \sin\theta$, $d = 1-\cos\theta$, and $e_x = 1-u_x^2$, $e_y = 1-u_y^2$, $e_z = 1-u_z^2$. In order to rotate the z-axis into the location described by the radial vector

$\hat{r}$, we imagine the equivalent process of rotating the original coordinate system by an angle θ around an axis perpendicular to the unit vector $\hat{\rho}$ and contained within the xy-plane. Such an axis of rotation is described by the unit vector $\hat{u} = \hat{x}\sin\varphi - \hat{y}\cos\varphi$. Substituting these values to the rotation matrix above, the form given by (4.20) and (4.21) follows readily.

4.7.5 The Time-Dependent Interaction Poynting Theorem

Taking the inverse Fourier transform of equations (4.48) and (4.49), the following sets are obtained

$$\begin{gathered}\nabla \times \bar{\mathcal{E}}_{\text{ev}} = -\mu \frac{\partial}{\partial t}\bar{\mathcal{H}}_{\text{ev}},\ \nabla \times \bar{\mathcal{H}}_{\text{ev}} = \varepsilon \frac{\partial}{\partial t}\bar{\mathcal{E}}_{\text{ev}}, \\ \nabla \cdot \bar{\mathcal{E}}_{\text{ev}} = 0,\ \nabla \cdot \bar{\mathcal{H}}_{\text{ev}} = 0,\end{gathered} \tag{4.106}$$

$$\begin{gathered}\nabla \times \bar{\mathcal{E}}_{\text{pr}} = -\mu \frac{\partial}{\partial t}\bar{\mathcal{H}}_{\text{pr}},\ \nabla \times \bar{\mathcal{H}}_{\text{pr}} = \varepsilon \frac{\partial}{\partial t}\bar{\mathcal{E}}_{\text{pr}}, \\ \nabla \cdot \bar{\mathcal{E}}_{\text{pr}} = 0,\ \nabla \cdot \bar{\mathcal{H}}_{\text{pr}} = 0\end{gathered} \tag{4.107}$$

Take the dot product of the first curl equation in (4.106) by $\bar{\mathcal{H}}_{\text{pr}}$ and the second curl equation in (4.107) by $\bar{\mathcal{E}}_{\text{ev}}$, subtract the results. It is found that

$$\begin{aligned}\bar{\mathcal{H}}_{\text{pr}} \cdot \nabla \times \bar{\mathcal{E}}_{\text{ev}} - \bar{\mathcal{E}}_{\text{pr}} \cdot \nabla \times \bar{\mathcal{H}}_{\text{pr}} \\ = -\varepsilon\bar{\mathcal{E}}_{\text{pr}} \cdot \frac{\partial}{\partial t}\bar{\mathcal{E}}_{\text{pr}} - \mu\bar{\mathcal{H}}_{\text{pr}} \cdot \frac{\partial}{\partial t}\bar{\mathcal{H}}_{\text{ev}}.\end{aligned} \tag{4.108}$$

Similarly, by taking the dot product of the second curl equation in (4.106) by $\bar{\mathcal{E}}_{\text{pr}}$ and the first curl equation in (4.107) by $\bar{\mathcal{H}}_{\text{ev}}$, subtracting the results, we obtain

$$\begin{aligned}\bar{\mathcal{H}}_{\text{ev}} \cdot \nabla \times \bar{\mathcal{E}}_{\text{pr}} - \bar{\mathcal{E}}_{\text{ev}} \cdot \nabla \times \bar{\mathcal{H}}_{\text{ev}} \\ = -\varepsilon\bar{\mathcal{E}}_{\text{ev}} \cdot \frac{\partial}{\partial t}\bar{\mathcal{E}}_{\text{ev}} - \mu\bar{\mathcal{H}}_{\text{ev}} \cdot \frac{\partial}{\partial t}\bar{\mathcal{H}}_{\text{pr}}.\end{aligned} \tag{4.109}$$

Applying the vector identity, $\nabla \cdot (\mathbf{A} \times \mathbf{B}) = \mathbf{B} \cdot (\nabla \times \mathbf{A}) - \mathbf{A} \cdot (\nabla \times \mathbf{B})$, equations (12.31) and (4.109) become

$$\nabla \cdot \left(\bar{\mathcal{E}}_{\text{ev}} \times \bar{\mathcal{H}}_{\text{pr}}\right) = -\varepsilon\bar{\mathcal{E}}_{\text{ev}} \cdot \frac{\partial}{\partial t}\bar{\mathcal{E}}_{\text{ev}} - \mu\bar{\mathcal{H}}_{\text{ev}} \cdot \frac{\partial}{\partial t}\bar{\mathcal{H}}_{\text{pr}}, \tag{4.110}$$

$$\nabla \cdot \left(\bar{\mathcal{E}}_{\text{pr}} \times \bar{\mathcal{H}}_{\text{ev}}\right) = -\varepsilon\bar{\mathcal{E}}_{\text{ev}} \cdot \frac{\partial}{\partial t}\bar{\mathcal{E}}_{\text{ev}} - \mu\bar{\mathcal{H}}_{\text{ev}} \cdot \frac{\partial}{\partial t}\bar{\mathcal{H}}_{\text{pr}}. \tag{4.111}$$

Adding (12.33) and (12.36), and observing the Leibniz product rule in handling contributions of the RHS, equation (4.76) immediately follows.

4.7.6 On the Divergence of the Total Evanescent Field Energy with Fixed Axis of Decomposition

Expand the dyadic Green's function into evanescent mode along the z-direction by using (**??**) and then substituting the result into (16.12). The following generalized electromagnetic field expansion can be obtained

$$\begin{aligned}\mathbf{E}(\mathbf{r}) = \tfrac{-\omega k\mu}{8\pi^2}\int\limits_V d^3r'\int_{-\infty}^{\infty}\int_{-\infty}^{\infty}dpdq \\ \times\bar{\boldsymbol{\Omega}}(\mathbf{K})\cdot\mathbf{J}(\mathbf{r}')\,e^{ik\left[p(x-x')+q(y-y')+m|z-z'|\right]},\end{aligned} \tag{4.112}$$

where

$$\mathbf{K} = \hat{x}kp + \hat{y}kq + \hat{z}\mathrm{sgn}(z-z')\,km. \tag{4.113}$$

That is, we don't here interchange the order of the spectral and source integrals because the exterior region will generally contain points within the antenna horizon. By decomposing the field into evanescent and propagating parts, it is found that

$$\begin{aligned}\mathbf{E}_{\mathrm{ev}}(\mathbf{r}) = \tfrac{-\omega k\mu}{8\pi^2}\int_V d^3r'\int_{p^2+q^2>1}dpdq \\ \times\bar{\boldsymbol{\Omega}}(\mathbf{K})\cdot\mathbf{J}(\mathbf{r}')\,e^{ik\left[p(x-x')+q(y-y')+m|z-z'|\right]},\end{aligned} \tag{4.114}$$

$$\begin{aligned}\mathbf{E}_{\mathrm{pr}}(\mathbf{r}) = \tfrac{-\omega k\mu}{8\pi^2}\int_V d^3r'\int_{p^2+q^2<1}dpdq \\ \times\bar{\boldsymbol{\Omega}}(\mathbf{K})\cdot\mathbf{J}(\mathbf{r}')\,e^{ik\left[p(x-x')+q(y-y')+m|z-z'|\right]}.\end{aligned} \tag{4.115}$$

Next, a spherical region enclosing the antenna is introduced and denoted by $V(r_0)$, where r_0 is the radius of the sphere. The total evanescent (nonpropagating) energy is calculated using (4.79) with fixed direction of decomposition chosen along the z-axis, which gives after using (4.114)

$$\begin{aligned}W_{\mathrm{ev}}^e = \tfrac{\omega^2k^2\mu^2\varepsilon}{256\pi^4}\int_{V_{\mathrm{ext}}}d^3r\int_V d^3r'\int_V d^3r'' \\ \times\int_{p^2+q^2>1}dpdq\int_{p'^2+q'^2>1}dp'dq' \\ \times\bar{\boldsymbol{\Omega}}(\mathbf{K})\cdot\mathbf{J}(\mathbf{r}')\cdot\bar{\boldsymbol{\Omega}}^*(\mathbf{K}')\cdot\mathbf{J}^*(\mathbf{r}'') \\ \times e^{ik\left[p(x-x')+q(y-y')+m|z-z'|\right]} \\ \times e^{-ik\left[p'(x-x'')+q'(y-y'')+m'^*|z-z''|\right]},\end{aligned} \tag{4.116}$$

where $V_{\mathrm{ext}} = V_\infty - V(r_0)$ is the region exterior to the sphere $V(r_0)$. We still don't know if this integral will converge, so expression (4.116) should be considered a tentative formula. From physical grounds, it is expected that the calculation will face the problem of dealing with waves along a plane perpendicular to the z-axis. In such domains, the electromagnetic field expansion into evanescent modes along the z-axis consists actually of only *pure* propagating modes. As will be seen below, when the spherical coordinate system is employed in performing the space integral, there is indeed a convergence problem when the evaluation of the total energy approaches the critical xy-plane. In explicating this difficulty, it will be explicitly shown now that the limit of the total energy when $\theta \rightarrow \pi/2^{\pm}$ does not exist.

Assuming that the order of integrations in (4.116) can be interchanged (a justification of this assumption will be given later), we write after expressing the space cartesian coordinates in terms of spherical coordinates

$$
\begin{aligned}
W_{\text{ev}}^{e} = & \frac{\omega^2 k^2 \mu^2 \varepsilon}{256\pi^4} \int\limits_V d^3 r' \int\limits_V d^3 r'' \\
& \times \int_{p^2+q^2>1} dp dq \int_{p'^2+q'^2>1} dp' dq' \\
& \times e^{ik\left(p'x''+q'y''-px'-qy'\right)} \\
& \times \int\limits_0^{2\pi} \int\limits_0^{2\pi} \int\limits_{r_0}^{\infty} r^2 dr d\theta d\varphi \sin\theta \\
& \times e^{ik\left[\zeta r \sin\theta + m\left|r\cos\theta - z'\right| + m'\left|r\cos\theta - z''\right|\right]} \\
& \times \bar{\Omega}\left(\mathbf{K}\right) \cdot \mathbf{J}\left(\mathbf{r}'\right) \cdot \bar{\Omega}^{*}\left(\mathbf{K}'\right) \cdot \mathbf{J}^{*}\left(\mathbf{r}''\right),
\end{aligned}
\tag{4.117}
$$

where

$$
\zeta = \left(p - p'\right)\cos\varphi + \left(q - q'\right)\sin\varphi. \tag{4.118}
$$

We focus our attention now on radial integrals in the form

$$
I = \int\limits_{r_0}^{\infty} dr r^2 e^{ik\left[\zeta r \sin\theta + m\left|r\cos\theta - z'\right| + m'\left|r\cos\theta - z''\right|\right]}. \tag{4.119}
$$

In Figure 4.5(a), we illustrate the geometry of the problem needed in computing this integral. Here, two source points z' and z'' are required in the evaluation around which a change in the definition of the integrand occurs. The angle θ will determine the exact location of z' and z'' with respect to r_0. Also, implicit here is the angle φ which will generate the 3D pattern out of this plane.

To simplify the calculation, the integral (4.119) will be evaluated for the special case $z' = z''$. Also, it will be assumed that $r_0 \cos\theta < z'$. The motivation behind these assumptions is the anticipation of the result that the limit $\theta \rightarrow \pi/2$ does not exist. In this case, it is evident that in such a limit the radial vector $\hat{r}$ will meet the circle $r = r_0$ before any z'.

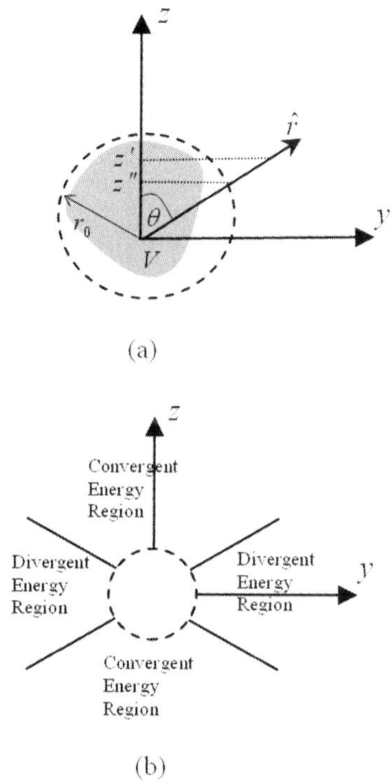

Figure 4.5 (a) The geometry behind the calculation of the space integral in (4.116). Here, the shaded region V refers to an arbitrary antenna current distribution enclosed within a fictitious sphere with radius r_0. In the figure, the two source points z' and z'' are chosen randomly. (b) The differentiation of the space V_{ext} exterior to sphere $V(r_0)$. The upper and lower regions correspond to convergent evanescent energy integrals while the left and right regions contain divergent evanescent energy. The z-axis can be freely rotated and hence the resulting total evanescent mode energy in the convergent two regions can acquired for the purpose of attaining a deeper analysis of the antenna near field structure. In both figures we show only the zy-plane section of the problem.

Integral (4.119) can be evaluated using the identity

$$\int x^2 e^{cx} dx = e^{cx} \left[\frac{x^2}{c} - \frac{2x}{c^2} + \frac{2}{c^3} \right]. \tag{4.120}$$

For $\theta = \pi/2$, (4.119) is divergent. For other angles, it is convergent since the radial dependence involves exponential decay through the imaginary constants m and m'.

Therefore, the integral with respect to θ in the tentative energy expansion (4.117) is ill-defined. The best we can do is to introduce an exclusion region $\pi/2 - \delta < \theta < \pi/2 + \delta$, and compute the evanescent field energy in the exterior regions, that is, the upper and lower regions $0 \le \theta \le \pi/2 - \delta$ and $\pi/2 + \delta \le \theta \le \pi$, both with $r \ge r_0$. In such a case, which is depicted in Figure 4.5(b), it is easy to prove that the energies computed in the upper and lower regions are finite. This follows from the fact that the fields in such regions are exponentially decaying with respect to r. Using (4.120), the corresponding infinite radial integral (4.119) is convergent. Moreover, by using an argument similar to Appendix 4.7.1, the same integral can be shown to be uniformly convergent. It follows then that the order of integrations with respect to the source and space variables can be interchanged because the former is finite. Also, since the Weyl expansion is uniformly convergent for $|z - z'| \neq 0$, the integrals with respect to the space variables and the spectral variables can be interchanged except at the plane $\theta = \pi/2$, which we have already excluded.[41] This formally justifies the general expression for the evanescent field energy, which now can be written as

$$\begin{aligned} W^e_{\text{ev}}(\hat{u}, \delta) = {} & \tfrac{\omega^2 k^2 \mu^2 \varepsilon}{256\pi^4} \int\limits_V d^3 r' \int\limits_V d^3 r'' \\ & \times \int_{p^2+q^2>1} dp dq \int_{p'^2+q'^2>1} dp' dq' \\ & \times e^{ik\left(p' x'' + q' y'' - p x' - q y'\right)} \\ & \times \left(\int\limits_0^{2\pi} \int\limits_0^{\pi/2-\delta} \int\limits_{r_0}^{\infty} dr d\theta d\varphi \sin\theta + \int\limits_0^{2\pi} \int\limits_{\pi/2+\delta}^{\pi} \int\limits_{r_0}^{\infty} dr d\theta d\varphi \sin\theta \right) \\ & \times r^2 e^{ik\left(\zeta r \sin\theta + m\left|r\cos\theta - z'\right| + m'\left|r\cos\theta - z''\right|\right)} \\ & \times \bar{\boldsymbol{\Omega}}(\mathbf{K}) \cdot \mathbf{J}(\mathbf{r}') \cdot \bar{\boldsymbol{\Omega}}^*(\mathbf{K}') \cdot \mathbf{J}^*(\mathbf{r}''). \end{aligned} \tag{4.121}$$

41 The shrewd reader will observe that in evaluating the integral (4.119), the integrand will meet with the singularities $|z - z'| = 0$ and $|z - z''| = 0$, at which the Weyl expansion is not uniformly convergent. However, since the radial integral clearly exists, its value is unchanged by the actual value of the integrand at the two *discrete* locations mentioned above. This is in contrast to the situation of radial integration at the plane $\theta = \pi/2$. In the latter case, the singularity $|z - z'| = 0$ is enforced at a *continuum* of points and so the interchange of integrations, together with all subsequent evaluations, are not justified.

It should be kept in mind that the last factor in this equation is function of z and z' as can be seen from (4.113). Here, we have emphasized the dependence of this energy expression on the exclusion angle δ. Also, since this energy depends on the direction of the axis of decomposition (in this particular example, it was chosen as the z-axis for simplicity), the dependance on this orientation is retained explicitly. The structure of an antenna near field can be analyzed by calculating the total evanescent energy for full azimuthal and elevation angle scan, with a suitable choice for δ. In this way, we have introduced what looks like a "near-field pattern," in analogy with the far-field radiation pattern.[42]

42 The expression (4.121) is complicated by the fact that the source and spectral integrals cannot be interchanged. In particular, rotation of the axis of decomposition $\hat{u}$ by a matrix $\bar{\mathbf{R}}$ cannot be simplified by effectively rotating the spectral vector $\mathbf{K}$ through the inverse operation. For this reason, it does not appear possible to gain further quick insight into the rotation effect on the evanescent energy as given above.

Chapter 5

The Scalar Antenna Near Field

5.1 INTRODUCTION

There has been a growing interest throughout the last few years in the topic of antenna near fields, motivated by both theoretical and applied concerns, where there appears to be a convergence toward more compact systems working at various spatial scales, for example nanoscale applications, metamaterials, miniaturized microwave and millimeter technology, etc [43], [35], [44]. On the other hand, the design and devolvement of large and complex antenna arrays force us to decrease the spacing between the elements, resulting in strong mutual coupling, a phenomenon that involves the near field of the radiating sources. The topic of electromagnetic energy has also been studied extensively in the last few decades, especially in connection with reactive energy, e.g., see the comprehensive paper [53]. In the new approach to the near field developed in Chapters 3 and 4, it was proposed that one of the most important aspects of the antenna field is its dynamic tendency to propagate or not propagate differently in different directions in space. The analysis there was based on the classic plane wave spectrum (Weyl) expansion (see [56] for background material) combined with a dynamic rotation of the local coordinate system in order to study the physical structure of the radiation problem. The analysis revealed the complexity of energy aspects in antennas, especially those related to localized and stored energy. However, it is possible using this approach to quantify many important features in the localization of the radiation field by computing the total nonpropagating part and comparing it with the total field. Due to the complexity of the analysis, no numerical study was presented in Chapter 4 for simple and exact special cases. We attempt here to focus on such smaller problems but within the perspective of the general insights obtained previously in the researches conducted by the authors.

The topic of this chapter is an investigation of the fundamental physical aspects of the near field produced by radiators comprised of scalar sources. The generalization to the full-wave vectorial case is more involved but relies on the insights developed after the study of the scalar case presented here. The overall structure of this chapter is the following. In Section 5.2, we first provide a general motivation for the study of the scalar theory. In Section 5.3, we formulate the problem starting from the vantage point of Chapter 4, which is reviewed here for self completeness. In particular, we show how the traditional Weyl expansion is combined with our dynamic approach to shed some light on the subtle manner in which the radiated field tend to propagate in different radial directions. In Section 5.4, we show that the radial field generated by a point scalar source, i.e., the radial Green's function, can be evaluated in simple analytical form when the origin of the coordinate system is located at the source itself. In Section 5.5, we provide in-depth analysis of the analytical results just obtained with comparison to numerical evaluation of the spectral integrals. Several interesting features are observed, for example the total vanishing of the propagating parts at certain spheres even in the far zone. The important topic of the interaction energy between the propagating and nonpropagating parts is also discussed. In Section 5.6, we give a brief indication of how to deal with the case of multiple sources, for example antenna arrays or continuous source distributions. Finally, we end with conclusions. The theory proposed in this chapter is applied extensively in Chapter 17 to develop new numerical analysis tools suitable for antenna design and development.

5.2 MOTIVATION FOR THE STUDY OF STUDY OF SCALAR NEAR-FIELD THEORY

Although the electromagnetic problem is strictly speaking never scalar, the scalar case provides a very attractive viewpoint to the topic at large and presents some very compelling advantages, which we briefly review here. To begin with, the vector Helmholtz equation, which governs the propagation of electromagnetic fields in the space surrounding the source, is obtained from the scalar Helmholtz equation, which governs the scalar problem, by updating the scalar field of the latter to the three

Cartesian components of the former [28]. Therefore, although the vector problem is considerably more complex, it is actually based on a the scalar case. Understanding the simpler theory provides therefore good foundations for working with the full-wave vectorial case. This approach has been favourite for many throughout the history of electromagnetic theory.

Another important advantage for studying the scalar theory is provided by the fact that many fundamental *vectorial* electromagnetic problems can be reformulated in a form that resembles the scalar case. That does not mean that a full-wave problem can be made equivalent to a scalar one, which is contradiction in terms, but that certain complex aspects in the vectorial case can be put in the easier form of the scalar one. We will give here a general example that illustrates this idea.

Consider a generic interaction problem between two current distributions $\mathbf{J}_1(\mathbf{r})$ and $\mathbf{J}_2(\mathbf{r})$ radiating in free spaces within volume supports V_1 and V_2. Suppose we are interested in studying the phenomenon of energy exchange between the two sources, for example, the problem of mutual coupling in antenna arrays. Any such a study will sooner or later involve the consideration of basic integrals of the following form

$$I = \int_{V_1} d^3r \int_{V_2} d^3r' \, \mathbf{J}_1(\mathbf{r}) \cdot \left(\bar{\mathbf{I}} + \frac{1}{k^2} \nabla\nabla \cdot \right) g(\mathbf{r}, \mathbf{r}') \cdot \mathbf{J}_2(\mathbf{r}'), \tag{5.1}$$

where $g(\mathbf{r}, \mathbf{r}')$ is the free-space Green's function (17.17). Physically, the direct meaning of the expression (5.1) reduces to understanding energy exchange between the antennas represented by $\mathbf{J}_1(\mathbf{r})$ and $\mathbf{J}_2(\mathbf{r})$ as mediated by the "channel" $\bar{\mathbf{G}}(\mathbf{r}, \mathbf{r}') = \left(\bar{\mathbf{I}} + 1/k^2 \nabla\nabla \cdot\right) g(\mathbf{r}, \mathbf{r}')$, i.e., the full-wave vectorial Green's function. However, it is possible to apply integration by parts in order to convert (5.1) into an expression containing only integrals of the form

$$I' = \int_{V_1} d^3r \int_{V_2} d^3r' \, \mathbf{J}_1(\mathbf{r}) \cdot g(\mathbf{r}, \mathbf{r}') \cdot \mathbf{J}_2(\mathbf{r}'), \tag{5.2}$$

$$I'' = \int_{V_1} d^3r \int_{V_2} d^3r' \, \nabla \cdot \mathbf{J}_1(\mathbf{r}) \, g(\mathbf{r}, \mathbf{r}') \, \nabla' \cdot \mathbf{J}_2(\mathbf{r}'). \tag{5.3}$$

Technically speaking, the trick is to move the differential operators in (5.1) from the free-space Green's function to the source. This is, for example, done routinely in method of moment formulations, for instance see [35].

From the physical point of view, the problem has not changed. However, from the computational aspect, the integrals (5.2) are much easier to work with. Indeed, for examining the problem of the near field, as will be done below, the analysis of the free-space Green's function $g(\mathbf{r},\mathbf{r}')$ into propagating and nonpropagating modes using (5.2) is much easier to perform than the original (5.1). *Therefore, understanding the scalar near-field problem can provide considerable help in tackling the interaction problem in general electromagnetic systems both computationally and theoretically.* For example, it will be shown in Section 5.4 that the radial total propagating and nonpropagating parts can be put in simple analytical forms and that there is here no need to perform numerical integration.

Finally, a strong motivation for the study of the scalar case is the relation with the far-field problem in the full-wave vectorial case. Indeed, the scalar Green's function (17.17) is simply the spherical wave representing the far field of any antenna [28]. While we are mainly interested here in studying the near field behavior of this wave, we will have occasions to say something interesting about its behavior in the far zone. On the other hand, the spherical wave (17.17) contributes to the near field of any antenna and becomes important even before reaching the far zone, for example in the intermediate field zone (see Chapter 3). Therefore, the insight developed here for the scalar can be considered a direct contribution to studying one aspect of the full-vectorial near field of any antenna.

A direct utilization of the proposal motivated by the expression (5.2) can be found in Chapter 17, where extensive numerical results and more examples beyond this chapter can be found there.

5.3 DEVELOPMENT OF THE RADIAL LOCALIZED NEAR-FIELD GREEN'S FUNCTION

The scalar problem is the one connected with establishing the scalar field $\psi(\mathbf{r})$ produced by a scalar source density $\rho(\mathbf{r})$ defined on a compact support V. The fields can be rigorously defined everywhere, including the source region, but in the present we are concerned mainly with the problem in the exterior region $\mathbb{R}^3 - V$. Therefore, we work with infinite, homogeneous, and isotropic space containing only a single source enclosed within the (possibly multi-connected) region V. In the time-harmonic case, the wave equation reduces to the scalar Helmholtz equation $\nabla^2\psi(\mathbf{r}) + k^2\psi(\mathbf{r}) = 0$. The scalar Green's function of this equation is given by the well-known expression [33]

$$g(\mathbf{r},\mathbf{r}') = \frac{e^{ik|\mathbf{r}-\mathbf{r}'|}}{|\mathbf{r}-\mathbf{r}'|}. \tag{5.4}$$

The importance of this Green's function arises from the fact that the radiated fields can now be expressed bt the following intuitive form

$$\psi(\mathbf{r}) = \int_V d^3r\, g(\mathbf{r}, \mathbf{r}')\, \rho(\mathbf{r}) . \tag{5.5}$$

We would like to further decompose the Green's function into two parts, one pure propagating and the other evanescent. This task can be accomplished by using the Weyl expansion [35]

$$\frac{e^{ikr}}{r} = \frac{ik}{2\pi} \int_{-\infty}^{\infty} \int_{-\infty}^{\infty} dpdq \frac{1}{m} e^{ik(px+qy+m|z|)}, \tag{5.6}$$

where

$$m(p,q) = \begin{cases} \sqrt{1-p^2-q^2} & ,p^2+q^2 \le 1 \\ i\sqrt{p^2+q^2-1} & ,p^2+q^2 > 1 \end{cases} . \tag{5.7}$$

The Weyl expansion shows that the total scalar Green's function can be divided into the sum of two parts, one as pure propagating waves and the other as evanescent, hence nonpropagating part. Explicitly, we write [35]

$$g(\mathbf{r}, \mathbf{r}') = g_{\mathrm{ev}}(\mathbf{r}, \mathbf{r}') + g_{\mathrm{pr}}(\mathbf{r}, \mathbf{r}'), \tag{5.8}$$

where the propagating and nonpropagating (evanescent) parts are given, respectively, by the expressions

$$g_{\mathrm{ev}}(\mathbf{r}, \mathbf{r}') = \frac{ik}{8\pi^2} \int_{p^2+q^2>1} dpdq \frac{1}{m} e^{ik\left[p(x-x')+q(y-y')\right]} e^{im|z-z'|}, \tag{5.9}$$

$$g_{\mathrm{pr}}(\mathbf{r}, \mathbf{r}') = \frac{ik}{8\pi^2} \int_{p^2+q^2<1} dpdq \frac{1}{m} e^{ik\left[p(x-x')+q(y-y')\right]} e^{im|z-z'|}. \tag{5.10}$$

By transforming the double integrals into cylindrical coordinates, making use of the integral representation of the Bessel function, we write the Weyl expansion in the following form [35]

$$g_{\mathrm{ev}}(\mathbf{r}, \mathbf{r}') = \frac{k}{4\pi} \int_0^{\infty} du J_0\left(k\rho_s\sqrt{1+u^2}\right) e^{-k|z-z'|u}, \tag{5.11}$$

$$g_{\mathrm{pr}}(\mathbf{r}, \mathbf{r}') = \frac{ik}{4\pi} \int_0^{1} du J_0\left(k\rho_s\sqrt{1-u^2}\right) e^{ik|z-z'|u}, \tag{5.12}$$

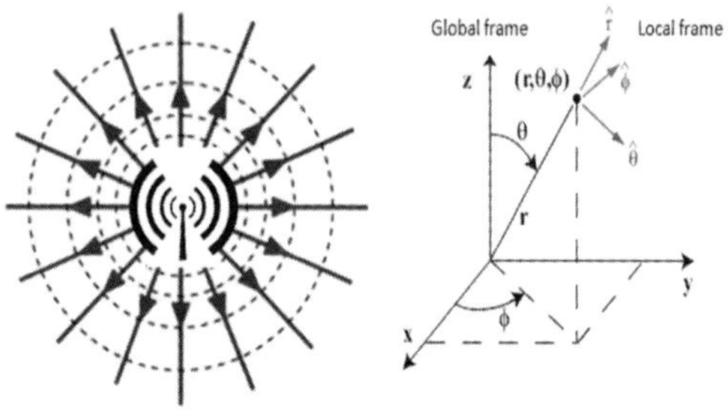

Figure 5.1 Radial Near Field. A generic antenna is shown in the left with the radial direction field along which a decomposition of the total field into propagating and nonpropagating modes is enacted. To the right, we show the notation of the coordinate systems used. The global coordinate system is the Cartesian xyz system. The local frame is given the notation of the spherical coordinates and the use of double primes is avoided for convenience in the notation.

where $\rho_s = \sqrt{(x-x')^2 + (y-y')^2}$.

We now introduce the radial localized near field of the radiation problem first proposed in Chapter 4. The main idea is to distinguish between three coordinate systems in the radiation problem:

1. Global coordinate system.
2. Source coordinate system.
3. Local coordinate system.

The global and source frames are the conventional ones used in electromagnetic theory, and are usually denoted by the unprimed and primed notation, e.g., $\mathbf{r}$ and $\mathbf{r}'$ respectively. In Chapter 4, it was found after a critical examination of the traditional approach (carried out in Chapter 3) that in order to adequately describe the engineering aspects of the radiation problem in the near field, there is a need to introduce a *third* coordinate system, the *local* frame, which was denoted there by double prime notation $\mathbf{r}''$. The physico-engineering significance of this new frame is quite straightforward and can be explained briefly as follows.

Consider a generic antenna system as shown in the left of Figure 5.1. As we now know, the field in the near zone can no longer be considered a pure propagating wave. Instead, it has the ambiguous character of being composed of both propagating and nonpropagating parts.[1] A Fourier analysis (i.e., via the Weyl expansion) of the spatial field is then required in order to decompose radiation into propagating and nonpropagating modes along any arbitrary chosen direction in space. This analysis, therefore, requires the specification of a direction at each point in the exterior region of the antenna problem along which the above mentioned decomposition into propgating/nonpropagating parts will be effectuated. The *radial* local frame is one natural choice that was pointed out in the general analysis of Chapter 4. As can be seen from Figure 5.1, it consists of choosing the direction field of the Fourier decomposition to coincide with the familiar radial field, i.e., at each location **r** in the exterior region, we attach the direction $\hat{r}$ and analyze the total field into propagating and nonpropagating modes along this vector. In this way, instead of the total field (which does not say anything about the inner dynamic structure of the near field), we obtain *two* different fields, the propagating radial near field, and the nonpropagating radial near field. The sum of the two will obviously give back the total field; however, we now obtain through this decomposition a firm grasp of how the original field structure splits dynamically as we move the local observation frame around the source.

The engineering motivations of this process are evident: We need to know how the field behaves around specific parts of the physical structure supporting the current distribution. By fixing the global frame on this rigid physical structure, and then rotating the local frame *with respect to the global frame*, we obtain in essence the relational dynamics of the problem in the sense that the tendency of the field to split into propagating and nonpropagating modes is described in reference to the physical body of the antenna itself. The three coordinate systems are then tightly connected with each other in a process that eventually leads to the explication of a physically real phenomena: the dynamical production of the propagating field with respect to the geometry of the source. It is the opinion of the authors that the conventional approach to energy and propagation in antenna systems using the methods of input impedance, reactive energy, and radiation density, are not adequate to deal with the problem in the way described above.

In Figure 5.1, we use the unprimed notation xyz to refer to the global frame. The local frame will be referred to by the usual spherical coordinates r, θ, and ϕ. However, *no double prime is used here*. The reader should be aware that the angles θ and ϕ refer to the *direction field* along which the dynamic decomposition into propgating/nonpropgating modes will be performed. No semantic confusion

1 In literature, this division is usually termed "static" and "radiated" fields. The difficulties of this approach, which is based on a hasty generalization of phenomena applicable only to small dipoles antennas, were criticized in Chapter 3.

between the typical spherical angles used to define position in $\mathbf{r}$ and these two angles above should be made, although their numerical values happen to coincide only in this case of radial localized near field.[2]

5.4 DERIVATION OF THE RADIAL LOCALIZED NEAR-FIELD GREEN'S FUNCTION FOR THE SCALAR PROBLEM

To proceed further, we need to write down the local frame coordinates explicitly in terms of the global frame. To do this, the rotation matrix (4.20) introduced in Chapter 4 is used. Using this matrix, we can express the local frame coordinates in terms of the global frame's using the following relations

$$\mathbf{r}'' = \bar{\mathbf{R}}(\theta,\phi)\cdot\mathbf{r},\ \ \mathbf{r}'_s = \bar{\mathbf{R}}(\theta,\phi)\cdot\mathbf{r}'. \tag{5.13}$$

It should be immediately stated that this rotation matrix will also rotate the $x''y''$-plane around the z''-axis with some angle. However, it was shown in Chapter 4 that the total propagating and nonpropagating modes are independent of rotation of the local frame around its z'-axis.

Since the scalar Green's function (17.17) is invariant with respect to rotation of the coordinate system, we replace $\mathbf{r}$ in (17.23) with $\mathbf{r}''$ using (5.13). This leads to

$$\frac{\exp(ikr)}{r} = \frac{ik}{2\pi}\int_{-\infty}^{\infty} dpdq\frac{1}{m}e^{i\mathbf{K}\cdot\left(\bar{\mathbf{R}}\cdot\mathbf{r}\right)}, \tag{5.14}$$

where the spectral variable (wavevector) is given by

$$\mathbf{K} = \hat{x}kp + \hat{y}kq + \hat{z}\mathrm{sgn}(z-z')\,km. \tag{5.15}$$

Here, sgn stands for the signum function. From the expressions of the rotation matrix elements (4.21)), we easily find that

$$\frac{\exp(ikr)}{r} = \frac{ik}{2\pi}\int_{-\infty}^{\infty} dpdq\frac{1}{m}e^{ikmr}. \tag{5.16}$$

We next further simplify this integral by converting the double integral into a single integral and then removing the singularity of the integrand. To achieve this,

2 The more general decomposition theorem, of which the radial localized field is only a special case, was discussed extensively in Chapter 4.

we work in the polar coordinates of the spectral domain, i.e., we employ $p = v\cos\varphi$ and $q = v\sin\varphi$. The integral (5.16) becomes then

$$\frac{\exp(ikr)}{r} = \frac{ik}{2\pi}\int_0^{2\pi}\int_0^{\infty} dv d\varphi \frac{v}{\sqrt{1-v^2}} e^{ik\sqrt{1-v^2}r}. \tag{5.17}$$

Since the integrand is independent of φ, we evaluate the outer integral and then separate the Green's function into two parts as in (5.8). The outcome is

$$g_{\text{ev}}^{\text{rad}} = ik\int_1^{\infty} dv \frac{v}{\sqrt{1-v^2}} e^{ik\sqrt{1-v^2}r}, \tag{5.18}$$

$$g_{\text{pr}}^{\text{rad}} = ik\int_0^{1} dv \frac{v}{\sqrt{1-v^2}} e^{ik\sqrt{1-v^2}r}. \tag{5.19}$$

Finally, perform the transformations of variable $u = -i\sqrt{1-v^2}$ and $u = \sqrt{1-v^2}$ in (5.18) and (5.19), respectively. We obtain

$$g_{\text{ev}}^{\text{rad}} = k\int_0^{\infty} du\, e^{-kur} = \frac{1}{r}, \tag{5.20}$$

$$g_{\text{pr}}^{\text{rad}} = ik\int_0^{1} du\, e^{ikur} = \frac{e^{ikr}-1}{r}. \tag{5.21}$$

These are the main results of the scalar theory of the radial propagating-evanescent Green's functions. In the remaining parts of this section, we explore some of their salient consequences.

5.5 DISCUSSION OF THE RESULTS AND THEIR PHYSICAL CONSEQUENCES

Before discussing the nature of the analytical expressions of the radial localized Green's functions derived above, we first present some numerical studies justifying the physical importance of the radial direction in describing the near field of the source. We consider a simple example comprised of a single point source located at the origin and ask how the radiate field tend to propagate at the point $(0, d, 0)$. Figure 5.2 presents the propagating scalar Green's function numerically when the rotation matrix has $\theta = \pi/2$. The variation with respect to ϕ are recorded at position $(0, 10\lambda, 0)$, hence in the far field of the source. It is interesting that at exactly the

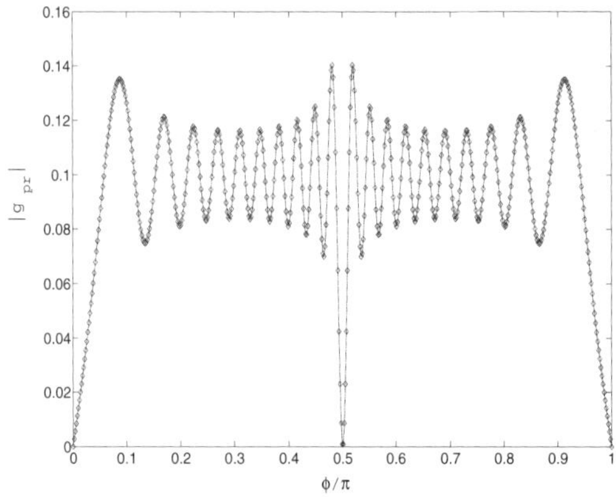

Figure 5.2 Spectral content of the propagation scalar Green's function observed at $(0, 10\lambda, 0)$ with $\theta = \pi/2$.

radial direction, in this case $\phi = \pi/2$, the total propagating part is zero! On the other hand, for directions around the radial orientation the propagating field has significant value. At a slightly different position $(0, 10.25\lambda, 0)$, Figure 5.3 shows that, contrary to the previous case, the propagating part has its most significant value concentrated at the radial direction. (The same results are also obtained for the position $(0, 9.75\lambda, 0)$.) This phenomena, the vanishing of the radial propagating part at some locations in space, will be explained below based on the analytical expressions (5.18) and (5.19). For the time being, we notice that most of the time the field tends to propagate maximally in the radial direction. For example, consider Figure 5.4 where this time we observe the spectral composition of the radiated field in the near-field zone. It is clear that energy can be found propagating in all directions, but the highest concentration of energy is located in the radial direction. *Our conclusion is that even though at certain discrete locations the radiated field has zero total propagating part, almost everywhere in space energy tends to flow maximally in the radial direction.* This provides some further justification for the theoretical and practical importance of the radial Green's function besides their obvious intuitive motivation.

Let us now examine carefully the structure of the radial Green's functions (5.18) and (5.19). Our purpose is to gain some insight into the physical settings of the problem by exploiting the analytical simplicity of the derived expressions.

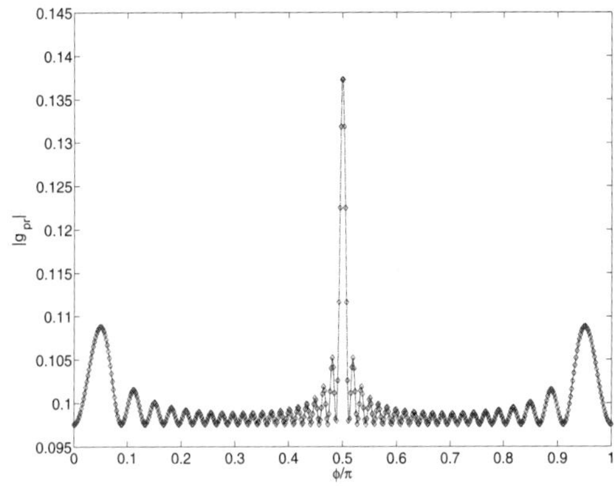

Figure 5.3 Spectral content of the propagation scalar Green's function observed at $(0, 10.25\lambda, 0)$ with $\theta = \pi/2$.

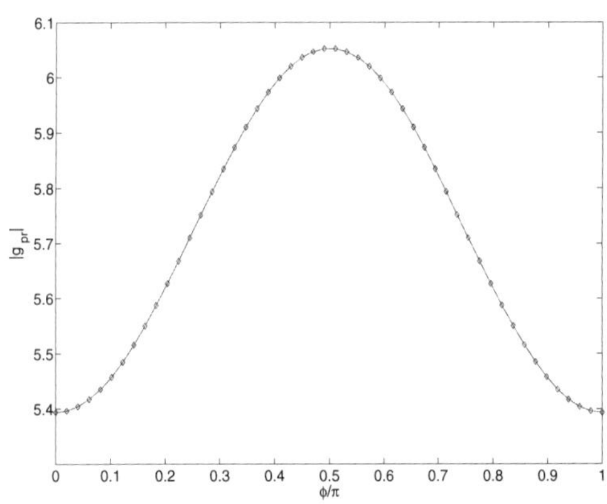

Figure 5.4 Spectral content of the propagation scalar Green's function observed at $(0, 0.15\lambda, 0)$ with $\theta = \pi/2$.

Spherical symmetry. Both scalar radial Green's functions are independent of θ and ϕ. This is expected since a point source in the scalar problem has no structure. Therefore, the problem is spherically symmetric. However, note that this applies only to the *radial* decomposition chosen in this chapter.

Independence of the wavenumber. A striking feature in the expression (5.20) is the fact that the dependence of the total radial evanescent field on k cancels out in the final formula. In particular, the localized near field is independent of frequency in the scalar problem. This shows that all scalar radiation problems possess a single universal structure of the nonpropagating field. This is connected with the next observation.

Connection between the evanescent field and electrostatics. The evanescent radial Green's function is simply (numerically speaking) the electrostatic Green's function. This suggests that there is deep connection between the electromagnetic near-field theory and potential theory.

Singularity of the radial propagating-evanescent Green's functions. The evanescent Green's function is singular at $r = 0$. This is, however, not the case with the propagating Green's function, which takes a nonzero value at the origin. Indeed, by L'hospital rule we have

$$\lim_{r\to 0} g_{\mathrm{pr}}^{\mathrm{rad}}(r) = \lim_{r\to 0} \frac{e^{ikr} - 1}{r} = ik \lim_{r\to 0} \frac{e^{ikr}}{1} = ik. \tag{5.22}$$

This shows that *the radiation field is already propagating right at the source location itself.* In other words, the field is created already in a propagating state! This result is surprising since one would expect intuitively that if the propagating part is finite at the source, it would approach the value zero there. Instead, the calculation (5.22) proves that the propagating field behaves differently in the neighborhood of the source.

Figure 5.5 shows the radial propagating scalar Green's functions. It is clear that the propagating part attains its greatest magnitude at $r = 0$, implying that not only the propagating field is nonzero at the source location, but actually it is maximal there.

Rate of decay. Another very interesting observation is that the rate of decay of both the propagating and evanescent scalar Green's functions is the same. Indeed, since $\exp(ikr) - 1$ is bounded, the function $g_{\mathrm{rv}}(r)$ is asymptotically like $1/r$. Therefore, both the propagating and evanescent fields behave in the far field as $1/r$. *The evanescent field contribute to the far field by the same level of magnitude contributed by the pure propagating part.*

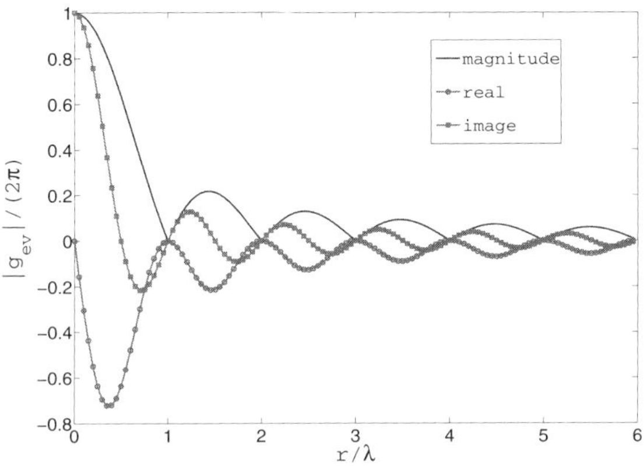

Figure 5.5 Radial propagating scalar Green's function.

Vanishing of the total propagating part. From (5.21) we see that the propagating field can vanish at any sphere

$$r = 2n\lambda, \ n \in \mathbb{N}. \tag{5.23}$$

This means that at point on any sphere with radius equals exactly an even multiple of λ, the totality of radial propagating modes all cancel each other. *On such a sphere, the radiation field is completely nonpropagating (evanescent) along the radial directions.*

Energy ratio. Consider now the energy density of the propagating and evanescent parts of the radiation field. We define the energy ratio Γ as

$$\Gamma := \frac{\left|g_{\mathrm{pr}}\left(r\right)\right|^{2}}{\left|g_{\mathrm{ev}}\left(r\right)\right|^{2}} = \left|e^{ikr} - 1\right|^{2}. \tag{5.24}$$

Figure 5.6 illustrates this energy ratio. It is clear that at the source the ratio is zero although the propagating part is not zero as we found above. This is possible because the evanescent field energy is infinite at the source. The fact that the limit of the ratio is zero restores some of the intuitive picture expected in radiation problem. *That is, although the individual total propagating field is nonzero at the source, its relative strength compared with the total evanescent field is zero.*

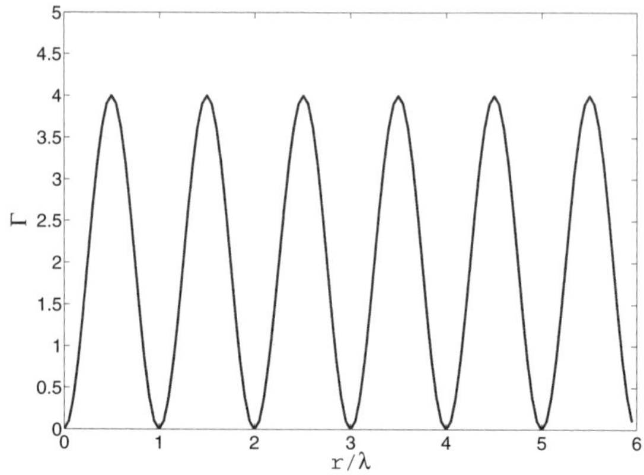

Figure 5.6 The energy ratio between the radial propagating and evanescent energy densities for the scalar problem.

The truly interesting feature of the energy ratio Γ is that the vanishing of the total propagating part extends to the far field zone as was already observed numerically in Figure 5.2. Indeed, any even multiple of λ satisfies the condition (5.23) and hence even deep in the far zone, one still encounters spheres where there is no total propagating mode. This is somehow contrasting with the naive common sense picture we usually draw about radiation problems where it is expected that the field is "totally propagating" in the far zone. Notice that the total field on the spheres $r = n\lambda$ is not zero. Actually, the radiation field of the scalar problem is never zero. Only the total pure propagating part is zero. This is again connected with the fact established above that in the far zone, both the propagating and evanescent modes are at the same level of magnitude.

Interaction energy. Consider now the total energy density in terms of the propagaing and evanscent parts. Simple calculations shows that

$$|g(r)|^2 = \underbrace{\frac{1}{r^2}}_{\substack{\text{self}\\ \text{evanscent}\\ \text{energy}}} + \underbrace{\frac{2}{r^2}\mathrm{Re}\left\{e^{ikr}-1\right\}}_{\text{interaction energy}} + \underbrace{\frac{\left|e^{ikr}-1\right|^2}{r^2}}_{\substack{\text{self propgating}\\ \text{energy}}} . \tag{5.25}$$

Therefore, the interaction energy between the propagating and evanescent parts is given by the simple expression

$$\mathcal{E}_{\text{int}} := \frac{2}{r^2}\text{Re}\left\{e^{ikr} - 1\right\} = \frac{2}{r^2}\left(\cos kr - 1\right). \tag{5.26}$$

Figure 5.7 shows the variation of the interaction energy (5.26) with distance. It is immediately observed that $\mathcal{E}_{\text{int}} \leq 0$ for all r. This is not surprising, since the self energy density of the radial evanescent part is actually equal to the total field energy, necessitating that when the propagating and evanescent parts mutually interact with each other, they both loss energy.

Notice that $\mathcal{E}_{\text{int}} = 0$ for all $r = 2n\lambda$, where n is a natural number. This is in harmony with the fact that at these spheres, the propagating part is zero so the interaction energy is trivially also zero. However, at the spheres $r = (2n+1)\lambda$, the interaction energy reaches its smallest value at $\mathcal{E}_{\text{int}} = -2/r^2$. In this case, the total propagating field energy density is equal to twice the total energy $1/r^2$.

Such observations suggest that care should be taken when attempting to interpret the evanescent part of the total field. It is usually assumed that the energy "stored" in the field is located in the nonpropagating part of this field. However, the simple calculation above points to opposite conclusions, for total energy seems to be distributed not only among the propagating and evanescent modes, but also in the interaction zone between the two. Now, although the evanescent modes for arbitrary (extended or nonpoint) source distribution will tend to be localized around the antenna, the propagating modes obviously exist everywhere. The fact that part of the total energy is always allocated to the *interaction* between purely propagating field and a localized one seems to imply that part of the *stored* energy may be related somehow to propagating waves escaping to infinity! This is another evidence indicating the difficulty in dealing with stored energy using a time-harmonic theory.[3]

Total radial evanescent energy. From the energy density expressions we can compute the total energy in any finite volume by simple integration

$$W_{\text{ev, pr}} = \int_V d^3r\, \mathcal{E}_{\text{ev, pr}}(r), \tag{5.27}$$

Consider a spherical volume bounded between the radii $r = a$ and $r = b$. The total electric radial evanescent energy in this region can be readily calculated as

$$W_{\text{ev}} = 4\pi\,(b - a). \tag{5.28}$$

3 In Chapter 4, the authors suggested that stored energy cannot be properly understood without a transient (time-dependent) analysis of the antenna system.

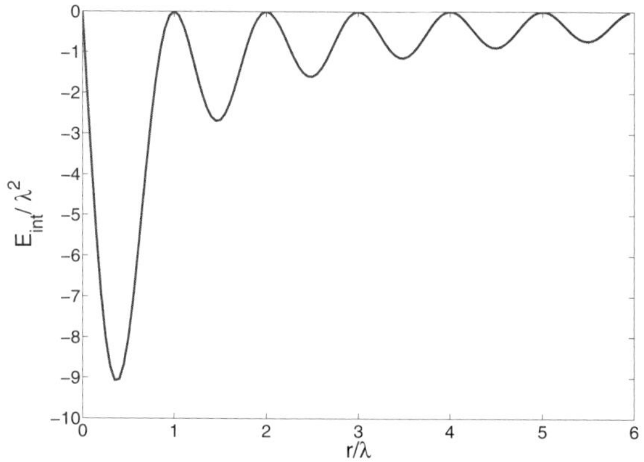

Figure 5.7 The interaction energy between the radial propagating and evanescent energy densities for the scalar problem.

We note that $W_{\rm ev} = 4\pi b$ at $a = 0$, which is finite, although the evanescent field itself is infinite at the source. However, as $b \to \infty$, the total energy diverges. Therefore, *the total radial evanescent energy is infinite, even though the individual modes are pure evanescent and decays exponentially at infinity.*

The total radial propagating energy is computed in the following form

$$W_{\rm pr} = 8\pi \left[b - a - \frac{1}{k} \left(\sin kb - \sin ka \right) \right]. \tag{5.29}$$

The limit does not exist in an infinite domain as should be the case with pure propagating modes (i.e., total propagation energy is always infinite in any radiation problem). The total interaction energy is simply the negative of the total propagating energy (5.29)

5.6 THE CASE OF MULTIPLE SCALAR SOURCES

Knowledge of the scalar radial Green's functions in the closed analytical form (5.18) and (5.19) derived in Section 5.4 permits us to compute the near field decomposition due to an arbitrary source. However, there is a subtlety in the multiple point source case that need to be clarified first. We noted in Section 5.4 that the derivation of the

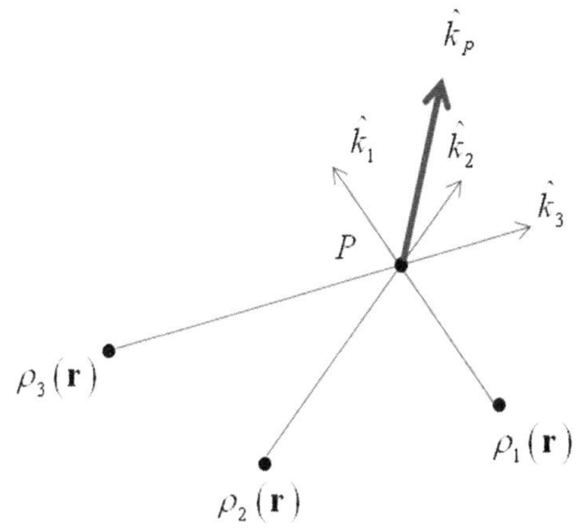

Figure 5.8 A diagram of three sources producing a near field at point P. Each point source will contribute a total propagating field along the direction specified by the line joining it with the observation point. A rough estimate of the direction of the resultant sum is the vector $\hat{k}_P$, which can be considered an average direction.

closed form expressions above is possible only when the origin of the coordinate system is chosen to coincide with the point source location. Obviously, when there are two or more point sources, this is not possible simultaneously. Therefore, the physical situation appear a bit more complicated in this case.

Before giving the mathematical expressions of the near field decompositions, we first consider the problem as illustrated graphically in Figure 5.8. We have three point sources $\rho_1(\mathbf{r})$, $\rho_2(\mathbf{r})$, and $\rho_3(\mathbf{r})$ and it is desired to know how the near field behaves at point P. When the method of Section 5.4 is used to compute the the propagating and evanescent parts of the field due to each source taken individually, the radial contributions of the sources $\rho_1(\mathbf{r})$, $\rho_2(\mathbf{r})$, and $\rho_3(\mathbf{r})$ will appear as $\hat{k}_1$, $\hat{k}_2$, and $\hat{k}_3$, respectively. By invoking the physically intuitive assumption that the net near field tendency to propagate or not propagate is the superposition of the separate sources taken individually, we may assume that the total propgating/nonpropgating parts of the near field at point P will be described with respect to the vector $\hat{k}_P$, which is the unit vector parallel to $\hat{k}_1 + \hat{k}_2 + \hat{k}_3$. This is to be understood, however, only as an *average* direction of energy flow.

If we have a continuous source in a volume V, the expressions of the total propagating and nonpropagating parts follow from the expressions (5.18) and (5.19)

$$\psi_{\mathrm{ev}}(\mathbf{r}) = \int_V d^3r' \frac{\rho(\mathbf{r}')}{r}, \tag{5.30}$$

$$\psi_{\mathrm{pr}}(\mathbf{r}) = \int_V d^3r' \frac{\rho(\mathbf{r}')\left(e^{ikr}-1\right)}{r}. \tag{5.31}$$

The unit vector of "most likely propagation" is estimated as

$$\hat{k}_P = \frac{\int_V d^3r' \,(\mathbf{r}-\mathbf{r}')}{\left\| \int_V d^3r' \,(\mathbf{r}-\mathbf{r}') \right\|}. \tag{5.32}$$

Therefore, the direction along which the near field tends to "maximally" propagate (or localize) depends on the geometric configuration of the relative positions of the radiating point sources with respect to each other. Evidently,, *each source current distribution will produce a near field localization pattern that reflects the shape of the radiator.*

Finally, we mention that it is possible to improve the estimation (5.32) by inserting a "weight" at each propgation/nonpropagation direction proportional to the magnitude of the total propagating part at the point under consideration. In this case, we can replace (5.32) with

$$\hat{k}_P = \frac{\int_V d^3r' \,|\psi_{\mathrm{pr}}(\mathbf{r})|\,(\mathbf{r}-\mathbf{r}')}{\left\| \int_V d^3r' \,|\psi_{\mathrm{pr}}(\mathbf{r})|\,(\mathbf{r}-\mathbf{r}') \right\|}. \tag{5.33}$$

The use of the absolute value of $\psi_{\mathrm{pr}}(\mathbf{r})$ in (5.33) is intentional and aims at eliminating any cancellation from the total sum. The reason is that from the energy viewpoint, a negative or positive total propagating parts are the same. A negative amplitude of a plane wave does not imply that the direction of propagation is reversed. Consequently, in computing the average direction of energy flow, it is more appropriate to use only positive values.

5.7 CONCLUSION

The chapter presented a study of the near-field structure of the basic problem of scalar point sources radiating in free space. The topic was approached from the new perspective of the dynamic structure of the near field based on careful spectral analysis of radiation using the plane wave spectrum combined with a rotating coordinate frame of reference. The analysis focused on the radial flow of energy given its intuitive appeal. We motivated this choice by some numerical examples indicating that most of the energy tend to move maximally in the radial direction. The radial Green's function of the scalar problem was then derived in simple analytical closed form for both the propagating and nonpropagating parts. We studied in depth the physical structure of the radiation problem revealed by these expressions and documented some interesting observations. Finally, a quick formulation for the array or multiple source problem was outlined, which may be used in the future for near-field shaping and focusing applications.

Chapter 6

Morphogenesis of Electromagnetic Radiation in the Near-Field Zone

6.1 INTRODUCTION

6.1.1 Philosophical Resume

The guiding thread in our investigation is the search for a suitable theoretical level at which general knowledge about the antenna can be summarized in the most economical form. By observing the pattern of the evolution of the physical and mathematical sciences throughout the last one hundred year, it appears to us that the most fundamental level at which a scientific investigation can be enacted is the *topological* one.

There are two levels of structures that may be exploited in formulating a general theory of the antenna near field. The first is the *topological layer*. It consists of systematic abstraction from the real number system used in representing electromagnetic phenomena within the framework of Maxwell's equations. The second level is the *physical layer*. This consists of any input arising from observing the way physical quantities actually behave when represented by quantities selected from the topological layer. For example, we mention here the invariance properties of the electromagnetic fields, which are expressions of the *continuous* symmetry of the underlying topological base structure when equipped with additional structures like differentiability and geometry.[1]

Our theoretical program here is to start from the physical layer and build the way back to the highest abstract level, that of the topological layer. This task is accomplished in the following manner. First, the rotational invariance of scalar sources in electromagnetic theory are observed.[2] We exploit this symmetry in the interesting case where there are several point sources co-existing next to each other. The rotational symmetry of each source is broken by choosing a particular orientation of the coordinate system along which we attempt a mathematical description of the electromagnetic near field.[3] We therefore move from perfect (rotational) symmetry to superposition of a multiplicity of broken symmetries, all coexisting and hence resonating with each other. This picture was described in detail in Chapter 4 from the *engineering* point of view, while the fundamental level will be treated comprehensively elsewhere. This chapter will provide an introduction to this latter more fundamental way of looking at the near field.

In such a more advanced stage, we will derive a system of nonlinear ordinary differential equations describing the actual dynamic structure of the field. We then search for the *singularities* of this system (critical points in the phase space of the problem). It is the topological structure of these singularities that is, in our opinion, the broadest abstract level in the scientific description of antennas. Since singularities are eminent points in the phase space structuring *virtual* solutions of the problem starting from given initial conditions, a topological description of the distribution of such singularities is naturally the best general way so far to encode all the relevant information of a given antenna. Such an encoding provides a deeper understanding of how the current distribution may vary in order to effect certain structural changes in the near fields, or how sensitive is the whole antenna system's radiation field to a continuous change in some antenna parameters of interest, etc. Once this topological understanding of the antenna problem is attained, we can work our way again toward the physical layer, and then back again to the topological layer in light of new questions, problems, and proposals. *It is this immanent resonant interrelation that communicates back and forth between the topological and physical layers what should be taken as the ultimate object of*

1 The rotational and translational symmetries of the electromagnetic fields are consequences of modeling physical phenomena by the topology of Euclidean spacetime. However, it is *only* through *observations* that we come to know that phenomena appearing in the real world are indeed well-behaved with respect to certain general symmetry principles.

2 Notice again that this is an expression of a structure at the physical layer.

3 As will appear later, this 'mathematical description' is chosen to be a decomposition of the total field into propagating and nonpropagating parts.

any scientific theory. A physical theory sensitive to both the fundamental structural principles and the operational behavior of radiating electromagnetic systems must address this dual mode of theoretical description.

6.1.2 General Scope of this chapter

The purpose of this chapter is presenting a brief outline of a general mathematical model of electromagnetic radiation and energy flow in the source region viewed as a *space-time process*. The familiar Poynting flow can be best interpreted as a *hydrodynamic* phenomenon occurring in some yet undiscovered foundational fluid dynamic formulation of Maxwell's theory. It is well known that following the advent of special relativity in 1905, there has been a general shift away from models involving a "substantial ether" serving as substrates underlying electromagnetic phenomena, mainly due to the justified objection that such an invisible media are clearly incapable of being uniquely singled out by any known experimental apparatus. However, by accepting the electromagnetic field as simply a mathematical function on manifolds, we may study the problem of energy flow by resorting only to considerations imposed by the structure of Maxwell's equations themselves. Such an approach will be adopted in this work, where we model space-time flow as the dynamic splitting of the near field into propagating and nonpropagating modes as observation moves progressively away from the source region.

The definition of 'propagation' here is based on the most simple consideration of what constitutes a moving wave. More specifically, by performing a Fourier analysis on the field, each resulting spectral component is seen to be either pure propagating mode $\exp(ix - iut)$ or a nonpropagating mode in the form of an exponential function $\exp(x - ut)$, where x plays the role of space, t the time, and u the speed, all real quantities. As was discovered in Chapter 4, the freedom allowed in choosing the orientation of the coordinate system in the free space Green's function (what was termed *local* frame) is the essence of the *dynamic* structure of the electromagnetic field. The reason is that this local orientation, fully captured by the Lie group SO(3), (the 3-dimensional rotation group), is the "space of additional forms" that appears to emerge in the far zone although *apparently* absent in the near zone. Our analysis suggests that for the purpose of characterizing the dynamic structure of radiation, every electromagnetic field must not only be labeled by a position in the 3-dimensional Euclidean space $\mathbb{R}^3$, but also by an element of SO(3) as much. Viewed in this form, the effective configuration space of the electromagnetic radiation problem becomes $\mathbb{R}^3 \times \mathrm{SO}(3)$, which is a 6-dimensional non-Euclidean manifold with a nontrivial but well-understood topology.

The empirical content of the proposed theory lies in the discovery of the existence of a *scalar* field on the configuration space $\mathbb{R}^3 \times \mathrm{SO}(3)$ characterizing the dynamic splitting of the radiation field into propagating and nonpropagating modes. This scalar field is what we call the *propagation potential* of the antenna. It is simply the relative ratio of the energy of the propagating field to the energy of the total field. Its precise construction will be detailed in this chapter. By computing the *gradient* of this potential function with respect to a suitable Riemannian metric, a new *vector* field is obtained, the *morphogenetic field*, and from this it follows the ordinary differential equations on the 6-manifold $\mathbb{R}^3 \times \mathrm{SO}(3)$ controlling the associated gradient flow.

6.2 THE FUNDAMENTAL PROBLEM OF MORPHOGENESIS IN ANTENNA THEORY

What is precisely the sense of that "profound mystery" associated with the creation of the antenna's radiation pattern? Why do we need to undertake a painful fundamental investigation within the framework of *morphogenesis* in order to understand the production of radiation fields in general antenna systems? Is there, to begin with, a genuine problematic underlying our understanding of the radiation field aside from the common knowledge of the general expressions of the antenna's field, which have been available for long time?[4] The answer to such questions requires some preliminary reflection on the terms implicitly employed for framing these and similar inquires. We start by noticing that the radiation *pattern* is essentially a *form*. Furthermore, it is a *spatial* form, amenable to mathematical description, in particular using the geometric language of differential manifolds [59]. The 'form' under consideration here is simply the mere fact that the radiation field, the physical content of the electromagnetic field in the antenna's far zone, exists mostly in the state of *propagating* modes, i.e., the mathematical expression of the fields tend to appear as a *traveling wave*.[5]

It has been well known for a long time that the near field, the field generated by the antenna observed at an electrically small distance from the antenna, does not behave as a propagating mode [33], [37], [35]; instead, it appears "static," to use an inaccurate term but adequate for our present purpose, i.e., describing the antenna near field as *non*-propagating or simply "not moving." By attributing 'form'

4 Indeed, available in the form of the *dyadic Green's function*, which *is* known in analytical closed form.

5 For an elementary introduction to the mathematical concept of traveling waves, see [91]. For a classical exposition, see [89].

to the existence of propagating modes in the spectral content of the field, we can now say that the near field *lacks form*, while the far field possesses a plenty of an unambiguous and definite form. It is then legitimate to ask the following question: If form was lacking in the early stage of the field (the near field), from where did it emerge later in the far field? Was it "imposed" from the "outside," or was it somehow originally hidden invisibly in the near field? And in any case, what are the mechanisms responsible for the creation of this form? The fundamental problem of morphogenesis is finding satisfactory scientific answers to these questions. *The task is to provide an immanent* (in contrast to *transcendent*) *explanation of the genesis of form, i.e., an objectively scientific explication of the mechanisms of the formative processes unfolding the near field into the far field.* In the language of the antenna system, we need to figure out the mechanism of the production of propagating modes from nonpropagating modes.

Throughout our investigations, it will turn out that the research program of morphogenesis is extremely fruitful for *both* the theoretical *and* applied points of view. Indeed, the epigenetic theory to be developed in this chapter appears to be naturally attuned to *practical* considerations while at the same abstract level where it expresses itself as a *fundamental* theory. This should not be a great surprise given that the engineering sciences are directly concerned with the problem of *form*, probably in a way more striking that the situation with the physical sciences. In the latter, form manifests itself as the mathematical harmony disclosed in a newly established law of nature. On the other hand, engineering is the science of creating, manipulating, transporting, processing, and even destroying various forms, shapes, structures, and so on. Communication theory is concerned with the transmission of form as information. Power engineering is interested in transporting energy in the organized form of electric current (rather than the disorganized from of, for example, heat). Computer engineering builds complex machines capable of processing and manipulating information in multiple forms. Construction engineering builds highly organized structures by arranging matter in rigid form. And so on. Physical sciences strive to *understand* form, but engineering sciences aim to *control* it. However, it is obvious from the history of science that the best way to control something is to understand how it works. This probably explains why technology has been in a sense *prior* to basic science in the historical context of not only the modern era, but even that dating backs to antiquity [169].[6]

Deeper knowledge of how a form is produced can be utilized in a variety of methods designed later by practitioners to channel the process of devolvement (the genesis of form) such that desirable particular end results may be obtained. This view can be better appreciated when reflecting on how difficult it is to change a particular feature in a mature organism by crude manipulation of its early embryo. If the process of biological devolvement is not fully understood, the final outcome of such a manipulation cannot be anticipated in an accurate fashion. This observation goes actually beyond a mere analogy; later in the chapter, we will propose a tentative scheme of "biological development" as the most natural framework for casting a reasonable theoretical interpretation of the mathematical equations governing the growth of the antenna radiated fields.

6.2.1 Propagation Model for the Antenna Fields

We need now to move toward a concrete implementation of the general program outlined above. The concept of *form*, with its essentially wider semantic scope, must now be narrowed down to a simpler level amenable to mathematization. By *form* we now focus our attention on that aspects pertaining to *propagation of the field*, i.e., the character of the field as a *moving function over space in time*. Mathematically speaking, we describe propagation by giving a a vector $\mathbf{k}$, with direction providing information about the direction of propagation, while its magnitude supplies information about the "speed of motion." The idea we suggest here is a generalization of the approach developed in Chapter 4, where we studied the structure of the electromagnetic field in the spectral domain by means of disclosing for the distribution of wavevectors k in the field structure. This implies that a Fourier analysis of the three-dimensional field function over space has to be performed. It is well known in literature that this analysis, when applied to functions satisfying the Helmholtz equation, will lead to the Weyl expansion of the field generated by a scalar point source [94], [35]. We are not going to repeat the basic derivation here, but instead refer the reader to the standard literature on the subject. In what follow, the general conceptual essence of the subject will be

6 A classical demonstration of this observation in the modern era can be looked up in any good book on the history of 19th-century science. The practical need to design better steam engines in order to sustain the rising demands of the Industrial Revolution was *the* stimulus leading to the creation of fundamental theory of modern thermodynamics. Also, the prospects opened up by the electrical sciences for achieving long-distance communication was noticed very early, as can be found in the famous "magnetic communication" experiment conducted by Gauss and Weber.

presented here in a rather simplified form, that pertaining to the *scalar* theory, not the fully electromagnetic complete picture.

To be more precise, let us assume that we are given a scalar field $\psi(\mathbf{r})$ generated by a point source and satisfying the scalar Helmholtz equation. For the well-behaved fields of mathematical physics, it is always possible to perform a Fourier analysis by writing the Fourier integral

$$\psi(\mathbf{r}) = \int_{-\infty}^{\infty} d^3k\tilde{\psi}(\mathbf{k})e^{-i\mathbf{k}\cdot\mathbf{r}}. \tag{6.1}$$

Here, the orientation of the coordinate system along which the spectral analysis is performed is chosen at random. Of course, by changing this orientation, the Fourier transform $\tilde{\psi}(\mathbf{k})$ itself will change, but the value of the field observed at the same spatial location $\mathbf{r}$ remains the same, and this applies to any well-behaved field, i.e., not only the electromagnetic and acoustic fields. There is nothing remarkable about this observation, which we consider to be devoid of any original physical content. However, the situation changes dramatically when we take into consideration the fact that the scalar field *does* satisfy Helmholtz equation. In this case, it is possible to show that the scalar Green's function Fourier integral can be reduced into a two-dimensional integral (the classic plane-wave spectrum of Weyl's expansion). The reduced integral, moreover, has the curious property that it is the sum of two parts, one in which only *propagating* modes in the z-direction are admissible, and the other where only *evanescent* (exponentially decaying) modes can be included. When the total radiated field produced by arbitrary source is expressed in terms of the actual source and the scalar Green's function, it follows at once that the total field is written as

$$\psi(\mathbf{r}) = \psi_{\text{pr}}(\mathbf{r}) + \psi_{\text{ev}}(\mathbf{r}), \tag{6.2}$$

where $\psi_{\text{pr}}(\mathbf{r})$ and $\psi_{\text{ev}}(\mathbf{r})$ stand for the propagating and evanescent (nonpropagating) parts, respectively.

Up to now, this is a standard knowledge in classic field theory (the mathematical formalism of acoustic and electromagnetic fields). What has to be emphasized here is the *choice* of the orientation of the observation frame along which the decomposition (6.2) was conducted. If the observation frame in the plane wave spectrum analysis of the scalar Green's function was rotated, the respective values of the total propagating and nonpropagating parts will change too, while of course their sum will always give back the original total value of the field. Now, we notice that the propagating part represents the total sum of all plane waves propagating along the z-direction. The evanescent part, being an exponentially decaying field, represents clearly the nonpropagating field along the same direction. There is, therefore, an unambiguous physical meaning to be attached to both fields. *They serve to delimitate the propagation structure of the field with respect to a given observation frame.* What about other observation frames? Here we should immediately mention that it is not the field itself what is being "observed," but instead the scalar Green's function itself. Indeed, there is only one global frame used to write the position vector $\mathbf{r}$ of the field, but a *new* observation frame has to be introduced in order to perform the spectral analysis leading to the decomposition of the fields into propagating and nonpropagating modes similar to (6.2). The details where provided in Chapter 4, but the general idea is simple and states that the field decomposition can now be written as

$$\psi(\mathbf{r}) = \psi_{\mathrm{pr}}\left(\mathbf{r}, \bar{\mathbf{R}}\right) + \psi_{\mathrm{ev}}\left(\mathbf{r}, \bar{\mathbf{R}}\right). \tag{6.3}$$

Therefore, the quantities $\psi_{\mathrm{pr}}\left(\mathbf{r}, \bar{\mathbf{R}}\right)$ and $\psi_{\mathrm{ev}}\left(\mathbf{r}, \bar{\mathbf{R}}\right)$ represent the propagating and nonpropagating parts, respectively, while the local observation frame is oriented using the rotation matrix $\bar{\mathbf{R}} \in \mathrm{SO}(3)$. The explicit functional dependence of the propagating and nonpropagating parts on $\bar{\mathbf{R}}$ serves to indicate that these quantities are not unique and that the total field can be Fourier analyzed in different ways according to the orientation of the observation frame with respect to the source distribution. Indeed, what attaches a special physical significance to this fact is that the aforementioned dependence is completely controlled by the spatial distribution of the source. In other words, the non-uniqueness of the Fourier analysis, in the sense understood above as decomposition into propagating and nonpropagating modes along a certain axis, is not a mathematical artefact, but a consequence based on the physical distribution of the source responsible of producing the radiated field and applies only to certain class of fields, i.e., those satisfying Helmholtz equation.[7]

It is a good idea to summarize the particular propagation model we chose in our work as the *form* of the antenna radiated fields under study. We provide the following procedure

1. Fix a global coordinate frame $\mathbf{r} = \langle x, y, z\rangle$.
2. Generate a rotated local frame $\mathbf{r}'' = \langle x'', y'', z''\rangle = \bar{\mathbf{R}} \cdot \mathbf{r}$. Here, $\bar{\mathbf{R}}$ is a general rotation matrix.
3. Along the z''-axis of the local frame, expand the fields into a sum of propagating and nonpropagating modes.

7 There is the interesting question of whether the idea of field decomposition into propagating and nonpropagating modes can be developed for other equations of mathematical physics, which we leave open here.

This process is illustrated graphically in Figure 6.1a. There we show the rotated local frame $x''y''z''$ sharing the same origin with the global frame xyz. The source frame $x'y'z'$ is assumed without loss of generality to coincide with the global frame and is suppressed in the figure. A general current distribution $\mathbf{J}(\mathbf{r})$ represents the mathematical model of the antenna as an electromagnetic source function in the form of a vector field on the Euclidean manifold. As usual, we always assume that this function has a compact support given by the shaded area in the figure.

It is worthy mentioning that the *total* propagating and nonpropagating parts of the field, namely $\psi_{\rm pr}\left(\mathbf{r}, \bar{\mathbf{R}}\right)$ and $\psi_{\rm ev}\left(\mathbf{r}, \bar{\mathbf{R}}\right)$, where the spectral analysis is conducted along the z''-axis, do *not* vary if the orientation of the local frame $x''y''z''$ is changed by only a rotation around the z''-axis.[8] This interesting observation was exploited in the formulation of the concept of radial streamlines, which the authors believe to be the most natural object of study in connection with the field spectral structure of general antenna systems from the *engineering* point of view. Although in this chapter, we go beyond the streamline picture, it is good to keep in mind the observation mentioned above about the invariance of the decomposition to local rotation around the main axis of the propagation model. It appears that even though the total propagating and nonpropagating field parts have this property, for the full and most fundamental theory the complete rotation matrix as an element of SO(3) has to be retained in its entirety. That is, in contrast to Chapter 4, here we don't introduce a spherical parameters like θ and φ in order to describe the direction of the z''-axis. Instead, the full rotation matrix, which is a *three*-dimensional non-Euclidean differential manifold, is retained and used as part of the extended configuration space of the problem. This will lead to a theory with a mathematical structure very different from the formalism of Chapter 4. For example, we argue that it is not possible to fully comprehend the structure of the antenna near field without working with the geometrical structure of a 6-dimensional differential manifold with the form $\mathbb{R}^3 \times \mathrm{SO}(3)$ as will be explained in detail later. At this stage, it is probably helpful to recall the precise meaning of the term 'propagating.' Even though we focus in our propagation model on the decomposition of the field into a propagating and nonpropagating modes along a certain axis, it is not possible mathematically to ignore the specific directions of the remaining two orthogonal directions. If we say that $\psi_{\rm pr}\left(\mathbf{r}, \bar{\mathbf{R}}\right)$ is the total propagating field along the z-axis of a frame obtained by rotating the global coordinate system by the matrix $\bar{\mathbf{R}}$, then the the other two axes, the x- and y- axes, are part of the mathematical description of the way in which the field is written as a sum of propagating modes. This is a fundamental difference between a particle (mechanistic) model and a

8 Cf. Chapter 4.

genuine field problem. In the former, the motion of a material or point particle is fully described mathematically by giving a vector in the configuration space of the problem, namely the velocity of the particle. In the latter case, the field problem, we find that although there *is* a preferred direction in space, the main axis along which we enact the spectral decomposition into propagating and nonpropagating parts as explained in detail above, still it is *not* possible to ignore the functional dependence of the field on the other orthogonal directions when writing the actual expansion.[9] The total propagating part as a number is invariant to the rotation around the main decomposition axis, but the *dynamics* of the problem, in a sense to be developed carefully later, requires working with the full rotation group, which is a *three* dimensional manifold, rather than two.

Although the local frame has to share the same origin with the global frame,[10] it is very instructive to introduce *another* local frame $x_l y_l z_l$, which has the same orientation of the original local frame $x''y''z''$, but this time with its origin O_l translated to the general location $\mathbf{r}$ outside the source region. The situation is depicted in Figure 6.1b. The second local frame has an orientation described in terms of the rotation matrix $\bar{\mathbf{R}}$, in exactly the same way as the first local frame of Figure 6.1a. We show also a trajectory or path spanned by the vector $\mathbf{r} = \mathbf{r}(\tau)$, where τ is a real parametrization of the three-dimensional curve. At this stage, it is not quite clear yet what is the physical motivation for introducing such a path in the antenna near zone, but let us pursue the idea now mathematically. (The concept of trajectory will emerge immanently at a later stage of the formulation.)

As the point $\mathbf{r}(\tau)$ changes with τ, we imagine that the rotation matrix $\bar{\mathbf{R}}$ itself also changes with τ, i.e., we write explicitly $\bar{\mathbf{R}} = \bar{\mathbf{R}}(\tau)$. Everywhere, we allow an arbitrary but smooth dependence on the motion parameter τ. In this picture, "motion" in the antenna near field zone can be visualized as a change of the spatial location, specified by $\mathbf{r}(\tau)$, accompanied with an arbitrary smooth change of the orientation of the local frame $x_l y_l z_l$ generated by $\bar{\mathbf{R}}(\tau)$. Physically, this means that the electromagnetic field is being decomposed at each spatial location $\mathbf{r}(\tau)$ into a sum of propagating and nonpropagating parts along the z-axis of a *local* frame oriented according to the rotation matrix $\bar{\mathbf{R}}(\tau)$, i.e., the z''- and z_l- axes. As mentioned before, for the actual computation of this spectral decomposition, the first local frame $x''y''z''$ has to be used; however a translated second local frame can be used to picture the process graphically. We notice that the second picture,

9 This is probably connected with the fact that 3D rotation cannot be completely parameterized by only two parameters. Three parameters and several coordinate patches are necessary to full describe the rotation space viewed as a manifold.

10 Because the scalar Green's function is invariant only to rotation, while translation will change the function because the distance to the source point is now modified.

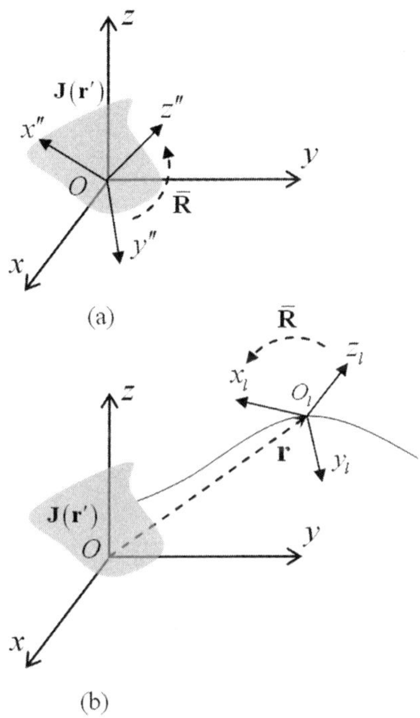

Figure 6.1 An illustration of the propagation model of a general antenna system specified by the current distribution $\mathbf{J}(\mathbf{r})$. Here, the coordinate system (frame) xyz is the global system of the problem. The source frame is given by $x''y''z''$ and is assumed without loss of generality to coincide with the global frame. In addition to these two frames of reference, a third coordinate system is introduced, the local frames xyz and $x_l y_l z_l$. (a) The local frame $x''y''z''$ has as its origin the same location of the origin of the global frame. It is used in the actual computation of the antenna propagation potential. (b) Here, another local frame is introduced, the system $x_l y_l z_l$ is descried mathematically by two objects, the position vector $\mathbf{r}$ of the local frame's origin O_l, and the rotation matrix $\bar{\mathbf{R}} \in \mathrm{SO}(3)$. The latter is used to generate the orientation of the local frame with respect to the global coordinate system. (The same rotation matrix is also used to describe the orientation of the first local frame $x''y''z''$.) Shown here also is a general path in the antenna near zone. The path is mathematically described by the trajectory spanned by the vector $\mathbf{r} = \mathbf{r}(\tau)$, i.e., the locus of the origin O_l of translated local frame $x_l y_l z_l$.

that of Figure 6.1b, is formally identical to the motion of a rigid body in classical mechanics. Indeed, the position vector $\mathbf{r}(\tau)$ plays the role of the body's center of gravity, while the rotation of the body in the three-dimensional Euclidean space is determined by the orientation of the local frame $x_l y_l z_l$ [76]. As is well known in this case, the configuration space of the mechanical problem is $\mathbb{R}^3 \times \mathrm{SO}(3)$. The second depiction of motion in the near zone of the antenna, i.e., Figure 6.1b, is formally identical to the motion of rigid body in three-dimensional Euclidean space. Throughout this chapter, we will adopt this fundamental analogy by working with the six-dimensional manifold $\mathbb{R}^3 \times \mathrm{SO}(3)$ as the (extended) configuration space of the antenna near-field problem.

6.2.2 The Idea of the Antenna Propagation Potential

Armed with the mathematical model of propagation in the antenna near zone described in Figure 6.1, we now move one step forward and introduce the most fundamental concept in this chapter, the idea of *the antenna propagation potential.* This will be a direct devolvement of the observation we made before about the peculiarity shared by fields satisfying the Helmholtz equations, namely that they admit a spectral decomposition into propagating and nonpropagating parts along a chosen axis in freely rotating local observation frame of reference.

We will continue to work with the scalar problem in this non-technical summary of the mathematical content of the present theory. That is, we now refer to the spectral expansion (6.3). Consider a general point in the configuration space $\mathbb{R}^3 \times \mathrm{SO}\,(3)$ described as $\left(\mathbf{r}, \bar{\mathbf{R}}\right)$. Imagine first that the position is fixed. Now, one can change the orientation of the local frame in an arbitrary fashion and obtain generally varying spectral decompositions of the same total field. As was mentioned before, only the direction of the z-axis of the local frame will change the spectral composition of the field since rotation of the xy-plane around the z-axis does not change the total propagating and nonpropagating parts.[11]

We would like to express this overall varying spectral composition in a simple quantitative fashion. Since we endorse in this chapter the viewpoint of morphogenesis, where *form* was identified with the propagating part of the field, we find that the most natural approach for defining the form of the antenna near field is to express the energy of the propagating part as a *fraction* or *relative ratio* of

11 But we repeat again, the orientation of the z-axis results from a 3D rotation, and hence cannot be successfully built into the mathematical expressions of the Weyl expansion without resort to the *full* rotation group, i.e., by parameterizations involving *three* variables, not only two parameters as in Chapter 4.

the total field. That is, we define a new scalar field, the antenna propagation potential $V = V\left(\mathbf{r}, \bar{\mathbf{R}}\right)$ as

$$V\left(\mathbf{r}, \bar{\mathbf{R}}\right) = \frac{\left|\psi_{\mathrm{pr}}\left(\mathbf{r}, \bar{\mathbf{R}}\right)\right|^2}{\left|\psi\left(\mathbf{r}\right)\right|^2} = \frac{\left|\psi_{\mathrm{pr}}\left(\mathbf{r}, \bar{\mathbf{R}}\right)\right|^2}{\left|\psi_{\mathrm{pr}}\left(\mathbf{r}, \bar{\mathbf{R}}\right)\right|^2 + \left|\psi_{\mathrm{ev}}\left(\mathbf{r}, \bar{\mathbf{R}}\right)\right|^2}, \tag{6.4}$$

where it is understood in this definition that the potential is defined only in regions where $\psi\left(\mathbf{r}\right) \neq 0$. That is, the antenna propagation potential is a positive real function bounded from above by one and is defined over a subregion $U \subset \mathbb{R}^3 \times \mathrm{SO}\left(3\right)$, i.e., $0 < V\left(\mathbf{r}, \bar{\mathbf{R}}\right) \leq 1$, $\forall\left(\mathbf{r}, \bar{\mathbf{R}}\right) \in U \subset \mathbb{R}^3 \times \mathrm{SO}\left(3\right)$. It can be proved that this function is smooth on U under quite general conditions on the antenna current distribution.

The antenna propagation potential measures the *relative* amount of propagating field at a particular location when the orientation of the observation frame is specified. Being a ratio, it serves as an indicator of the "degree" of form (=propagation) attained by the field at a given point in the configuration space $\mathbb{R}^3 \times \mathrm{SO}\left(3\right)$. From this, we can see at once that the function V is an *intensive* quantity, similar to temperature, pressure, electrostatic potential, gravitational potential, etc. In this case, a corresponding physical phenomenon is naturally associated with *changes* in the values of the potential, not its actual absolute value as with extensive quantities. This is clear from the definition of the antenna propagation potential given in (6.4).

In order to describe how the field is actually propagating in a given small region, one starts by a certain location in the Euclidean space $\mathbb{R}^3$, then examines the behavior of the potential V at all orientations $\bar{\mathbf{R}}$. The value of the propagation potential in itself does not help much in shedding light on how the field will move from that location to another. This latter depends on the *difference* between the value of the potential at the initial location, and its values at other nearby locations. If the difference is large, then one expects a significant amount of the field to flow into the new location with the larger value of V. The implicit assumption we make here is that electromagnetic flow, like all physical processes encountered in classical field theory, is *local*, i.e., motion occurs at infinitesimal steps in contiguous regions in the configuration space of the problem. According to this picture, *the overall flow of electromagnetic energy is governed by* ***differences*** *in the propagation potential.*

The truth of this proposition does not hinge on an additional empirical law, but is a consequence of the *definition* of the propagation potential itself. Indeed, the most remarkable feature exploited here is the existence of such a potential, which is a result deduced from the Weyl expansion. The capacity of the near field to split into propagating and nonpropagating parts by changing the orientation of the local observation frame is considered by us as something on bar with an "empirical discovery" in the sense that it provides the necessary foundations for defining the antenna propagation potential as an intensive quantity. The flow of electromagnetic waves is driven by the field of intensive differences associated with this potential.

In mathematical physics, the precise content of this 'field of intensive differences associated with a given potential' is encoded in the idea of the *gradient* of the potential. Mathematically speaking, the *gradient* is a vector field pointing in the directions of maximum increase in the associated potential.[12] Therefore, the gradient is exactly the quantity capturing the true sense of intensive difference since it supplies the field of directions along which flow or motion will "most likely" tend to form and develop, where the "most likely" here is the extremum of the increase in the potential V mentioned above. The gradient field, conventionally denoted as ∇V, is then postulated here as the fundamental force driving the formation of form or propagation structure in the near field.

Before moving to the next step, we would like to come back to the interpretation of the scalar field V as a potential. We notice that although we postulate a physical significance mainly for the difference of potential, an individual value of the potential, computed at some location in the extended configuration space, is not totally devoid of meaning. Indeed, since V is defined as the relative strength of the propagating field, it can serve as a quick indicator of the "degree of maturity" of the field in the form of propagation structure. Although terms like 'maturity' may strike the reader as a bit metaphorical in a physical investigation, we would like to indicate that the biological interpretation of this mathematical theory to be developed in Section 6.5 will make the use of such a jargon less objectionable as it may appear on first look. At this stage of our development, we expect that when the position $\mathbf{r}$ is close to the antenna current distribution, most the field will exist in its early underdeveloped form of nonpropagating fields. Therefore, for all values of $\bar{\mathbf{R}}$, the corresponding propagation potential at the location $\mathbf{r}$ is small. As we move away from the antenna, say along the direction of radiation beam, the value of V tends to grow in magnitude. Notice that our main objective in this chapter is to describe the mechanism of the formation of the near field, not to decide whether a given field is propagating or not. Therefore, our study will examine the pattern of *variations* in the antenna propagation potential, most simply and intuitively as captured by the idea of the gradient ∇V. Deciding whether the obtained field is "mature" enough

12 Conceptually speaking, Faraday introduced this idea to electromagnetism in his early analysis of electrostatics. The mathematical details were worked out later by mathematical physicists like William Thomson, Tait, and Maxwell. In our personal opinion, it is *Faraday* who deserves most of the credits for the concept of *dynamics driven by potential gradients*.

or not is a secondary result that can be obtained as a bonus from the fundamental evolution equations to be presented next.

6.2.3 The Fundamental Evolution Equations of the Antenna Near Fields

Given a new vector field, what is the best method to understand its structure? When Faraday was trying to answer this question for the magnetic field, he performed the now famous experiment of iron filings, in which the small iron pieces interacted with the field by forming a characteristic spatial distribution in the space surrounding the magnet, the 'lines of forces' of the field, as they were termed later by Faraday. Mathematically, what happens here is exactly the following: Search for a bundle of trajectories in the configuration space of the problem such that at each point on a given trajectory, the tangent to the path points into the direction of the vector field assigned to that point. The fundamental theorem of the existence of solutions to first-order ordinary differential equations guarantees the existence (and uniqueness) of this bundle of trajectories if the vector field is smooth enough [74]. Assuming that our gradient field ∇V is smooth (we will prove this later), the trajectories filling the configuration space $\mathbb{R}^3 \times \mathrm{SO}(3)$ associated with this vector field are solutions to the following system of ordinary differential equations

$$\frac{d\mathbf{r}}{d\tau} = \nabla_{\mathbf{r}} V, \quad \frac{d\bar{\mathbf{R}}}{d\tau} = \nabla_{\bar{\mathbf{R}}} V, \tag{6.5}$$

where $\nabla_{\mathbf{r}} V$ and $\nabla_{\bar{\mathbf{R}}} V$ are the gradients evaluated with respect to $\mathbf{r}$ and $\bar{\mathbf{R}}$, respectively. The variable τ is a real parameter for the path. The equations (6.5) has to be augmented with initial conditions as points located in the configuration space $\mathbb{R}^3 \times \mathrm{SO}(3)$ such that ∇V is smooth. Notice also that the equations are coupled because each of $\nabla_{\mathbf{r}} V$ and $\nabla_{\bar{\mathbf{R}}} V$ is a function of both $\mathbf{r}$ and $\bar{\mathbf{R}}$ at the same time.

We call equations (6.5) *the fundamental evolution equations of the antenna near field*. This is a system of six nonlinear coupled first-order ordinary differential equations on the manifold $\mathbb{R}^3 \times \mathrm{SO}(3)$. They supply, in our opinion, the entire mathematical content of the problem of genesis of form in the near zone, where 'form' is defined as the propagation structure of the field. They are written as evolution equations in terms of the real parameter τ. As the value of this variable increases, the field flows away from the antenna and tends generally to propagate toward the far zone. The reason behind using the term 'evolution' is that the solution of these equations, as is well-known from the general theory of ordinary differential equations, can be expressed as one-parameter transformation group [74]. Mathematically speaking, the group transforms any initial region in the configuration space of the problem into another region by generating a flow from the former ending in the latter. Since the vector field appearing on the RHS of (6.5) is a gradient of the antenna *propagation* potential, the use of the term 'evolution' suggested by 'flow' and 'propagation' is doubly motivated.

A qualitative (topological) analysis of the solution of equations (6.5) will be supplied in Section 6.4. An analytical solution of these equations does not exist, and only direct numerical integration can provide quantitative results. For each antenna current distribution, the entire system (6.5) has to be integrated and the trajectories in a region in the configuration space under interest can be displayed. Similar to the iron filling experiment, we expect that the geometrical structure of these trajectories will shed significant light on the electromagnetic structure of the field in the antenna near zone.

We now find that the concept of 'path in the near field,' introduced in Figure 6.1b, is an essential part of the theoretical analysis of the near field structure. Indeed, the integration of (6.5) provides as with the functions $\mathbf{r} = \mathbf{r}(\tau)$ and $\bar{\mathbf{R}} = \bar{\mathbf{R}}(\tau)$ postulated there. Notice that although the two sets of equations in (6.5) are coupled, once integrated, their solutions can be displayed not only as trajectories in the 6-manifold of the extended configuration space $\mathbb{R}^3 \times \mathrm{SO}\,(3)$, but also in the suggestive form of Figure 6.1b; i.e., as a normal spatial trajectory in the Euclidean space $\mathbb{R}^3$, and superimposed on it the change of the orientation of the local observation frame $x_l y_l z_l$ as the point $\mathbf{r} = \mathbf{r}(\tau)$ moves along its spatial trajectory. Our strategy will be to work with both representations of the solution of the fundamental evolution equations. Indeed, depending on the context and the intention, one geometrical model is sometimes far more convenient than the other.

There is another distinguishing features in equations (6.5). This is the fact that the manifold on which the fundamental evolution equations are defined, namely $\mathbb{R}^3 \times \mathrm{SO}(3)$, in contrast to the Euclidean space $\mathbb{R}^3$, cannot be covered completely by a single coordinate chart. As is well known from Lie theory, the rotation group SO(3) is a non-Euclidean three-dimensional manifold that requires a set of overlapping coordinate patches in order to cover it completely [81], [80], [60], [61]. Therefore, equations (6.5) can be written in a coordinate-free manner by refereing to such a global covering of the underlying configuration space $\mathbb{R}^3 \times \mathrm{SO}(3)$ by a the suitable differential structure.[13]

6.3 COMPARISON WITH THE POYNTING FLOW

It is interesting to write down the differential equations describing the flow of electromagnetic energy in terms of the classical Poynting flux flow density $\mathbf{S} = (1/2)\,\mathrm{Re}\,\{\mathbf{E} \times \mathbf{H}^*\}$. These are readily given as the coupled three equations in $\mathbb{R}^3$

$$\frac{d\mathbf{r}}{d\tau} = \frac{1}{2}\mathrm{Re}\,\{\mathbf{E}(\mathbf{r}) \times \mathbf{H}^*(\mathbf{r})\}\,. \tag{6.9}$$

Comparison between (6.9) and (6.5) reveals that the two systems of equations cannot be isomorphic (i.e., topological equivalent, c.f., [74]) since the dimensions of the corresponding configuration spaces are different. Moreover, careful asymptotic analysis (not given here) demonstrates that the energy flow in the far zone as predicted by (6.5) is *not* identical to the Poynting flow. These surprising results strongly suggests that *the unquestioned reliance on the conventional interpretation of the Poynting flux as a mathematical representation of electromagnetic energy flow in space and time should be reconsidered*.

13 To give more details, consider an open set $U \subset \mathbb{R}^3 \times \mathrm{SO}\,(3)$. Let $\varphi : U \to D \subset \mathbb{R}^6$ be a coordinate map with coordinates given by $x^n, n = 1, ..., 6$. The gradient field ∇V can be written in terms of this chart as [60]

$$\nabla V = \sum_{n=1}^{6} f^n(x) \frac{\partial}{\partial x^n}, \tag{6.6}$$

where the $f^n(x)$ are smooth functions in $x := \langle x^1, x^2, x^3, x^4, x^5, x^6 \rangle$. Define $f(x) := \langle f^1(x), f^2(x), f^3(x), f^4(x), f^5(x), f^6(x) \rangle$. In terms of this chart, we can write the fundamental evolution equation locally as

$$\frac{dx}{d\tau} = f(x), \tag{6.7}$$

or component-wise in detail as

$$\frac{dx^n}{d\tau} = f^n\left(x^1, x^2, x^3, x^4, x^5, x^6\right), \quad n = 1, ..., 6. \tag{6.8}$$

As is well known from basic differential manifold theory, it is possible to state the fundamental evolution equations in a coordinate-free manner. This requires the notion of *integral curves*, which can be defined in terms of the vector field ∇V without reference to a specific coordinate chart (U, φ) on $\mathbb{R}^3 \times \mathrm{SO}(3)$ [60], [61].

At this stage, it is convenient to refer to another interesting aspect implied by (6.5). The reader may have noticed the emergence of a new concept of time, the τ-time, beside the familiar t-time of electromagnetic (relativistic) phenomena. Such a multi-layering of the temporal structure is not new in physics, although it is usually ignored. We mention here that, for example, the photon concept is obtained in quantum electrodynamics by the second-quantization procedure of t-time-harmonic classical field modes [110]. On the other hand, the *finite* "time period" through which a photon is created at the source and subsequently annihilated at the detector obviously cannot refer to the same harmonic t-time. Therefore, it appears that the "development time" of the photon is more properly related to the τ-time of (6.5) than the formal t-time. This observation suggests the need for quantizing the theory presented in this chapter.

We also notice that τ plays a role similar to the physical time t in the sense that the near field tends to *evolve* and *flow* along trajectories in its configuration space parameterized by τ. However, the physical t-time has been already employed implicitly in constructing the antenna propagation potential V using the time-harmonic excitation condition $\exp(-i\omega t)$ in writing the Weyl expansion [35]. On the other hand, the parameter τ is intimately connected with the *spatial* structure of the configuration space $\mathbb{R}^3 \times \mathrm{SO}(3)$ in the sense that as τ increases, the field tends to convert from its embryonic "early" stage, in which most of its spectral content is nonpropagating (regions spatially close to the antenna current in the near-field zone,) to its mature phase, where most of the spectral composition of the field consists of propagating modes (the antennas far-field zone). Therefore, the parameter τ, which possesses a straightforward mathematical semantics, appears to play a double role in the physical sense by conveying *both* temporal and spatial meanings. We would like to introduce, vaguely, a new time we call the τ-time. We notice immediately that this new time is frequency dependent; actually, for each frequency ω there is a new set of fundamental evolution equations because the antenna propagation potential (6.4) is obviously a function of frequency. The τ-time is simply the mathematical content of the real parameter used to describe the solutions of the fundamental evolution equations (6.5). The ultimate connection between the physical t-time and the τ-time requires performing time-dependent plane wave spectrum analysis [35] in which the complete temporal dynamics of the radiated fields is determined by all the τ-times in the entire frequency range. This process is extremely complicated and it is not clear at the present stage wether it is warranted or not. In this chapter, we restrict our analysis fully to the time-harmonic excitation scenario.

It might be interesting to make some few additional observations that are somehow speculative in nature. Although the entire theory developed here is applicable only to *classical* fields, in which no quantization of the field itself is attempted, the trajectory picture emerging from the fact that the fundamental evolution equations (6.5) are *ordinary* differential equations strongly suggests a *"particle"* interpretation of the field. Indeed, one may argue that an integral curve of equations (6.5) represents motion of a "particle of radiation" in the antenna near zone. It is well known that photons, the only genuine particle of electromagnetic radiation, are obtained by first Fourier analyzing the classical fields then quantizing the obtained spectral expansion [111], [110]. (In this sense, each photon with energy E is associated unambiguously with a frequency ω through the Planck-Einstein equation $E = \hbar\omega$.) However, every physicist or engineer working with phenomena essentially involving the photon concept, e.g., quantum optics, acts with some mental imaginations in which the photon is visualized as *moving* somehow from region to another. (This remains true even with the fact that the concept of photon wavefunction is problematic [112].) When one tries to interpret the experiment in which a photographic plate registers the impact of a photon, it is this t-time-harmonic entity – which presupposes an *infinite*-time-harmonic waveform – that moves and hits spatially localized regions, where the impact process occurs in (another?) physical time. The concept of τ-time may help resolve this difficulty. One can speculate that the *definition* of photon requires a t-time, while the genetic process of photon generation, transfer/movement, and subsequent absorbtion/destruction are all based ultimately on the τ-time. This interpretation is not fundamental to our main theory because in the antenna problem, where the fields are classical (non-quantized) fields, there is no essential need for introducing the concept of particles of radiation.

6.4 THE ROLE OF SINGULARITIES

Our approach to the antenna near field through the fundamental equations (6.5) will be based on the qualitative theory of ordinary differential equations, which was originally proposed by Poincare and Lyabonov at the end of the 19th century [74]. The topological aspects of the flow generated by (6.5) can be completely described by the distribution of special points in the phase space of the problem. These points are usually termed *critical* or *fixed* points, but we adopt here the other commonly used terminology of *singularities*. They can be defined as the points in $\mathbb{R}^3 \times \mathrm{SO}(3)$ at which the gradient field ∇V is identically zero. The entire solution of (6.5) will be structured both locally and globally by the location and types of these singularities in the phase space. The local study begins by computing the differential of the field ∇V and obtaining its eigenvalues at each singularities. Depending on the nature of these, e.g., whether real (positive, negative, zero), imaginary (pure imaginary, etc), one can deduce valuable information about the behavior of the electromagnetic flow around the fixed point [74]. There has been an enormous effort in the mathematical community throughout the 20th century to classify all types of singularities and tabulate the resulting qualitative behavior. This literature can be directly utilized by the antenna community in the quest to understand the structure of electromagnetic flow in the near zone based on the equations (6.5).

Instead of reviewing this literature, we will propose here a new concept that appears to the authors natural in gradient-like flows. This is the idea of a *near-field wavefront*, which we define as the set of points in $\mathbb{R}^3 \times \mathrm{SO}(3)$ satisfying the equation $V = V_0$, for constant V_0. Therefore, this is the "level surface" of the propagation potential (6.4) at V_0. It is a generalization of the familiar idea of a equipotential surface in electrostatics. The near-field wavefront can be thought of as the moving structure of electromagnetic energy that recedes away from the source into the far-zone. The actual radiation field will carry the signature of this structure in the form in which it was initiated near the antenna current distribution. The question is now the following: How does this near-field wavefront change while moving away from the source?

Let M_{τ_1} and M_{τ_2} be two near-field wavefronts obtained from each other by the flow generated by (6.5), i.e., starting from time τ_1 and ending at τ_2. It can be shown based on a local analysis of the propagation potential that if no singularity is encountered during the motion of wavefront M_τ in the interval $[\tau_1, \tau_2]$, then the two fronts M_{τ_1} and M_{τ_2} will be diffeomorphic to each other, i.e., have essentially the same topological or qualitative shape. As it turns out, it is only when the wavefront passes through a singularity that a topological (qualitative) change occurs in the shape of the moving field.

The specific analysis of the case when singularities are present requires some more technical treatment. We will classify singularities as is customarily in differential topology into degenerate and nondegenerate points. The former is the case when the Hessian matrix of the propagation potential is singular, while the latter is the other possibility [87]. Any antenna propagation potential which has all of its singularities belonging to the nondegenerate type are called *Mores functions*. However, there is no reason why the antenna potential (6.4) must be always a Mores function, but it can be shown that the set of Morse functions are dense and hence any generic antenna potential can be safely assumed to be a Morse function (the details of this argument will be given elsewhere). This brings into the antenna problem the powerful machinery of differential topology in general, and Morse theory in particular, for our quest to understand the structure of the near field flow generated by (6.5).

Using Morse theory [87], it is possible to completely determine how the antenna near-field wavefront will change when passing through each possible type of singularities. The effect of the singularities is here described locally as the topological change enacted by the singularity in a small neighborhood surrounding it on the part of the near-field wavefront passing this region. The nature of this topological change turns out to depend fully on two factors, the nature of the antenna configuration space $\mathbb{R}^3 \times \mathrm{SO}(3)$ (e.g., dimension, homotopy and homology groups, etc.), and the *index* of the nondegenerate singularity (this quantity is defined in [87]).

The program of analysis of the structure of the antenna near field is now summarized as follows:

1. Using the antenna propagation potential, identify all singularities and classify them according to their indices.
2. Identify initial near-field wavefronts at any location of interest near the antenna.
3. Study how the near-field wavefront will undergo a fundamental qualitative change when passing through a singularity.
4. If no singularity is encountered by the near-field wavefront, then its essential qualitative form as originated in the near zone will pass unchanged into the far zone.

Based on this overall picture, we suggest that the specific form of the fields generated by the antenna device is completely governed by the locations and the indices of the antenna propagation potential singularities. Therefore, answering fundamental questions about electromagnetic radiation, in our view, may be attacked using the topological program sketched here. Notice that the task of implementing this procedure is now completely computational after the derivation of (6.5), although not an easy one and may require a considerable and sustained group effort.

6.5 FROM THEORETICAL BIOLOGY TO ANTENNA ENGINEERING

In the theory to be presented in this work, we make use of ideas inspired by the study of development and ontogenesis as observed in the biological world. That does not imply that our intention is to suggest that the antenna fields are a "living organism," although, as will become clear toward the end of this chapter, there is indeed a striking similarity between the manner in which an initial germ in the animal kingdom grows to form a mature organism and the mechanism of the emergence of well-defined radiation pattern in general antenna systems. Our objective of spelling out this remarkable similarity is twofold. First, we make a progress in the purely scientific study of electromagnetic radiation by probing in depth its mechanisms of morphogenesis in a mathematically rigorous fashion. On the other hand, this new understanding is shown to lead in a natural way to a significant insight into the prospects of attaining a finer control of the antenna fields by supplying clues for how to modify the current distribution.

Morphogenesis is the coming of form into being [161], [162], [163], [164]. The first comprehensive mathematical approach to the problem of the genesis of form can be traced back to the classic work by Thompson [155], which was taken up again by Thom in his catastrophe theory [163], [164]. Without going into much details about the subject matter of these works, we first make notice of the fact that at a very general level, the framework of morphogenesis involves fundamentally two pillars: 1) an emphasis on *form* as the subject proper of the scientific investigation, and 2) the *genesis* of this form, its creation as a historical process in which an initial germ unfolds and "evolves" into a complete form. The sense of the term 'complete' here remains ambiguous till we delimit its meaning as essentially refereing to the emergence of form; i.e., the final (mature) object unfolding from the initial germ possesses an unequivocal *new* form that was *absent* in the germ itself.

In antenna systems, the field close to the physical support of the current distribution is "devoid of radiation" in the sense of being consisting mainly of nonpropagating modes.[14] On the other hand, the field away from the antenna contains the "well-developed" state of radiation in the form of propagating or traveling modes. Therefore, the natural signification of the word 'form' is being either propagating or nonpropagating. In this sense, the near-field part is "less-developed" than the far-field parts, since the latter contains rich amount of "complete" forms or propagating modes. In our opinion, this simple picture is fundamental in achieving proper understanding of the inner working of antenna systems. It explains why it has been extremely difficult in antenna practice during the last six or seven decades to come up with a perspicuous methodology for antenna design. Indeed, the recurrent reference to antenna design as being more of an "art" than science stems from the fact that the process of electromagnetic radiation is a complex genetic process in which there is roughly speaking a plethora of pathways of growth and development taking each part of an initial germ into a complete form in the far field. Think of an early

14 Cf. Chapter 4.

animal embryo. It is well known that modification of this initial germ may lead to all kinds of complications in the final attained form. One thing that was observed very early in embryology is that small modifications in the initial germ may lead to enormous variations in the morphology of the mature organism [156]. Something very similar occurs in antenna practice. Modification of the antenna parameters leads to changes in the current distribution, affecting directly the near field shell, which we now take as the initial morphogenetic germ of the problem of electromagnetic radiation. Therefore, it is possible for all kinds of changes to occur in the final form, the far field. *We would like then to visualize the antenna system as a device with the main function – ontologically speaking – producing form, where 'form' here means electromagnetically propagating fields along certain directions.*

The meaning of propagation is a that a field form moves from one location in space to another. Like any motion, this requires knowing the direction of movement and its speed or amount. We proposed in Chapter 4 a mathematical model for the description of propagation in the near-field zone. The model relies on exploiting the freedom of orienting the observation frame in the expression of the Weyl expansion. And as it turned out in this chapter in the section on propagation potentials, a quantity can be computed to give the "amount of propagation" along various directions in space. within this framework, the form of the radiation field will be determined by assigning to each location in space, and an orientation of the observation coordinate system, a number or a group of numbers describing the "amount of propagation" we spoke about above. *Form* then can be mathematically quantified via our model as will become clearer later. In the remaining parts of this chapter, we will attempt to integrate the previous mathematical theory with the embryology or biological development.

6.5.1 Epigenesis

What is the difference between morphogenesis and epigenesis?[15] We would like to consider morphogenesis as the proper name to be given for the *theoretical* program of the scientific description of the antenna electromagnetic near field. In this sense, morphogenesis is the overall frame of study, while epigenesis refers to a particular emphasis laid down by the theorist on his *empirical* observation and subsequent reflections and speculations. We will show later that the discovery of

15 The technical term 'epigenetics' was introduced by Waddington in 1947 [160]. It is adapted from the Aristotelian term 'epigenesis.' It appears that from Waddington's term is derived the now standard adjective 'epigenetic.' We are not concerned here with the labyrinth of the etymological uses that this term has elicited in biology and philosophy. Instead, we try as much as possible to provide our own definitions and understandings of the various terms used.

the propagation potential of the antenna system represents an instance of empirical finding. Later on, the postulation of a six-dimensional configuration space (6-manifold), the construction of the associated morphogenetic field, and the derivation of the fundamental evolution equations, are all to be considered as theoretical constructs proposed by the theorist within the general outline of the program of morphogenesis.

We can go farther beyond this philosophical introduction and define epigenesis as the emergence of a form that was not present in the previous stage. In this sense, the use of the term epigenesis suggests that 1) a temporal evolution is under consideration, 2) there is a clear demarcation between a past and new stages, and 3) in the new stage there is a well-defined form that was absent in the old stage. Now this can be contrasted to the case of morphogenesis, which was defined roughly as the coming of form into being. A form may already exist in some past stage of the developmental process, and yet continues to be present in the new stage. For example, as we will explained in Section 6.4, the near field wave front is homeomorphic to all evolved fronts as long no Morse singularity is passed over. In this example, the form is merely the *topology* of the 5-manifold comprising the near-field wave front. If we define form as the (metric) *geometry* of this 5-manifold, then there is always a potential creation of new form, i.e., change in the geometric isomorphism as we move along the flow. In this case, all flows involves epigenesis if form is defined geometrically, but only special cases involve epigenesis if form is defined topologically.

Epigenesis can then be defined as the generation of a discontinuity. A change of form is itself a form viewed at the meta-linguistic level. Morphogenesis is the science of the creation of all kinds of forms, but epigenesis narrows the focus down to empirical findings of concrete cases. The distribution of the singularities in the antenna near-field zone is a kind of "concrete universal" event, to borrow the concept from the philosopher Gellis Deleuze. It is concrete because involved is a concrete spatial distribution of the singularities in the phase space. But also universal because it governs the behavior of not a single trajectory around the singularity, but an entire continuum of all possible (later we will better say *virtual*) trajectories that pass near the particular location of that singularity.

Finally, we reinstate one of the main terminological issues adopted by this chapter. The propagation potential of the antenna is a real-valued, smooth function defined on the six-dimensional manifold of the morphogenesis configuration space. As such, this function produces a new six-dimensional manifold, the "graph" of the function itself now viewed as a differential submanifold embedded in an higher-dimensional ambient Euclidean manifold [84], [86]. Following a similar idea proposed by Waddington's concept of 'canalization' in his theory of genetic assimilation [158], we call this 6-manifold, the "graph" of the propagation potential, the *epigenetic landscape*. In this sense, the term 'epigenetic' is used liberally to generally indicate the mere morphogenetic nature of the particular potential. As we mentioned before, the existence of this scalar field, the propagation potential, is taken as an empirical discovery about electromagnetic radiation.

6.5.2 Waddington's Canalization

The concept of *canalization* was proposed by the biologist Waddington in the 1940s in response to a series to experimental studies suggesting that living organisms are capable of responding to external environmental stimuli by an adaptation that can become inheritable after several generations [158], [159], [160]. We are not interested here in the biological context itself, which is basically the idea of genetic assimilation. Instead, we would like to focus on the abstract theoretical scheme that lies behind Waddington's proposal. This is the ability of an embryological system to "buffer" random stimuli that are not exceeding certain threshold; the development process appears to resist random changes in the sense that the evolution of the system manages to persists in reaching the same final objective goal after enough interval to time. In other words, developmental mechanisms in the biological world posses *robustness* with respect to both external (environment) and internal (genetic) small fluctuations. Figure 6.2 illustrates the basic idea. The motion of the ball represents a possible developmental history undergone by the organism. If we take the case depicted in one scenario, say A, we notice that if the initial stimuli managed to push development in this trajectory, then any random small disturbing effect affecting the rolling ball while in its trajectory has little impact on the final destination reached. The reason is that the trajectory, viewed as a "steep valley," compensates for slight deviations by pushing the ball back to the main track. This ability for spontaneous automatic control or self-correction was observed empirically by embryologists. Waddington's proposal is that there is indeed something in the development system that acts like the trajectory, or what he named later the *chreod*, refereing by this new term to to a possible well-defined developmental trajectory in the epigenetic landscape.

It is now well-known that Waddington himself had not gone enough beyond the orthodoxy of neo-Darwinism by proposing the concept of canalization. Rene Thom went further than Waddington by first giving the concept of chreod a precise mathematical definition [163], and then by postulating the existence of the chreod on a firm level of reality, at least in the sense of applied ontology [164]. The concept

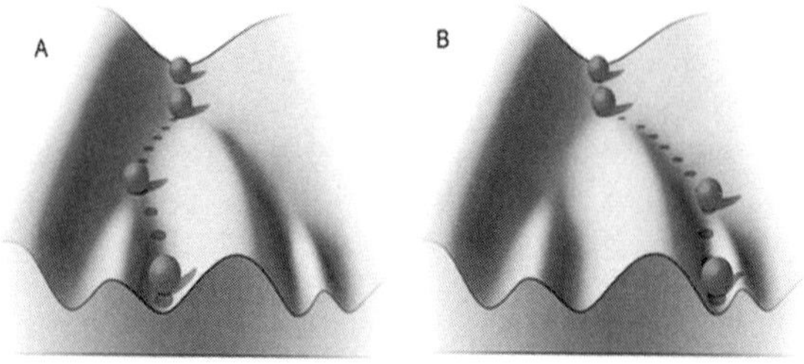

Figure 6.2 An illustration of Waddington's concept of canalization in biological devolvement. The "graph" of the antenna propagation potential is interpreted by us as the epigenetic landscape proposed originally by Waddington. Electromagnetic flow is here homologous to embryological evolution and both are depicted in the figure as the movement of the ball along a potential trajectory in the landscape or what Waddington later named 'chreod.' The two scenarios A and B represents two potential developmental trajectories. In the biological case, it is up to environmental effects (through selection pressure) or genetic stimuli to switch the devolvement from A to B or vice versa. In the antenna near field zone, the emergence of such a clear cut two trajectories is an indication of the formation of possibly multiple beams if this two-canalization chreod persists up to the far zone. One question remains: what plays in the antenna case the role of "random" stimulus capable of switching the evolution from one chreods to another? We suggest that the answer to this question in the context of a morphogenetic field theory necessitates the use of the concepts of the *virtual* and the *actual*.

of canalization as robustness or persistence of behavior is now well-understood mathematically in terms of the concept of *attractor* that Thom utilized in his precise definition. The fact that a behavior tends to robustly manifests itself in time is not new, and can be found in many textbooks on dynamics [78].

What we would like to investigate further in this position is not the idea of chreod as an attractor, but rather the deeper implications that this analogy have for a *field* theory. Indeed, if the configuration space of the field problem is the classical Euclidean space-time, then we immediately face the difficulty of explaining what exactly we mean by an "initial condition" at a particular location. There are no "particles" that may move embryology like the case in dynamics, and the question remains whether the whole discussion of canalization is essential to our morphogenetic theory or not.

We believe that one of the main contributions of morphogenesis in understanding the nature and the mechanism of the formation of electromagnetic radiation lies in clarifying the inadequacy of the conventional Euclidean space-time to the mathematical description of the problem. It is interesting that Waddington himself had noticed early enough that embryological devolvement cannot be properly studied except in *higher*-dimensional space [160]. Here, the *phenotypes* that are relevant to the evolvement process are taken into account as "extra dimensions" in the configuration space of the problem.

We will implement this idea in the theory of the morphogenesis of electromagnetic radiation by choosing a new "phenotype" and appending it to the standard 3D configuration position space in order to make the *extended* configuration space. This new "phenotype" is the orientation of the local observation frame needed in the mathematical model of propagation in the electromagnetic field. According to this model, each element of the rotation group SO(3) corresponds to a possible phenotype. Development concerns the historical unfolding of series of spatial locations and the phenotypes adjoined to them. Since the rotation group is itself a 3-manifold, the space $\mathrm{SO}(3) \times \mathbb{R}^3$ is the 6-manifold or the extended configuration space of our morphogenetic theory. The fact that the phenotype space is the 3-manifold SO(3) suggests that one way of thinking about the extended configuration space is viewing the latter as an annexion of a full group SO(3) to each spatial location. This interpretation will prove to be powerful when the solutions of the fundamental evolution equation are processed to draw physical conclusions about the behavior of the antenna fields. However, while solving the equations themselves, it is important to retain the full "wholeness" of the 6-manifold as a unified configuration space for the entire problem. Indeed, the six fundamental differential equations are all generally coupled.

6.5.3 The Virtual and the Actual

It appears to us that the difference between morphogenetic theories of field structures, like the electromagnetic and acoustic fields, and traditional morphogenetic theories considered as refinement of dynamics [163], [164], is the increasing necessity for introducing the fundamental concepts of the *virtual* and the *actual*. If one looks back at Waddington's canalization, it can be noticed that "randomness in the initial conditions" is one of the main factor triggering the developmental processes by setting it to take course in one chreod instead of others. This seems to be a legitimate and comprehensible assumption in such theories since the model 'chreod = attractor' is inspired by the theory of dynamic systems, which is, in certain sense, the modern mathematical theory of dynamics in classical mechanics. In this context, there is always an unambiguous conception of what constitutes an initial condition. In the extended configuration space of our morphogenetic theory, the "initial position" involves an initial position in space *and* a given element of SO(3). While one can easily think of spatial positions as forming initial points for the morphogenetic evolution, it is less obvious what kind of meaning should be allotted to "initial phenotype," here mathematically given as elements of SO(3). The ambiguity can be resolved by noticing that an initial region in the extended configuration space $\mathbb{R}^3 \times \mathrm{SO}(3)$ is better viewed as a "germ," an embryonic stage that is still in the early developmental phase. As was noticed early on by Driesch [156], one fundamental characteristic of such an early "germs" is their "equi-potentiality" for future developments. This means that each part of the embryo is capable of developing, if forced or stimulated to do so, to any possible final form. We think that the term 'equi-potential' is not very convenient here because in Section 6.4 we introduced the concept of near-field wave front as an equipotential surface (actually, a 5-manifold) where 'potential' here refers to the propagation potential of the antenna. We propose to call Driesch's embryologic property *equi-virtuality*. It still means the same thing implied by Driesch's equi-potentiality, which is the equal possibility for each spatial part of the germ to develop into a mature part of the full organism actualizing the full phenotype or form. In this sense, we define the *morphogenetic germ* U as any open subset $U \subset \mathbb{R}^3 \times \mathrm{SO}(3)$ of the extended configuration space taken as an initial region for the future evolution under the fundamental (morphogenetic) evolution equations. As expected, this definition losses its importance if the spatial locations in U are already in the far zone. In this case, one is already talking about the "full or mature organism," the far-field radiation pattern. In order for U to play an important, nontrivial part in the morphogenetic theory of electromagnetic radiation, it is necessary to take the spatial part of U to be in the near zone, as close as possible to the antenna current distribution. Here, the rotation-group part of U, i.e., the elements of SO(3), give the virtual phenotype or electromagnetic form, the spectral expansion at the considered spatial location. In the morphogenetic germ, all possible forms (phenotypes) are legitimate initial conditions for the fundamental morphogenetic equations. However, the appearance of a well-defined form, propagation for a continuous domain of spatial points in which the propagation vectors are coherent, will give rise to the emergence of the radiation pattern. Away from the antenna, flow is in concrete (we say *actual*) directions. Very close to the antenna, there is still no well-developed field of directions in which there is coherence in the propagation vectors associated with neighboring points. All directions of propagation or flow are equally legitimate, and we say that all these directions are *virtual*. The virtual is still real, exactly as the actual, but it is different from the latter in being "not realized," not "selected" or "implemented" by the mechanism of morphogenesis. A noncoherent mixture of things, where coherence is assigned the value of a given form, is still real, but with respect to the form under consideration has all forms virtually built into it without being able to actualize a particular or concrete one.

In order to clarify these abstract ideas, we come back to the antenna propagation potential introduced in (6.4). The mathematical function $V = V\left(\mathbf{r}, \bar{\mathbf{R}}\right)$ is defined in a domain $D \subset \mathbb{R}^3 \times \mathrm{SO}(3)$ in which the potential is a smooth function on this manifold.[16] Consider the 2-sphere $S_a^2 = \{r = a\}$ chosen such that the entire antenna current distribution is contained inside its interior. For simplicity, we work with a single antenna, instead of the most general case in which an entire array is considered. We are interested in understanding the genesis of the antenna radiation field by starting from this sphere. The question of how did we get from the interior region of the antenna to this sphere is interesting but is beyond the scope of this book. Therefore, our focus is concentrated in understanding the structural emergence of form starting from this enclosing sphere that already lies in the exterior region of the antenna system, but very close to the boundary of this region. If we consider the fundamental evolution equations of the antenna near field (6.5) in order to solve this problem, we notice that knowledge of the spatial locations on the sphere S_a^2 is not enough to integrate the equations uniquely since it is still required to specify the location in the SO(3) part of the full 6-manifold of the configuration space. Indeed, the morphogenetic germ of this particular configuration is the 5-submanifold $U_a^S = S_a^2 \times \mathrm{SO}(3)$. This submanifold will be called the *standard morphogenetic germ for the antenna problem* in our investigation, and will be refereed to repeatedly as U_a^S.

At this stage, the antenna propagation potential V does not distinguish between any orientation of the local observation frame attached to a given point on the enclosing sphere S_a^2; indeed, for $\mathbf{r} \in S_a^2$ the potential reduces to a smooth function on the submanifold U_a^S. In general, the value of V changes for different orientations $\bar{\mathbf{R}} \in \mathrm{SO}(3)$; although one may view through such evaluations how the potential restricted into U_a^S is behaving, it is not possible theoretically to identify unambiguously a specific orientation to serve as the sole legitimate initial point in the integration of the evolution equations (6.5). Our strategy for resolving this difficulty is based on the interpretation of the morphogenetic germ inspired by Driesch's concept of equi-virtuality. That is, we declare the full range of possible forms, mathematically encoded by *all* values of the orientations $\bar{\mathbf{R}} \in \mathrm{SO}(3)$, as legitimate initial points for integrating (6.5). The *entire* 5-submanifold U_a^S is then considered as the initial region in the full 6-manifold configuration space D from which we compute the flow (diffeomorphism group) generated by the evolution equations (6.5). The result from this process is a new region in the far zone, coinciding with the

16 For a definition of smoothness of real-valued functions defined on a general abstract manifolds, see [60], [84], [85], [86].

radiation pattern, and giving in the details of the transition via the aforementioned flow itself the mechanism of the formation of the radiation field.

The question now is this: What really performs the real transition from the arbitrariness of taking into account *all* possible orientations in the initial region – the standard morphogenetic germ U_a^S – into the emergence of well-defined radiation pattern in the far zone? Notice that this definiteness of the radiation field in the far zone translates mathematically, according to our propagation model in Figure 6.1, into the selection of certain orientations, i.e., *specific* elements of SO(3), as "realized" in the spectral composition of the field in the far zone; if the propagation potential $V\left(\mathbf{r}, \bar{\mathbf{R}}\right)$ is evaluated in the far zone along *other* directions, it will yield small values, while along the selected directions its value is large, possibly approaching unity in some cases (e.g., at full peaks in the radiation pattern). Our answer to the question raised above is that some sort of an *attractor* in the phase space of this problem has "selected" certain orientations by forcing the flow initialized in the full (sub)manifold U_a^s to be channeled or canalized in certain pathways. Otherwise, there will be no uniquely specified propagation in that region in the far zone. An attractor acts like a "focal point" in the phase space of the problem organizing the behavior of all trajectories originating in the surrounding regions [78]. In Section 6.4, the exact meaning of these attractors will be studied with more care within the context of the singularities (critical points) of the gradient field of the antenna problem. What is relevant to the present discussion is that the attractor is merely the mechanism in which all the virtual states taken into full consideration in the standard morphogenetic germ U_a^S, namely the entire (compact) rotation group SO(3) attached to each location on S_a^2, are *actualized* in the far zone. This explain, in our opinion, how a definite form, a phenotype, a preferred direction for propagation/nonpropagation was generated from an early embryonic state, the germ U_s^S, where more states (the virtual states) were present than found in the later (mature) states of the far zone, say on the radiation sphere S_∞^2.

The term 'virtual' was consistently emphasized in our presentation while the deceptively similar term 'potential' was deliberately avoided. The reason is that the latter got mixed up from early on with Aristotle's use of this key term in his work. Our use of the key couple of terms 'virtual' and 'actual' is more in line with the modern philosophical tradition of Bergson [166], [167] and the postmodern work of Deleuze [168]. However, the idea of the 'actualization of the virtual' seems to be present in the writings of many open-minded biologists, especially embryologists. In the ontological context of our theory, which is not the topic of this chapter, we think of the actual as the product resulting from the virtual when the latter has undergone a process of ontological epigenetic devolvement. The actualization of the virtual is the production of being by working on a "raw material" consisting of virtual beings. One way to visualize the epigenetic process leading from the virtual into the actual is to imagine the actualized being as the net outcome of a "condensation" of a multiplicity of virtual states. Of course, the mechanism of "condensation" here has to be explicated by a reasonable model in order to avoid falling back into metaphorical traps. The actualization of the virtual through the formative power of the morphogenetic field is proposed in our work as an extension of Deleuze's ideas [168] where singularises plays the essential part in forming the transition from the virtual to the actual through the ontological operator of difference. *Difference* is incorporated in our work through the *intensive* nature of the morphogenetic field as the *gradient* field associated with the antenna propagation potential.

6.5.4 The Concept of the Morphogenetic Field

According to the standard story, the idea of a field organizing and orchestrating the process of development was introduced independently in the 1920s by three biologists, Hans Spemann in 1921, Alexander Gurwitsch in 1922, and Paul Weiss in 1923 [162]. The sense of the fundamental importance of these postulated fields was allotted to their value in organizing normal growth and devolvement and also directing the processes of regulation and regeneration, for example, if the normal devolvement is disturbed for some external reason.

It is interesting, however, to notice that the conceptual basis of the field concept as a guiding *force* in morphogenesis can be found in the earlier work of Driesch, in particular, in the semi-mathematical concept of the 'prospective potency' of a given part of the embryo to develop into several forms [156]. This concept appears to be a more sophisticated adaptation of the framework of potential functions in mechanics. What is of direct interest to us here is that Driesch's 'prospective potency' was defined as a *function over space*, making it resembling something like a scalar field, another strong indication of its conceptual analogy to the potential function in mathematical physics. In a later stage of his theory, Driesch introduced (non-mathematically) the main force of biological development as a form of "intensive manifoldness." The concept of *intensity* was cautiously used by Driesch who is also a philosopher. The potential function is related to the idea of *intensity* in the sense that force is generated by *difference* in potential. For example, if we take the most intuitive scalar potential field, temperature, then we immediately recall that heat flows under the intensive force of temperature differences (or more precisely, heat *gradient*). Similarly, in classical mechanics, the gravitational force is generated by the gradient of the gravitational potential, a scalar field. The electric (Columbian) force is generated by the gradient of the electrostatic potential. And so on. The idea of the intensive versus the extensive is well known in both physics and philosophy.[17] Modelling the force of biological devolvement by Driesch as '*intensive* manifoldness' suggests, in our opinion, that the concept of morphogenetic field can be traced back to the work of this biologist and theorist.

In this chapter, we endorse the framework of *morphogenetic fields* as the appropriate theoretical framework for understanding the problem of morphogenesis in the antenna near field zone. Our formulation is more sophisticated than the typically non-mathematical models found in biology, especially in the early literature, but less speculative and ambitious that the later catastrophe theorists [163], [164]. We adopt the proposition that *the antenna morphogenetic field is the gradient of the antenna propagation potential.* This gradient is computed with respect to the 6-manifold of the extended configuration space of the problem. It describes the directions[18] in this phase space along which the field tends "most strongly" to transform from its poor-form state of nonpropagating modes to a rich-form state of propagating modes. The meaning of the vague expression 'most strongly' is unambiguously fixed through our definition by fact that the gradient points to the direction of *maximum* increase in the scalar field from which it is derived.

6.6 CONCLUSION

We would like now to start a summary of the main ideas presented in this chapter. The efforts undertaken her focus on bringing attention to the richness and complexity of the problem of morphogenesis in general, and the formation of electromagnetic radiation in the antenna near zone in particular. The theoretical framework of morphogenesis has been singled out by us as a crucial step in understanding the genesis of radiation in a quite general and hence necessarily abstract manner. The abstractness here is not merely a byproduct of the formulation of the evolution equations of the near field, but, we believe, the inevitable consequence of the effort to understand the actual physical content of these equations. The derivation of the evolution law (6.5) does not require directly any biological input for its completion. However, the purely empirical fact of the existence of the antenna propagation

17 For a penetrating philosophical analysis of the extensive and intensive, see Bergson's early work [165]. For a modern, actually postmodern, treatment, the reader is referred to the work of Deleuze, for example, see [168].

18 'Direction' is a geometric, not topological concept, and therefore a metric for the configuration space of the antenna problem has to be chosen.

potential (which in tern is a consequence of the Weyl expansions) cannot be understood fully by just performing numerical integration of the differential equations. A deeper understanding of the unique way in which the new degrees of freedom associated with the rotation group SO(3) are distributed in the extended phase (configuration) space of the problem calls for a radically new manner of dealing with evolution questions in physics and engineering. However, the proposed interpretation sketched out very briefly in this present chapter is actually a devolvement of ideas originally developed long time ago by thinkers like Driesch and Waddington [156], [158]. Our main contribution is in putting all these scattered ideas together in a coherent whole. A quite well-developed feeling of the dynamic content of these theories applied to the understanding of electromagnetic radiation will be attained gradually in time when further theoretical and empirical research on the morphogenesis of electromagnetic radiation is performed.

Part II

The Antenna Current Green's Function (ACGF) Formalism

Chapter 7

The Antenna Current Green's Function Formalism as a Paradigm

7.1 GENERAL INTRODUCTION: THE PROGRAM OF ELECTROMAGNETIC SYSTEMS

7.1.1 Explanatory Adequacy in Scientific Research

Chief among the closely guarded ideals of science is the attainment of what has become known as *explanatory* adequacy, in contrast to *descriptive* adequacy [1], [2]. The latter concerns the objective depiction of the observations under investigation conducted in a manner as faithful as possible to the phenomena themselves, and without introducing doubtful elements of subjective standards. As such, a given phenomenon can be considered 'adequately described' when an unequivocal procedure for reproducing the main features of this phenomenon exists. In antenna and circuit engineering, this ideal appears to have been achieved. Indeed, given any antenna or circuit structure, it is possible now, at least in principle, to either perform a full-wave numerical simulation or collect measured data such that both procedures describe the system very well, and the two sets of results can be shown to be approximately identical.

The situation, however, appears to be different with respect to the other ideal of science, the *explanatory* function of applied electromagnetic theory. Informally, explanation is the explication of the unfamiliar in terms of the familiar. 'Adequacy' of an explanation means that the corresponding explication is rich enough to encompass most of the details of the phenomena under investigation judged by typical practitioners as important. In natural science, and in particular electromagnetic theory, we may define explanatory adequacy formally as the explication of detailed mechanisms articulating the intricate web of basic processes leading to the production of the observed phenomena. This definition in turn depends on the term 'basic processes.' By this we mean very simply something like a familiar process. The "familiar" components include things like geometrical transformations (e.g., rotation, translation), harmonic modes, resonance, atoms, field-oscillator interaction, and so on. These are all part of the daily bread and butter of the scientist and engineer, and consequently any theory that makes recourse only to such familiar processes is deemed by our presentation as explanatory adequate.

We will give an example illustrating the fundamental distinction between descriptive and explanatory adequacy. A basic problem in electromagnetic theory is this. Two antennas A and B are placed in close vicinity to each other such that coupling between them occurs in the near-field zone. Let antenna A functions as a source and antenna B as a receiving system. We can observe the voltage developed at the receiving port of antenna B. To *adequately describe* this problem, it is possible either to directly measure the voltage at the physical port or to perform a full-wave simulation of the combined system A + B. Both approaches lead to answers within the limits of the now universally accepted numerical and empirical procedures. Next, suppose that antenna A is kept fixed, but now we rotate antenna B with respect to some frame of reference. In general, the received voltage will change. Again, it is possible either to measure or compute the new voltage. However, while such a procedure is perfectly adequate to describing any new situation, it is totally at loss when it comes to the question *why* and *how* this voltage changes. Indeed, neither the empirical nor the numerical procedures are capable of even formulating this question in a mathematically precise manner. What we encounter here is a fundamental shortcoming in descriptive methods of mainstream science in general. They are, by design, incapable of probing deeply *beneath* the routinely measured/computed in order to provide alternative views about the rich tapestry of mechanisms buried under the surface of observed phenomena.

In this part, we will develop a formalism for the analysis of general electromagnetic systems, mainly antennas but also including circuits, such that some clues about the issue of explanatory adequacy defined above can be provided. The *antenna current Green's function formalism* includes the previous stage of descriptive adequacy,[1] and therefore does not sacrifice the accuracy and precision of the traditional descriptive procedures. However, it is further developed here in order

1 Because currently the mentioned Green's function can be obtained practically only using full-wave numerical solutions, though we hope lab-based methods will be developed in the near future.

to supply a deeper insight into electromagnetic energy exchange and transfer in general systems, particularly the core phenomena of near-field coupling and interaction. The ultimate purpose of providing explanatory adequacy is the attainment of acceptable rationalization of the topic under study, allowing us consequently to fully control the phenomena and hence exploiting some of the subject matter's future unseen potentials.

7.1.2 Context and Motivations

Core problems in electrical engineering, at least if attention is restricted to the physical layer, revolve around the existence of well-defined systems, commonly named *circuits*, responsible for the implementation of functions originally conceived by the designer. Deciding whether such functions involve radiation, reception, filtering, generation, etc, depends on the actual anatomical details of these circuits, but the overall abstract concept, that of a well-defined, isolable module, the circuit *element* as such, remains the key theoretical presupposition upon which the entire edifice of electrical engineering is based. Therefore, the total complex of a given engineering system can be analyzed into separate basic building blocks variously interconnected with each other (usually in a very complicated manner) while situated in a global platform that may itself enters in the future into the community of other more or less equally complicated systems.

It is important to appreciate the hidden assumptions implicit in any *circuit* formulation of problems with electrical engineering purport. The circuit paradigm is based on the existence of units satisfying some or all of the following conditions:

1. **Axiom 1**. Each unit is well defined. That is, when embedded within the overall system, there are no great ambiguities in delimitating the boundaries formed by the unit, or the physical domain occupied by the device, or its material constitution, or the nature of its connection with other units existing in its neighborhood.
2. **Axiom 2**. Each unit possesses some sort of independent identity. By this we mean that a unit has a character (electromagnetic in nature) that persists in relative autonomy with respect to the existence of other nearby elements. For example, a circuit that operates as an oscillator should not turn to functioning as something else, say a mixer, because a transmission line is placed in its vicinity. Although the existence of a transmission line, or any other device, will theoretically affect the electromagnetic character of the oscillator, this

effect should not be so vast to the degree of turning the oscillator into something completely different.

3. **Axiom 3**. The focus of the engineer or the designer is not on the entire performance measure of the unit, but is limited instead only to certain *terminal-like* quantities, such as voltage, impedance, current, power flow, etc. These quantities of interest usually give rise to a *finite* set. In contrast, the full range of the information relevant to the operation of any electromagnetic unit is *infinite* (because electromagnetic fields, being smooth functions on the space-time continuum, are infinite "lists of data" by definition).

When an electromagnetic unit satisfies the three conditions above, we usually refer to it by the term 'circuit.' These three conditions then are fundamental and form the axiomatic basis of any rational discourse dealing with circuits and networks in electrical engineering.

The main contents of this circuit paradigm has been definitively consolidated during the early years of the birth of electrical engineering, a process that was stimulated by war-related activities but later took off as an independent discipline in the period immediately following the end of World War II, mostly in the United States. The two early pioneers of electrical engineering, Norbert Wiener and Y. M. Lee, formulated the mathematical apparatus needed in setting up the stage for a comprehensive and adequate description of general electrical networks. Although practical research in wireless technology dates earlier by almost half a century (Marconi, Rayleigh, Heaviside, etc), the systematic integration of radiating elements (antennas) within larger systems started to gather momentum only after the maturation of the *impedance concept* and the establishment of *network theory* as a paradigm for the scientific study of electrical systems.

Probably the main feature in the Wiener–Lee approach was the reliance on the spectral analysis technique (Fourier methods) to deal with the circulation of information within an existing electrical network. Therefore, we see that the now *de facto* concepts of *transfer function* and *filtering* play a fundamental role in rationalizing our conception of large-and-complex systems. The reasons motivating this trend are not hard to find. The Fourier method has been one of the main tools at the disposal of theoretical physicists since its inception by Fourier during the first two decades of the 19th century. The method, being already a mathematically powerful tool, proved also to be a extremely fertile general "thinking platform" for reflecting on nature in broad terms. The idea that a general physical effect can be expanded as a linear sum of oscillations coheres very nicely with the intuitive understanding of interactions as a sort of mechanical responses to certain applied or forced vibrating modes. (The metaphor of the ringing of a vibrating fork is buried deep in the memory of most people.) It was not difficult therefore to imagine an extension of the Fourier method, especially after its rigorous justification by 19th-century mathematicians, such as Cauchy, Reimann, Dirichlet, in order to apply it to all natural phenomena that happen to obey linear differential equations. Indeed, the use of Fourier analysis in electrical engineering during the last six or seven decades proved to be the single most pervasive and influential theoretical idea at the core of the entire discipline. But it should be kept in mind that the deployment of the Fourier method, similar to what happened with the stochastic methods of statistical mechanics, was not created from scratch by Wiener and Lee, but consciously appropriated from an already existing mature tradition in theoretical physics.

However, we observe that *the systematic application of the Fourier methods described above was focused almost exclusively on time, rather than space*. That is, the Fourier expansion was performed with respect to the temporal variation of the various signals involved. This is quite understandable when viewed from the historical context of the developments that took place immediately after 1945. Indeed, the majority of early industrial applications involved low-frequency circuits, which can be modeled very accurately as lumped-element circuit elements. Moreover, for the majority of other systems that cannot be modeled in this way, a distributed-element model was developed for basic circuits and transmission lines. This circuit model was compared directly to classical electromagnetics using the theory of electromagnetic wave propagation in waveguide structures, which had been already developed much earlier during the last century. For radiating elements, i.e., antennas, whether operational in the transmitting or receiving modes, the strategy was to characterize them in the near-field perspective using the concept of *equivalent impedance*, i.e., by replacing the antenna itself by a simple lumped-element circuit object. In this way, the antenna part can be analyzed when connected with other circuit elements using the same mathematical formalism. The resulting microwave network theory was quite successful (and remains so) because it is solely concerned with few terminal quantities like voltages and currents at identifiable physical ports. *However, when the analysis is focused on the spatial variations of the fields involved at these ports, a situation that is created and sustained by modern developments in the applied field and the monotonic growth of the complexity of existing and projected systems, the standard network theory becomes inadequate.*

In recent years, the need to know more about the spatial distribution of the fields in complex antenna and circuit systems mobilized the use of brute-force full-wave numerical solution of Maxwell's equations through one of the popular numerical methods available today, e.g., FEM, MoM, FDTD. We believe that although knowledge of certain numerical ratios or data is in principle possible by either full-wave simulation or direct measurement, the theoretical understanding of the nature of electromagnetic interactions and the mechanisms of information transfer within general electromagnetic systems is still lacking. This gap, however, cannot be filled by simply improving the efficiency of numerical or experimental methods. The reason is that such methods can provide information about only a *concrete* system, i.e., the particular system that is being simulated or measured at the time. Instead, such a gap in our knowledge can be filled only by developing a general theoretical framework for analyzing and studying electromagnetic flow and interactions in general electromagnetic systems at a broad level. This study, we believe, has to answer in particular to the demand for constructing a Fourier formalism for *space*, not time.

The electromagnetic problem is inherently more complicated because of the following peculiarities. While we can expand the spatial variation of the signal applied to a given electromagnetic device in terms of the Fourier transform, the device itself is described in spatial terms, say via the geometrical details of the device's physical boundaries. There is a direct connection, generally very complex, between the spatial information content of the fields circulating within and around a given electromagnetic system and the spatial geometry of the system itself. It is the analysis and the understanding of such subtle interrelations what constitutes a major strand in our contributions in this book.

For instance, it will be shown in Chapter 9 that by Fourier analyzing the input signal into spatial-harmonic modes, and by also performing the same operation on the *Green's function* describing the device interacting with this input signal, it will be possible to propose a mechanism explaining the overall interaction picture in terms of simple basic field–oscillator interaction processes familiar in atomic physics. This picture was attained only after performing the *spatio*-spectral analysis of the right degrees of freedom. However, note that although this analysis involves considerable field-theoretic considerations, it is still a *circuit*-based analysis in the sense that only *localized* parts of the systems, the input and the output terminals, are concerned. The mode of the analysis we are proposing here is therefore intermediate in complexity between a conventional microwave circuit on one hand, and antenna theory and an exact full-wave field analysis on the other, although it is as accurate and rigorous as the latter.

Few comments must be inserted here regarding how we would like to place out theory with regard to the rest of the electromagnetic community, particularly the optical field of research. Optics is basically relevant to the short wavelength limit $\lambda \rightarrow 0$. That is, with the exception of the very recent development of near-field nano-optics, all of classical optics deals with far-field phenomena. One manifestation of this fact is that optical devices placed in the vicinity of each other do not couple electromagnetically as is the case in the microwave regime where the operating wavelength is comparable with the physical dimension of the devices.[2] Finally, we mention that in optics there has been from the beginning a considerable interest in the spatial variations of the fields, in contrast to the case in microwave engineering. For example, beam shaping, focusing, spatial filtering, are all spatial operations that are analyzed using Fourier methods. However, they are still in the main *far*-field phenomena. Although in this work we will make use of spatial Fourier transform methods, the theoretical framework, near-field interactions, is different from classical disciplines such as Fourier optics.

7.1.3 Overview of the Present Work

The prime objective of the present investigations will be the construction of a complete mathematical formalism suitable for the analysis of energy–information flow and interactions in general systems, composed of both circuit and antenna parts (this will become the main interest of Chapter 9, while this chapter is concerned with establishing the theoretical foundations of the concept of the antenna current Green's function itself). Our arguments will demonstrate that such analysis of general electromagnetic systems is reducible to the antenna problem. More precisely, we treat pure circuit problems as manifestations of antenna-antenna interactions where coupling occurs in the near-field zone. The antenna problem itself is analyzed into three modes of operations, Modes A, B, and C. We later (Chapter 9) employ suitable reciprocity theorems to eliminate Mode C by reducing it to Mode A. The two remaining modes, Mode A and Mode B, are essentially distinct and one cannot be reduced to the other. We then provide in Chapter 9 a theory for the interrelations between all these three modes and study various practical configurations of possible interactions. The net outcome is what we believe a general formalism suitable for handling the analysis of arbitrary metallic electromagnetic systems understood ultimately in terms of the foundational problem of antenna theory, i.e., the theory of the three modes mentioned above.[3] Applications of the ACGF formalism will be developed in several independet chapters. Chapter 14 apply the formalism to

2 By 'electromagnetic coupling' we understand change in the electromagnetic responsitivity of one device because of the presence of another object near to it. The most general definition of electromagnetic coupling is developed in Chapter 16.

3 The extension of the same ideas to other types of interacting media, for example mixed metallic and dielectric materials, is possible in principle but quite lengthy and so will not be attempted in the present book.

receiving antenna arrays, with special focus on far-field excitations. The more difficult and substantially undevloped topic in the literature, i.e., arrays responding to generic near-field illumination, will be briefly dealt with in Chapter 15. Applications to mutual coupling analysis are presented in Chapter 16. Finally, the utlization of the ACGF formalism for the analysis and design of MIMO systems is discussed in Chapter 18.

One of the fundamental insights that the antenna current Green's function formalism can provide, especially when combined with the theory of the antenna near fields, is the crucial relevance of the way various local and global frames of reference are being interpreted when closer examination of the dynamic content of the phenomena under consideration is attempted. For example, it will be shown that no preferred local coordinate system can be chosen in the mathematical description of the antenna current Green's functions of Modes A and C, which is a consequence from the fact that the configuration space of this problem is a general 2D surface (2-manifold) with a curvature (or Reimannian metric) that cannot be fixed in advance for general antennas. Multiple coordinate systems can be defined and introduced locally. The actual content of the theory, for example the 2D Green's tensor, is expressed only in terms of one frame of reference but its transformation into the language of other frames is readily available using the methods of differential geometry.

On other hand, the core of this chapter will be the explication of the *relations* between the modes, particularly Mode B and Mode C, as we endeavor do in Chapter 9. For this problem, we will show that adequate understanding of the mechanisms of electromagnetic coupling in the near-field zone can be attained by paying special attention to the geometrical transformations (rotation and/or translation) of certain local coordinate systems that are directly connected with the basic physical processes in the interaction picture. It will be demonstrated later in detail how such basic geometrical transformations, such as rotations, lie behind the complex mechanism generating the actually observed change in the received voltage.

Consequently, it is appropriate to reinstate here that a fundamental theoretical theme recurring in various forms throughout the entire book is the particular attention we pay to the intricate issues surrounding the use of coordinate systems (or frames of reference). Although it has been often emphasized in the literature of electromagnetic theory that Maxwell's equations can be given a "coordinate-free" formulation, for example using the calculus of differential forms, we believe that coordinate systems are not dispensable, and that even a coordinate-free formalism presupposes more or less a concept of coordinate systems. It is precisely the manner in which coordinate systems are defined with respect to distributions of sources and physical boundaries that will dictate the way in which a process of electromagnetic interaction will unfold. Even though no preferred coordinate frame is singled out, for each concrete problem a set of global and local frames of references have to be established unambiguously. Our subsequent analysis suggests that a careful study of the geometrical transformations involved in the electromagnetic phenomena now described in terms of those chosen frames will make a rational understanding of the physical problem ready at hand.

7.1.4 Fundamental Assumption about General Electromagnetic Systems

This book aims at studying electromagnetic systems at the most general level. The motivations for this research, beside the original incentive of the search for new and deeper scientific knowledge, also include some practical issues. One of the most important of these issues is the search for alternative methods for analyzing large-and-complex general systems, which will probably dominate many of the advanced applications of the future.[4] As such, it is important to develop some mathematical procedures capable of providing a platform for thinking about these systems. Our main fundamental ontological postulate about general systems will be the following [3]

> Every system, except the universe, interacts with other systems in certain respects, and is isolated from other systems in other respects. (7.1)

This axiom will be tacitly presupposed in our approach below. Actually, the same axiom can be shown to be implicitly assumed in conventional antenna theory. If one is studying the radiation produced by a single element, then the antenna under consideration is taken to be electromagnetically isolated from the rest of the universe. On the other hand, when radiation by an *array* of closely spaced elements is considered, then interactions between various objects is taken into account. In the case of *general* systems, the decision concerning which elements are to be considered isolated and which are strongly interacting cannot be answered in advance, but must be investigated *empirically* in connection with a particular system. In this book, we discuss theoretically the various possible scenarios in general system analysis and show that even at such an abstract level some concrete knowledge about electromagnetic interactions can be attained. More specific examples will also be given whenever necessary, e.g., see Chapters 14, 15, 16, 18.

4 Indeed, although the future is in principle an open-ended process, the only thing an engineer can claim about the future with certainty is that it will deal with systems vastly more complex than those known at the present. A glance at the history of technology immediately confirms this claim.

Indeed, without treating some parts of a complex circuit as relatively independent entities, no hope can be attained in analyzing the complex whole. At the same time, in electromagnetic systems, more than any other subsystems like low-frequency circuits, it is extremely important to remain always alert about the latent channels of interactions. Electromagnetic systems in general operate as an integral whole, and this is one of the reasons why recourse to full-wave simulation of the entire system, say an antenna array with strong mutual coupling, has been very frequent in the last two decades. The art of analyzing large-and-complex systems will rely on striking a certain balance between both the relative autonomy of a circuit element and its electromagnetic coupling with the surrounding elements. The analysis to be developed in this book will pay attention to these issues.[5]

7.1.5 The Circulation of Information in General Electromagnetic Systems

There is another theme that will appear repeatedly throughout our investigations. This is the issue of the flow and circulation of information within an existing electromagnetic system. Unfortunately, this topic is seldom treated in the vast literature on applied electromagnetics. The subject matter of our remarks goes to the manner in which unknown field variations move from one location in the system to another. Notice that once a given circuit or antenna system is assembled, usually no modification in the geometry is possible.[6] However, the actual *form* of the field that is being injected into the system throughout the latter's operational duration is contingent on the practical circumstances of the system's function and environment. This is akin to speaking about 'information' in communication theory where the form of the signals cannot be anticipated in advance. Usually, in order to analyze the performance of such systems, a stochastic technique is introduced. In the case of conventional circuit theory developed with spectral methods in time, the analysis can be executed in the now traditional manner.[7] However, for the case

5 However, we immediately remind the reader that this chapter does not replace empirical studies of concrete systems. Our approach is abstract and general in the sense that it supplies a broad mathematical scheme and a physical interpretation of this scheme. We don't provide rules of thumbs or empirical prescriptions for things like, for example, when to consider two elements strongly coupled or not. Such a practical insight can be generated by using our mathematical formalism together with its physical content in direct empirical studies of specific systems. Examples of such an empirical information obtained using our theory can be found in the applications chapters of the present work.

6 With the exception of electro-mechanical systems.

7 That is, the mathematics of linear system theory and its extension to include stochastic processes, time series, data correlations, etc.

of general electromagnetic system, where the analysis must be based on *spatial* Fourier methods, it is not clear how to perform such a statistical study.

What happens in practice is this. The antenna engineer and the microwave circuit engineers design their devices by ensuring that they meet certain standard operational measures, such as directivity, gain, polarization, input impedance, etc. After assembling, the system is handed over to other specialists. The communication and system engineers in turn characterize the system in terms of their own measures, such as signal-to-noise ratio, intersymbol interference, probability of error, sensitivity analysis, budget analysis, etc. What is not performed is an actual analysis of how the information dealt with by the latter engineers is really circulated in the existing electromagnetic system designed by the former engineers. Indeed, we find here that the accumulated experience of network theory, which is based on dealing with Fourier analysis with respect to time, fails to measure up with the need to study electromagnetic coupling and flow, a processes that involves essentially *spatial* considerations.

The mathematical methods developed in this part will pay attention to the issue of the circulation of information within a general system by clearly indicating how the spatial variations of the injected signal affect the responses produced at distant parts of the system. Since our whole approach is based on frequency-domain formulation, only the spatial content of the information flow will be analyzed. The full-fledged study of *both* temporal and spatial dynamics is beyond the scope of this book; however, any such a study, we believe, will probably involve a a formalism more or less resembling the one we are proposing here.

7.1.6 The Interrelation Between Synthesis and Analysis

Notice that although the ultimate objective in electrical engineering is the design or *synthesis* of complex systems in order to perform certain preconceived tasks, the procedure of *analysis* is fundamental and cannot be dispensed with. In order to understand the reason behind this, it may be helpful to picture the actual development of any design process. Initially, the engineer implements a design as an *idea* in the most economic means available to him: By *imagining* a very rough approximative mental blueprint, a kind of primitive initial prototype. He then needs either to verify the imagined idea or to explore its consequences (practically both alternatives turn out to be necessary). For this the engineer proceeds by making use of the next available *analytical* tool, say, a circuit simulator. Notice that even if the ultimate devolvement will depend on the use of full-wave simulators, all of the next-in-line methods awaiting the engineer in research and development depend on pure analysis of existing prototypes. There is, however, an indirect way in which analysis participates crucially in the synthesis process. By going back to the first step in the design process, the creation (by the imagination of the engineer) of an initial prototype, a question comes out naturally: how did the designer come up with the basic idea after all? Did the blueprint comes "out of the blue"? Obviously, there is no unequivocal answer to this kind of queries, which belong more conveniently to the field of the psychology of creativity. The celebrated moment of mentally creating the initial prototype is essentially (ontologically speaking) synthetic, i.e., it is irreducible in its ownmost being to an analytic element. But there is no denying that the *ability* of the engineer to create a good initial design hinges on his own personal grasp of the problem. If the only means available to him are either numerical or experimental, then little can be achieved aside from common sense and practiced intuition. However, the existence of a comprehensive mathematical theory that fully explicates the physics behind the working of electromagnetic systems greatly enhances the ability of the engineer to think a new design. We conclude then from all of the above that *deep understanding of analysis is an essential step in any actual design process*. Therefore, although this book is tends to focus on the analytical in the sense that it develops a comprehensive theoretical understanding of the mathematics and physics of *existing* systems, we believe the theory presented here will be of value for future engineering studies of large-and-complex systems, nano-scale problems, and artificial media. As history has been teaching us up to date, a profound comprehension of the inner workings of a complex system is the best incentive for the invention of better systems that can surpass the old design in ways usually totally unseen without the mediation of theory.

7.2 OUTLINE OF THE GENERAL PROBLEMS OF ANTENNA THEORY

Antennas can operate in two different modes, the transmitting and receiving modes. The choice of this particular dichotomy is purely pragmatic, being ultimately reducible to the concrete and specific ways in which scientists and engineers need to deal with radiating structures functioning as parts of larger and more complex systems [28]. For research purpose, we will attempt in this part to study the two separate (transiting and receiving) operational modes of general antenna systems. The goal is to understand

1. their principles of operation when viewed separately from each other,

2. and the mode of interrelation between these different modes of operation when combined within a single system.

For studying the first point, we take the analysis of the antenna in the transmitting mode for granted and focus on the receiving mode. That does not imply that the transmitting mode operation is now a completed or finished research topic, for nothing is farther from truth than this. However, as will be shown later, the first stage of the transmitting mode antenna system, i.e., solving for the induced current using, for example, the method of moment, is a relatively speaking well mature subject now.[8]

Here, Part II, we will study the second point, the interrelational aspect of the transaminating and receiving mode, from the perspective of the near field. This will be achieved by showing that the antenna in the receiving mode can be viewed as a collection of "atomic resonators," i.e., well-defined objects interacting resonantly with the illuminating field. The latter is analyzed using the recently developed theory of the antenna near-field structure in the spectral domain (Part I). It will be demonstrated in that rotating a given antenna, which is immersed in the near field of another antenna, will change the received voltage by varying the contributions of the illuminating field's propagating and nonpropagating (evanescent) parts interacting with the aforementioned atomic resonators.[9]

In our approach, we will systematically demarcate the two modes of operation. Although a reciprocity theorem relates some of the performance measures of the two modes to each other, the mechanisms of operation of the antenna system in the two situations are generally quite different.[10] As we will strive to clarify throughout, one of the main contributions of this research program is the correction of some widely held misunderstanding regarding such an issues.

We present the overall study of the antenna system in the a threefold process as shown in Figure 7.1, where the full path leading from a given voltage source in one antenna to the receiving port of another antenna illuminated by the field produced by the former is illustrated. The starting point of any practical PEC

8 The analysis of the second stage (see the second block in Figure 7.2a) is undertaken in Part I of this book.

9 As will be shown later, this analysis was considerably simplified by assuming that this rotation of the receiving mode antenna immersed in the field of a transmitting mode antenna does not affect the current distribution on both antennas. That is, the *current-current* interaction is weak, while our focus is concentrated on the *field-voltage* interaction, the *field* being the illumination one radiated by the transmitting mode antenna, while the *voltage* is the signal developed at the physical port of the receiving mode antenna.

10 For an extended discussion and analysis of this issue not invoking spectral considerations, see Chapter 14.

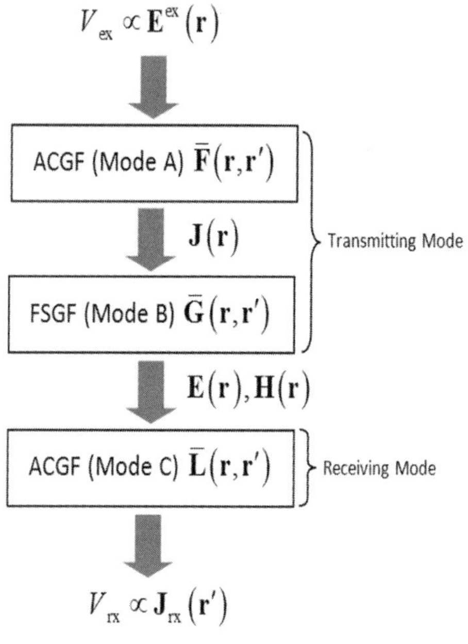

Figure 7.1 Combined general description of antenna system. The system is complete in the sense that the full chain of processes starting from the voltage excitation of a transmitting antenna up to the physical port of a receiving antenna is included in the analysis.

antenna is a source excitation V_{ex} proportional to an externally applied electric field $\mathbf{E}^{\text{ex}}$.[11] The antenna responds by generating an electric current distribution $\mathbf{J}(\mathbf{r})$ such that the re-radiated field plus the applied fields satisfy the boundary condition on the antenna surface [28]. The electromagnetic problem, being linear, admits a Green's function $\bar{\mathbf{F}}(\mathbf{r}, \mathbf{r}')$ that connects the output current with the input field excitation, which we call here the *antenna current Green's function* (ACGF). That is, we write[12]

$$\mathbf{J}(\mathbf{r}) = \int_S ds' \bar{\mathbf{F}}(\mathbf{r}, \mathbf{r}') \cdot [\hat{n} \times \mathbf{E}^{\text{ex}}(\mathbf{r}')]. \tag{7.2}$$

Here, S represents the (orientable) antenna surface (all interfaces presenting material discontinuities in the antenna system). The unit vector $\hat{n}$ is the normal vector on S pointing in the outward direction.[13]

If the impressed field is localized, then the surface integral in (7.2) can be reduced to the region of influence of this input.[14] We call this step in the transmitting phase Mode A. That is, Mode A is the stage in the overall transmitting phase in which a locally applied external field will produce a current on the entire surface of the antenna system.[15]

Next, we introduce Mode B of the overall transmitting phase. Here, the antenna current distribution $\mathbf{J}(\mathbf{r})$ will radiate fields everywhere in space through the free space Green's function (FSGF) $\bar{\mathbf{G}}(\mathbf{r}, \mathbf{r}')$. This can be written at once as

$$\mathbf{E}(\mathbf{r}) = \int_S ds' \bar{\mathbf{G}}(\mathbf{r}, \mathbf{r}') \cdot \mathbf{J}(\mathbf{r}'). \tag{7.3}$$

11 For general boundary conditions, the externally applied magnetic field has to be included. This does not affect any of the general conclusions to be drawn henceforth since the antenna current Green's function formalism is valid for any (macroscopic) electromagnetic system.

12 Throughout this book, an implicit time-harmonic dependence $\exp(-i\omega t)$ is assumed and suppressed everywhere. Also, we have the homogeneous space wavenumber given by $k_0 = \omega\sqrt{\varepsilon_0\mu_0}$. Here ε_0 is the electric primitivity and μ_0 is the magnetic permeability of free space.

13 Obtaining the ACGF $\bar{\mathbf{F}}(\mathbf{r}, \mathbf{r}')$ in (7.2) is the most difficult part in the overall process of the transmitting mode. Generally speaking, the whole "art" of antenna engineering can be reduced to the problem of knowing how to tailor the antenna specifications in order to produce a desired current distribution in response to a known source. For now, we obtain this Green's function only through numerical solvers in conjunction with the distribution-theoretic construction in Section 8.2.3.

14 Cf. equation (8.11). Moreover, for simplicity, we restricted ourselves here to only metallic antennas. The generalization to dielectric and metallic antennas is possible but a little bit more involved.

15 It is also possible to formulate the whole problem in terms of volume equivalence theorem. In this case, the antenna surface S should be replaced by some effective volume V. However, here we analyze the antenna problem using the concept of surface equivalence only. It appears to the author that the fundamental understanding of Mode A does not require dealing in an essential way with the consideration of configuration spaces possessing dimensionality larger than two.

Here, the FSGF is given by (18.36). The radiated field $\mathbf{E}(\mathbf{r})$ represents the final output of the antenna system in the transmitting phase. We therefore managed to explicate this phase as composed of two modes cascaded in series, first Mode A followed by Mode B, as shown in Figure 7.2a.

The second phase, i.e., Mode B, which involves the free space Green's function, is completely different. Here the Green's function $\bar{\mathbf{G}}(\mathbf{r},\mathbf{r}')$ is fully known in closed-form analytical solution.[16]

We now consider the final stage in Figure 7.1, the antenna receiving part. The previously produced fields $\mathbf{E}(\mathbf{r})$ and $\mathbf{H}(\mathbf{r})$ impinge on the receiving antenna. The latter may be located in the near or far zone of the transmitting antenna. The illuminating fields interact with the entire surface of the receiving antenna and produce a voltage that can be observed at the receiving part. In general, an induced surface current is generated on *both* the PEC part of the port and the antenna surface. However, for practical applications, where PEC ports are used to collect the developed voltage, only this voltage is pertinent to applications. Therefore, here we restrict our focus in the study of receiving antenna systems to observation of voltages developed at unique locations in the illuminated antennas, i.e., the physical port.

The receiving antenna problem is still governed by Maxwell's equations where the incident field plays the excitation or forcing term. Similar to the situation in the transmitting mode, the existence of external fields disturbs the boundary condition so a current distribution is generated on the antenna surface in order to re-establish the correct electromagnetic transition condition. The linearity of the problem allows us to express the relation between the input (illumination field) and output (antenna current distribution) by *another* Green's function operator, say $\bar{\mathbf{L}}(\mathbf{r},\mathbf{r}')$. This mode, which we call Mode C, is similar to Mode A in the sense of being described by an antenna current Green's function. The produced current is therefore given by

$$\mathbf{J}_{\mathrm{rx}}(\mathbf{r}) = \int_S ds \bar{\mathbf{L}}(\mathbf{r},\mathbf{r}') \cdot \mathbf{E}(\mathbf{r}') . \tag{7.4}$$

Again, we ignore here the interaction with the magnetic field since the antenna system is assumed for simplicity to be PEC system. The integration surface S is that of the receiving antenna.

16 The produced field $\mathbf{E}(\mathbf{r})$ can be roughly classified into far field and near field. The far field has a simple well-known analytical form (spherical outgoing wave). An outline of a general theory of the antenna near field providing in-depth analysis of its spatial structure have been developed in Chapters 3, 4, and 5.

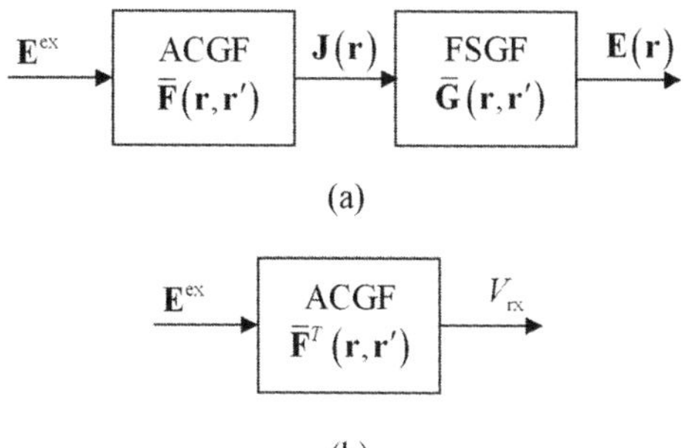

Figure 7.2 General description of antenna system. (a) Transmitting mode model. (2) Receiving mode model.

In Chapter 9, it was proved that Mode C receiving ACGF $\bar{\mathbf{L}}(\mathbf{r},\mathbf{r}')$ is related to the Mode A transmitting ACGF $\bar{\mathbf{F}}(\mathbf{r},\mathbf{r}')$ by simple transpose relation. It is possible then to determine the voltage observed at the receiving port if the ACGF, of which the excitation voltage is also placed in the same port location, is known. As will be seen later, if the port location is changed, a new ACGF has to be calculated. Therefore, it is possible to divide the main problems in antenna theory into two types as shown in Figure 7.2 where we treat the transmitting and receiving modes as separate problems. The transmitting part consists of Mode A followed by Mode B. The input is some external field $\mathbf{E}^{\text{ex}}(\mathbf{r})$ while the output is fields $\mathbf{E}(\mathbf{r})$ and $\mathbf{H}(\mathbf{r})$ radiated everywhere in the exterior region. This is shown in Figure 7.2a. The receiving mode problem, Mode C above, is given by the transpose of same Green's function of the Mode A. Figure 7.2b shows this problem where the illuminating field is a general external field $\mathbf{E}^{\text{ex}}(\mathbf{r})$, not necessary the one mentioned in Figure 7.2a. We would like to add that although there seems to be a simple relation between the transmitting Mode A and the receiving Mode C, the mechanisms of electromagnetic interactions in the case of general arrays problems are very different.[17]

The general pictures presented by either the combined transmitting/recieving system of Figure 7.1, or the individual models of Figure 7.2, serve to delimitate the main problems in antenna engineering as they usually arise in practice. As has been stated repeatedly so far, the fact that all modes of electromagnetic interactions

17 For example, the existence of variable loads connected to the ports of the receiving modes introduces additional complexity that was absent in the transmitting mode.

can be described by linear operators naturally suggests representing each mode by a proper Green's function.

Chapter 8

Foundations of the Antenna Current Green's Function Formalism

8.1 DEFINITION OF THE ANTENNA CURRENT GREEN'S FUNCTION

8.1.1 Definition of the Antenna System

We first make a precise definition of what we mean by an *antenna system*.[1] Consider a region D in the Euclidean 3-dimensional space $\mathbb{R}^3$ inside which an antenna is located. By the term 'antenna' we usually understand a composition of various materials with varying electromagnetic properties such that the entire structure, when excited by placing a localized electric field at a certain location (excitation port), becomes capable of producing electromagnetic fields propagating efficiently away from the structure into the exterior region $\mathbb{R}^3 \setminus D$. For the mathematical formulation of our problem, we will adopt the vantage point of the surface equivalence theorem in which only changes of the electromagnetic character of the medium through surfaces are taken into account. Indeed, we will assume that there is a finite number N of surfaces composing the antenna system such that it is associated with each surface S_n an electromagnetic boundary condition BC_n describing the exact manner in which the field must undergo a change when moving across this surface. The total surface comprising the antenna system will be defined simply as the union of all these individual surfaces

$$S = \bigcup_{n=1}^{N} S_n, \; \mathrm{BC} = \{\mathrm{BC}_n\}_{n=1}^{N} . \tag{8.1}$$

1 We anticipate that for the extension of the antenna current Green's function formalism to general dielectric-PEC systems, this definition will prove vital in securing the correctness of the derivation.

That is, the boundary condition BC of the antenna system is the set of all individual boundary conditions associated with the individual surfaces S_n. It is seen then from (8.1) that the definition of a general antenna system possesses two components. The first is purely *geometrical* and consists of the internal structure of the combined surface S as the sum (set-theoretic union) of the individual surfaces S_n. The second component is *electromagnetic* and consists of the corresponding data $\mathrm{BC}_1, \mathrm{BC}_2, ..., \mathrm{BC}_N$. In this book we will treat only the PEC boundary condition; i.e., here BC will be simply $\{\mathrm{PEC}\}$. However, note that *all the concepts to be developed under the rubric of the antenna current Green's function formalism can be generalized to arbitrary electromagnetic boundary conditions.* From now on, by the 'antenna physical body' we understand the total surface S.[2]

The surfaces composing S are described mathematically using the rigorous theory of two-dimensional *differential manifold*, that is, a geometrical surface with an additional structure allowing us to perform calculus on this surface.[3] Although manifold theory is now standard in the mathematical literature, we believe some important conceptual issues relevant to the 2-manifold theory are in need for further clarifications in connection with electromagnetic theory. It is well known that a general surface is not topologically equivalent to a Euclidean patch. Technically, this translates into the fact that it is not possible to cover a general surface with only one coordinate patch. Instead, one has to revert into a collection of coordinate neighborhoods overlapping in a specific technical manner described in detail in [61]. Although this observation may appear of rather technical import, it actually implies some considerations that are crucial for the appreciation of the subtleties involved in the concept of the antenna current Green's function and how it differs from the classical Green's functions of mathematical physics, where the later arise from *differential* equations. As we will show later, in order to obtain the ACGF, one must solve an *integral* operator equation to obtain a vector field (the current) defined on the global surface S. The geometric details of this surface determines completely the nature of the operator at question.

The topology of S is the *global qualitative* features encoded in the geometry of S. There is an organic connection between this topology and the solution of the integral operator equation mentioned above, a connection that is still not well understood due to the fact that no analytical solution of the general surface integral

2 At an even more abstract level, one may prefer to define the antenna system as the ordered triple $\langle \mathcal{S}, \mathrm{BC}, \vartheta \rangle$, where $\mathfrak{S} = \{S_1, S_2, ...\}$, and the function ϑ is defined as $\vartheta : \mathfrak{S} \rightarrow \mathrm{BC}$, mapping every surface $S_n \in \mathfrak{S}$ to the corresponding boundary condition $\mathrm{BC}_n \in \mathrm{BC}$. Notice the difference between $\mathfrak{S}$ defined above and S in (8.1).

3 The basic definitions and results needed for our aims here can be found in any book dealing with modern differential geometry. For example, see [61].

equation exists. However, even with the absence of such solutions, we believe it may be possible to infer some information about the ACGF by applying suitable topological techniques in order to pin out the relation between the ACGF and the shape of S. Such a study is beyond the scope of this book and therefore will be avoided. However, the fact that several coordinate patches are needed to cover a general surface is at the heart of the ACGF method as will be shown in detail later. Indeed, it will be demonstrated that there is no unique way for writing the 2D tensor representing the ACGF. In general, any collection of coordinate patches can do the job. The 3D tensor representing the ACGF in the language of a global frame of reference can be derived from the local representation of the 2D tensor. One manifestation of the topology of S will then be seen as a restriction on the available coordinate patches and the manner of their overlap. Since such coordinate patches enter directly into the structure of the 2D ACGF tensor, and therefore, into the derived 3D tensor, we begin, although in at a relatively primitive level, to catch a glimpse of the impact of the topology of the antenna system surface S on the electromagnetic characterization of the device under consideration.

The subtle topological issues in antenna theory can probably be best illustrated by the following simple example. Consider a closed loop and a curved open wire. Topologically, these two structures are different because there is no continuously invertible bijective map that can take one into the other. (On the other hand, a square loop and a circular loop are topologically equivalent because such a transformation can be easily constructed.) Now, we know that wire antennas act like *electric* dipole while loop antennas behave like *magnetic* dipoles. The two electromagnetic types of dipoles are qualitatively, and hence fundamentally, different. We may now suspect that the topological difference between the loop and wire structures plays some role in this important dissimilarity in the electromagnetic performance. To our knowledge, this topological structure of antenna theory has never been seriously discussed in literature. From the viewpoint of the present work, we notice that the special topology of closed loops entails that it is impossible to cover the entire loop, regardless to its shape, by a single coordinate patch. This is not the case with an open wire where such a coordinate map can be immediately constructed. To describe *all* the points of the closed loop, one must make resort to *multiple* coordinate patches overlapping in the manner described in [61]. We may then begin to appreciate the fundamental importance of paying the utmost attention to the full machinery of two-manifold theory. Indeed, the ACGF of a general antenna system can *not* be described by a single unique global coordinate system. Only a *local* representation is possible. The rich topology of 2D surfaces comprising the structure of the antenna

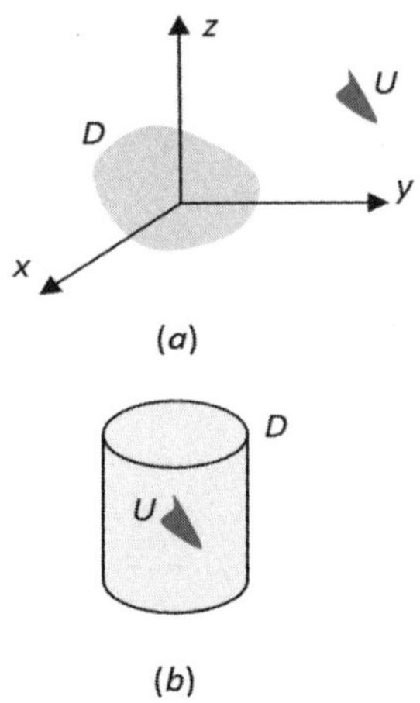

Figure 8.1 Illustration of the interaction between between a source (externally applied field) localized in a compact region U and an antenna system contained within another compact set D. a) $U \cap D = \emptyset$. b) $U \subset D$.

system as defined in (8.1) are encoded in the manner of the distribution and overlap of these coordinate patches used in the construction of the ACGF.

In order to better grasp this definition of the antenna system, let us consider the mechanism of the excitation of an arbitrary antenna. According to the definition (8.1), it is only the boundary condition BC that is responsive to an externally applied field. The reason is that if a given field is nonzero at a location in S, the total surface of the antenna system, then the BC applicable there is disturbed. In order to remain valid, a nonzero current distribution is generated on the *entire* surface S which will radiate electromagnetic fields everywhere, but most importantly the radiated fields, when combined with the external fields, will satisfy the global boundary condition BC. In Figure 8.1, two possible scenarios are presented. We would like to consider externally applied electromagnetic fields that are 1) smooth[4] and 2) vanish outside a compact support U.[5] Assume that the entire antenna system is located in a region D, which can be taken also to be compact without loss of generality. In Figure 8.1a, we consider first the situation when the compact support of the source U and the antenna domain D are disjoint, i.e., when $U \cap D = \emptyset$. In this case,

4 By 'smooth' function we mean a function with continuous partial derivatives of all order.

5 It is possible to show explicitly that such functions exist using basic techniques in elementary differential topology.

even with the usual time-harmonic excitation condition $\exp(-i\omega t)$, the source, being completely confined within the region U, cannot disturb any of boundary conditions BC, and therefore the antenna will not be excited.[6] Consider now the situation depicted in Figure 8.1b. In this case, we assume for simplicity that the antenna system consists of merely a homogeneous dielectric cylinder with some relative permittivity different from one. Inside this homogeneous regions, we apply an externally controlled field with compact support U as in Figure 8.1a. In this case, the localized source, although located *within* the antenna domain D itself, still does *not* disturb any boundary condition[7] and therefore the antenna is *not* excited and no radiated fields will be observed. As should become clear now, it is the fact that the material filling the antenna domain is homogeneous throughout the entire source domain U what prevents any interaction between the localized source and the actual antenna system. These simple facts should be kept in mind as they figure prominently in the derivation of the ACGF in the receiving mode, especially in the case when more complicated boundary conditions than the PEC are taken into account.[8]

Before leaving this point, let us comment briefly on the issue of electromagnetic interaction between two objects placed in the vicinity of each other. Imagine that the first object is a PEC part while the second is a dielectric object such that the regions occupied by both are disjoint. As is well known, if we excite the first object, the PEC part of the antenna system, the presence of the other object will affect the radiated field. One may wonder how could this be explained in light of our observation that the mechanism of antenna excitation consists solely of the disturbance caused by the external field at a boundary surface right at the location where the source is applied. If we excite the PEC part by a localized source, then no direct disturbance of the boundary condition of the second object is immediately pertinent. What happens actually is the following. The field generated by the PEC object is unlikely under general circumstances to decay rapidly enough away from the excited PEC object. Therefore, since the dielectric object is located at the

6 The situation, of course, is different with plane wave excitation. In the latter case, such an external field is be definition nonvanishing everywhere in $\mathbb{R}^3$, in particular at D.

7 In this particular example, the electromagnetic boundary condition is the continuity of the tangential electric and magnetic fields along the cylindrical surface separating the homogeneous dielectric region from the surrounding infinite free space.

8 A careful reader may wonder why we assume the localized source to be defined over a *compact* support U instead of just assuming a Dirac delta source. The reply is that although the Dirac delta source works perfectly well in this argument of Figure 8.1, some other readers may object that it is not a "physical" field since it is not smooth. To avoid such philosophical argument, we made use only of smooth functions defined on a compact support.

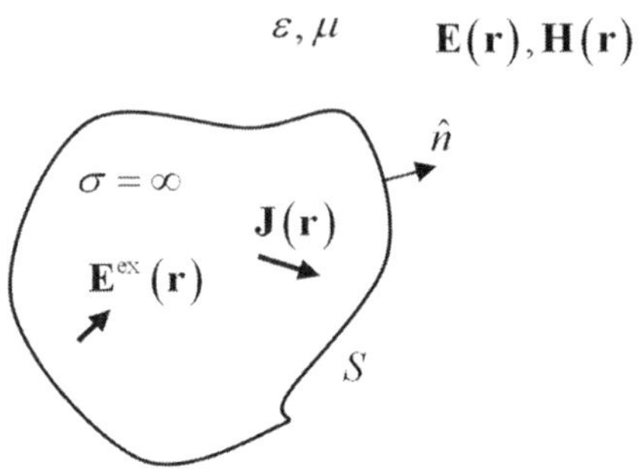

Figure 8.2 A metallic antenna with arbitrary shape excited by an external electric field $\mathbf{E}^{\text{ex}}(\mathbf{r})$ producing a current distribution on S that in turn generates the fields $\mathbf{E}(\mathbf{r}), \mathbf{H}(\mathbf{r})$ throughout an infinite homogeneous space surrounding the antenna described by a dielectric constant ε and permeability μ. For simplicity, we set $\varepsilon = \varepsilon_0$ and $\mu = \mu_0$ for the rest of this book.

vicinity of the metallic one, the field radiated by the latter will in turn disturb the boundary condition of the former. The actual radiated field has obviously to satisfy simultaneously both boundary conditions in order to serve as the self-consistent unique field produced by the antenna system. The details of this analysis belongs to the study of Mode A, and is in our opinion still poorly understood.[9]

8.1.2 An Intuitive Approach

Figure 8.2 illustrates the general geometry of the problem. An PEC antenna is enclosed within the volume inscribed by a closed surface S.[10] A unit normal vector $\hat{n}$ points outward of the surface S. The structure is excited by a localized electric field distribution $\mathbf{E}^{\text{ex}}(\mathbf{r})$, i.e., a field with a compact support contained in S. Since only

9 Notice that although nowadays it is routine to compute or measure the self-consistent field radiated by an antenna array under the condition of strong mutual coupling, very little is known about the detailed mechanisms giving rise to the actually observed field. One reason for this is perhaps the excessive reliance on numerical simulations, which don't always provide the right framework for a deeper analysis of radiation problems.

10 By 'closed surface' we mean a compact 2-manifold with no boundary [61]. Therefore, since in practice any surface PEC has a finite nonvanishing thickness, this model can deal with arbitrary PEC physical shapes. It is possible however to formulate the problem with respect to a mathematical surface that is open. In this case, the ideal situation of exactly zero thickness can be modeled. In both cases, the mathematical theory of 2-manifolds or surfaces is the same and therefore we restrict our treatment in the text to closed surfaces.

the tangential component of this field is relevant to the PEC boundary condition, we work only with the surface (vector field) $\mathbf{E}_t^{\text{ex}} := \hat{n} \times \mathbf{E}^{\text{ex}}$. Subsequently, the antenna responds by generating an electric current distribution $\mathbf{J}(\mathbf{r})$ with a compact 2-dimensional support consisting of the entire antenna physical body surface S.[11]

The relation between the impressed field $\mathbf{E}_t^{\text{ex}}$ and the generated current $\mathbf{J}(\mathbf{r})$ is given by the electric field integral equation (EFIE) [70]

$$\mathbf{E}_t^{\text{ex}}(\mathbf{r}) = -i\omega\mu_0\, \hat{n} \times \int_S ds' \left[\bar{\mathbf{I}} + \frac{1}{k_0^2}\nabla\nabla\cdot \right] \frac{e^{ik_0|\mathbf{r}-\mathbf{r}'|}}{4\pi\,|\mathbf{r}-\mathbf{r}'|} \cdot \mathbf{J}(\mathbf{r}'), \tag{8.2}$$

which is a Fredholm equation of the first kind. Equation (8.2) can be put in the form $\mathbf{E}_t^{\text{ex}} = \mathcal{L}\mathbf{J}(\mathbf{r})$, where $\mathcal{L}$ is a linear (but unbounded) operator [70]. Therefore, mathematically the problem of exciting the antenna can be understood in terms of a linear map $\mathcal{L}$ defined on the linear space of vector fields on the compact support S. Indeed, this function space can be given the structure of Banach space. (See [70].) Theoretically speaking, Mode A and Mode C in antenna systems can be studied completely in a reduced-dimensional space represented by the 2-manifold S. Mode B, however, essentially involves the full Euclidean space $\mathbb{R}^3$ in order to describe electromagnetic radiation into free space.

It can be proved using the basic theory of linear operators that the electric-field exterior problem is uniquely solvable provided that k_0 is not an eigenvalue of the interior problem [70].[12] Since the integral operator $\mathcal{L}$ defined by (8.2) is linear, its inverse $\mathcal{L}^{-1}$ is also linear [58]. Therefore, we expects that $\mathcal{L}^{-1}$ admits a Green's function. However, it is not immediately clear how this function should be written. In general, since the antenna excitation problem in Figure 8.2 is embedded in $\mathbb{R}^3$, one expects that the ultimate Green's function must be written as a 3D tensor $\bar{\mathbf{F}}(\mathbf{r}, \mathbf{r}')$ as in (7.3).[13] While this is obviously correct, care must be taken in interpreting the components of this tensor. As will be shown below, even before

11 All of the theoretical devolvements to be presented below in connection with Figure 8.2 applies to the case when multiple antenna elements are located in the vicinity to each other with arbitrary geometries and positions. For reasons of brevity, we will refer to the total antenna array surface by S and refrain from using multiple indices to distinguish different elements within the array. A detailed analysis for the general array case can be found in Chapter 14.

12 However, for the numerical solution of surface integral equations using methods that depend on mesh descretizations, problems may occur at frequencies matching the eigenmodes of the interior problems [40]. This difficulty can be dealt with by using combined integral equations [40], [70]. In this part, which is not concerned with the numerical method used in obtaining the ACGF, we don't worry about conditions related to convergence or actual numerical accuracy.

13 For a simplified review of the dyadic calculus needed in electromagnetic theory, see [34]. For more comprehensive introduction to tensor calculus in general, see [58].

enforcing reciprocity considerations, it is not true that the components of this tensor are all independent. The reason turns out to be in the nature of the operator equation $\mathbf{E}_t^{\text{ex}}(\mathbf{r}) = \mathcal{L}\mathbf{J}(\mathbf{r})$, which is really in a function space on two-dimensional non-Euclidean manifold S, not the three-dimensional Euclidean space $\mathbb{R}^3$. It will be demonstrated shortly that the Green's tensor of such a problem is properly a 2D tensor $F_{nm}, n, m = 1, 2$, defined by piecing together functions defined *locally* on S. This remains true even though the operator $\mathcal{L}$, being an integral rather than differential operator, is actually a global operator.

In order to mathematically describe the tensor F_{nm}, one has to choose suitable local coordinate systems on S. Let us consider the problem of exciting the antenna at location $\mathbf{r}'$ and observing the induced current generated at $\mathbf{r}$. There are no preferable coordinate frames in the atlas defining S, and therefore we pick any two patches, one (U_l, x_l) such that $\mathbf{r} \in U_l$, while the other $(U_{l'}, x_{l'})$ satisfies $\mathbf{r}' \in U_{l'}$. (Notice that in general it is not necessary that the two patches overlap, even if both $\mathbf{r}'$ and $\mathbf{r}$ belong to a connected surface.) Here, x_l and $x_{l'}$ are in $\mathbb{R}^2$ and represent the local coordinates of points located in U_l and $U_{l'}$, respectively. In order to facilitate our construction of the tensor $F_{nm}(\mathbf{r}, \mathbf{r}')$, we choose two orthogonal unit vectors $\hat{\alpha}_l^1(\mathbf{r}), \hat{\alpha}_l^2(\mathbf{r})$ at $\mathbf{r}$ in the language of U_l, and $\hat{\alpha}_{l'}^1(\mathbf{r}'), \hat{\alpha}_{l'}^2(\mathbf{r}')$ at $\mathbf{r}'$ in the language of $U_{l'}$.[14] We can then formally construct all possible combinations of tensor products $\hat{\alpha}_l^n(\mathbf{r})\,\hat{\alpha}_{l'}^m(\mathbf{r}')$ for $n, m = 1, 2$ as will be used below.

In the standard literature on differential manifold theory, tensor products are defined only with respects to vectors defined at the *same* point, i.e., vectors belonging to the same tangent space TM_p, where p is some point in the (2-dimensional) manifold under consideration. The tensor product $\hat{\alpha}_l^n(\mathbf{r}) \otimes \hat{\alpha}_{l'}^m(\mathbf{r}')$ involves vectors belonging to two different tangent spaces, i.e., $TM_{\mathbf{r}}$ and $TM_{\mathbf{r}'}$. This difficulty can be overcome in several ways. Probably the most direct strategy is to exploit the fact that the vectors involved can all be embedded in the larger space $\mathbb{R}^3$. In this case, it is always possible to perform parallel transport of any vector from a given point to an arbitrary point. Since the inner product is invariant with respect to permutations involving vectors that are obtained from each other by the process of parallel transport, the tensor product above can always be construed as a 3D cartesian tensor product.

This solution may not be found in harmony with our earlier claim that Modes A and C can be studied, at least theoretically, as processes occurring completely in a 2D space, rather than $\mathbb{R}^3$. In this case, it is desirable not to refer to the larger

14 Notice that two linearly independent vectors do not necessary constitute *coordinate* basis. In particular, the two orthonormal vectors chosen above may not be coordinate basis. This, however, has no important bearing on what follows.

space in the definition of the antenna current Green's function operator. It is enough for our purpose to *define* the operator entity $\hat{\alpha}_l^n(\mathbf{r}) \otimes \hat{\alpha}_{l'}^m(\mathbf{r}') : TM_{\mathbf{r}'} \to TM_{\mathbf{r}}$ by the equation $\hat{\alpha}_l^n(\mathbf{r}) \otimes \hat{\alpha}_{l'}^m(\mathbf{r}') \cdot \hat{\alpha}_{l'}^q(\mathbf{r}') := \hat{\alpha}_l^n(\mathbf{r}) \left[\hat{\alpha}_{l'}^m(\mathbf{r}') \cdot \hat{\alpha}_{l'}^q(\mathbf{r}')\right]$, which is possible since we have a natural inner product on S (i.e., the Euclidean inner product inherited from the ambient space $\mathbb{R}^3$ into which S is embedded) available for computing $\hat{\alpha}_{l'}^m(\mathbf{r}') \cdot \hat{\alpha}_{l'}^q(\mathbf{r}')$. Finally, we mention that by reverting to the techniques of tangent bundles on abstract manifolds, this argument can be made fully rigorous. However, we don't pursue further such a formalization in the present work.

Denote by $\delta_{\mathbf{r}'}$ a Dirac-like delta (point) source located at $\mathbf{r}' \in S$. (Detailed examination of this Dirac delta functions is given in Section 8.2.3, particularly in the discussion around (8.30).) The antenna response to a point source polarized along the unit vector $\hat{\alpha}_{l'}^1(\mathbf{r}')$ at $\mathbf{r}'$ is itself a vector but located this time at $\mathbf{r} \in S$. In particular, it can be written as

$$\mathcal{L}^{-1}\left\{\delta_{\mathbf{r}'}\hat{\alpha}_{l'}^1(\mathbf{r}')\right\} := F_{11}\hat{\alpha}_l^1(\mathbf{r}) + F_{21}\hat{\alpha}_l^2(\mathbf{r}). \tag{8.3}$$

Similarly, the response to point excitation directed along $\hat{\alpha}_{l'}^2(\mathbf{r}')$ is expressed in the form

$$\mathcal{L}^{-1}\left\{\delta_{\mathbf{r}'}\hat{\alpha}_{l'}^2(\mathbf{r}')\right\} := F_{12}\hat{\alpha}_l^1(\mathbf{r}) + F_{22}\hat{\alpha}_l^2(\mathbf{r}). \tag{8.4}$$

Now, since an arbitrary tangent vector at $\mathbf{r}'$ can be written as a linear combination of $\hat{\alpha}_{l'}^1(\mathbf{r}')$ and $\hat{\alpha}_{l'}^2(\mathbf{r}')$, it follows from the linearity of the operator $\mathcal{L}^{-1}$ that the antenna current produced in response to any input can be written completely in terms of the four functions F_{nm} defined above. Furthermore, by using the inner product inherited by S from $\mathbb{R}^3$, we can express the Green's tensor as

$$\bar{\mathbf{F}}(\mathbf{r},\mathbf{r}') = \sum_{n=1}^{2}\sum_{m=1}^{2} F_{nm}(\mathbf{r},\mathbf{r}')\,\hat{\alpha}_l^n(\mathbf{r})\,\hat{\alpha}_{l'}^m(\mathbf{r}'). \tag{8.5}$$

From the basic identity $\left(\hat{a}\hat{b}\right) \cdot \hat{c} = \hat{a}\left(\hat{b} \cdot \hat{c}\right)$ [58], which is true for any three vectors $\hat{a}$, $\hat{b}$, and $\hat{c}$, the expansion (8.5) is justified by formally observing its role in the expression (7.2).

The four functions $F_{nm}(\mathbf{r},\mathbf{r}')$, $n, m = 1, 2$, determine *completely* the antenna response to an arbitrary electric field excitation construed as a vector field on S. As was emphasized explicitly in (8.5), this expansion of the ACGF depends on the two local coordinate systems containing the observation and source points, i.e., on the choices (U_l, x_l) and $(U_{l'}, x_{l'})$, respectively.[15] In general, it is possible to transform this tensor from one local frame to another using the formula (8.57) derived in Section 8.2.6. Therefore, in practice a particular choice of coordinate frames is made, e.g., the two patches U_l and $U_{l'}$, and then the four corresponding functions F_{nm} are computed. Afterwards, a translation to arbitrary other coordinate languages can be made using the transformational calculus developed in Section 8.2.6. In effect, any suitable local frame can be chosen since all local coordinate systems on S are in principle equivalent.

It is now an easy matter to deduce the form of the ACGF 3D tensor, that is, the expression of the 2D tensor (8.5) but this time written in terms of the global (i.e., the one associated with the ambient space $\mathbb{R}^3$) 3D Cartesian frame xyz. Define $x_1 := x$, $x_2 := y$, and $x_3 := z$. Each tangential vector $\hat{\alpha}_l^n(\mathbf{r})$ can be expanded in terms of the unit vectors $\hat{x}_m$ as

$$\hat{\alpha}_l^n(\mathbf{r}) = \sum_{m=1}^{3} \beta_{lm}^n(\mathbf{r})\, \hat{x}_m, \tag{8.6}$$

where the functions $\beta_{lm}^n(\mathbf{r})$ are easily determined using the inner product on the manifold S by the relation $\beta_{lm}^n(\mathbf{r}) = \hat{\alpha}_l^n(\mathbf{r}) \cdot \hat{x}_m$. Substituting (8.6) into (8.5), we obtain

$$\bar{\mathbf{F}}(\mathbf{r}, \mathbf{r}') = \sum_{p=1}^{3}\sum_{q=1}^{3} F_{x_p x_q}(\mathbf{r}, \mathbf{r}')\, \hat{x}_p \hat{x}_q, \tag{8.7}$$

where

$$F_{x_p x_q}(\mathbf{r}, \mathbf{r}') := \sum_{n=1}^{2}\sum_{m=1}^{2} F_{nm}(\mathbf{r}, \mathbf{r}') \beta_{lp}^n(\mathbf{r})\, \beta_{l'q}^m(\mathbf{r}'). \tag{8.8}$$

The form of the antenna current Green's function given by (8.7) will be refereed to as the 3D ACGF tensor. We must notice that, in contrast to the components of the 2D tensor appearing in (8.5), *the components of the 3D ACGF are in general not independent of each other*. This can be easily inferred from the expression (8.8), a fact that is not really surprising if we keep in mind that the genuine problems of Modes A and C possess a configuration space of the dimension two. Now, the Green's function, being properly a distribution or a generalized function, cf. Section

15 In fact, for a fully rigorous notation one has to indicate explicitly in the functional form of F_{nm} their dependence on the choice of the coordinate patches U_l and $U_{l'}$, for example by writing $F_{nm}^{ll'}(\mathbf{r}, \mathbf{r}')$ instead of $F_{nm}(\mathbf{r}, \mathbf{r}')$. However, the latter version was adopted in order to simplify the notation. The reader has to bear in mind all the time that any reference to specific components of a tensor necessarily involves a choice of a coordinate frame.

8.2.3, is never dealt with outside its defining (7.2). In such equations (and also in (8.9) below), the Green's function is never multiplied by an arbitrary vector, but always appears associated with a vector field *tangential* to S, the configuration space of the physical and mathematical problem. For example, it is not appropriate to interpret the components of the 3D ACGF individually, say by calling $F_{x_1 x_1}$ 'the x_1-component of the response of the antenna to an applied electric field polarized in the x_1 direction.' Such an interpretation is rejected from the outset because it is not always the case that the x_1-component of this applied field is tangential to the antennas surface at the location under consideration.

We are now ready to write down the integral equation satisfied by the ACGF. This can be achieved from (8.2) with the help of (8.7). The result is

$$\begin{aligned}\hat{n} \times \hat{a}\, \delta\left(\mathbf{r} - \mathbf{r}'\right) = -i\omega\mu_0\, \hat{n} \times \int_S ds'' \left[\bar{\mathbf{I}} + \tfrac{1}{k_0^2}\nabla\nabla\cdot\right] \\ \times \tfrac{e^{ik_0|\mathbf{r}-\mathbf{r}''|}}{4\pi|\mathbf{r}-\mathbf{r}''|} \cdot \left[\bar{\mathbf{F}}\left(\mathbf{r}'', \mathbf{r}'\right) \cdot \left(\hat{n} \times \hat{a}\right)\right],\end{aligned} \tag{8.9}$$

where $\hat{a}$ is unit vector describing the polarization of the impressed field point source. It is clear that since the problem in (8.2) is vectorial in $\mathbb{R}^3$, the appropriate Green's function describing the relation between the input excitation (electric field) and the output (surface electric current density) should be the 3D tensor (8.7).

In (8.9), the vector $\mathbf{r}'$ gives the position of the impressed excitation $\mathbf{E}^{\mathrm{ex}} = \hat{a}\, \delta\left(\mathbf{r} - \mathbf{r}'\right)$. The ACGF will represent the response to this particular excitation in the form of the *Cartesian* (i.e., a 3-dimensional) vector $\bar{\mathbf{F}}\left(\mathbf{r}'', \mathbf{r}'\right) \cdot \left(\hat{n} \times \hat{a}\right)$, which gives the electric current surface density at the location of the vector $\mathbf{r}''$. The free space dyadic Green's function $\bar{\mathbf{G}}\left(\mathbf{r}, \mathbf{r}''\right)$ defined in (18.36) will generate the radiated field due to a point current source located at $\mathbf{r}''$ observed at the position $\mathbf{r}$. This radiated field will cancel exactly the tangential component (to the surface S) of the externally supplied impulsive excitation $\mathbf{E}^{\mathrm{ex}}$ defined above. Therefore, it is important for consistency to keep in mind the semantic differences between the three position vectors $\mathbf{r}$, $\mathbf{r}'$, and $\mathbf{r}''$ while reading (8.9). This integral equation is interesting from the theoretical point of view because it encodes in a concise manner the interplay between two very different types of electromagnetic Green's functions, the traditional free-space function (18.36) (forward problem) and the antenna current Green's function (a kind of "reverse" problem). Finally, the relation (8.9) can be justified rigorously using the special distribution theory developed in Section 8.2.3. However, the details are not given here for brevity.

8.1.3 Some General Remarks

The problem solved by obtaining the antenna current dyadic Green's function $\bar{\mathbf{F}}(\mathbf{r}, \mathbf{r}')$ is the *inverse* of the classical problem solved by physicists long time ago, which is described by the free space dyadic Green's function (FSGF)

$$\bar{\mathbf{G}}(\mathbf{r}, \mathbf{r}') = \left[\bar{\mathbf{I}} + \frac{1}{k_0^2}\nabla\nabla\cdot\right] \frac{e^{ik_0|\mathbf{r}-\mathbf{r}'|}}{4\pi|\mathbf{r}-\mathbf{r}'|}, \tag{8.10}$$

where subsequently the superposition principle is used to compute the antenna radiated field once the current distribution on the antenna is known as shown in (7.3). Therefore, the latter problem, Mode B, was already solved in a closed analytical form, while the former one, the *engineering* problem of designing antennas, i.e., Mode A, cannot be solved exactly in terms of simple known function because integral equations in general do not admit such solutions [70].

Informally, equation (7.2) may be verified by substituting it into (8.2) and using (8.9) after interchanging the (finite) integral operator with the dyadic operator $\bar{\mathbf{F}}$. But a rigorous procedure for the verification of expressions like (7.2) must ultimately be based on distribution theory since the dyadic Green's functions are generalized functions [58]. The rudiments of such a formulation is provided in Section 8.2.3, where the existence of the ACGF is established based on first principles, i.e., distribution theory and differential manifold theory.

It is important to mention that the ACGF $\bar{\mathbf{F}}(\mathbf{r}, \mathbf{r}')$ defined above is in general *not* a function of $\mathbf{r} - \mathbf{r}'$. This also applies to the ACGF $\bar{\mathbf{L}}(\mathbf{r}, \mathbf{r}')$ of the receiving mode. *Therefore, the ACGFs of Mode A and Mode C are not spatially shift-invariant, while the free space Green's function is shift invariant.* This considerably complicates the mathematical treatment of Modes A and C compared with standard Green's function (FSGF) methods in electromagnetics, in which the latter have been developed mostly in connection with the investigation of Mode B. The reason behind this vital difference between the ACGF and the FSGF can be explained as follows. The *configuration space* of a given mode is defined as the space on which the output field (= the field produced by the physical process associated with the mode under consideration) is defined. Mode B, as can be seen from (7.3), has the full 3D Euclidean space $\mathbb{R}^3$ as its configuration space. However, from (7.2) and (7.4) we clearly see that the configuration space of Modes A and C is the 2D space (surface) S; here, in $\mathbf{J}(\mathbf{r})$, the output field of these two modes, the current distribution, is defined only on this lower dimensional space, i.e., $\mathbf{r} \in S$. The configuration space of Mode B, then, being the Euclidean manifold $\mathbb{R}^3$, is a *homogeneous* space.[16] On the other hand, the surface S is obviously *not* homogeneous in general. This is why the ACGFs of Modes A and C are not only dependent on the distance between the source and observation points, but also on the relative locations of these points with respect to other points on the surface, for example the boundary. If we take very special antennas in which S is a homogeneous space, for example an infinite plane PEC sheet or a perfect conducting sphere, then the ACGF of these structures *is* shift invariant, and in these cases we can write, for instance, $\bar{\mathbf{F}}(\mathbf{r}, \mathbf{r}') = \bar{\mathbf{F}}(\mathbf{r} - \mathbf{r}')$ because it can be shown that (8.9) is invariant to a uniform shift of both the observation and source points. *For other nonsymmetric antenna shapes, this conclusion need not be true in general.*

8.1.4 Response to Arbitrary Excitation using the ACGF Method

In Mode A, the antenna is usually excited by a *localized* electric field, say defined on a small compact support $U \subset S$. In this case, (7.2) is reduced to

$$\mathbf{J}(\mathbf{r}) = \int_U ds \bar{\mathbf{F}}(\mathbf{r}, \mathbf{r}') \cdot [\hat{n} \times \mathbf{E}(\mathbf{r}')] . \tag{8.11}$$

Notice that this equation is not valid for Mode C, where the antenna in the receiving mode is illuminated by an external field that is generally nonvanishing everywhere, and therefore in the latter case the full integration region S has to be retained as in (7.2). For the case of Mode A, we observe that equation (8.11) suggests that when an antenna is used as a radiator in a given system in which the excitation area U is already fixed, then if the data $\bar{\mathbf{F}}(\mathbf{r}, \mathbf{r}')$, $\mathbf{r}' \in U$, is stored, one can compute the current on the antenna for *arbitrary* excitation field $\mathbf{E}(\mathbf{r}')$. This situation is expected in practice since the actual excitation, the function $\mathbf{E}(\mathbf{r}')$, $\mathbf{r}' \in U$, with fixed U, depends on the information being processed or the overall mode along which the entire system is operating.

The antenna element or array, together with their excitation (feeding) mechanism which determines U, are usually fixed after design and installation, while the information content, the variations of the fields being transferred and processed throughout the system, is not known a priori for obvious reasons. For example, in communication systems, the actual variations of the field depend on the specific information being transferred, processed, detected, etc. Equation (8.11) then is of

16 By a homogeneous space we understand a space in which given any point located inside this space, all of its neighborhoods look the same compared with other neighborhoods containing any other point. The expression 'look the same' can be made precise in differential geometry, see [61].

great value for system engineers since *once the ACGF is computed and stored as explained above, the operation of the system can be predicted without the need to perform a new full-wave simulation for every new excitation.*

8.2 DISTRIBUTIONAL FOUNDATIONS FOR THE ANTENNA CURRENT GREEN'S FUNCTION

8.2.1 Resume of the Distributional Theory

In Sections 8.2.2-8.2.5, we develop the mathematical foundations of the ACGF formalism using distribution theory. There are two main motivations behind this formulation. The first is to provide a rigorous justification of the major theorem (7.2). This we achieve in Section 8.2.3 where the ACGF is proved to exist by actually constructing one. The inverse reciprocity theorem Chapter 9 is then utilized to prove the relation (7.2).

The second motivation is providing an approach to actually compute the ACGF using numerical methods. It is clear that no closed-form analytical expression exists for the general electromagnetic boundary-value problem and therefore the use of full-wave numerical methods to approximate the antenna Green's function is inevitable. Our construction of the ACGF in the form of a suitable limit of a sequence of distributions, given in (8.33), can serve as a foundation for testing and developing methods aiming at utilizing the ACGF in the role of a transfer function model for various concrete applications, for example receiving antenna arrays (for a brief overview, see Chapter 9).

The structure of our approach to the distributional theory can be summarized very briefly in the following manner. We first provide in Section 8.2.2 a direct construction of the Green's function for a scalar problem. This fictitious problem is dealt with first in order to motivate two technical proposals: 1) approximating the ACGF as a limit (in a suitable sense) of responses to special delta sequence, and 2) constructing a *local* delta sequence that reflects the differential geometry of the antenna system surface S. After solving the scalar problem, the construction of the actual electromagnetic ACGF is attempted in Section 8.2.3 using a chain of carefully chosen definitions. We then prove the main relation (7.2) rigorously using the inverse reciprocity theorem, which will be derived in detail in Chapter 9. Finally, anticipating the major applications of the ACGF formalism in Chapter 9, we develop in Appendices 8.2.4 and 8.2.5 the mathematical foundations of the Fourier transform of the ACGF using tempered distribution theory. In particular, we show that the ACGF as a surface function can be replaced by a sequence of functions with 3-dimensional compact support in $\mathbb{R}^3$. This idea is essential for securing the validity of the main formulation of Chapter 9, where interaction between two antennas in the near zone is addressed.

8.2.2 Direct Construction of the ACGF Using Distribution Theory: Scalar Theory

In this section and Section 8.2.3, we will provide an outline for the construction of the ACGF of a general antenna system specified by a 2-manifold S using the theory of distributions or generalized functions (we don't distinguish between the two terminologies). For simplicity, we will work out the construction first for a scalar problem. The extension to the electromagnetic case will be taken up in Section 8.2.3.

The ACGF is defined as the response of the antenna system to a local[17] delta-source excitation $\delta(\mathbf{r}-\mathbf{r}')$ placed in S, i.e., for $\mathbf{r}' \in S$. The ACGF is then defined as $F(\mathbf{r},\mathbf{r}') := \mathcal{L}^{-1}\delta(\mathbf{r}-\mathbf{r}')$, where again $\mathcal{L}$ is the (surface) electromagnetic operator of the problem. Let $\mathcal{D}(M)$ be the standard space of test functions in distribution theory [58], [67], i.e., the space of smooth (infinity differentiable) functions with bounded support in M, where M is either $\mathbb{R}^n$ for some $n \geq 1$, or a smooth manifold embedded in some $\mathbb{R}^n$.[18] The *topological dual* of $\mathcal{D}(M)$ is the space of continuous linear functionals on $\mathcal{D}(M)$, which is customarily denoted by $\mathcal{D}'(M)$.[19] Let $\varphi(\mathbf{r}) \in \mathcal{D}$. The distribution F is then defined tentatively as the continuous linear functional given by the relation

$$\begin{aligned}\langle F(\mathbf{r},\mathbf{r}'),\varphi(\mathbf{r})\rangle &:= \left\langle \mathcal{L}^{-1}\delta(\mathbf{r}-\mathbf{r}'),\varphi(\mathbf{r})\right\rangle \\ &= \left\langle \delta(\mathbf{r}-\mathbf{r}'),\mathcal{L}^{-1}\varphi(\mathbf{r})\right\rangle.\end{aligned} \tag{8.12}$$

Here the scalar interaction is as defined by

$$\langle \phi(\mathbf{r}),\varphi(\mathbf{r})\rangle := \int_S ds\,\phi(\mathbf{r})\,\varphi(\mathbf{r}), \tag{8.13}$$

which is well defined for any two functions ϕ and φ that are locally integrable on S. The foundations of the definition (8.12) rest on the fact that the operator $\mathcal{L}$ is self

17 The precise definition of 'local' used here will be given later in (8.28).

18 In the case that M is such a smooth manifold, we set $M := S$ throughout the rest of this work.

19 A *functional* g on $\mathcal{D}$ is a function $g : \mathcal{D} \to \mathbb{R}$. The suitable definition of continuity here is technical and can be found in the standard literature, for example [58], [67].

adjoint, a fact that is intimately connected with the inverse reciprocity theorem of the antenna current Green's function formalism proved in Chapter 9.

In the present formulation, choosing space of test functions as $\mathcal{D}$ will present an immediate problem, which is the fact that for $\varphi(\mathbf{r}) \in \mathcal{D}$, in general $\mathcal{L}^{-1}\varphi(\mathbf{r}) \notin \mathcal{D}$, i.e., the current distribution obtained by inverting the operator $\mathcal{L}$ is not necessarily smooth, although still possessing a compact support (since S is compact) [70]. Therefore, the standard distribution theory associated with the space $\mathcal{D}$ cannot be used directly in constructing the ACGF F as a distribution.

The basic idea in our construction is the observation that the ACGF is always used within an integral in order to compute the induced current on S due to some external excitation. Therefore, we always encounter F in expressions like $\langle F(\mathbf{r},\mathbf{r}'), E(\mathbf{r})\rangle$, where $E(\mathbf{r})$ is the applied electric field excitation, which plays the role of the test function $\varphi(\mathbf{r})$ defined above. Now, although $J(\mathbf{r}) = \mathcal{L}^{-1}E(\mathbf{r})$ need not be smooth, it is certainly continuous [70]. Therefore, it will be enough for most practical considerations to examine the provisional definition (8.12) in the case where the function $\mathcal{L}^{-1}\varphi(\mathbf{r})$ is required to be at least continuous.

We now present a sequence of functions belonging to the delta family, i.e., a sequence of sufficiently well-behaving functions approximating the Dirac delta function [71]. The crucial details lie in requiring only the continuity of the test function, rather than the much stronger condition of infinite differentiability. The following is one possible candidate of such a delta family [71], [67]

Theorem 8.2.1. *Consider the family $\{f_n(x)\}_{n\in\mathbb{N}}$ of nonnegative locally integrable functions with $x \in \mathbb{R}^m, m \geq 1$, such that the following two conditions are satisfied:*

1. *For some $A > 0$, $\lim_{n\to\infty} \int_{|x|<A} f_n(x)\,dx = 1$.*
2. *For every $A > 0$, we have $\lim_{n\to\infty} f_n(x) = 0$ uniformly for every $|x| \geq A$.*

Then for every continuous function $\varphi(x)$ satisfying $\int_{\mathbb{R}^m} \varphi(x)\,dx < \infty$, we have

$$\lim_{n\to\infty} \int_{\mathbb{R}^m} f_n(\mathbf{r})\,\varphi(\mathbf{r})\,dx = \varphi(0)\,. \tag{8.14}$$

Actually, even continuity of $\varphi(x)$ only at $x = 0$ is enough.

Let us select a delta family satisfying the conditions of Theorem 8.2.1. One possible delta family is provided by the following theorem [71]

Theorem 8.2.2. *Consider a sequence of functions $\{f_n(x)\}_{n\in\mathbb{N}}$, $x \in \mathbb{R}^m, m \geq 1$, where $f_n(x) := n^m f(nx)$, and $f(x)$ is a nonnegative locally integrable function in $\mathbb{R}^m$ satisfying $\int_{\mathbb{R}^m} f(\mathbf{r})\,dx = 1$. Then relation (8.14) holds for the sequence f_n.*

It is not difficult to show that the conditions of Theorem 8.2.1 are satisfied for the sequence defined in Theorem 8.2.2 and hence this delta family can be used in computing the ACGF as will be demonstrated later below. Moreover, by simply choosing f to be any smooth function with bounded support, the resulting sequence will also be smooth with bounded support, and thus satisfying a stronger requirement enforced in Section 8.2.3.

Our strategy will be to employ such delta families in order to insure that the last equality in (8.12) is well defined by replacing it with a limit modeled on (8.14). To achieve this rigorously, we will use the concept of weak convergence as follows.

The *distributional limit* is defined as [71]

Definition 1. *Let* $\{f_n\}_{n\in\mathbb{N}}$ *be a family of distributions in* $\mathcal{D}'(M)$. *We say that* f_n *converges distributionally to the distribution* $g \in \mathcal{D}'(M)$ *as* $n \to \infty$ *if*

$$\lim_{n\to\infty} \langle f_n, \varphi \rangle = \langle g, \varphi \rangle, \tag{8.15}$$

for each $\varphi \in \mathcal{D}(M)$.

We then have the following important theorem in distribution theory [71]

Theorem 8.2.3. *If* $\lim_{n\to\infty} \langle f_n, \varphi \rangle$ *exists for every* $\varphi \in \mathcal{D}(\mathbb{R}^m)$, $m \geq 1$, *then there exists a unique distribution* $g \in \mathcal{D}'(\mathbb{R}^m)$ *such that* $\lim_{n\to\infty} f_n = g$ *in the sense of Definition 1, that is*

$$\langle g, \varphi \rangle = \lim_{n\to\infty} \langle f_n, \varphi \rangle, \tag{8.16}$$

for each $\varphi \in \mathcal{D}(\mathbb{R}^m)$.

Consider any point $\mathbf{r}' \in S$. Our first goal is to construct a delta sequence defined totally in S, that is, a definition of the delta function centered at $\mathbf{r}'$ *intrinsic* in S. Pick any local coordinate system (U_l, x_l) such that $\mathbf{r}' \in U_l$. Consider a delta family $\{f_n(x_l)\}_{n\in\mathbb{N}}$, $x_l \in \mathbb{R}^2$, satisfying the conditions of Theorem 8.2.2. We will make a stronger demand by requiring these functions to be smooth and with bounded support.[20] (This will insure that $\mathcal{L}^{-1} f_n(\mathbf{r})$ is well defined for all $n \in \mathbb{N}$ and that it is continuous [70].)[21]

20 A differentiable function is continuous, and consequently locally integrable. A smooth function on bounded support has bounded derivatives and hence uniformly Lipschitz continuous, and consequently uniformly Holder continuous. For integral equations formalisms making use of uniform Holder spaces, see [70].

21 By assuming that the delta sequence is smooth, we impose a quite strong condition that will most likely meet the demands of all types of boundary-value problems. However, for a specific given operator, probably less stringent conditions can be imposed.

Now let V_l be the coordinate image in $\mathbb{R}^2$ of the region U_l, i.e., let $\mathbb{R}^2 \supset V_l = x_l(U_l)$. Next, choose the functions $f_n(x_l)$ such that for $n = 1$, the support of $f_1(x_l)$ is contained in V_l. Since we have from the specific delta sequence of Theorem 8.2.2 $\text{supp}\{f_n(x_l)\} \subset \text{supp}\{f_1(x_l)\}$ for all $n > 1$, we conclude that $\text{supp}\{f_n(x_l)\} \subset V_l$ for all $n \geq 1$. In other words, all the functions belonging to this delta family in $\mathbb{R}^2$ have supports within the local coordinate patch V_l.

The interaction between a locally integrable function $f(\mathbf{r})$ confined in U_l and a function $\varphi(\mathbf{r})$ continuous on S is given by the following integral

$$\int_{U_l} ds f(\mathbf{r})\varphi(\mathbf{r}) = \int_{V_l} dx_l \sqrt{g(x_l)} f(x_l)\varphi(x_l), \tag{8.17}$$

where the second integral is obtained from the first by using the well-known area element in differential geometry $ds = \sqrt{g(x_l)}dx_l$ [61]. Here $g(x_l)$ stands for the determinant of the Riemannian metric tensor of the 2-manifold S. Next, define the following sequence of functions on V_l

$$f_n^S(x_l, x_l') := \frac{f_n(x_l - x_l')}{\sqrt{g(x_l)}} \tag{8.18}$$

This definition does work because we know that $g \neq 0$ in any U_l and for arbitrary 2-manifold S [61]. From (8.17) and (8.18) we find after applying Theorem 8.2.2

$$\begin{aligned} \lim_{n\to\infty} \langle f_n^S(\mathbf{r},\mathbf{r}'), \varphi(\mathbf{r})\rangle &= \lim_{n\to\infty} \textstyle\int_{U_l} ds f_n^S(\mathbf{r},\mathbf{r}')\varphi(\mathbf{r}) \\ &= \lim_{n\to\infty} \textstyle\int_{V_l} dx_l f_n(x_l - x_l')\varphi(x_l) = \varphi(x_l'). \end{aligned} \tag{8.19}$$

The important thing to remember is that for the validity of (8.19) above the test function $\varphi(\mathbf{r})$ is required to be at least continuous at $\mathbf{r}'$ and integrable throughout U_l.[22] Therefore, the family of functions $\{f_n^S(x_l, x_l')\}_{n\in\mathbb{N}}$ represents a l*ocal delta sequence* in S at $\mathbf{r}' = \mathbf{r}'(x_l')$. These functions are defined *intrinsically* in S, and so no need arises for functional embedding in $\mathbb{R}^3$.

Let $F_n(\mathbf{r},\mathbf{r}') := \mathcal{L}^{-1} f_n^S(\mathbf{r},\mathbf{r}')$. Each ordinary function $F_n(\mathbf{r},\mathbf{r}')$ on S generates a *regular*[23] distribution in $\mathcal{D}(S)$, but a *singular* distribution in $\mathcal{D}(\mathbb{R}^3)$,

22 Since we require the delta sequence functions f_n to be of bounded support, only integrability on V_l is needed, not the entire domain $\mathbb{R}^2$ as required in Theorem 8.2.1, the latter being much more general than what is needed here.

23 We associate with each locally integrable function $f(x)$, $x \in M$, its *regular* distribution f. (That is, we don't distinguish in notation between an ordinary function and its associated regular distribution.) The regular distribution f is defined as the continuous linear functional $\langle f, \varphi\rangle := \int_M f(x)\varphi(x)\,dx$, $\forall\varphi \in \mathcal{D}(M)$. It can be proved that any locally integrable function generates a regular distribution in $\mathcal{D}'(M)$. For details, see [67], [71].

which we also denote by F_n.[24] These distributions F_n will now be shown to represent an *approximation* of the exact ACGF distribution F to be defined below.

From the fact that $\mathcal{L}$ is self-adjoint [13], it follows that $\left\langle \mathcal{L}^{-1} f_n^S(\mathbf{r},\mathbf{r}'), \varphi(\mathbf{r})\right\rangle = \left\langle f_n^S(\mathbf{r},\mathbf{r}'), \mathcal{L}^{-1}\varphi(\mathbf{r})\right\rangle$ for every $\varphi \in \mathcal{D}(\mathbb{R}^3)$. The function $\mathcal{L}^{-1}\varphi(\mathbf{r})$ is continuous for each $\varphi \in \mathcal{D}(\mathbb{R}^3)$ [70] and since $\mathcal{L}^{-1}\varphi(\mathbf{r})$ is of bounded support on S, it is absolutely integrable. It follows then from (8.19) that the $\lim_{n\to\infty}\langle F_n, \varphi(\mathbf{r})\rangle$ exists. Consequently, by Theorem 8.2.3 there exists a unique distribution $F \in \mathcal{D}'(\mathbb{R}^3)$ such that

$$\langle F, \varphi \rangle = \lim_{n\to\infty} \left\langle \mathcal{L}^{-1} f_n^S(\mathbf{r},\mathbf{r}'), \varphi(\mathbf{r})\right\rangle, \quad \forall \varphi \in \mathcal{D}(\mathbb{R}^3). \tag{8.21}$$

Therefore, The scalar antenna Green's function F is now formally defined by the basic relation (8.21) as a singular distribution in $\mathcal{D}'(\mathbb{R}^3)$. By replacing $\mathcal{D}(\mathbb{R}^3)$ in (8.21) by $\mathcal{D}(S)$ and using the version of Theorem 8.2.3 valid in $\mathcal{D}'(S)$, we obtain the definition of F as a distribution in $\mathcal{D}'(S)$. The singular distribution $F \in \mathcal{D}'(\mathbb{R}^3)$ corresponds to the 3D ACGF while the distribution $F \in \mathcal{D}'(S)$ can be compared with the 2D ACGF.

Our approach is based on choosing *any* (smooth and with bounded support) local delta family that insures the existence of the limit (8.19). In this sense, one way to interpret the definition (8.21) is that it associates the ACGF F with the *equivalence class* of all (smooth and with bounded support) local delta families satisfying (8.19). As long as the operator $\mathcal{L}$ is invertible, one can insure the existence of the antenna current Green's function through the definition (8.21). For the particular operator we chose to work with in this chapter, namely the standard electric-field integral operator, it can be shown using functional-analytic methods that the problem is uniquely solvable if and only the free-space wavenumber k_0 is not an eigenvalue of the interior problem [70]. The extension of this basic idea to the full dyadic case and the proof of the Green's function theorem are the subject of Section 8.2.3.

24 Remember that the definition of the singular distribution in $\mathcal{D}'(\mathbb{R}^3)$ generated by a function $\sigma(\mathbf{r})$ integrable on S is the linear continuous functional given by

$$\langle \sigma(\mathbf{r}), \varphi(\mathbf{r})\rangle := \int_S ds\, \sigma(\mathbf{r})\,\varphi(\mathbf{r}), \quad \forall \varphi(\mathbf{r}) \in \mathcal{D}\left(\mathbb{R}^3\right). \tag{8.20}$$

8.2.3 Direct Construction of the ACGF Using Distribution Theory: Electromagnetic Theory

Choose two local coordinate systems containing the source location $\mathbf{r}'$ and the observation point $\mathbf{r}$ and select two orthonormal unit vectors $\hat{\alpha}_n, n = 1, 2$ at each location. We then have

$$\mathcal{L}_{\mathbf{r}}^{-1}\delta\left(\mathbf{r}-\mathbf{r}'\right)\hat{\alpha}_1\left(\mathbf{r}'\right) = \hat{\alpha}_1\left(\mathbf{r}\right)F_{11}\left(\mathbf{r},\mathbf{r}'\right) + \hat{\alpha}_2\left(\mathbf{r}\right)F_{21}\left(\mathbf{r},\mathbf{r}'\right), \tag{8.22}$$

$$\mathcal{L}_{\mathbf{r}}^{-1}\delta\left(\mathbf{r}-\mathbf{r}'\right)\hat{\alpha}_2\left(\mathbf{r}'\right) = \hat{\alpha}_1\left(\mathbf{r}\right)F_{12}\left(\mathbf{r},\mathbf{r}'\right) + \hat{\alpha}_2\left(\mathbf{r}\right)F_{22}\left(\mathbf{r},\mathbf{r}'\right). \tag{8.23}$$

Here the subscript $\mathbf{r}$ in $\mathcal{L}_{\mathbf{r}}^{-1}$ serves to indicate that the operator is applied to the $\mathbf{r}$-vector function $\delta\left(\mathbf{r}-\mathbf{r}'\right)\hat{\alpha}_n\left(\mathbf{r}'\right), n = 1, 2$. Similar consideration applies when we write $\mathcal{L}_{\mathbf{r}'}^{-1}$ with the obvious modifications. Based on the notation of (8.22) and (8.23), we can write down at once the full form of the ACGF tensor in operator form

$$\begin{aligned}\bar{\mathbf{F}}\left(\mathbf{r},\mathbf{r}'\right) = &\left[\mathcal{L}_{\mathbf{r}}^{-1}\delta\left(\mathbf{r}-\mathbf{r}'\right)\hat{\alpha}_1\left(\mathbf{r}'\right)\right]\hat{\alpha}_1\left(\mathbf{r}'\right)\\ &+\left[\mathcal{L}_{\mathbf{r}}^{-1}\delta\left(\mathbf{r}-\mathbf{r}'\right)\hat{\alpha}_2\left(\mathbf{r}'\right)\right]\hat{\alpha}_2\left(\mathbf{r}'\right).\end{aligned} \tag{8.24}$$

Our goal now is give the expression (8.24) a rigorous meaning using distribution theory by following the train of thought developed in Section 8.2.2.

We define the interaction of two vector fields on S by the formula [13]

$$\left\langle \mathbf{X}_1\left(\mathbf{r}'\right), \mathbf{X}_2\left(\mathbf{r}'\right)\right\rangle := \int_S ds'\, \mathbf{X}_1\left(\mathbf{r}'\right)\cdot\mathbf{X}_2\left(\mathbf{r}'\right). \tag{8.25}$$

Also, we make the following definition

$$\left\langle \bar{\mathbf{F}}\left(\mathbf{r},\mathbf{r}'\right), \mathbf{X}\left(\mathbf{r}'\right)\right\rangle := \int_S ds'\, \bar{\mathbf{F}}\left(\mathbf{r},\mathbf{r}'\right)\cdot\mathbf{X}\left(\mathbf{r}'\right). \tag{8.26}$$

Therefore, the ACGF theorem (7.2), which gives the current induced on the antenna due to excitation field $\mathbf{E}(\mathbf{r})$, can be expressed compactly as

$$\mathbf{J}\left(\mathbf{r}\right) = \left\langle \bar{\mathbf{F}}\left(\mathbf{r},\mathbf{r}'\right), \mathbf{E}\left(\mathbf{r}'\right)\right\rangle. \tag{8.27}$$

We notice again that the electric field can be thought of as formally playing the role of "test function" in distribution theory. This observation was exploited in [13] in order to prove the ACGF theorem (8.27).

The space $\mathcal{D}(\mathbb{R}^3)$ was defined in terms of scalar functions but we can easily extend this function space to vector fields $\mathbf{T}(\mathbf{r})$ on S. In this case, we need a relation like the following

$$\langle \delta\left(\mathbf{r}' - \mathbf{r}\right)\hat{\alpha}\left(\mathbf{r}'\right), \mathbf{T}\left(\mathbf{r}'\right)\rangle = \hat{\alpha}\left(\mathbf{r}\right)\cdot\mathbf{T}\left(\mathbf{r}\right). \tag{8.28}$$

Note that the integration is with respect to $\mathbf{r}'$. Here, $\mathbf{T}(\mathbf{r}')$ is required to be only a *continuous* vector field on S. On the other hand, $\hat{\alpha}(\mathbf{r}')$ is to be taken as a smooth unit vector field also on S in line with the geometric formulation of the ACGF introduced previously.

In (8.22), (8.23), and (8.28), although the delta functions are written in the language of $\mathbb{R}^3$, for the actual approximation of the ACGF using distribution theory we must use the special *local* delta family in S introduced formally by (8.18). Let such a family be denoted by $\left\{f_n^S\left(\mathbf{r}', \mathbf{r}\right)\right\}_{n\in\mathbb{N}}$. Using (8.25), (8.13), then (8.19), we find

$$\begin{aligned}&\lim_{n\to\infty}\left\langle f_n^S\left(\mathbf{r}', \mathbf{r}\right)\hat{\alpha}\left(\mathbf{r}'\right), \mathbf{T}\left(\mathbf{r}'\right)\right\rangle \\ &= \lim_{n\to\infty}\left\langle f_n^S\left(\mathbf{r}', \mathbf{r}\right), \hat{\alpha}\left(\mathbf{r}'\right)\cdot\mathbf{T}\left(\mathbf{r}'\right)\right\rangle = \hat{\alpha}\left(\mathbf{r}\right)\cdot\mathbf{T}\left(\mathbf{r}\right).\end{aligned} \tag{8.29}$$

Therefore, we can define the *local* vector delta function $\delta\left(\mathbf{r}' - \mathbf{r}\right)\hat{\alpha}\left(\mathbf{r}'\right)$ on S as the distributional limit[25]

$$\delta\left(\mathbf{r}' - \mathbf{r}\right)\hat{\alpha}\left(\mathbf{r}\right) := \lim_{n\to\infty} f_n^S\left(\mathbf{r}', \mathbf{r}\right)\hat{\alpha}\left(\mathbf{r}'\right). \tag{8.30}$$

In this case, the desired relation (8.28) is obtained by means of (8.29). The reader should carefully note how the argument of the function $\hat{\alpha}$ changes from $\mathbf{r}$ to $\mathbf{r}'$ when moving from the LHS to the RHS of (8.30).

We have now all the tools necessary to justify (8.24) in the sense of distribution theory. When the electromagnetic operator $\mathcal{L}^{-1}$ is invertible, it is always possible to construct the exact ACGF $\mathcal{L}_{\mathbf{r}}^{-1}\delta\left(\mathbf{r} - \mathbf{r}'\right)\hat{\alpha}_n\left(\mathbf{r}'\right), n = 1, 2$ using the definition (8.21) by performing two constructions corresponding to the two cases $l = 1, 2$. In details, we have

$$\begin{aligned}&\lim\nolimits_{n\to\infty}\left\langle \mathcal{L}_{\mathbf{r}'}^{-1} f_n^S\left(\mathbf{r}', \mathbf{r}\right)\hat{\alpha}_l\left(\mathbf{r}'\right), \mathbf{T}\left(\mathbf{r}'\right)\right\rangle \\ &= \lim\nolimits_{n\to\infty}\left\langle f_n^S\left(\mathbf{r}', \mathbf{r}\right)\hat{\alpha}_l\left(\mathbf{r}'\right), \mathcal{L}_{\mathbf{r}'}^{-1}\mathbf{T}\left(\mathbf{r}'\right)\right\rangle \\ &= \lim\nolimits_{n\to\infty}\left\langle f_n^S\left(\mathbf{r}', \mathbf{r}\right), \hat{\alpha}_l\left(\mathbf{r}'\right)\cdot\mathcal{L}_{\mathbf{r}'}^{-1}\mathbf{T}\left(\mathbf{r}'\right)\right\rangle.\end{aligned} \tag{8.31}$$

But $\mathcal{L}_{\mathbf{r}'}^{-1}\mathbf{T}\left(\mathbf{r}'\right)$ is continuous [70] while $\hat{\alpha}_l\left(\mathbf{r}'\right)$ by definition is smooth. Consequently, $\hat{\alpha}_l\left(\mathbf{r}'\right)\cdot\mathcal{L}_{\mathbf{r}'}^{-1}\mathbf{T}\left(\mathbf{r}'\right)$ is continuous. Hence, from (8.19) we conclude that the limit exists. Therefore, we define the *vector* distribution $\mathcal{L}_{\mathbf{r}'}^{-1}\delta\left(\mathbf{r}' - \mathbf{r}\right)\hat{\alpha}_l\left(\mathbf{r}\right)$ as the

25 Cf. Definition 1.

distributional limit

$$\mathcal{L}_{\mathbf{r}'}^{-1}\delta\left(\mathbf{r}'-\mathbf{r}\right)\hat{\alpha}_l\left(\mathbf{r}\right) := \lim_{n\to\infty}\mathcal{L}_{\mathbf{r}'}^{-1}f_n^S\left(\mathbf{r}',\mathbf{r}\right)\hat{\alpha}_l\left(\mathbf{r}'\right), \tag{8.32}$$

which is valid for $l = 1, 2$.

The procedure should be obvious by now. We define the *dyadic* functions $\left\{\bar{\mathbf{F}}_n\left(\mathbf{r},\mathbf{r}'\right)\right\}_{n\in\mathbb{N}}$ by the relation

$$\begin{aligned}\bar{\mathbf{F}}_n\left(\mathbf{r},\mathbf{r}'\right) := {}& \left[\mathcal{L}_{\mathbf{r}}^{-1}f_n^S\left(\mathbf{r},\mathbf{r}'\right)\hat{\alpha}_1\left(\mathbf{r}\right)\right]\hat{\alpha}_1\left(\mathbf{r}'\right)\\ &+\left[\mathcal{L}_{\mathbf{r}}^{-1}f_n^S\left(\mathbf{r},\mathbf{r}'\right)\hat{\alpha}_2\left(\mathbf{r}\right)\right]\hat{\alpha}_2\left(\mathbf{r}'\right).\end{aligned} \tag{8.33}$$

Therefore, from (8.33) we have

$$\begin{aligned}\bar{\mathbf{F}}_n^T\left(\mathbf{r}',\mathbf{r}\right) := {}& \hat{\alpha}_1\left(\mathbf{r}\right)\left[\mathcal{L}_{\mathbf{r}'}^{-1}f_n^S\left(\mathbf{r}',\mathbf{r}\right)\hat{\alpha}_1\left(\mathbf{r}'\right)\right]\\ &+\hat{\alpha}_2\left(\mathbf{r}\right)\left[\mathcal{L}_{\mathbf{r}'}^{-1}f_n^S\left(\mathbf{r}',\mathbf{r}\right)\hat{\alpha}_2\left(\mathbf{r}'\right)\right].\end{aligned} \tag{8.34}$$

Now, consider

$$\bar{\mathbf{F}}\left(\mathbf{r},\mathbf{r}'\right) := \lim_{n\to\infty}\bar{\mathbf{F}}_n\left(\mathbf{r},\mathbf{r}'\right), \bar{\mathbf{F}}^T\left(\mathbf{r}',\mathbf{r}\right) := \lim_{n\to\infty}\bar{\mathbf{F}}_n^T\left(\mathbf{r}',\mathbf{r}\right), \tag{8.35}$$

where the limit is the distributional convergence of Definition 1. By construction, the delta family $\left\{f_n^S\left(\mathbf{r}',\mathbf{r}\right)\right\}_{n\in\mathbb{N}}$ will insure that the two limits on the RHS of the equations (8.35) exist in a manner identical to the way we proved that the definition (8.32) is well posed but will not give the details.[26]

By the inverse reciprocity theorem, we have $\bar{\mathbf{F}}^T\left(\mathbf{r},\mathbf{r}'\right) = \bar{\mathbf{F}}\left(\mathbf{r}',\mathbf{r}\right)$. Therefore, we may write

$$\lim_{n\to\infty}\left\langle\bar{\mathbf{F}}_n\left(\mathbf{r},\mathbf{r}'\right),\mathbf{E}\left(\mathbf{r}'\right)\right\rangle = \lim_{n\to\infty}\left\langle\bar{\mathbf{F}}_n^T\left(\mathbf{r}',\mathbf{r}\right),\mathbf{E}\left(\mathbf{r}'\right)\right\rangle. \tag{8.36}$$

Next, using the relation $\mathbf{E}\left(\mathbf{r}'\right) = \mathcal{L}_{\mathbf{r}'}\mathbf{J}\left(\mathbf{r}'\right)$, it is possible to rewrite the RHS of (11.55) as

$$\lim_{n\to\infty}\left\langle\bar{\mathbf{F}}_n^T\left(\mathbf{r}',\mathbf{r}\right),\mathbf{E}\left(\mathbf{r}'\right)\right\rangle = \lim_{n\to\infty}\left\langle\bar{\mathbf{F}}_n^T\left(\mathbf{r}',\mathbf{r}\right),\mathcal{L}_{\mathbf{r}'}\mathbf{J}\left(\mathbf{r}'\right)\right\rangle. \tag{8.37}$$

Since the operator $\mathcal{L}$ is self-adjoint, so is its inverse [67], [13]. Therefore, after substituting the expression of $\bar{\mathbf{F}}_n^T\left(\mathbf{r}',\mathbf{r}\right)$ in (8.34) into (8.37), we can move the operator $\mathcal{L}_{\mathbf{r}'}^{-1}$ in (8.34) to act on the function $\mathcal{L}_{\mathbf{r}'}\mathbf{J}\left(\mathbf{r}'\right)$,[27] resulting in $\mathcal{L}_{\mathbf{r}'}^{-1}\mathcal{L}_{\mathbf{r}'}\mathbf{J}\left(\mathbf{r}'\right) = \mathbf{J}\left(\mathbf{r}'\right)$.

26 In the proof, the definition (8.26) should be used first, then (8.25) and (8.13), successively, to reduce a form like $\left\langle\bar{\mathbf{F}},\mathbf{E}\right\rangle$ to the form where (8.19) can be used to establish the convergence of the ordinary limit. Distributional limit existence then follows by the force of Theorem 8.2.3.

27 That is, for a self-adjoint operator $\mathcal{L}$, we have for any two vector fields $\mathbf{X}_1$ and $\mathbf{X}_2$ belonging to the domain of definition of $\mathcal{L}$, the following equality holds $\left\langle\mathcal{L}\mathbf{X}_1\left(\mathbf{r}'\right),\mathbf{X}_2\left(\mathbf{r}'\right)\right\rangle = \left\langle\mathbf{X}_1\left(\mathbf{r}'\right),\mathcal{L}\mathbf{X}_2\left(\mathbf{r}'\right)\right\rangle$.

Consequently, the RHS of (8.37) becomes

$$\begin{aligned}&\hat{\alpha}_1(\mathbf{r})\lim_{n\to\infty}\left\langle f_n^S(\mathbf{r}',\mathbf{r})\,\hat{\alpha}_1(\mathbf{r}'),\mathbf{J}(\mathbf{r}')\right\rangle\\&\quad+\hat{\alpha}_2(\mathbf{r})\lim_{n\to\infty}\left\langle f_n^S(\mathbf{r}',\mathbf{r})\,\hat{\alpha}_2(\mathbf{r}'),\mathbf{J}(\mathbf{r}')\right\rangle\\&=\hat{\alpha}_1(\mathbf{r})\left[\hat{\alpha}_1(\mathbf{r})\cdot\mathbf{J}(\mathbf{r})\right]+\hat{\alpha}_2(\mathbf{r})\left[\hat{\alpha}_2(\mathbf{r})\cdot\mathbf{J}(\mathbf{r})\right]\\&=\mathbf{J}(\mathbf{r}),\end{aligned}\tag{8.38}$$

where use was made of (8.28) in the form of (8.29). Therefore, we reach

$$\mathbf{J}(\mathbf{r})=\lim_{n\to\infty}\left\langle\bar{\mathbf{F}}_n(\mathbf{r},\mathbf{r}'),\mathbf{E}(\mathbf{r}')\right\rangle.\tag{8.39}$$

This is the main relation we are looking for. It states that the exact ACGF can be obtained by a certain limit of converging sequences. For any desired error ϵ, we can find a sufficiently large integer N such that for each $n > N$, $\left|\mathbf{J}(\mathbf{r})-\left\langle\bar{\mathbf{F}}_n(\mathbf{r},\mathbf{r}'),\mathbf{E}(\mathbf{r}')\right\rangle\right| < \epsilon$. The convergence in (8.39) is pointwise in $\mathbf{r}$, *not* uniform. That is, for each $\mathbf{r}$, one must look for the corresponding integer $N(\mathbf{r},\epsilon)$, such that

$$\mathbf{J}(\mathbf{r})\simeq\left\langle\bar{\mathbf{F}}_{N(\mathbf{r},\varepsilon)}(\mathbf{r},\mathbf{r}'),\mathbf{E}(\mathbf{r}')\right\rangle.\tag{8.40}$$

In contrast, for *uniform* convergence, N must be dependent only on the error level ϵ. Such a stronger type of convergence is desirable but in general cannot be guaranteed according to our construction.

8.2.4 On the Fourier Transform of the ACGF Tensor

Working in the 3D Euclidian space $\mathbb{R}^3$, the ACGF is not a classical function, but a distribution, and that is for *two* reasons. The first is the obvious fact that $\bar{\mathbf{F}}(\mathbf{r},\mathbf{r}')$ is a response to a distribution, namely the Dirac delta source. The second is that the ACGF in $\mathbb{R}^3$ represents a *surface* function on S. Therefore, its Fourier transform should be ultimately grounded on a suitable formalism, namely *tempered* distribution theory [67].

We will propose a solution in line with the basic construction of the ACGF presented by (8.21). Let us denote by $\mathcal{P}'(\mathbb{R}^3)$ the space of *rapid decay* functions on $\mathbb{R}^3$ [67], [68], [71]. The space of tempered distributions is the topological dual and is denoted by $\mathcal{P}'(\mathbb{R}^3)$. As is well known, every distribution $g\in\mathcal{P}'$ possesses a Fourier transform $\mathcal{F}g$ also in $\mathcal{P}'(\mathbb{R}^3)$ [67]. Moreover, the exact corresponding version to Theorem 8.2.2 holds for the tempered distribution space $\mathcal{P}'(\mathbb{R}^3)$, i.e., if $\langle f_n,\varphi\rangle$ converges for every $\varphi\in\mathcal{P}'(\mathbb{R}^3)$, then there exists a unique distribution $F\in\mathcal{P}'(\mathbb{R}^3)$ such that $\langle F,\varphi\rangle=\langle f_n,\varphi\rangle$ for all $\varphi\in\mathcal{P}'(\mathbb{R}^3)$ [68].

It is very natural then to consider the sequences of functions $\{F_n^{pq}(\mathbf{r},\mathbf{r}')\}_{n\in\mathbb{N}}$, $p,q=1,2$, defined by (8.33), and ask whether they generate tempered distributions in $\mathcal{P}'(\mathbb{R}^3)$. The answer is in the affirmative. Indeed, in a certain sense, the functional $\left\langle \bar{\mathbf{F}}_n(\mathbf{r},\mathbf{r}'), \mathbf{T}(\mathbf{r})\right\rangle = \int_S ds \bar{\mathbf{F}}_n(\mathbf{r},\mathbf{r}') \cdot \mathbf{T}(\mathbf{r})$ define a distribution in $\mathcal{P}'(\mathbb{R}^3)$. Linearity is obvious so we will prove continuity. Let $\mathbf{T}_m(\mathbf{r}) \to \mathbf{T}(\mathbf{r})$ in the space $\mathcal{P}(\mathbb{R}^3)$.[28] We find

$$\begin{aligned} &\left|\left\langle \bar{\mathbf{F}}_m(\mathbf{r},\mathbf{r}'), \mathbf{T}(\mathbf{r}) - \mathbf{T}_m(\mathbf{r})\right\rangle\right| \\ &\qquad \le \int_S ds \|\bar{\mathbf{F}}_m(\mathbf{r},\mathbf{r}')\| \cdot \|\mathbf{T}(\mathbf{r}) - \mathbf{T}_m(\mathbf{r})\| \\ &\qquad = \sup_{\mathbf{r}\in S} |\mathbf{T}(\mathbf{r}) - \mathbf{T}_m(\mathbf{r})| \int_S ds \|\bar{\mathbf{F}}_m(\mathbf{r},\mathbf{r}')\|, \end{aligned} \tag{8.41}$$

where we define the norm of the tensor as $\|\bar{\mathbf{F}}\| := \sum_{p=1}^{2}\sum_{q=1}^{2}\|F^{pq}\|$. However, convergence in $\mathcal{P}'(\mathbb{R}^3)$ implies uniform convergence on S, i.e., $\sup_{\mathbf{r}\in S}|\mathbf{T}(\mathbf{r}) - \mathbf{T}_m(\mathbf{r})| \to 0$. Therefore, from (8.41) we find $\left|\left\langle \bar{\mathbf{F}}_m(\mathbf{r},\mathbf{r}'), \mathbf{T}(\mathbf{r}) - \mathbf{T}_m(\mathbf{r})\right\rangle\right| \to 0$ and hence conclude that the functional $\bar{\mathbf{F}}_m$ is continuous in $\mathcal{P}'(\mathbb{R}^3)$, which establishes that it is indeed a distribution in $\mathcal{P}'(\mathbb{R}^3)$.

We next need the following theorem [68]

Theorem 8.2.4. *Consider a sequence of distributions in $\mathcal{P}'$ such that $f_n \to f$ distributionally. Then, we have $\mathcal{F}f_n \to \mathcal{F}f$ also distributionally in $\mathcal{P}'$.*

That is, if a sequence converges distributionally to a distribution in $\mathcal{P}'$, then the Fourier transforms of the converging distributions converge distributionally (also in $\mathcal{P}'$) to the Fourier transform of the former limit. Since each of the scalar functions comprising F_n^{pq} is in $\mathcal{P}'(\mathbb{R}^3)$, they possess Fourier transforms $\{\mathcal{F}F_n^{pq}\}_{n\in\mathbb{N}}$ also in $\mathcal{P}'(\mathbb{R}^3)$ [67]. Similarly, the components of the limit function $\bar{\mathbf{F}} \to \bar{\mathbf{F}}_n$ are in $\mathcal{P}'(\mathbb{R}^3)$ and so they possess distributional Fourier transforms $\mathcal{F}F_n^{pq} \in \mathcal{P}'(\mathbb{R}^3)$. From Theorem 8.2.4, we conclude that $\mathcal{F}\bar{\mathbf{F}} \to \mathcal{F}\bar{\mathbf{F}}_n$ because the limit holds for each of the four functions F^{pq}, $p,q=1,2$. In detail, we may write

$$\langle \mathcal{F}F^{pq}, \varphi(\mathbf{r})\rangle = \lim_{n\to\infty} \langle \mathcal{F}F_n^{pq}(\mathbf{r},\mathbf{r}'), \varphi(\mathbf{r})\rangle, \quad \forall\varphi \in \mathcal{P}\left(\mathbb{R}^3\right). \tag{8.42}$$

This establishes the Fourier transform of the ACGF in the rigorous sense of tempered distribution theory. In practice, especially in numerical computations, we will seldom work with the distribution itself, but rather with an approximation in terms of test functions. This topic will be dealt with next in Section 8.2.5.

28 Here, we understand that each component of the vector field $\mathbf{T}(\mathbf{r})$ converges to the corresponding component of $\mathbf{T}_m(\mathbf{r})$ in the sense of convergence in the space $\mathcal{P}'(\mathbb{R}^3)$.

8.2.5 On the Inverse Fourier Transform of the ACGF Tensor

Our goal here is to understand the inverse Fourier transform of the ACGF in terms of a *sequence* of relations converging in some acceptable sense. The theoretical basis for this construction is the following important theorem

Theorem 8.2.5. *Every distribution in $\mathcal{D}'(\mathbb{R}^3)$ can be written as the distributional limit of a sequence of regular distributions generated by test functions in $\mathcal{D}(\mathbb{R}^3)$. In other words, the space $\mathcal{D}(\mathbb{R}^3)$ is dense in $\mathcal{D}'(\mathbb{R}^3)$.*

Hence, we can approximate the distribution $\bar{\mathbf{F}}(\mathbf{r},\mathbf{r}')$ by a sequence of regular distributions $\{\bar{\mathbf{F}}_n^d(\mathbf{r},\mathbf{r}')\}_{n\in\mathbb{N}}$ in the sense that $F_n^{d,pq}(\mathbf{r},\mathbf{r}') \to F^{d,pq}(\mathbf{r},\mathbf{r}')$, $p,q = 1,2$, distributionally in $\mathcal{D}'(\mathbb{R}^3)$. Indeed, Theorem 8.2.5 allows us to write

$$\lim_{n\to\infty}\int_{\mathbb{R}^3} d^3r\bar{\mathbf{F}}_n^d(\mathbf{r}',\mathbf{r})\cdot\mathbf{E}(\mathbf{r}) = \langle\bar{\mathbf{F}}(\mathbf{r}',\mathbf{r}),\mathbf{E}(\mathbf{r})\rangle = \mathbf{J}(\mathbf{r}'). \tag{8.43}$$

Notice that the support of the integral on the LHS is in $\mathbb{R}^3$. The fact that the integral on the RHS is on S indicates again that the ACGF is a singular distribution in both $\mathcal{D}'(\mathbb{R}^3)$ and $\mathcal{P}'(\mathbb{R}^3)$.

Each of the functions $F^{d,pq}(\mathbf{r},\mathbf{r}')$ is both smooth and of bounded support in $\mathbb{R}^3$. Therefore, they possess an *ordinary* Fourier transform and also an inverse Fourier integral that is absolutely and uniformly convergent.[29] That is, we have

$$\bar{\mathbf{F}}_n^d(\mathbf{r},\mathbf{r}') = \frac{1}{(2\pi)^3}\int_{\mathbb{R}^3} d^3k\,\bar{\mathbf{F}}_n^d(\mathbf{k},\mathbf{r}')\,e^{i\mathbf{k}\cdot\mathbf{r}}, \tag{8.44}$$

Now, since each smooth function $\bar{\mathbf{F}}_n^d(\mathbf{r},\mathbf{r}')$ is ordinary, it defines a *regular* distribution in $\mathcal{D}'(\mathbb{R}^3)$ [67]. Moreover, since it has a compact support, it is automatically a function of slow growth and consequently its associated regular distribution is also in $\mathcal{P}'(\mathbb{R}^3)$ [67]. Consequently, it can be inferred from Theorem 8.2.4 that in the space of tempered distributions, $\bar{\mathbf{F}}_n^d(\mathbf{r}',\mathbf{r}) \to \bar{\mathbf{F}}(\mathbf{r}',\mathbf{r})$ entails $\bar{\mathbf{F}}_n^d(\mathbf{r}',\mathbf{k}) \to \bar{\mathbf{F}}(\mathbf{r}',\mathbf{k})$.

It can be easily proved that the regular tempered distribution generated by $\bar{\mathbf{F}}_n^d(\mathbf{r}',\mathbf{k})$, and the distributional Fourier transform of the regular distribution $\bar{\mathbf{F}}(\mathbf{r}',\mathbf{r})$ are identical [67]. Therefore, we don't distinguish in notation between distributional transforms and Fourier transforms of regular distributions in $\mathcal{P}'(\mathbb{R}^3)$. As an example of how to move freely between distributions and regular functions,

29 Cf. Theorem 8.2.6.

consider the following manipulation

$$\begin{aligned}\lim_{n\to\infty}\left\langle \bar{\mathbf{F}}_n^d\left(\mathbf{r}',\mathbf{r}\right),\mathbf{E}\left(\mathbf{r}\right)\right\rangle &= \lim_{n\to\infty}\left\langle \mathcal{F}^{-1}\bar{\mathbf{F}}_n^d\left(\mathbf{r}',\mathbf{k}\right),\mathbf{E}\left(\mathbf{r}\right)\right\rangle \\ = \lim_{n\to\infty}\left\langle \bar{\mathbf{F}}_n^d\left(\mathbf{r}',\mathbf{k}\right),\mathcal{F}^{-1}\mathbf{E}\left(\mathbf{r}\right)\right\rangle &= \left\langle \bar{\mathbf{F}}\left(\mathbf{r}',\mathbf{k}\right),\mathcal{F}^{-1}\mathbf{E}\left(\mathbf{r}\right)\right\rangle \\ = \left\langle \mathcal{F}^{-1}\bar{\mathbf{F}}\left(\mathbf{r}',\mathbf{k}\right),\mathbf{E}\left(\mathbf{r}\right)\right\rangle &= \left\langle \bar{\mathbf{F}}\left(\mathbf{r}',\mathbf{r}\right),\mathbf{E}\left(\mathbf{r}\right)\right\rangle,\end{aligned} \tag{8.45}$$

or in short, $\mathcal{F}^{-1}\bar{\mathbf{F}}_n^d\left(\mathbf{k},\mathbf{r}'\right)\to\bar{\mathbf{F}}\left(\mathbf{r}',\mathbf{r}\right)$, which is the same as (8.43). The first equality in (8.45) follows from (8.44); the second equality from the fact the operator $\mathcal{F}^{-1}$ is self-adjoint; the third equality from $\bar{\mathbf{F}}_n^d\left(\mathbf{r}',\mathbf{k}\right)\to\bar{\mathbf{F}}\left(\mathbf{r}',\mathbf{k}\right)$ just established above; and the last equality again follows from the self-adjoint property of $\mathcal{F}$.[30]

The relation (8.44) states that one may approximate both the Fourier transform and the inverse Fourier integral of the ACGF by the corresponding transforms of the smooth approximating functions $\left\{\bar{\mathbf{F}}_n^d\left(\mathbf{r},\mathbf{r}'\right)\right\}_{n\in\mathbb{N}}$, all possessing 3-dimensional bounded supports in $\mathbb{R}^3$. There is an enormous advantage in working with such functions in both theory and applications. Some of the nice behavior of these functions is captured by the following theorems [67], which we will have occasions to use throughout this chapter, cf. Chapter 9.

Theorem 8.2.6. *Let f be a function of C^2 class such that f, f', and f'' are all integrable. Then $\mathcal{F}f$ is also integrable and the Fourier inversion formula holds.*

Theorem 8.2.7. *If $f\in L^1$ is a function of bounded support, then its Fourier transform is of C^∞ class.*

Theorem 8.2.8. *Let $f(\boldsymbol{r})$ be a function differentiable p times. Then its Fourier transform $\mathcal{F}f := \tilde{f}(\boldsymbol{k})$ satisfies*

$$\left|\tilde{f}\left(\mathbf{k}\right)\right| \leq \left|2\pi\mathbf{k}\right|^{-p}\int_{\mathbb{R}^3}\left|\tilde{f}^{(p)}\left(\mathbf{r}\right)\right| d^3r. \tag{8.46}$$

Note that the three theorems above can be generalized to arbitrary spaces $\mathbb{R}^m, m\geq 1$.

8.2.6 Transformation Calculus for the 2D ACGF Tensor

First, consider the four functions $F_{nm}, n, m = 1, 2$, forming the components of the 2D ACGF tensor (8.5). They can be interpreted as the scalar Green's function defined by (8.21). Now, the local delta sequence introduced in (8.18) does depend on

30 In (8.45), we assume that the components of $\mathbf{E}(\mathbf{r})$ belong to $\mathcal{P}'(\mathbb{R}^3)$.

the choice of the local coordinate system, say (U_l, x_l). Indeed, in a new coordinate system, say (U_q, x_q), the quantity $\sqrt{g}$ transforms as [61]

$$\sqrt{g\left(x_q\right)} = \left|\frac{\partial x_l}{\partial x_q}\right| \sqrt{g\left(x_l\right)}, \tag{8.47}$$

where $|\partial x_l/\partial x_q| := \det\left(\left[\partial x_l^n/\partial x_q^m\right]\right)$, i.e., the determinant of the transformation matrix $\partial x_l^n/\partial x_q^m, n, m = 1, 2$. From (8.18), the local delta sequences transforms then in the new coordinate system as

$$f_n^{S,q}\left(x, x'\right) = \left|\partial x_l/\partial x_q\right|^{-1} f_n^{S,l}\left(x, x'\right). \tag{8.48}$$

This implies that for each choice of the local coordinate system containing $\mathbf{r}'$, there exists a *different* sequence of approximating functions F_n. In other words, we should write F_n^l in order to explicitly refer to the local coordinate system used in constructing (8.18). However, a careful examination of the definition (8.21) reveals that the *distribution* F does *not* depend on the choice of the local coordinate system. In other words, $F_n^l \rightarrow F$ as $n \rightarrow \infty$ for all possible values of l. The fact that the approximating sequence F_n does depend on the choice of the local coordinate system is not a problem because in practice the Green's function, as a distribution, always appears inside a surface integral on S. In all these cases, from the relation (8.19) it can be seen at once that the final result of any such an integration does *not* depend on the choice of the local coordinate system.

It remains then to examine the dependence of the ACGF on the local coordinate system that arises from the orthonormal vectors $\hat{\alpha}_l^n$. Suppose we are given a 2D Green's tensor as in (8.5). The explicit components of this tensor are the four functions $F_{nm}, n, m = 1, 2$, and they are dependent on the choice of the source and observation frames of references, $(U_{l'}, x_{l'})$ and (U_l, x_l), respectively, and also upon the choice of corresponding two sets of orthonormal vectors $\hat{\alpha}_{l'}^1\left(\mathbf{r}'\right), \hat{\alpha}_{l'}^2\left(\mathbf{r}'\right)$ and $\hat{\alpha}_l^1\left(\mathbf{r}\right), \hat{\alpha}_l^2\left(\mathbf{r}\right)$. We are given another two coordinate patches $(U_{q'}, x_{q'})$ and (U_q, x_q) such that both $U_{q'} \cap U_{l'}$ and $U_q \cap U_l$ are nonempty. Furthermore, let $\mathbf{r}' \in U_{q'} \cap U_{l'}$ and $\mathbf{r} \in U_q \cap U_l$. In this case, it is possible to express the coordinates and the tangent vectors defined at $\mathbf{r}$ and $\mathbf{r}'$ in terms of the language of the new coordinate patches $(U_{q'}, x_{q'})$ and (U_q, x_q). The purpose of the present Section is detailing how this purely geometrical problem is to be properly handled. To achieve this task, we will rely on the standard machinery of smooth manifolds [61].

Let us expand the two sets of orthonormal vectors chosen at $\mathbf{r}$ and $\mathbf{r}'$ in terms of coordinate basis with respect to the frames (U_q, x_q) and $(U_{q'}, x'_{q'})$ as follows

$$\hat{\alpha}_l^n(\mathbf{r}) = \sum_{m=1}^{2} \zeta_l^{mn} \frac{\partial}{\partial x_l^m}, \ \hat{\alpha}_{l'}^n(\mathbf{r}') = \sum_{m=1}^{2} \zeta_{l'}^{mn} \frac{\partial}{\partial x_{l'}^m}, \tag{8.49}$$

with $n = 1, 2$. In terms of the new patches (U_q, x_q) and $(U_{q'}, x'_{q'})$, we have similarly

$$\hat{\alpha}_q^n(\mathbf{r}) = \sum_{m=1}^{2} \zeta_q^{mn} \frac{\partial}{\partial x_q^m}, \ \hat{\alpha}_{q'}^n(\mathbf{r}') = \sum_{m=1}^{2} \zeta_{q'}^{mn} \frac{\partial}{\partial x_{q'}^m}. \tag{8.50}$$

The relation between the coordinate basis of the two representations can be written easily as [61]

$$\frac{\partial}{\partial x_q^n} = \sum_{m=1}^{2} \frac{\partial x_l^m}{\partial x_q^n} \frac{\partial}{\partial x_l^m}, \ \frac{\partial}{\partial x_{q'}^n} = \sum_{m=1}^{2} \frac{\partial x_{l'}^m}{\partial x_{q'}^n} \frac{\partial}{\partial x_{l'}^m}. \tag{8.51}$$

The inverse relation is given by

$$\frac{\partial}{\partial x_l^n} = \sum_{m=1}^{2} \frac{\partial x_q^m}{\partial x_l^n} \frac{\partial}{\partial x_q^m}, \ \frac{\partial}{\partial x_{l'}^n} = \sum_{m=1}^{2} \frac{\partial x_{q'}^m}{\partial x_{l'}^n} \frac{\partial}{\partial x_{q'}^m}. \tag{8.52}$$

Substituting (8.49) into (8.5), we find

$$\bar{\mathbf{F}}(\mathbf{r}, \mathbf{r}') = \sum_{mm'} \sum_{nn'} F_{mm'} \zeta_l^{nm} \zeta_{l'}^{n'm'} \frac{\partial}{\partial x_l^n} \frac{\partial}{\partial x_{l'}^{n'}}. \tag{8.53}$$

This represents the expansion of the 2D ACGF tensor, originally given in terms of two sets of orthonormal vectors, now in terms of the coordinate basis $\partial/\partial x_l^n, n = 1, 2$. Note that the tensorial basis in turns are $(\partial/\partial x_l^n)(\partial/\partial x_{l'}^m), n, m = 1, 2$. Now, substituting (8.52) into (8.53), we arrive at

$$\bar{\mathbf{F}}(\mathbf{r}, \mathbf{r}') = \sum_{mm'} \sum_{nn'} \sum_{pp'} \frac{\partial x_q^p}{\partial x_l^n} \frac{\partial x_{q'}^{p'}}{\partial x_{l'}^n} F_{mm'} \zeta_l^{nm} \zeta_{l'}^{n'm'} \frac{\partial}{\partial x_q^p} \frac{\partial}{\partial x_{q'}^{p'}}. \tag{8.54}$$

This gives the representation of the 2D ACGF tensor in terms of the coordinate basis of the *new* coordinate patches.

In order to give the final transformational rule, we first need to invert (8.50) to obtain

$$\frac{\partial}{\partial x_q^n} = \sum_{m=1}^{2} \kappa_q^{mn} \hat{\alpha}_q^m(\mathbf{r}), \quad \frac{\partial}{\partial x_{q'}^n} = \sum_{m=1}^{2} \kappa_{q'}^{mn} \hat{\alpha}_{q'}^m(\mathbf{r}'), \tag{8.55}$$

where the inverse matrix is defined as

$$\left[\kappa_q^{mn}\right] := \left[\zeta_q^{mn}\right]^{-1}, \quad \left[\kappa_{q'}^{mn}\right] := \left[\zeta_{q'}^{mn}\right]^{-1}. \tag{8.56}$$

Finally, substituting (8.50) into (8.54), we reach the most general result desired

$$\begin{aligned}\bar{\mathbf{F}}(\mathbf{r},\mathbf{r}') = \sum_{vv'} \Bigg[\sum_{mm'} \sum_{nn'} \sum_{pp'} \frac{\partial x_q^p}{\partial x_l^n} \frac{\partial x_{q'}^{p'}}{\partial x_{l'}^n} \zeta_l^{nm} \zeta_{l'}^{n'm'} \\ \times \kappa_q^{vp} \kappa_{q'}^{v'p'} F_{mm'} \Bigg] \hat{\alpha}_q^v(\mathbf{r}) \hat{\alpha}_{q'}^{v'}(\mathbf{r}').\end{aligned} \tag{8.57}$$

This represents the transformational law connecting the 2D ACGF tensor, written in the language of orthonormal vectors in terms of coordinate patches with indices in the atlas of S given by l and l', to a corresponding new tensor, i.e., that expressed in terms of the language of new coordinate patches at q and q', respectively.[31] As we can see after inspecting (8.57), given the four components of the 2D ACGF tensor computed in (8.5), one can transform this tensor to any other sets of orthonormal vectors and with respect to any other admissible coordinate patch in the atlas describing the differential geometry of the antenna surface S. The transformation law can be stated as

$$F_{vv'}^{\text{new}} = \sum_{mm'} \sum_{nn'} \sum_{pp'} \frac{\partial x_q^p}{\partial x_l^n} \frac{\partial x_{q'}^{p'}}{\partial x_{l'}^n} \zeta_l^{nm} \zeta_{l'}^{n'm'} \kappa_q^{vp} \kappa_{q'}^{v'p'} F_{mm'}^{\text{old}}. \tag{8.58}$$

where $F_{vv'}^{\text{new}}$ represent the components of the new tensor expressed in terms of the old components $F_{mm'}^{\text{old}}$.

8.2.7 Remarks on the Tensorial Character of the ACGF

In the standard literature on differential manifold theory, tensor products are defined only with respects to vectors defined at the *same* point, i.e., vectors belonging to the

31 In (8.57), only the explicit dependence of $\hat{\alpha}_q^v$ and $\hat{\alpha}_{q'}^{v'}$ on their spatial locations was mentioned. However, all the remaining factors contain in turn their own spatial dependence. Only F_{nm} depends jointly on $\mathbf{r}$ and $\mathbf{r}'$. On the other hand, the functions $\partial x_q^p / \partial x_l^n$, $\zeta_l^{n,m}$, $\kappa_q^{v,p}$ depend on $\mathbf{r}$, while the functions $\partial x_{q'}^{p'} \big/ \partial x_{l'}^n$, $\zeta_{l'}^{nm}$, $\kappa_{q'}^{v',p'}$ depend on $\mathbf{r}'$.

same tangent space TM_p, where p is some point in the (2-dimensional) manifold under consideration. The tensor product $\hat{\alpha}_l^n(\mathbf{r})\,\hat{\alpha}_{l'}^m(\mathbf{r}')$ involves vectors belonging to two different tangent spaces, i.e., $TM_{\mathbf{r}}$ and $TM_{\mathbf{r}'}$. This difficulty can be overcome in several ways. Probably the most direct strategy is to exploit the fact that the vectors involved can all be embedded in the larger space $\mathbb{R}^3$. In this case, it is always possible to perform parallel transport of any vector from a given point to an arbitrary point. Since the inner product is invariant with respect to permutations involving vectors that are obtained from each other by the process of parallel transport, the tensor product above can always be construed as a 3D Cartesian tensor product.

This solution may not be found in harmony with our earlier claim that Modes A and C can be studied, at least theoretically, as processes occurring completely in a 2D space, rather than $\mathbb{R}^3$. In this case, it is desirable not to refer to the larger space in the definition of the antenna current Green's function operator. It is enough for our purpose to *define* the operator entity $\hat{\alpha}_l^n(\mathbf{r})\,\hat{\alpha}_{l'}^m(\mathbf{r}') : TM_{\mathbf{r}'} \to TM_{\mathbf{r}}$ by the equation $\hat{\alpha}_l^n(\mathbf{r})\,\hat{\alpha}_{l'}^m(\mathbf{r}') \cdot \hat{\alpha}_{l'}^q(\mathbf{r}') := \hat{\alpha}_l^n(\mathbf{r})\,[\hat{\alpha}_{l'}^m(\mathbf{r}') \cdot \hat{\alpha}_{l'}^q(\mathbf{r}')]$, which is possible since we have a natural inner product on S (i.e., the Euclidean inner product inherited from the ambient space $\mathbb{R}^3$ into which S is embedded) available for computing $\hat{\alpha}_{l'}^m(\mathbf{r}') \cdot \hat{\alpha}_{l'}^q(\mathbf{r}')$. Finally, we mention that by reverting to the techniques of tangent bundles on abstract manifolds, this argument can be made fully rigorous. However, we don't pursue further such a formalization in the present work.

Chapter 9

Interrelationships Between Operational Modes of General Antenna Systems

9.1 GENERAL INTRODUCTION

At the present point of Part II, we start developing some of the major applications made available by the systematic deployment of the antenna current Green's function (ACGF) formalism introduced in chapter 8. Following our proposal of studying general electromagnetic systems through three operational modes, i.e., Modes A, B, and C, efforts in this Chapter will be focused on the various schemes of interrelations between the modes, rather than the modes themselves. This choice of focus is based on our belief that the dynamic content of electromagnetic theory can be best captured by paying special attention to *differential* (in the sense of *relational*) structures holding between basic processes isolated and identified earlier as the main constituents of the phenomena under investigation. Accordingly, our interest is not restricted solely to obtaining certain amount of data pertaining to one given mode, but rather we aim at understanding how one mode interrelates with another, *regardless of what the particular state of each of these modes is at the time.*[1]

To achieve this task, it is necessary to dig deep enough into the connective structure combining two given modes with each other. This connective structure,

1 For example, the dynamic content of classical mechanics can be captured by Newton's second law $\mathbf{F} = m\mathbf{a}$, where $\mathbf{F}$ is the force, m the mass, and $\mathbf{a}$ the acceleration. As such, the law does not give us direct information on what is the particular applicable force or how to measure the acceleration. Instead, the law states that whatever is the force $\mathbf{F}$ and the acceleration $\mathbf{a}$ in a material system with mass m, the *relation* between the two vectors is given by such and such equation.

especially as developed in Sections 9.4 and 9.5 below, will take the form of a *mechanism of interaction* explicated with the help of a theory of the electromagnetic near field of Part I. An idea of *spatial bandwidth* is captured in our formulation and highlighted within the context familiar in system engineering. Moreover, it is found that there are two regimes in electromagnetic interactions, resonant and nonresonant, and each has its own dominant features. Some other interesting findings include the concept of *generalized transfer function*, which turns out to be the most general invariant structure for the antenna system (as a spectral concept). In a certain sense, it goes beyond the ACGF itself as a transfer function. We also propose a general design methodology that can help in constructing approximations of antenna shapes capable of meeting arbitrary desired spectral characteristics specified in terms of the generalized transfer function mentioned above. The chapter ends by discussing the dynamic genesis of the received voltage in an antenna spatially displaced in the near-field zone of the source, with expected repercussions regarding the understanding of how arrays work, besides its relevance for foundational research.

Section 9.2 will develop in detail an argument first presented in [13] but expanded this time in the light of the subsequent devolvements of Chapter 8. The goal is to situate the inverse reciprocity theorem, which relates Mode A to Mode C, into the context of the 2D and 3D ACGF tensors defined rigorously in the distributional construction of Chapter 8. In Section 9.4, we introduce the near-field perspective and start exploring the structure of the relation between Mode A and Mode B. This aims at gaining some general knowledge concerning how the spatial variations of the excitation field control the radiated fields. Finally, Section 9.5 contains the bulk of the contribution of the present chapter. It aims at studying the relation between a current source of a nearby antenna and a general receiving antenna. The various findings of this part, we hope, will contribute to the final synthesis of the overall theoretical understanding attained via the use of the ACGF formalism. In particular, we focus on two dual contexts, the static and dynamic genesis of the received voltage in near-field zone interactions. The static genesis will present the substructure upon which the the dynamic picture can be erected. The latter explicates the (geometrical) mechanism of how motion of the receiving antenna changes the observed voltage. This analysis, which is quite general, may form the basis for future investigations of particular systems, such as antenna arrays and near field measurement.

9.2 SYMMETRY OF THE ANTENNA CURRENT GREEN'S FUNCTION: THE INTERRELATIONSHIP BETWEEN MODE A AND MODE C

9.2.1 Introduction

The antenna current induced by a delta source can seldom by represented in a closed-form analytical expansion, and hence it is difficult to see in advance what kind of properties it has. However, we prove in this section a very general symmetry property possessed by the dyadic function $\bar{\mathbf{F}}(\mathbf{r}, \mathbf{r}')$. The upshot of the argument is that in the case where the focus in the receiving mode is on certain terminal-like quantities, e.g., received voltage, current, power, etc, then it is possible to relate the ACGF of Mode C to the ACGF of Mode A in a quite simple manner. The basic idea of reciprocity applied to transfer functions appears to have been already formulated in [25] and [26] within the context of circuit-theoretic models of antenna systems. The idea was recently applied numerically to antenna arrays to predict the received voltages with strong mutual coupling using transmitting mode data. However, in all these works, the approach was worked out informally in a close reading of the numerical method adopted for use in performing the computation without explicitly refereing to the antenna current Green's function concept itself.[2] The derivation of the receiving ACGF (Mode C) in this section is constructed directly using field theory, not a circuit or numerical approximation.

9.2.2 Derivation of the Symmetry Relation of the ACGF

The starting point is Lorentz reciprocity theorem in one of its standard integral forms [28]. We assume that two sources $\mathbf{J}_1(\mathbf{r})$ and $\mathbf{J}_2(\mathbf{r})$ produce the fields $\mathbf{E}_1(\mathbf{r}), \mathbf{H}_1(\mathbf{r})$ and $\mathbf{E}_2(\mathbf{r}), \mathbf{H}_2(\mathbf{r})$. The magnetic sources are set to zero since the problem involves only PEC objects. It can then be shown using Maxwell's equations that [28]

$$\int_V dv\,(\mathbf{J}_1 \cdot \mathbf{E}_2 - \mathbf{J}_2 \cdot \mathbf{E}_1) = \oint_{S_0} d\mathbf{s} \cdot (\mathbf{E}_1 \times \mathbf{H}_2 - \mathbf{E}_2 \times \mathbf{H}_1). \tag{9.1}$$

Here, the surface S_0 incloses the antenna surface S.[3] If the surface S_0 is taken to infinity, then it can be shown that, provided the surrounding medium is assumed to be infinite and homogeneous, the surface integral on the RHS of (9.1) vanishes [28].

2 For further details about the ACGF method applied to empirical contexts with some comparison with existing reciprocity methods in literature, see the following applications-oriented chapters.

3 Since we assume perfect electric conductor, the terms involving the magnetic field on the LHS of (9.1) has already been dropped out.

We are left with

$$\int_V dv\,(\mathbf{J}_1\cdot\mathbf{E}_2-\mathbf{J}_2\cdot\mathbf{E}_1)=0. \tag{9.2}$$

The integration volume V is the volume inclosed by the metallic antenna body S.

Now consider the situation described by the electric field integral equation in the form stated in Chapter 8, equation (8.2). Here, we have two sets of electromagnetic data, each containing three types of quantities: 1) The externally applied electric field $\mathbf{E}^{\rm ex}$, 2) the current induced by this field $\mathbf{J}$, and 3) the field radiated by the induced current $\mathbf{E}^{\rm rad}$.[4] At the antenna surface, the following relation holds $\hat{n}\times\mathbf{E}^{\rm rad}=-\hat{n}\times\mathbf{E}^{\rm ex}$. Consider now the configuration

$$\mathbf{E}_1^{\rm ex}(\mathbf{r})=\hat{a}\,\delta\,(\mathbf{r}-\mathbf{r}_1)\,,\;\hat{n}\times\mathbf{E}_1^{\rm rad}(\mathbf{r})=-\hat{a}\,\delta\,(\mathbf{r}-\mathbf{r}_1)\,, \tag{9.3}$$

$$\mathbf{E}_2^{\rm ex}(\mathbf{r})=\hat{b}\,\delta\,(\mathbf{r}-\mathbf{r}_2)\,,\;\hat{n}\times\mathbf{E}_2^{\rm rad}(\mathbf{r})=-\hat{b}\,\delta\,(\mathbf{r}-\mathbf{r}_2)\,, \tag{9.4}$$

$$\mathbf{J}_1(\mathbf{r})=\bar{\mathbf{F}}(\mathbf{r},\mathbf{r}_1)\cdot\hat{a},\;\mathbf{J}_2(\mathbf{r})=\bar{\mathbf{F}}(\mathbf{r},\mathbf{r}_2)\cdot\hat{b}. \tag{9.5}$$

The first equations in (9.3) and (9.4) give the form of the externally applied field polarized according to the arbitrary unit vectors $\hat{a}$ and $\hat{b}$ *tangential to S* and located at arbitrary two positions $\mathbf{r}_1$ and $\mathbf{r}_2$. The second equations in (9.3) and (9.4) give the response of the antenna system to these excitations.[5] The two currents appearing in (9.5) finally provide us with the electric current induced on the antenna by the externally applied field. They are written in terms of the ACGF tensors.[6] However, the manner in which the relation between the ACGF tensor $\bar{\mathbf{F}}$ and the induced currents was expressed in (9.5) should be studied carefully by the reader. It is crucial in the present derivation to realize that the statement of Lorentz reciprocity theorem relates sets of sources (currents) to the fields produced by these sources. Our goal is to reach a statement regarding the exact reverse of this relation: we would like to learn something about the receiving mode ACGF in terms of the transmitting mode ACGF but for the physical situation when the sources are fields while the outputs are currents. Strictly speaking, the electromagnetic operator of this

4 We just happen to be not interested in the present derivation in the current producing the externally applied field, which can be computed directly from Maxwell's equation. It is part of the idealization of any scientific model that irrelevant quantities are dropped out of consideration when they do not affect the parameters of interest in the theory. However, this source current will be taken directly into consideration in Section 9.5.

5 For the rigorous definition of the surface Dirac delta function in the context of the antenna current Green's function, see Chapter 8.

6 The local sifting property derived in Chapter 8 is assumed and used freely throughout the present derivation.

problem is the EFIE operator $\mathcal{L}$ defined by 8.2, which is wholly different from the forward electromagnetic operator to which the classical Lorentz reciprocity theorem applies. However, we can bypass the difficult problem of working directly with $\mathcal{L}$ by noticing the mechanism of PEC objects: They simply work by generating a current such that the radiated field is canceled by the applied field. In order to apply Lorentz reciprocity to this situation, we must still operate with sources in the form of currents and outputs in the form of fields. The appropriate sets will be then the currents in (9.5) as the sources and the *second* fields in (9.3) and (9.4), i.e., the fields $\mathbf{E}_1^{\mathrm{rad}}(\mathbf{r})$ and $\mathbf{E}_2^{\mathrm{rad}}(\mathbf{r})$. These two fields are strictly produced by the currents $\mathbf{J}_1$ and $\mathbf{J}_2$ according to the typical forward electromagnetic operator to which the Lorentz reciprocity theorem applies. We now can appreciate the peculiar way in which the arguments of the ACGF were written in (9.5). Recalling the clarification given in Chapter 8 regarding the semantics of the various arguments of the Green's functions involved there, we find that in the functional form $\bar{\mathbf{F}}(\mathbf{r},\mathbf{r}')$, the agrement $\mathbf{r}'$ stands for location of the *impressed* field (now should be replaced by *radiated* fields), while $\mathbf{r}$ stands for the spatial location of the current on the antenna surface. This latter location is simply a position vector ranging on the total surface S. In this way, the two sets of sources and radiated fields to be related to each other in the next step have the correct functional form required by the Lorentz reciprocity theorem. Therefore, we obtain from (9.2), (9.5), and the second equations in (9.3) and (9.4) the result

$$\begin{aligned}-\int_V dv\Big[\big(\bar{\mathbf{F}}(\mathbf{r},\mathbf{r}_1)\cdot\hat{a}\big)\cdot\hat{b}\,\delta(\mathbf{r}-\mathbf{r}_2)\\ -\Big(\bar{\mathbf{F}}(\mathbf{r},\mathbf{r}_2)\cdot\hat{b}\Big)\cdot\hat{a}\,\delta(\mathbf{r}-\mathbf{r}_1)\Big]=0.\end{aligned} \tag{9.6}$$

In writing this equation, we made use of the fact that the currents $\mathbf{J}_1(\mathbf{r})$ and $\mathbf{J}_2(\mathbf{r})$ have nonvanishing values only on the antenna surface S, which allowed us to restrict the value of the radiated field $\mathbf{E}_{\mathrm{rad}}(\mathbf{r})$ to the tangential components only as given by (9.3) and (9.4). This integral can now be immediately evaluated to give

$$\big[\bar{\mathbf{F}}(\mathbf{r}_2,\mathbf{r}_1)\cdot\hat{a}\big]\cdot\hat{b}=\Big[\bar{\mathbf{F}}(\mathbf{r}_1,\mathbf{r}_2)\cdot\hat{b}\Big]\cdot\hat{a}. \tag{9.7}$$

Since $\mathbf{r}_1$ and $\mathbf{r}_2$ were assumed arbitrary in our derivation, they can be replaced by $\mathbf{r}$ and $\mathbf{r}'$. We now notice that the proper unit vectors $\hat{a}$ and $\hat{b}$ are tangential to S. Therefore, in order to spell out the meaning of (9.7) in terms of the components of the ACGF, only the 2D tensor can be used. Moreover, since $\hat{a}$ and $\hat{b}$ are arbitrary, any choice of the coordinate patches of the source and observation points, together with a choice of two orthogonal directions in each patch, can be matched by proper corresponding selections of $\hat{a}$ and $\hat{b}$. In this case, we immediately arrive from (9.7) at our main result

$$[F_{nm}(\mathbf{r},\mathbf{r}')] = [F_{nm}(\mathbf{r}',\mathbf{r})]^T, \tag{9.8}$$

where T denotes the transpose operation applied to the matrix representation $[F_{nm}]$ of the 2D dyadic operator. In component form, equation (9.8) is expanded as

$$\begin{aligned} F_{11}(\mathbf{r},\mathbf{r}') &= F_{11}(\mathbf{r}',\mathbf{r}), F_{12}(\mathbf{r},\mathbf{r}') = F_{21}(\mathbf{r}',\mathbf{r}), \\ F_{21}(\mathbf{r},\mathbf{r}') &= F_{12}(\mathbf{r}',\mathbf{r}), F_{22}(\mathbf{r},\mathbf{r}') = F_{22}(\mathbf{r}',\mathbf{r}). \end{aligned} \tag{9.9}$$

The same symmetry relation is also enjoyed by the 3D ACGF tensor, a fact proved formally in the next section.

9.2.3 On the Symmetry of the 3D ACGF Tensor

We show here that the 3D ACGF tensor also satisfies the simple symmetry relation (9.8) enjoyed by the 2D ACGF tensor. To see this, we study the components of the Mode C 3D ACGF tensor as given in Chapter 8 by (8.7). By swapping the source and observation variables, we obtain for Mode C the following components

$$F_{x_p x_q}(\mathbf{r}',\mathbf{r}) = \sum_{n=1}^{2}\sum_{m=1}^{2} F_{nm}(\mathbf{r}',\mathbf{r})\beta_{l'p}^{n}(\mathbf{r}')\,\beta_{lq}^{m}(\mathbf{r}). \tag{9.10}$$

In order to show that the 3D tensor is indeed symmetric, we swap the indices p and q in (9.10) and then use (9.8) to obtain

$$F_{x_p x_q}(\mathbf{r}',\mathbf{r}) = \sum_{n=1}^{2}\sum_{m=1}^{2} F_{mn}(\mathbf{r},\mathbf{r}')\beta_{l'p}^{m}(\mathbf{r})\beta_{lq}^{n}(\mathbf{r}'). \tag{9.11}$$

By relabeling the dummy indices in the sums of (9.11) as $n \to m$ and $m \to n$, it follows at once by comparing (9.11) above with (8.7) in Chapter 8 that

$$F_{x_p x_q}(\mathbf{r}',\mathbf{r}) = F_{x_q x_p}(\mathbf{r}',\mathbf{r}). \tag{9.12}$$

That is, we can safely write in general

$$\bar{\mathbf{F}}(\mathbf{r},\mathbf{r}') = \bar{\mathbf{F}}^T(\mathbf{r}',\mathbf{r}), \tag{9.13}$$

regardless to whether the 2D or 3D form of the ACGF is used. In any case, Mode C ACGF can still be related to Mode A ACGF through the relation (9.10) together with (9.8) or (9.9).[7]

9.2.4 Discussion of the Results

We now put the mathematical theorem (9.8) proved above in the physical context of antenna applications. That is, we consider Mode C as defined in terms of its characteristic Green's function $\bar{\mathbf{L}}(\mathbf{r},\mathbf{r}')$. The situation encountered in practice is this. It is usually the case that we solve numerically for the ACGF in the transmitting mode, typically in connection with calculations similar to that outlined in Chapter 8, i.e., for *localized* excitation region U corresponding to the physical *port* used to inject information into the antenna system. This is Mode A. For Mode C, the same antenna is used as a *receiving* structure. The received voltage is usually collected at the same physical port used in Mode A, i.e., the receiving "area" will be also U [28].[8] Now a question naturally arises regarding whether the two modes of operation can be related to each other, which can now be easily answered with the help of (9.8). From (7.2) and (7.4) and (9.8) above, we find

$$\bar{\mathbf{L}}(\mathbf{r},\mathbf{r}') = \bar{\mathbf{F}}^T(\mathbf{r}',\mathbf{r})\,. \tag{9.14}$$

It is significant for the derivation of this relation to carefully observe the reference of each argument belonging to the various Green's functions involved as they appear in their defining relations (7.2) and (7.4).

The distinguishing feature of the ACGF formalism, we believe, is that the Lorentz reciprocity theorem (9.1) is *not* the core of the method. Instead, it is the antenna current *Green's function* what is at stake here. In this sense, the method of expressing the received current using (7.4) is unique and represents the signature of the formalism adopted in this book. Indeed, the method is valid for *arbitrary* illuminating field in the receiving mode. The simple relation (9.14) allows us to predict the current induced in the receiving port occupying the area U, when the

7 The reader may consider (9.10) as an effective method to compute the *receiving* Green's function $\bar{\mathbf{L}}(\mathbf{r},\mathbf{r}')$ defined in Chapter 8. This fact will be heavily exploited in the forthcoming applications chapters.

8 There is no loss of generality in the subsequent analysis if the receiving port is different from the transmitting antenna port.

antenna is functioning in Mode C, using some data computed from Mode A.[9] *However, it should be kept in mind that it is only the received signals at U that can be predicted, and not the induced current distribution on the entire antenna system surface S.* Since the focus of the engineer is on terminal-like quantities, and hence the limited inverse relation implied by (9.14) ("limited" in the sense of being applicable in practice to a spatially *common* area U in both Modes A and C) is adequate for most of the applications of electromagnetic systems involving both antennas and circuits.[10]

9.2.5 Comparison with Traditional Reciprocity Methods

Probably the closest formulations to the ACGF to be found in the literature are some special remarks in [26], although the concept of the current Green's function itself for general antenna systems was never explicitly formulated there. The majority of the works in applied electromagnetics, e.g., see [32], [31], [37], [28], [54], appears to be based on the direct application of the Lorentz reciprocity theorem to special cases tailored for the particular problem under consideration. In this section, we will pick up and highlight one characteristic traditional reciprocity method formulation to be found in [28], and compare it to the picture proposed by the ACGF formalism.

Consider an antenna system with a surface S supporting its electromagnetic boundary condition. The equivalent circuit of an antenna in the receiving mode consists of a current generator $I_{\rm rx}$ connected in parallel with the admittance $Y_{\rm in}$, which is simply the input admittance seen when looking into the input terminals of the antenna's physical port during the transmitting mode operation [26]. The main intention of the reciprocity method is to establish a formula for the current $I_{\rm rx}$ in terms of some transmitting mode data. Using Lorentz reciprocity theorem, such formula was found and is given by [28]

$$I_{\rm rx} = (-1/V_{\rm tx}) \int_{\mathcal{V}} dv \mathbf{E}_{\rm tx}(\mathbf{r}) \cdot \mathbf{J}_{\rm src}(\mathbf{r}). \tag{9.15}$$

9 That is, the reciprocity theorem is applied to the Green's function, and then the latter is used to enact the calculations. In literature, it is usually the reciprocity theorem what is applied directly to certain combinations of variables of interest. Such methods are inherently less general than the ACGF formalism. It is clear that any specific result obtained in one of the common reciprocity methods existing now in literature can be also obtained using the ACGF formalism. For more detailed comparison with literature, see Section 9.2.5.

10 Further and detailed investigation of the issues of mutual coupling, load and intrinsic impedance characterization, improved computational models, etc, in antenna arrays using the ACGF formalism armed with the reciprocity theorem will be dealt with elsewhere.

Here, $V_{\rm tx}$ stands for the input port's voltage in the transmitting mode. $\mathbf{J}_{\rm src}$ is the source current located in the source region $\mathcal{V}$ in the exterior of S; it produces the fields impinging on S in the receiving mode. The field $\mathbf{E}_{\rm tx}$ is the electric field produced by the antenna S in the region $\mathcal{V}$ when working in the transmitting mode. It is clear that (9.15) does achieve the intended goal stated above in the sense that it expresses the receiving mode current $I_{\rm rx}$ in terms of only transmitting mode data, namely $\mathbf{E}_{\rm tx}$ in the source region $\mathcal{V}$. We would like now to compare this formula with the ACGF formulas (7.4) and (9.14).

First of all, we need to be careful about the effect of the source region $\mathcal{V}$ on the operation of the antenna S. In general, the transmitting mode data $\mathbf{E}_{\rm tx}$ are computed or measured in the presence of $\mathbf{J}_{\rm src}$ [28]. The motivation for this restriction is that the antenna S may couple electromagnetically with the source. This coupling does *not* occur with the *current* $\mathbf{J}_{\rm src}$ itself, but with the *physical layer* supporting it, i.e., the region $\mathcal{V}$ and its associated boundary condition. In practice, however, we are very seldom interested in the details of the source generating the fields impinging on the antenna under consideration. In characterizing systems working in the receiving mode, we are interested in the fields impinging on S, which may be generated by infinite varieties of source distributions. We notice now that the ACGF formalism is well suited to this task since it requires as an in input to the antenna S only the illumination field $\mathbf{E}^{\rm inc}$. We will then assume for a while that the antenna S does not couple electromagnetically with V. A discussion of the implications of not posing this assumption will be presented at the end.

Suppose that the antenna S is excited by a field $\mathbf{E}^{\rm inc}$ generated by the source $\mathbf{J}_{\rm src}$. In this case, both the ACGF formula (7.4) and (9.15) are exact relations since they were both derived from field theory. There is a difference, however, in the nature of the transmitting data used in the two formulas. In (9.15), this data has the form of a field computed or measured in the source region $\mathcal{V}$. In (7.4), on the other hand, the Green's function $\bar{\mathbf{F}}(\mathbf{r},\mathbf{r}')$ (in the form of a current distribution) is our transmitting data. This difference may appear superficial at a first examination since eventually both formulas are indeed exact. However, the crucial contrast between the ACGF formula and all other traditional reciprocity methods emerges when we consider the quite practical scenario where other sources radiate in the vicinity of S. Suppose that there is another source region $\mathcal{V}'$ with current $\mathbf{J}'_{\rm src}$. In order to use (9.15) for the computation of the received current $I_{\rm rx}$, we must compute a *new* set of transmitting mode data, namely the field radiated by S in the transmitting mode evaluated in the new source region $\mathcal{V}'$. In the ACGF formula (7.4), the situation is quite different. In principle, there is no need to compute any new transmitting mode data. The ACGF $\bar{\mathbf{F}}(\mathbf{r},\mathbf{r}')$ used previously can be perfectly employed in computing the response of S to the new fields. Thus, if the new source region does not overlap with the first region, i.e., when $\mathcal{V} \cap \mathcal{V}' = \emptyset$, then none of the old transmitting mode data can be utilized in the computation of the received current using the formula (9.15), while the situation is exactly the opposite with the ACGF formula (7.4). The difference between the two methods becomes even graver when one considers an *arbitrary* number of fields impinging on S emanating from various source regions. For example, there may be a far field illumination due to multiple antennas in the far zone of S plus the fields generated by the sources $\mathcal{V}$, $\mathcal{V}'$, $\mathcal{V}''$, etc, in the near zone. In all such cases, the same formula (7.4) can be used with an integration restricted on a fixed region, the antenna's own surface S. Using (9.15), on the other hand, requires knowledge of fields radiated by S in all those source regions, with different integrations performed in the various source regions involved.[11]

It is important to be clear about the proper origin of the advantageous position of the ACGF formula (7.4) compared with (9.15). This is not because of some deficiency in the latter formula, or to suggest that we managed to use the same reciprocity result, the classical Lorentz theorem, in a way superior to the manner in which it was used in (9.15). What is actually the case is this. The ACGF formula (7.4), augmented by the inverse reciprocity theorem (9.14), involves not only the use of Lorentz reciprocity theorem as manifested in the derivation of (9.14), but also the use of *another* powerful theorem, namely the Green's function theorem (7.4). What we have achieved here is that the classical Lorentz reciprocity theorem was applied not to the inputs and outputs of the electromagnetic system S, but to the *transfer* function, the *system* function, the *Green's* function of S, i.e., $\bar{\mathbf{F}}(\mathbf{r}, \mathbf{r}')$. It is for this very reason that (7.4), when combined with (9.14), becomes capable of dealing with the arbitrary general situation when multiple illumination scenarios are prevalent, while traditional reciprocity methods fail to achieve the same result without deploying additional sets of transmitting mode data.

We would like now to say something about the assumption that the antenna under consideration S does not electromagnetically couple with a source region $\mathcal{V}$ located in the near zone. Assume that this is not the case. In such scenario, both the ACGF formalism and the method of (9.15) have to take into account explicitly the effect of the mutual interaction (between S and $\mathcal{V}$) on the data $\mathbf{E}_{\text{tx}}$ and $\bar{\mathbf{F}}(\mathbf{r}, \mathbf{r}')$. In both cases, the predictions of the two methods are exact. Consider another source in region $\mathcal{V}'$ to be introduced to the problem as before. If the new region also

11 The proper origin of the advantageous position of the ACGF formula (7.4) compared with (9.15) is that the ACGF formula (7.4), augmented by the inverse reciprocity theorem (9.14), involves not only the use of Lorentz reciprocity theorem as manifested in the derivation of (9.14), but also the use of *another* powerful theorem, namely the Green's function theorem (7.4).

couples electromagnetically with S, then again both the ACGF formalism and the method of (9.15) has to be modified to reflect the effect of this mutual interaction between the source and the antenna under consideration. However, we may notice that if additional source regions $\mathcal{V}''$, $\mathcal{V}'''$, etc, were introduced at positions in which they alone do *not* electromagnetically couple with S, while $\mathcal{V}$ and $\mathcal{V}'$ remains so, then obviously the ACGF formalism has the advantage over the formula (9.15) as explained in the previous paragraphs.

But the extreme situation in which *every* added source region will couple significantly with the antenna S is quite odd and in disharmony with the spirit of the usual investigations of antenna practice. If for some input X we compute the output Y by using the transfer function H via the relation $Y = HX$, then it is implicitly assumed that H is independent of X. Indeed, we don't want to have a system in which $H = H(X)$, making the computation of the output Y by the relation $Y = H(X)X$ quite awkward. The main underlying theme of system theory is that there exists in practice a characteristic function, the transfer function or the Green's function, *independent of the nature of the input*, in which the output due to any excitation can be readily computed. Now that does not mean that the opposite situation cannot be encountered in practice, for nothing is further from truth than this. In electromagnetic systems, especially in the microwave regime, the possibility of strong coupling in the near zone is real. But if the device under consideration is an antenna working in the receiving mode under various field illumination scenarios, then this idealized situation, antenna interacting with impinging *fields*, is the dominant theme expected in applications, and the ACGF formalism is the most natural setting for this kind of problems. In case the antenna performance is different for each new source of field illumination, then this is not a proper single antenna problem anymore; instead, one then has to consider the antenna S together with the source regions as a *combined* larger system since the component parts are strongly interacting with each other all the time. In this case, the ACGF formalism can be applied to an *array* problem, as explained in detail in Chapter 14.

9.3 INTERLUDE: THE SIGNIFICANCE OF THE NEAR-FIELD PERSPECTIVE IN APPLIED ELECTROMAGNETICS

We provide some general remarks about the importance of devoting considerable attention to the near-field perspective in a general analysis of electromagnetic systems, a theme that will dominate Sections 9.4 and 9.5 below. There is a need for fresh re-examination of the fundamental issues pertaining to the principles of operations of electromagnetic systems viewed mainly from the near field viewpoint developed in Part I. The reasons for this can be stated as follows. First, the radiation pattern of the array is itself the outcome of a complex genetic process in which the near field continuously differentiates and changes in form, giving rise to a determinate shape of radiation at the far zone. Consequently, understanding the structure of the near field allows us to better grasp the hidden potentials of controlling the radiated fields through educated modifications of the current distribution on the system. Second, electromagnetic interactions between array elements, especially elements placed very close to each other, is a direct coupling phenomenon between near fields shells. The physics of the near field then dictates the manner in which electromagnetic coupling will affect the performance of the array under consideration, e.g., see Chapter 16. Third, for future applications, which tends to increase dramatically in complexity, such as metamaterial environments (artificial media) and nano-structures, it is important to get a hold on the fine details of the spatial structure of the near field in order to theoretically pre-plan new devices and guide the research and devolvement process. The spectral analysis of the near field is one such perspective on the spatial structure (see Chapter 4). Although the utilization of the evanescent fields has found its way into near-field nano-optics, especially for imaging applications, very little research in this direction has been conducted for antennas working in the microwave regime. Consequently, we devote the remaining parts of this chapter to detailed theoretical investigations of the spectral aspects in electromagnetic interactions using the formalism of the ACGF developed above.

9.4 SPECTRAL ANALYSIS OF THE ANTENNA-SOURCE RADIATED FIELDS: THE INTERRELATIONSHIP BETWEEN MODE A AND MODE B

9.4.1 Introduction

Having established a precise mathematical formalism for describing antenna systems working in Mode A in Chapter 8, Mode C in Section 9.2, and the exact relation between these two modes, we direct our attention now toward the intermediate stage essential in any electromagnetic system, Mode B, i.e., the conversion of a current source into an electromagnetic field capable of interacting with far and near objects. Although Mode B is the most widely studied topic in classical electromagnetic theory, little is understood about the fundamental physics of the relation between

Mode A and Mode B on one side, and Mode B and Mode C on another. In the this section, we focus our attention on the former relation and leave the latter to the next section.

The complete antenna system comprised of Mode A and Mode B is shown in Figure 7.1. In order to bring a new insight into this system, we notice that the input excitation field $\mathbf{E}^{\mathrm{ex}}$ is a function of space defined on a compact support U. To understand the combined A + B system, one is naturally interested in mapping out the relation between the input $\mathbf{E}^{\mathrm{ex}}$ and the radiated fields $\mathbf{E}$ and $\mathbf{H}$. Due to the considerable complexity of the radiated fields, we should select certain features in the output variables that can be deemed the most interesting theoretically and practically within the present context of investigation. Now, the antenna system is a spatial object in the sense that its physics is dictated by the geometry of the surfaces supporting the electromagnetic boundary condition. The fields themselves are functions on space. We would like then to analyze with sufficient depth their spatial structure and the manner in which spatial degrees of freedom interact with the geometrical details of the antenna system total surface. However, for the interest of applied electromagnetic theory, we are most concerned about the *wave* phenomena associated with the radiated fields. This means that an adequate appreciation of the peculiar mechanism in which the field converts from nonpropagating to propagating modes, as the observation point is moved in the spatial domain surrounding the antyenna/circuit system, is the main general feature most likely to arouse our interest. From the perspective of the antenna current Green's function, the analysis of the near field of general antenna systems takes a new light as this section and the following one will make clear. Henceforth, the dynamic decomposition of the near field into propagating and nonpropagating modes will be taken up in the present work as the major thematic framework for the subsequent inquiries into the interrelation between the two "outer" modes, Mode A and Mode C, on one side, and the "inner" mode, Mode B, on the other side.

9.4.2 Derivation of the Main Relation Between Mode A and Mode B

The main method of investigation to be adopted in this work is phenomenological in the following sense. We will not force dogmatically any external *ad hoc* viewpoint, but the analysis will proceed according to the nature of the phenomena under investigation themselves. More particularly, we will focus on the *dynamic* picture of the radiated field as developed in Part I and try to relate it to the input excitation $\mathbf{E}^{\mathrm{ex}}(\mathbf{r})$. To accomplish this, we will perform a Fourier expansion of the ACGF, then plug its expression into the form of the dynamic decomposition of the near field (to propagating and nonpropagating modes). In Section 9.4.3, we discuss the physical significance of the results.

The main idea behind the dynamic picture presented in Part I was the utilization of two coordinate frames to describe the splitting of the electromagnetic fields into propagating and nonpropagating modes.[12] Mathematically, this was accomplished by using the well-known Weyl expansion that expands the scalar free space Green's function into propagating and evanescent modes. The local observation frame is described by a matrix $\bar{\mathbf{R}}$ that defines the 3D rotation of this frame with respect to a previously chosen global frame. The following general decomposition theorem was derived in Part I

$$\mathbf{E}_{\mathrm{ev}}(\mathbf{r};\hat{u}) = \int_{p^2+q^2>1} dpdq\, \mathbf{D}(\mathbf{K},\hat{u})\, e^{i\mathbf{K}\cdot(\bar{\mathbf{R}}\cdot\mathbf{r})}, \tag{9.16}$$

$$\mathbf{E}_{\mathrm{pr}}(\mathbf{r};\hat{u}) = \int_{p^2+q^2<1} dpdq\, \mathbf{D}(\mathbf{K},\hat{u})\, e^{i\mathbf{K}\cdot(\bar{\mathbf{R}}\cdot\mathbf{r})}, \tag{9.17}$$

where

$$\mathbf{D}(\mathbf{K},\hat{u}) := \bar{\mathbf{R}}^T(\hat{u}) \cdot \bar{\boldsymbol{\Omega}}(\mathbf{K}) \cdot \bar{\mathbf{R}}(\hat{u}) \cdot \mathbf{J}\left[\bar{\mathbf{R}}^T(\hat{u}) \cdot \mathbf{K}\right]. \tag{9.18}$$

Here, $\hat{u} = \hat{x}\sin\theta\cos\varphi + \hat{y}\sin\theta\sin\varphi + \hat{z}\cos\theta$ is a unit vector describing the orientation of the z-axis of the local frame.[13] The spectral variables p and q describe the Fourier expansion in the x and y directions, while the third variable m, pertinent to the corresponding expansion in the z-direction, is related to the first two by the relation $m = \sqrt{1-p^2-q^2}$. The spectral vector $\mathbf{K}$ is given by

$$\mathbf{K} := \hat{x}k_0 p + \hat{y}k_0 q + \hat{z}\,\mathrm{sgn}(z)\, k_0 m, \tag{9.19}$$

where sgn stands for the signum function and k_0 for the free space wavenumber. The spectral polarization dyad is defined as

$$\bar{\boldsymbol{\Omega}}(\mathbf{K}) := \frac{-\omega\mu}{8\pi^2 k_0 m}\left(\bar{\mathbf{I}}k_0^2 - \mathbf{K}\mathbf{K}\right). \tag{9.20}$$

We will call the vector function $\mathbf{D}(\mathbf{K})$ the *field spectral composition*. Its physical significance lies in the fact that through $\mathbf{D}(\mathbf{K})$ we can fully specify the dynamics

12 The reader is referred to Part I for full details. However, the background presented in this chapter is sufficient to follow the main lines of analysis and forthcoming conclusions.

13 Note that the functional dependence on $\hat{u}$ comes from the fact that the rotation matrix $\bar{\mathbf{R}}$ depends on the spherical angles φ and θ through relations that can be found in Chapter 4.

of the radiated fields in such a way that at each orientation of the local observation frame, determined by $\hat{u}$, it is possible to determine which spectral components are present in the fields and how they split into propagating and nonpropagating modes as in Chapter 4. The spatial Fourier transform $\mathbf{J}(\mathbf{K})$ is defined by

$$\mathbf{J}(\mathbf{K}) := \int_S ds\, \mathbf{J}(\mathbf{r})\, e^{-i\mathbf{K}\cdot\mathbf{r}}, \tag{9.21}$$

where the volume V originally used in Chapter 4 is replaced in (9.21) by the S appearing in the general definition of the antenna system presented in 8.1.1.[14]

The expansions (16.19) and (16.18) are valid only in the region exterior to the infinite slab $|z''| < L_0$, which is perpendicular to the (varying) direction $\hat{u}$, where z'' is the z-coordinate of the local (rotating) frame of reference.[15]

The physical content of the decomposition (9.16) and (9.17) boils down to this: When the observation point moves around the antenna current distribution such that $\hat{u}$ points along a direction specified by the spherical angles φ and θ, then the manner in which the field splits into propagating and nonpropagating modes along this direction *will vary from one location to another*.[16] This overall manner of variation is the essences of the *dynamic* description of the near field not only because it maps out the overall process in which the field converts from its "static" (nonpropagating) form into the dynamic (propagating) form, but more because *no fixed observation frame is singled out a priori for the description of the radiation process*.[17]

Now, by substituting (7.2) into (9.21), we obtain

$$\mathbf{J}(\mathbf{K}) = \int_S ds \int_U ds'\, \bar{\mathbf{F}}(\mathbf{r}, \mathbf{r}') \cdot \mathbf{E}_t^{\text{ex}}(\mathbf{r}')\, e^{-i\mathbf{K}\cdot\mathbf{r}}, \tag{9.22}$$

14 For an elementary method to eliminate this Fourier transform from the general decomposition theorem (16.19) and (16.18), see the appendix in Chapter 16. Note however that this method is approximative, not exact.

15 However, it was found in Part I that the concept of *radial streamlines*, which is the most interesting description of the the spectral structure of the antenna near field from the engineering point of view, is not restricted by this limitation on the region of validity of the general expansion (9.16) and (9.17). In the remaining parts of this chapter, we will work only with the general decomposition theorem (9.16) and (9.17).

16 At each location, propagation (nonpropagating) is understood in the sense of "instantaneous" velocity in mechanics.

17 For example, it is possible within the scope of the general theory developed in Part I to describe the exact mechanism in which the far-field radiation pattern develops from the near-field region as one moves continuously from the "source sphere" $r = L_0$ to the "radiation sphere" $r = \infty$. The essence of this work hinges on careful understanding of the role played by coordinate systems in the description of the electromagnetic field radiated by the source.

where $\mathbf{E}_t^{\text{ex}}(\mathbf{r}') := \hat{n} \times \mathbf{E}^{\text{ex}}(\mathbf{r}')$ is the tangential component of the excitation field. The two integrals in (9.22) are over compact supports S and U, which allows us to interchange the order of the s'- and s''- integrations [62]. Therefore, we reach the following result

$$\mathbf{J}(\mathbf{K}) = \int_U ds' \, \bar{\mathbf{F}}(\mathbf{K}, \mathbf{r}') \cdot \mathbf{E}_t^{\text{ex}}(\mathbf{r}'), \tag{9.23}$$

where

$$\bar{\mathbf{F}}(\mathbf{k}, \mathbf{r}') := \int_S ds \, \bar{\mathbf{F}}(\mathbf{r}, \mathbf{r}') \, e^{-i\mathbf{k}\cdot\mathbf{r}}. \tag{9.24}$$

For the justification of the Fourier transform formula (9.24) using distribution theory, see Chapter 8. Relations (9.18), (9.23), and (9.24) represent the main ingredients of the theory of the interrelation between Mode A and Mode B. Their significance and physical interpretation will be outlined next.

9.4.3 Discussion of the Results

We provide here a phenomenological reading of the results obtained in the previous section. We start from the decomposition theorem (16.19) and (16.18). The general form of this spectral expansion is the outcome of the analysis performed in Part I and which had led to the conclusion that the generator of the specific manner in which the near field of any antenna gives rise to the actually observed radiation pattern comes from the functional dependence of the integrands of the spectral expansions in (16.19) and (16.18) on the angles of the local frame φ and θ. Indeed, the entire content of the spectral theory of the antenna near field is encapsulated in the functional form $\mathbf{D}(\mathbf{K}, \hat{u})$, i.e., the spectral composition of the field is determined by the manner in which $\mathbf{D}$ depends *jointly* on $\mathbf{K}$ and $\hat{u}$. Given such determination, it is possible to fully reconstruct the history of how the near field dynamically splits into propagating and nonpropagating modes as the observation point moves away from the source.[18] What is needed in our present treatment is the following. Since Mode A is determined completely by the ACGF $\bar{\mathbf{F}}(\mathbf{r}, \mathbf{r}')$, full understanding of the interrelation between Mode A and Mode B requires acquiring conclusive knowledge about the way in which the ACGF transforms into the spectral function $\mathbf{D}(\mathbf{K}, \hat{u})$ for a given input excitation $\mathbf{E}_t^{\text{ex}}$. Note that once the spectral function $\mathbf{D}(\mathbf{K}, \hat{u})$ is found, it becomes routine to obtain the full picture of the physics of the antenna near fields using the machinery developed in Part I.

18 In this case, the particular concept of the radial streamline picture enters favorably into the picture but the details can be found in Part I.

For these reasons, *we claim that Mode B is sufficiently captured by the function* $\mathbf{D}(\mathbf{K}, \hat{u})$.

The reader may observe that there is some analogy here with the traditional approach to the analysis of linear systems using the temporal Fourier transform method. Indeed, if we write $y(t) = \int d\omega H(\omega) X(\omega) \exp(i\omega t)$ for the Fourier integral of the output $y(t)$ of some given system characterized by the transfer function $H(\omega)$ and is excited by the input $x(t)$, then we can easily see that the spectral composition of the output signal is determined by the functional dependence of the integrand given by $H(\omega) X(\omega)$, which corresponds in this interpretation to $\mathbf{D}(\mathbf{K}, \hat{u})$ in (16.19) and (16.18). The only difference concerning the application of the Fourier transform method to the analysis of the electromagnetic near field is that we encounter in the latter case a more complicated structure arising from the peculiarity of being able to expand Mode B radiation Green's function $\bar{\mathbf{G}}(\mathbf{r}, \mathbf{r}')$ in terms of propagating and nonpropagating modes, and the fact that this expansion depends on the choice of the observation coordinate system (see Chapter 4). In such case, we find that the spectral composition of the radiation field involves a new set of parameters, i.e., those encoded in $\hat{u}$ as φ and θ. Nevertheless, the conceptual basis of our approach is essentially the same as the traditional methods although with further refinement.

By the linearity of the electromagnetic problem we expect that each point excitation $\mathbf{E}_t^{\text{ex}}(\mathbf{r}') = \hat{a}\delta(\mathbf{r}' - \mathbf{r}'')$ will give rise to some contribution $\mathbf{E}(\mathbf{r}, \mathbf{r}'')$ to the radiated field. The total field will be then given by the sum of the contributions from all locations $\mathbf{r}''$, i.e., the integral $\int d^3r'' \, \mathbf{E}(\mathbf{r}, \mathbf{r}'')$. Now we can see from (16.19) and (16.18) that the radiated field $\mathbf{E}$ itself depends linearly on $\mathbf{D}$. Therefore, the spectral composition too $\mathbf{D}$ has to depend linearly on the excitation $\mathbf{E}_t^{\text{ex}}$. This explains the form found in (9.23). Therefore, it is enough to focus on the way in which a point source excitation generates the functional form $\mathbf{D}(\mathbf{K}, \hat{u})$. That is, the situation appears as if we are introducing a new Green's function connecting the input $\mathbf{E}_t^{\text{ex}}(\mathbf{r}')$ with the output $\mathbf{D}(\mathbf{K}, \hat{u})$ instead of the fields $\mathbf{E}(\mathbf{r})$ and $\mathbf{H}(\mathbf{r})$ themselves.[19]

The previous discussion motivates the formal introduction of the new Green's function $\bar{\mathbf{Y}}(\mathbf{K}, \hat{u}; \mathbf{r}')$ by the relation

$$\mathbf{D}(\mathbf{K}, \hat{u}) = \int_S ds' \, \bar{\mathbf{Y}}(\mathbf{K}, \hat{u}; \mathbf{r}') \cdot \mathbf{E}_t^{\text{ex}}(\mathbf{r}'). \tag{9.25}$$

19 The magnetic fields can be easily computed in a manner essentially following the general outline developed here for the electric field. However, the details are omitted.

Notice that there is nothing remarkable about the existence of such Green's operator, which follows trivially from the linearity of the various electromagnetic operators involved. However, we will be rather interested in the *internal structure* of the operator $\bar{\mathbf{Y}}(\mathbf{K}, \hat{u}; \mathbf{r}')$, which turns out to be both nontrivial and interesting.

From (9.18), (9.25), (9.23) we conclude that

$$\bar{\mathbf{Y}}(\mathbf{K}, \hat{u}; \mathbf{r}') = \Lambda(\mathbf{K}, \hat{u})\, \bar{\mathbf{F}}(\mathbf{k}, \mathbf{r}'), \tag{9.26}$$

where the Λ-operator is defined by the equation

$$\begin{aligned} \Lambda(\mathbf{K}, \hat{u})\, \bar{\mathbf{F}}(\mathbf{k}, \mathbf{r}') := \left[\bar{\mathbf{R}}^T(\hat{u}) \cdot \bar{\boldsymbol{\Omega}}(\mathbf{K}) \cdot \bar{\mathbf{R}}(\hat{u})\right] \\ \times \cdot\, \bar{\mathbf{F}}\left(\bar{\mathbf{R}}^T(\hat{u}) \cdot \mathbf{K}, \mathbf{r}'\right). \end{aligned} \tag{9.27}$$

Therefore, it is understood that a dyadic function of $\mathbf{k}$ is applied as an input to the operator $\Lambda(\mathbf{K}, \hat{u})$. The Λ-operator works by 1) evaluation of this input function at $\mathbf{K}$, 2) rotating the result by $\bar{\mathbf{R}}^T(\hat{u})$, and 3) *left* multiplying the thus obtained dyadic function by the tensor $\bar{\mathbf{R}}^T(\hat{u}) \cdot \bar{\boldsymbol{\Omega}}(\mathbf{K}) \cdot \bar{\mathbf{R}}(\hat{u})$.

We may call $\Lambda(\mathbf{K}, \hat{u})$ the *spectral AB-transfer function*, in which we understand the appropriate transfer function from Mode A to Mode B in the spectral domain. It constitutes the essential relation between Mode A and Mode B. Indeed, Mode A data $\bar{\mathbf{F}}(\mathbf{k}, \mathbf{r}')$ is transformed by the this Λ-operator into Mode B data $\mathbf{D}(\mathbf{K}, \hat{u})$. Consequently, we may visualize the operation of the antenna system as the interplay between Mode A and Mode B highlighted by the nature of the *geometrical* transformations enacted by the Λ-operator. The explication of the precise sense of "geometrical" here requires a closer examination of the structure of (9.27), which we provide briefly as follows.

Mode B presents the geometrical part of the electromagnetic process of radiation. The "genuine" electromagnetic part, that involving the electromagnetic boundary condition, belongs to Mode A and is stored in the ACGF $\bar{\mathbf{F}}(\mathbf{r}, \mathbf{r}')$. We have chosen to enact the analysis of the relation between Mode A and Mode B in the spectral domain, and hence the input data supplied in the first stage is given by the Fourier transform of the ACGF (9.24). Next, the Λ-operator (9.27) will transform the spectral Mode A ACGF tensor into the Y-tensor appearing in (9.25) by *purely geometrical operation*, mainly multiplication by a constant tensor (the transformed polarization dyad) and then rotating the result. Through this remarkably simple operation, the full spectral composition of the radiated fields, as exemplified by (9.18), reveals itself in the manifestation of the manner in which the electromagnetic fields dynamically split into propagating and nonpropagating modes according to the general decomposition theorem (16.19) and (16.18).

9.4.4 Conclusion

To begin with, we have achieved through (9.25) a *complete decoupling* of Modes A and B. The Λ-operator stands physically for the second stage in electromagnetic radiation following the production of the antenna current distribution in Mode A. In our opinion, that such decoupling between the modes via a linear operator (the Λ-operator) is possible is not an obvious *a priori* fact. Its relatively simple derivation, however, is an indication of 1) the fertility of the ACGF formalism, and 2) the correctness of the selection of the appropriate degrees of freedom in Mode B, i.e., via the choice of the field *spectral* composition $\mathbf{D}(\mathbf{K}, \hat{u})$ in (9.18).

The most interesting conclusion to be drawn from the foregoing analysis relates to expression (9.25), where it is found that the input excitation field $\mathbf{E}_t^{\mathrm{ex}}(\mathbf{r}')$ in the *spatial* domain controls directly the *spectral* composition of the radiated field in a simple and well-defined manner through the action of the Λ-operator. Indeed, *the Green's tensor* $\bar{\mathbf{Y}}(\mathbf{K}, \hat{u}; \mathbf{r}')$ *is independent of the excitation*, and can be computed and stored completely *prior* to the operation of the electromagnetic system. The nature of the excitation field is generally unknown in the system, and depends on the particular circumstances of the electromagnetic environment at the time. Yet, in any case, the manner in which the radiated fields structure themselves is fully determined physically by the Y-tensor. The actually observed field structure is simply the appropriate weighted sum of the $\mathbf{r}'$-indexed Y-tensors, each assigned a weight according to the value of the excitation $\mathbf{E}_t^{\mathrm{ex}}(\mathbf{r}')$ as can be seen from combining (9.25) with (16.19) and (16.18).

We also observe from the internal structure of the Λ-operator (9.27) that it divides into two factors. The first factor is the transformed spectral polarization dyad $\bar{\mathbf{R}}^T(\hat{u}) \cdot \bar{\mathbf{\Omega}}(\mathbf{K}) \cdot \bar{\mathbf{R}}(\hat{u})$, which is independent of both $\mathbf{r}'$ and $\mathbf{K}$. Such independence signifies that it is a genuinely *universal* element in Mode B. This fact was already observed in Chapter 4, however it acquires new light in the present formulation.[20] The second factor, $\bar{\mathbf{F}}\left(\bar{\mathbf{R}}^T(\hat{u}) \cdot \mathbf{K}, \mathbf{r}'\right)$ introduces the effect of the location of the actual excitation of the antenna system through the variable $\mathbf{r}'$; and the spectral shaping achieved through the rotation (by the matrix $\bar{\mathbf{R}}^T(\hat{u})$) of the Fourier transform of the antenna current Green's function but evaluated at the $\mathbf{K}$ defined by (16.17) instead of $\mathbf{k}$. It is then this second factor in the Λ-operator that will take care of the "individuality" of the particular antenna under consideration, in contrast to the first factor, which is essentially universal.

20 The origin of this constant tensor $\bar{\mathbf{R}}^T(\hat{u}) \cdot \bar{\mathbf{\Omega}}(\mathbf{K}) \cdot \bar{\mathbf{R}}(\hat{u})$ is the vectorial structure of the electromagnetic radiation operator, and hence the motivation for the use of the term 'polarization,'

Finally, notice that it is absolutely necessary in the analysis of the relation between Mode A and Mode B to use the 3D ACGF tensor (8.7) instead of the 2D tensor (8.5). Indeed, the spectral composition $\mathbf{D}(\mathbf{K}, \hat{u})$ in (9.18) is a 3D entity, and so is the spectral polarization dyad (9.20). This makes it mandatory to use the 3D ACGF in order to compute the correct transformation enacted by (9.25). Although Modes A and C are ultimately reducible to problems in 2D configuration space (the 2-manifold S), Mode B is essentially a process in the 3D configuration space $\mathbb{R}^3$, and hence its interaction with Mode A will make it necessary to revert the latter back to the full 3D problem. The same situation will emerge again later when we analyze the relation between Mode B and Mode C in Section 9.5.

We finally end this conclusion by a note on the functional dependence of the Y-tensor (9.25) on $\mathbf{r}'$, the position of the excitation field. For each different value of $\mathbf{r}'$, the ACGF has to be calculated anew, without being able to provide any definite general statement of the correlation between the various functions indexed by $\mathbf{r}'$. This can be seen at once from the defining integral equation of Mode A (8.2), in which the solution for a given ACGF depends crucially on the relative position of the excitation point $\mathbf{r}'$ within S. It is beyond the scope of the theory developed in this book to provide further insight on the numerical variation in the ACGF due to changes in $\mathbf{r}'$. Without delving deeper into this subject, we consider in the present work that each Y-tensor corresponding to a given value of $\mathbf{r}'$ is a new channel of interaction in variance with another Y-tensor evaluated at different $\mathbf{r}'$. However, we conjecture that further understanding of the effect of $\mathbf{r}'$ can be gleaned through a systematic investigation of the relation between the *topology* of S and the electromagnetic operators as represented by the ACGF.

9.5 SPECTRAL APPROACH TO ANTENNA-ANTENNA INTERACTIONS: THE INTERRELATIONSHIP BETWEEN MODE B AND MODE C

9.5.1 Introduction

So far our focus has been concentrated on various scenarios involving systems that can be deemed, in a certain sense, as *isolated* from other parts of the universe. In our case, we have been dealing with antennas working in pure Mode A, pure Mode C (Section 9.2), and the single antenna system comprised of Modes A + B (Section 9.4). This ideal situation already covers the ground up to most of the applications involving individual radiating antenna systems in which the device under consideration can be effectively enclosed inside a bounded surface. In reality, however, we seldom encounter such situation. Indeed, radiating structures are always in the process of mutual interaction with both near and far objects.[21] The full process, however, becomes extremely complex when all types of interactions are included once and for all. Our strategy will be a more pragmatic one, based in part on first assuming isolated system, as we have been doing so far, and then gradually introduce more complicated situations in which basic levels of interaction gradually become relevant and important. Understanding the more complicated case will be based, if we are fortunate, on some more primitive understanding of the less complicated situation.[22]

The most elementary situation that we will deal with is the interaction between two antennas, one operating as a source while the other works as a receiving system. The interaction of a given antenna with the *far* field produced by another antenna located in the latter's far zone represents the simplest possible scheme of interaction imaginable. Indeed, in this case the incident field is simple in form, being merely a plane wave. Moreover, due to the fact that the distance between the two objects is practically infinite, the presence of one is incapable of modifying the electromagnetic performance of the other.[23] In the this section, we deal with the more general case in which the two antennas interact with each other in the *near*-field zone.

Our approach to the problem of understanding the relational structure of Modes B and C will exploit a general thematization of a duality between *static* and *dynamic* genesis. The meaning of 'static genesis' is the determination of the general mechanism detailing how a *fixed* antenna located within the near-field radiation of another antenna is producing the physically observed received voltage. As will be shown in Section 9.5.8, this mechanism is developed after a thorough examination of the mathematical structure of the expression of the received voltage using the ACGF formalism. It explains how given electromagnetic and geometric configurations give rise to the actually observed port voltage by looking into the

21 It is probably worth the effort to compare this situation in applied electromagnetics with thermodynamics, in particular with the fundamental distinction in the latter field of study between *open* and *closed* systems. Closed systems can be considered as isolated assemblies of objects exchanging energy only within its well-defined boundary separating the system from the outside universe. Such idealization is obviously a fiction, but a very legitimate one.

22 In science, this can be seen in the process of developing *basic prototype systems*, i.e., extremely simple but yet very interesting systems that later prove essential in understanding the full physical picture. Examples of such prototypes include harmonic oscillator models in both classical and quantum mechanics, lower dimensional quantum field theories, Ising models in ferromagnetism, etc.

23 This situation has been dealt with theoretically and numerically in Chapter 14 and experimentally in [15].

process in the spectral domain. However, the discovered mechanism explains only how a *fixed* electromagnetic/geometric configuration is organically connected with the port voltage, leaving still open the question about how this particular received voltage would vary if the electromagentic/geoemtric configuration did change. In this chapter, we will focus only on the geometrical aspects of the above mentioned change, and the framework of such study will be undertaken through the theme of the *dynamic genesis* of Section 9.5.9. The basic philosophical motivation behind this kind of genesis, which we claim to be the most comprehensive investigation possible, is the fact that no privilege should be granted to a specific coordinate system in the endeavor to provide an explanatory model for a given electromagnetic phenomenon. Instead, focus must be directed on the dynamic interrelationship connecting various *situations* with each other through the web of an integral whole, in a way that treat all coordinate systems as equally significant in the explanatory roles they play in the model. This, in our opinion, can be attained only if a physical meaning is given to the process of changing the perspective of the observer, i.e., the coordinate system used to describe the problem. As will be shown in Section 9.5.9, by moving the received antenna immersed in the near zone of another antenna, we can exploit the dynamic splitting of the source's radiation field into propagating and nonpropagating modes in order to explain how exactly the received voltage will vary. This we will achieve again by working in the spectral domain and using the ACGF formalism.

9.5.2 Interaction Between Externally Applied Source and a Fixed Receiving Antenna in the Near-Field Zone

We will assume that the source antenna is given by *externally* controlled electromagnetic currents. This implies that the effect of the receiving system on the source will be neglected. Although it may appear to some readers that this is practically not feasible, we reply that within the framework of Maxwell's theory, this is a very legitimate assumption. Indeed, Maxwell's equations are a system of differential equations including forcing – or better called *independent*– terms representing the applied source. It is essential in understanding the mathematical boundary-value problem to realize that such forcing terms are truly independent of the actually 'dependent variables,' the electromagnetic fields themselves, and also the boundary

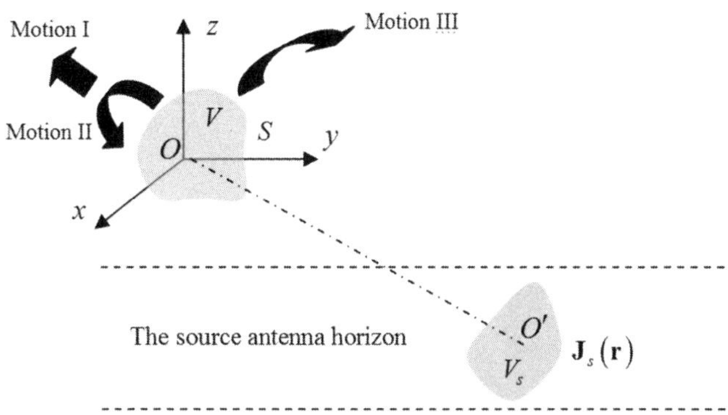

Figure 9.1 A general schematic diagram illustrating the geometry of the antenna-antenna interaction problem.

condition itself.[24] Therefore, by starting with modeling the source antenna as an externally applied current, we basically effect the first and most natural level in a hierarchy of theories with increasing complexity.

The general problem is geometrically illustrated in Figure 9.1. The externally applied source $\mathbf{J}_s(\mathbf{r})$ is assumed to be localized in the compact region V_s. The receiving antenna is located in the compact region V surrounded by a regular surface S (i.e., the 2-manifold S is the boundary of V). Without loss of generality, we restrict our attention to a connected surface S. For antenna arrays, the analysis can be generalized in a straightforward fashion to the case when S is a multiply-connected surface. A global coordinate system with origin at $O \in V$ is chosen as shown in Figure 9.1. The choice of this frame is made such that the source antenna horizon lies outside the receiving antenna region V. Aside from this restriction,

24 Sometimes it is necessary to relate the forcing terms and the boundary condition together under a common category, like being smooth (continuously differentiable). However, this is not an essential relation between the two entities. Indeed, the *functional* form of the forcing term is independent of the geometrical details of the boundary-condition surface or the numerical values of the various electromagnetic parameters of the media involved, etc.

this global frame of reference is completely arbitrary. Some origin for the source antenna, $O' \in V_s$, is also chosen randomly.[25]

The source antenna will produce a radiation field given by the following spectral expansion [35], [10]

$$\mathbf{E}(\mathbf{r}) = \int_{\mathbb{R}^2} dpdq\, \bar{\mathbf{\Omega}}(\mathbf{K}) \cdot \mathbf{J}(\mathbf{K})\, e^{i\mathbf{K}\cdot\mathbf{r}}. \tag{9.28}$$

Notice that there is no question of employing a rotating local frame as was done in Section 9.4. The reason is that in Figure 9.1 the choice of the coordinate system is fixed by the existence of another object, the antenna S, in the exterior region of the source antenna. This absence of rotating frames is the hallmark of the *static* genesis. Indeed, the goal of the dynamic genesis of Section 9.5.9 will be to find a way in which a local description of the interaction can be injected back into the picture.

The receiving antenna system S will be described by its exact ACGF $\bar{\mathbf{F}}(\mathbf{r}, \mathbf{r}')$. We introduce the inverse Fourier integral of the ACGF tensor

$$\bar{\mathbf{F}}(\mathbf{r}', \mathbf{r}) = \frac{1}{(2\pi)^3} \int_{\mathbb{R}^3} d^3k\, \bar{\mathbf{F}}(\mathbf{r}', \mathbf{k})\, e^{i\mathbf{k}\cdot\mathbf{r}}, \tag{9.29}$$

where the integration with the respect to the spectral variable $\mathbf{k}$ is carried throughout the entire Euclidian space $\mathbb{R}^3$. Throughout this section, we will understand the integral in (9.29) in the distributional sense developed in Chapter 8. That is, each spectral function $\bar{\mathbf{F}}(\mathbf{r}', \mathbf{r})$ will be replaced by a smooth approximation $\{\bar{\mathbf{F}}^d_m(\mathbf{r}', \mathbf{r})\}_{m\in\mathbb{N}}$ with bounded support on $v_m \subset \mathbb{R}^3$, a sequence of 3-dimensional volumes approximating S. The importance of this distributional approximation will become apparent shortly.

From (9.28) and (9.29), we obtain with the help of the ACGF of the antenna S the following expression for the current induced at $\mathbf{r}'$

$$\begin{aligned} \mathbf{J}(\mathbf{r}') = (2\pi)^{-3} \textstyle\int_S ds \int_{\mathbb{R}^3} d^3k \int_{\mathbb{R}^2} dpdq \\ \times \bar{\mathbf{F}}(\mathbf{r}', \mathbf{k}) \cdot \bar{\mathbf{\Omega}}(\mathbf{K}) \cdot \mathbf{J}_s(\mathbf{K})\, e^{i(\mathbf{K}+\mathbf{k})\cdot\mathbf{r}}. \end{aligned} \tag{9.30}$$

The next crucial step in our derivation involves changing the order of integrations in (9.30). We know that the pq-integral is uniformly convergent (Chapter 4) and

25 The figure also shows three different types of rigid motion of the receiving antenna S, which will be discussed in Section 9.5.9. Motion I consists of simple translation of S. Motion II is a simple local rotation of the antenna S around any axis passing through O. Motion III is a rotation of S around an axis passing through O' such that the antenna S will preserve its relative local orientation with respect to the source.

therefore it can be interchanged with s-integral since the latter is on compact region S. However, the third integral, i.e., the $\mathbf{k}$-integral, need *not* be uniformly convergent without further investigation. Since we aim at formulating the problem at the most general level, we will avoid imposing an ad hoc condition, like simply postulating the possibility of performing such exchange. To solve this problem, the reader is referred to Section 8.2.4, where we proposed a construction, using distribution theory, such that each Fourier transform of the ACGF $\bar{\mathbf{F}}(\mathbf{r}, \mathbf{r}')$ is absolutely and uniformly convergent. This was done by introducing a distributional approximation by a sequence of smooth functions (with bounded support) as mentioned in the text after (9.29) above. Based on this construction, we can first interchange the s- and $\mathbf{k}$-integrals, then interchange the s- and the pq- integrals, where the surface S will now be understood to refer implicitly to a sequence of volumes $v_m \subset \mathbb{R}^3$ approximating S as $m \to \infty$. Therefore, expression (9.30) can be written as

$$\mathbf{J}(\mathbf{r}') = (2\pi)^{-3} \int_{\mathbb{R}^3} d^3k \int_{\mathbb{R}^2} dpdq\, \bar{\mathbf{F}}(\mathbf{k}, \mathbf{r}')\, I(\mathbf{k}, \mathbf{K}) \;\cdot \mathbf{D}(\mathbf{K}), \tag{9.31}$$

where we define $\mathbf{D}(\mathbf{K}) := \bar{\mathbf{\Omega}}(\mathbf{K}) \cdot \mathbf{J}_s(\mathbf{K})$. The scalar function $I(\mathbf{k}, \mathbf{K})$ is defined as

$$I(\mathbf{k}, \mathbf{K}) := \int_S ds\, e^{i(\mathbf{K}+\mathbf{k})\cdot\mathbf{r}}. \tag{9.32}$$

We will call the function on the LHS of (9.32) *the spectral interaction kernel*. It plays a foundational role in the theory of electromagnetic interactions we are developing here and will be studied at great details shortly. It suffices for now to observe its role in the spectral interaction integral (9.31) where the function $I(\mathbf{k}, \mathbf{K})$ appears as a mediator linking the incoming field's spectral component at $\mathbf{K}$ and the receiving antenna ACGF's Fourier component at $\mathbf{k}$. In other words, *the interaction between the Fourier components of the illumination field and the receiving antenna Green's function is not direct, but has to be mediated by a new (third) function, namely the spectral interaction kernel* $I(\mathbf{k}, \mathbf{K})$.

Now let us divide the total receiving antenna surface S into an arbitrary number N of smaller surfaces $s_n, n = 1, 2, ..., N$, each with arbitrary shape, i.e., we may write

$$S = \bigcup_{n=1}^{N} s_n. \tag{9.33}$$

Therefore, we have $I(\mathbf{k}, \mathbf{K}) = \sum_{n=1}^{N} I_n(\mathbf{k}, \mathbf{K})$, where

$$I_n(\mathbf{k}, \mathbf{K}) := \int_{s_n} ds\, e^{i(\mathbf{k}+\mathbf{K})\cdot\mathbf{r}}. \tag{9.34}$$

Using this notation, we can express the total interaction integral (9.31) as

$$\begin{aligned} \mathbf{J}\left(\mathbf{r}'\right) = \left(2\pi\right)^{-3} \sum_{n=1}^{N} \int_{\mathbb{R}^3} d^3k \int_{\mathbb{R}^2} dpdq \\ \times \bar{\mathbf{F}}\left(\mathbf{r}', \mathbf{k}\right) I_n\left(\mathbf{k}, \mathbf{K}\right) \cdot \mathbf{D}\left(\mathbf{K}\right). \end{aligned} \tag{9.35}$$

The division of the antenna system surface into smaller arbitrary shaped regions as enacted by (9.33) is in line with the engineering perspective of the present work. That is, we don't search for ultimate fundamental constituents of the system under consideration, as for example in the search for new fundamental particles in the phenomena of subatomic interactions, but rather we leave it to the application considered to determine the exact manner in which a given antenna surface is to be split into various parts reflecting the physical aspects intended by the engineer, whether being for analysis or synthesis (design) considerations. In this way, the $n$nth spectral interaction kernel (9.34) represents the structure – associated with the geometric element s_n – mediating between the external illumination field and the internal Green's function of the *total* system S. The received signal, as a glance at (9.35) shows, is the sum of all contributions coming from the totality of the subsurfaces s_n. Therefore, by a careful study of the various terms appearing therein, the engineer will be able to gain some understanding of the inner mechanism through which each important geometrical part of the antenna system will unfold its contribution to the production of the actually observed received signal. Finally, we will refer to this scheme of thinking as the implicit *field of division* of an antenna system overall surface S into arbitrary parts. It is a field of division in the *dynamic* sense understood here as a constantly varying choice of the parts composing the antenna system which can always be modified according to the engineering need. In other words, it is a dynamic field of division, not a static one as in dividing a piece of crystals into fixed parts, namely atoms or molecules. Our antenna "atoms" s_n can always be redefined as long as (9.33) is satisfied.

9.5.3 The General Behavior of the Spectral Interaction Kernel

We start by observing that $I(\mathbf{k}, \mathbf{K})$ is actually a function of $\mathbf{k} + \mathbf{K}$, which motivates the name *kernel* since it appears as a multiplicative factor in the integrand of the spectral interaction integral (9.31). We will use the notations I_n, $I_n(\mathbf{k} + \mathbf{K})$, and $I_n(\mathbf{k}, \mathbf{K})$ interchangeably depending on the context.

The first important observation, however, is that the function $I_n(\mathbf{k}, \mathbf{K})$ is a a Fourier transform. To see this, the surface S is replaced by one of its approximating volumes v_m. Next, this region v_m is divided in turn as $v_m = \cup_{n=1}^{N} v_{mn}$. The resulting integral over v_{mn} is therefore seen as the usual 3-dimensional Fourier transform with respect to its $\mathbf{k}$ argument. However, note that with respect to the p and q arguments, it is *not* a 2-dimensional Fourier transform. We will not refer explicitly in the following analysis to the approximation volumes v_{mn} or v_m but this distributional context should always be understood implicitly, especially when performing numerical computations.

9.5.3.1 Short-Wavelength Behavior of the Spectral Interaction Kernel

Consider the following discontinues function

$$u_n(\mathbf{r}) = \begin{cases} 1, & \mathbf{r} \in s_n, \\ 0, & \text{otherwise.} \end{cases} \tag{9.36}$$

The spectral interaction kernel can therefore be written as the usual spatial Fourier transform $I(\mathbf{k}, \mathbf{K}) = \mathcal{F}\left\{u_n(\mathbf{r}) e^{i\mathbf{K}\cdot\mathbf{r}}\right\}$. We now make use of the classic Reimann-Lebesgue lemma, which states that the Fourier transform of an integrable function approaches zero as the magnitude of its spectral vector goes to infinity. Therefore, it immediately follows from this theorem that

$$\lim_{|\mathbf{k}|\to\infty} I_n(\mathbf{k}, \mathbf{K}) = 0. \tag{9.37}$$

This is our first important property of the spectral interaction kernel. The study of the behavior of I_n when both p and q goes to infinity is more difficult. We would like to prove that

$$\lim_{p^2+q^2\to\infty} I_n(\mathbf{k}, \mathbf{K}) = 0. \tag{9.38}$$

Equations (9.37) and (9.38) are valid for all $n = 1, ..., N$.

The main difficulty in interchanging the order of limit and integral in

$$\lim_{p^2+q^2\to\infty} \int_{s_n} ds \exp[i(\mathbf{k} + \mathbf{K}) \cdot \mathbf{r}]$$

is the following. For the most general problem, it is possible that the origin of the coordinate system be contained in s_n. In this case, the limit $\lim_{p^2+q^2\to\infty} \exp[i(\mathbf{k} + \mathbf{K}) \cdot \mathbf{r}]$ *will not be uniform* in s_n as can be gleaned after examining the behavior at $\mathbf{r} = 0$. Therefore, the classical theorem guaranteeing the permissibility of interchanging the order of integration and a limit of sequence of functions once the latter limit is uniform cannot be directly applied to our problem. Instead, use will be made of a stronger result in real analysis.

We say that a sequence of functions $f_n(x)$ is *uniformly bounded* on a closed interval A whenever there exists a nonnegative real number M such that $|f_n(x)| \leq M, \ \forall \ n \in \mathbb{N}, x \in A$. The following is the bounded convergence theorem for Riemann integrable functions [72]

Theorem 9.5.1. *If a uniformly bounded sequence of (Riemann) integrable functions f_n converges pointwise to a (Riemann) integrable function f on $[a, b]$, then the following relation holds*

$$\lim_{n\to\infty} \int_a^b f_n(x)\, dx = \int_a^b f(x)\, dx. \tag{9.39}$$

This theorem can be generalized to the space $\mathbb{R}^3$ where the interval $[a, b]$ is replaced by a compact set.

Define the function

$$g(\mathbf{r}) := \lim_{p^2+q^2\to\infty} e^{i(\mathbf{k}+\mathbf{K})\cdot\mathbf{r}} = \begin{cases} 1, & \mathbf{r} = 0, \\ 0, & \text{otherwise.} \end{cases} \tag{9.40}$$

We further notice that

$$\left| e^{i(\mathbf{k}+\mathbf{K})\cdot\mathbf{r}} \right| \leq 1, \ \forall p^2 + q^2 \in \mathbb{R}^+, \ \forall \mathbf{r} \in s_n. \tag{9.41}$$

Therefore, from the bounded convergence theorem (9.39) and the definition (9.40) we find

$$\begin{aligned} &\lim_{p^2+q^2\to\infty} \int_{s_n} ds\, e^{i(\mathbf{k}+\mathbf{K})\cdot\mathbf{r}} \\ &= \int_{s_n} ds \lim_{p^2+q^2\to\infty} e^{i(\mathbf{k}+\mathbf{K})\cdot\mathbf{r}} = \int_{s_n} ds\, g(\mathbf{r}) = 0. \end{aligned} \tag{9.42}$$

Hence, (9.38) is proved.

9.5.3.2 Boundedness Property

It is interesting to observe that all the functions are bounded. Indeed, from (9.34) we find

$$|I_n(\mathbf{k}, \mathbf{K})| < \int_{s_n} ds \left| e^{i(\mathbf{k}+\mathbf{K})\cdot\mathbf{r}} \right| < \int_{s_n} ds \left| e^{i\mathbf{K}\cdot\mathbf{r}} \right|. \tag{9.43}$$

However, from (16.17) we see that for real and complex $\mathbf{K}$, the quantity $\exp\left(i\mathbf{K}\cdot\mathbf{r}\right)$ is always less than or equal unity. Therefore, we find

$$|I_n\left(\mathbf{k},\mathbf{K}\right)| \leq I_n\left(\mathbf{k},\mathbf{K}\right)|_{\mathbf{k}+\mathbf{K}=0} = \int_{s_n} ds\,. \tag{9.44}$$

Therefore, the nth spectral interaction kernel is bounded from above by the area of s_n. Moreover, it attains its upper bound at $\mathbf{k}+\mathbf{K}=0$.

9.5.3.3 The Concept of Spatial Bandwidth in General Antenna systems

The results (9.37), (9.38), and (9.44) together help visualize the behavior of the spectral interaction kernels at infinity and strongly suggest that each function function I_n acts like a lowpass filter. Its input receives spectral contributions from the external illumination field (evaluated at $\mathbf{K}$) and the ACGF of the receiving antenna system (evaluated at $\mathbf{k}$). The two contributions are filtered by $I_n(\mathbf{k},\mathbf{K})$ and the total sum of all spectral variables $\mathbf{k}$ and p,q is formed. Moreover, because of the condition (9.44), it appears that this filter behaves more like a resonator, with its resonance peak centered at $\mathbf{k}+\mathbf{K}=0$. Consequently, it is very reasonable to guess that a phenomenon like bandwidth is at hand here. However, in order to substantiate this conclusion, we need to learn more about the rate of decay of the function $I_n(\mathbf{k},\mathbf{K})$.

We first observe that the function $I_n(\mathbf{k})$ cannot be absolutely integrable. The reason is that if it does, then its inverse Fourier integral will be uniformly convergent (by the Weierstrass M test), and hence the function $u_n\left(\mathbf{r}\right)\exp\left(i\mathbf{K}\cdot\mathbf{r}\right)$ would be continuous, contradicting the definition (9.36). Therefore, we must have

$$|I_n\left(\mathbf{k},\mathbf{K}\right)| \text{ decays at infinity slower than } |\mathbf{k}|^{-4}\,. \tag{9.45}$$

Now let us study the environment inside which the function $I_n(\mathbf{k},\mathbf{K})$ lives in the interaction integral (9.31). It is easy to show that $\lim_{p^2+q^2\to\infty}\Omega\left(\mathbf{K}\right)=\text{constant}.$[26] Moreover, from the definition of the Fourier transform of the source current given in (9.21) and the definition of of I_n in (9.34), we conclude that that both $\mathbf{J}_s(\mathbf{K})$ and $I_n(\mathbf{K})$ have roughly the same rate of decay as p^2+q^2 tends to infinity.[27] However, since the ACGF is smooth (see the construction in Section 8.2.4), it follows that $\left|\bar{\mathbf{F}}\left(\mathbf{k},\mathbf{r}'\right)\right|$ decays faster than $|\mathbf{k}|^{-p}$ for any positive integer p [67]. Therefore, by

26 This follows at once from the definition of the spectral polarization dyad given in (9.20).

27 That $\lim_{p^2+q^2\to\infty}\mathbf{J}_s\left(\mathbf{K}\right)=0$ can be proved by exactly the same procedure outlined inSection 9.5.3.1.

(9.45) we find that $I_n(\mathbf{k}, \mathbf{K})$ *has the slowest decay rate in the spectral interaction integral* (9.31).

On the basis of these findings, we introduce the idea of *spatial effective bandwidth* defined in terms of the spectral interaction function $I_n(\mathbf{k} + \mathbf{K})$. Its exact value will depend on the application but can always be properly defined as in resonant circuits and lowpass filters. We will denote the bandwidth associated with the antenna part s_n by BW_n. It is a function of the error level e below (with respect to which the amplitude of I_n is considered sufficiently small for the application at hand)

$$\left(|\mathbf{k}|^2 + k_0^2 p^2 + k_0^2 q^2\right)^{1/2} > BW_n \;\Rightarrow\; |I_n(\mathbf{k}, \mathbf{K})| < e. \tag{9.46}$$

This definition is always well posed because of the relations (9.37) and (9.38). Furthermore, the fact that $I_n(\mathbf{k} + \mathbf{K})$ has the slowest decay rate in the spectral interaction integral (9.31) makes the definition of BW_n independent of the particular illumination field and the ACGF. We will not always write explicitly $BW_n = BW_n(e_n)$ but use instead BW_n under the assumption that a suitable global implicit error level $e := \max\{e_1, e_2, ..., e_N\}$ has been fixed in advance. Based on these definitions, we can approximate the interaction integral (9.35) as

$$\begin{aligned}\mathbf{J}(\mathbf{r}') \simeq (2\pi)^{-3} \sum_{n=1}^{N} \int_{|\mathbf{k}|^2 + k_0^2(p^2+q^2) < BW_n} & d^3k\,dp\,dq \\ \times \bar{\mathbf{F}}(\mathbf{r}', \mathbf{k})\, I_n(\mathbf{k}, \mathbf{K}) & \cdot \mathbf{D}(\mathbf{K}).\end{aligned} \tag{9.47}$$

Therefore, we effectively replace the infinite integration domain in (9.31) by the 5-sphere $|\mathbf{k}|^2 + k_0^2(p^2 + q^2) < BW^2$.

9.5.4 Examples of the Spectral Interaction Kernel: Dipole and Patch Antennas

It is possible to compute the spectral interaction kernel in simple closed analytical forms for antennas like linear wires and patches. Consider a linear wire oriented along the z-axis with length $2L$ and origin at the middle point. The spectral kernel is readily evaluated as[28]

$$\begin{aligned}& I^{\text{dipole}}(\mathbf{k}, \mathbf{K}) \\ & = -i\frac{e^{i\left(k_z + k_0\sqrt{1-p^2-q^2}\right)L} - e^{-i\left(k_z + k_0\sqrt{1-p^2-q^2}\right)L}}{k_z + k_0\sqrt{1-p^2-q^2}}.\end{aligned} \tag{9.48}$$

28 We replace the surface S in (9.34) by a 1-dimensional line extended along the wire antenna.

On the other hand, for a square patch in the xy-plane with side length $2L$ and origin located at the center, the spectral interaction kernel is given by

$$I^{\text{patch}}(\mathbf{k},\mathbf{K}) = -\frac{e^{i(k_x+k_0p)L}-e^{-i(k_x+k_0p)L}}{k_x+k_0p} \times \frac{e^{i(k_y+k_0q)L}-e^{-i(k_y+k_0q)L}}{k_y+k_0q}. \tag{9.49}$$

It is not difficult to show by direct numerical computation that all the general properties of the $I(\mathbf{k},\mathbf{K})$ developed in Section 9.5.3 are applicable to the special forms given in (9.48) and (9.49). In particular, we mention that the spectral shape resembles a 1-dimensional sinc function for the case of the linear dipole and a 2-dimensional sinc function for the square patch. It is well-known that such shapes are resonant with approximate bandwidth given by $2\pi/L$ [67].

9.5.5 Miltipole Expansion of the Spectral Interaction Kernel

The goal of this Section is to study the inner structure of the spectral interaction kernel (9.34) within the framework of the classic multipole expansion. Our point of departure is expanding the integrand of (9.34) in Taylor series, which is valid for all complex arguments

$$\begin{aligned} e^{i(\mathbf{k}+\mathbf{K})\cdot\mathbf{r}} &= \sum_{m=0}^{\infty}\frac{i^m}{m!}\left[(\mathbf{k}+\mathbf{K})\cdot\mathbf{r}\right]^m \\ &= 1+\tfrac{i}{1}\mathbf{r}\cdot(\mathbf{k}+\mathbf{K})+\tfrac{i^2}{2}\left[\mathbf{rr}\cdot(\mathbf{k}+\mathbf{K})\right]\cdot(\mathbf{k}+\mathbf{K}) \\ &\quad+\tfrac{i^3}{6}\left\{\left[\mathbf{rrr}\cdot(\mathbf{k}+\mathbf{K})\right]\cdot(\mathbf{k}+\mathbf{K})\right\}\cdot(\mathbf{k}+\mathbf{K})+\cdots, \end{aligned} \tag{9.50}$$

where use was made of the identity $(\mathbf{a}_1\mathbf{a}_2...\mathbf{a}_n)\cdot\mathbf{b} = (\mathbf{a}_1\mathbf{a}_2...\mathbf{a}_{n-1})(\mathbf{a}_n\cdot\mathbf{b})$. We can write (9.50) compactly in the following form

$$e^{i(\mathbf{k}+\mathbf{K})\cdot\mathbf{r}} = \sum_{m=0}^{\infty}\frac{i^m}{m!}\underbrace{\mathbf{rr}\cdots\mathbf{r}}_{m\text{ times}}\cdot(\mathbf{k}+\mathbf{K})^m, \tag{9.51}$$

where the m-fold operation on $\mathbf{r}$ is tensor multiplication while the corresponding operation on $\mathbf{k}+\mathbf{K}$ is repeated inner product with the resulting mth-order tensor $\mathbf{rr}\cdots\mathbf{r}$.

Now the series (9.51), being a power series, is uniformly convergent in $\mathbf{r}$ on every compact region [72], in particular in s_n. Therefore, it can be integrated term by term [62], [72]. We obtain then after substituting (9.51) into (9.34) the following expansion

$$I_n(\mathbf{k}, \mathbf{K}) = \sum_{m=0}^{\infty} \widehat{Q}_{mn} \cdot (\mathbf{k} + \mathbf{K})^m, \tag{9.52}$$

where the mth *multipole moment* of I_n is defined as

$$\widehat{Q}_{mn} := \frac{i^m}{m!} \int_{s_n} ds \, \underbrace{\mathbf{rr} \cdots \mathbf{r}}_{m \text{ times}}. \tag{9.53}$$

It is important to pay due attention to the issue of the convergence of the multipole expansion (9.52). In the electromagnetic and physical literature, the expansion is usually truncated after few terms (sometimes without justification). This actually works because in the majority of the applications of such expansion, for example the interaction of light with matter, the wavelength of the electromagnetic radiation is large compared with the spatial extension of the characteristic atomic or molecular structure under illumination. In our problem, the main goal is the study of the received signal as given by the interaction integrals (9.31) and (9.35). As can be seen after a glance at these expressions, they involve infinite integrations in the space $\mathbb{R}^5$ over the spectral variables $\mathbf{k}$, p, and q. Therefore, the quantity $\mathbf{k} + \mathbf{K}$ appearing in (9.52) are by no means small, and in general it is not possible to terminate the multipole expansion after few terms. However, we will now show that an analytical approximation of I_n in the form of a finite polynomial in $\mathbf{k} + \mathbf{K}$ can be obtained provided we make use of the phenomenon of spatial bandwidth elaborated in (9.46) and (9.47).

We start by observing that the expansion (9.52) can be put in the form of a power series in three generally complex variables $X = k_x + k_0 p$, $Y = k_y + k_0 q$, $Z = k_y + k_0\sqrt{1 - p^2 - q^2}$, where we used (16.17) and the fact that the global coordinate system in Figure 9.1 was chosen *above* the source antenna horizon. That is, (9.52) can be written in the form $\sum_{mnl} a_{nml} X^n Y^m Z^l$. This is a power series in three variables, and in particular the third variable Z is generally complex. We will apply the following theorem [69]

Theorem 9.5.2. *Let the power series $\sum_{mnl} a_{nml} X^n Y^m Z^l$ be convergent at X_0, Y_0, and Z_0. It follows that the series is absolutely and uniformly convergent in the region $|X| < |X_0|, |Y| < |Y_0|, |Z| < |Z_0|$.*

Notice that the absolute convergence justifies our deducing the form

$$\sum_{mnl} a_{nml} X^n Y^m Z^l$$

from (9.52) by rearrangement of terms.

Now, choose $k_x = k_y = k_z = B$, $p = q = B/k_0$ where B is an arbitrarily large positive number. The multipole expansion (9.52) converges at this choice, and consequently from Theorem 9.5.2 the power series $\sum_{mnl} a_{nml} X^n Y^m Z^l$ converges uniformly in the region

$$\begin{aligned} &|k_x + k_0 p| < 2B, |k_y + k_0 q| < 2B, \\ &\left| k_z + k_0 \sqrt{1 - p^2 - q^2} \right| < 3B^2 - k_0^2. \end{aligned} \tag{9.54}$$

Notice that the last inequality in (9.54) follows from the fact that we choose B large enough to make the square root $\sqrt{1 - p^2 - q^2}$ imaginary.

Next, by the triangular inequality,[29] we may deduce from (9.54)

$$\begin{aligned} &|k_x| < 2B, |k_0 p| < 2B, |k_y| < 2B, |k_0 q| < 2B, \\ &|k_z| < 3B^2 - k_0^2, \left| k_0 \sqrt{1 - p^2 - q^2} \right| < 3B^2 - k_0^2. \end{aligned} \tag{9.55}$$

We now distinguish two possible scenarios. In the first, we have $p^2 + q^2 > 1$, in which case (9.55) give

$$k_x^2 + k_y^2 + k_z^2 + k_0^2 \left(p^2 + q^2 \right) < 8B^2 + 2 \left(3B^2 - k_0^2 \right)^2 + k_0^2. \tag{9.56}$$

In the second scenario, we assume $p^2 + q^2 < 1$. Combining this inequality with (9.55), we find

$$k_x^2 + k_y^2 + k_z^2 + k_0^2 \left(p^2 + q^2 \right) < 8B^2 + \left(3B^2 - k_0^2 \right)^2 + k_0^2. \tag{9.57}$$

From (9.56) and (9.57), we conclude that if we choose B large enough such that $BW^2 = \max\{BW_n^2\} < 8B^2 + \left(3B^2 - k_0^2 \right)^2 + k_0^2$, then we will insure that $|\mathbf{k}|^2 + k_0^2 \left(p^2 + q^2 \right) < BW_n^2$ for all $n = 1, 2, ..., N$. Therefore, *the multipole expansion is uniformly convergent in the 5-sphere* $|\mathbf{k}|^2 + k_0^2 \left(p^2 + q^2 \right) < BW^2$.

In particular, the uniform convergence of (9.52) in any spectral 5-sphere allows us to interchange the order of integration and summation in (9.47) and obtain

$$\begin{aligned} \mathbf{J}(\mathbf{r}') \simeq (2\pi)^{-3} \sum_{n=1}^{N} \sum_{m=0}^{\infty} \int_{|\mathbf{k}|^2 + p^2 + q^2 < BW_n} d^3k\,dp\,dq \\ \times \bar{\mathbf{F}}(\mathbf{r}', \mathbf{k}) \left\{ \widehat{Q}_{mn} \cdot [(\mathbf{k} + \mathbf{K})]^m \right\} \cdot \mathbf{D}(\mathbf{K}). \end{aligned} \tag{9.58}$$

29 I.e., $|x + y| < |x| + |y|, \forall x, y \in \mathbb{C}$.

On the other hand, the uniform convergence of the multipole expansion implies that we can approximate the received signal appearing in (9.47) to any degree of accuracy desired by including enough terms in the multipole series. The number of terms will depend on both the spatial bandwidth BW and the desired error. We will write this number as $M = M(BW)$ and drop reference to the error. Next, we can exploit the fact that the "atoms" s_n appearing in (9.33) may be chosen as small as possible and approximate each multipole moment in (9.53) as

$$\widehat{Q}_{mn} \simeq \frac{i^m}{m!} \underbrace{\mathbf{r}_n \mathbf{r}_n \cdots \mathbf{r}_n}_{m \text{ times}} \int_{s_n} ds, \tag{9.59}$$

where $\mathbf{r}_n \in s_n$. We will not spell out the details of how to choose s_n but outline the method. For simplicity, consider only polynomials in one variable x. For sufficiently small deviation in x, we can approximate any polynomial around a by $f(x) \simeq f(a) + f'(a)(x-a)$. However, if we choose $|x - a| << |f(a)|/|f'(x)|$, it may be assumed that the polynomial in this interval is practically a constant. After truncating the multipole series by $M(BW)$ terms, the integrand of (9.53) becomes a polynomial in a *finite* number of several variables. Therefore, the procedure outlined above can always be carried out in practice to choose a suitable (sufficiently small) s_n and the approximation (9.59) can be made as accurate as one wishes. This choice of N, however, will depend on the already chosen M because, as can be seen from (9.53), the nth multipole moment involves a polynomial of order m. The smallest size of s_n, which determines N, will then depend on the largest value of m, that is on M, which in turns depended in its determination on BW.

The reasoning above leads to the following analytical approximation of the spectral interaction kernel of the antenna system S

$$I(\mathbf{k}, \mathbf{K}) \simeq \sum_{n=1}^{N(M)} \sum_{m=0}^{M(BW)} \left(\frac{i^m}{m!} \underbrace{\mathbf{r}_n \cdot \cdot \mathbf{r}_n}_{m \text{ times}} \int_{s_n} ds \right) \cdot (\mathbf{k} + \mathbf{K})^m, \tag{9.60}$$

It is a polynomial in the cartesian components of $\mathbf{k} + \mathbf{K}$. We have explicitly indicated the dependence of M on BW and of N on the just chosen M. Notice that the validity of this approximation (9.60) depends crucially on the existence of an effective spatial bandwidth of the antenna system as suggested by (9.47).

9.5.6 Synthesis of Antenna Shapes to Meet Specific Spectral Characteristics

It is possible to gain further insight into the nature of the spectral interaction kernel by performing a multipole expansion as outlined in Section 9.5.5. The virtue of this derivation is that it situates the antenna-antenna interaction picture developed in this chapter in the familiar context of elementary light-matter interactions in physics. The multipole moments (9.53) provide a physically intuitive interpretation of the multipole expansion of I_n by thinking of the first-order term as a dipole interaction term, the second-order as a quadruple interaction, and so on. Since the spectral kernels I_n don't depend on the illumination field and the particular ACGF, this computation need be done once for a given antenna shape. Changes in the electromagnetic environment or the feeding mechanism will affect only the ACGF, while the interaction function I_n, mediating the coupling between the antenna and the exciting field, is fixed. Therefore, it is worth the effort computing and studying each term in the expansion (9.52) in the process of attempting to achieve deeper understanding of the principles of operations of a given class of antennas possessing the same geometrical shape.

The expansion (9.52) shows that the spectral function I_n is analytic in $\mathbf{k} + \mathbf{K}$, although not analytic in the five variables k_x, k_y, k_z, p, q pertinent to the antenna problem because of the nonlinear relation between m, in the z-component of $\mathbf{K}$, and p and q as evident in $m = \sqrt{1 - p^2 - q^2}$. However, although the multipole expansion (9.52) is not a power series, we were able to prove in Section 9.5.5 that it is uniformly convergent in any 5-sphere $|\mathbf{k}|^2 + k_0(p^2 + q^2) < BW^2$. Using this knowledge, together with the fact established in Section 9.5.3 regarding the existence of an effective spatial bandwidth of the antenna system, we managed to derive the analytical approximation of the spectral interaction kernel (9.60). It is important to notice the sense in which this analytical approximation is considered valid. The term 'analytical approximation' itself is meaningless unless one decides which quantity of interest the analytical expression is supposed to approximate. In this work, the spectral interaction function $I(\mathbf{k}, \mathbf{K})$ by itself is not important, but we are instead interested in its role as a mediator between the external field and the ACGF in the process of producing the physically observable received signal $\mathbf{J}(\mathbf{r}')$. Therefore, in the derivation of (9.60), we showed that the error in $\mathbf{J}(\mathbf{r}')$ can be made as small as one wishes by truncating the multipole expansion (9.52) by a finite number of terms. This depends on two crucial facts 1) the *uniform* convergence of (9.52) and 2) the existence of an effective *finite* spatial bandwidth of the antenna system. Since both facts have been established, we conclude that the analytical expression (9.60) can always be found provided one chooses sufficiently large M and N (see Section 9.5.5 for definitions and details).

The multipole moments of this expansion as given by (9.59) provide a simple means for reconstructing the geometry of the antenna as we now proceed to demonstrate. Suppose we are given a desired spectral function $I_d(\mathbf{k}, \mathbf{K})$ that we

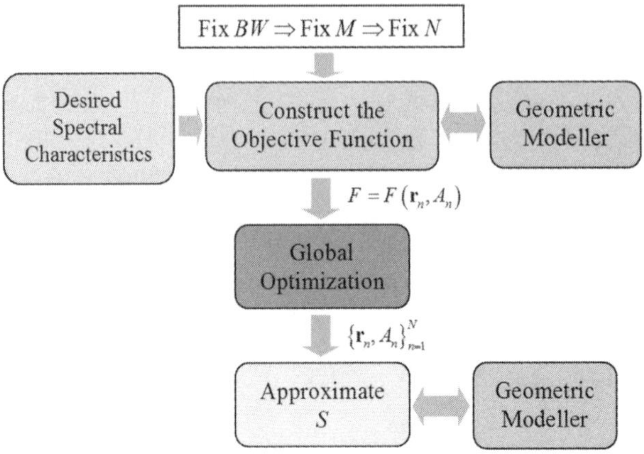

Figure 9.2 The general synthesis of an antenna shape S meeting a pre-given spectral characteristics.

would like to achieve by a certain antenna geometrical shape S_d. Suppose further that this spectral function has a spatial bandwidth BW_d. The analysis conducted in Section 9.5.5 demonstrated that it is always possible to obtain an analytical approximation of the spectral interaction function of an antenna system provided it has a finite bandwidth, namely the polynomial expression (9.60) as a function of $\mathbf{k}$, p, and q. Since the multipole expansion (9.52) was proved to be uniformly convergent in the 5-sphere $|\mathbf{k}|^2 + k_0(p^2 + q^2) < BW_d^2$, we can consider the minimization of the sup norm

$$F := \sup_{|\mathbf{k}|^2+k_0(p^2+q^2)<BW_d^2} \left| I_d(\mathbf{k}, \mathbf{K}) - \sum_{n=1}^{N(M)} \sum_{m=0}^{M(BW_d)} (i^m A_n/m) \underbrace{\mathbf{r}_n \cdots \mathbf{r}_n}_{m \text{ times}} \cdot (\mathbf{k} + \mathbf{K})^m \right|. \tag{9.61}$$

Here, $A_n := \int_{s_n} ds$ are the areas of the elements s_n. A powerful global optimization algorithm can be utilized to search for the best distribution of position vectors $\mathbf{r}_n$ and areas of s_n minimizing the objective function (9.61). From the knowledge of $\mathbf{r}_m$ and the rough sizes of the small "atoms" s_n, one can build a first approximation of the antenna shape that can meet the desired spectral characteristics I_d.

From the value of the effective bandwidth BW_d one can get a first estimate of the number of terms in the multipole expansion, i.e., the value $M(BW_d)$. Based on this choice of M, a rough estimate of N can be given as outlined in Section 9.5.5 provided we know the rough overall surface area of the antenna to be synthesized. The design methodology will then work for antennas with various total surface areas in the sense that for each such area the objective function (9.61) can be formed and an optimization search is conducted. The choice of the antenna size is certainly a matter to be left to the actual application at hand (which originally motivates the particular choice of the desired spectral characteristics I_d) and the concrete implementation of the design methodology suggested here.

Figure 9.2 demonstrates the proposed synthesis procedure. For a given spectral function $I_d(\mathbf{k}, \mathbf{K})$ and a spatial bandwidth BW, one can determine M and consequently N. An initial rough antenna shape is implemented by the geometric modeler and is fed into the process of constructing the objective function F (9.61), the latter now dependent on the main optimization parameters $\mathbf{r}_n$ and A_n. A global optimization strategy, for example genetic algorithms or the particle swarm optimization, may now be deployed in order to search for the best parameters minimizing F within the limits delimited by the allowable spatial region imposed through the geometric modeler. Once a best set of parameters $\mathbf{r}_n$ and A_n is found, their values are fed into the geometric modeler in order to construct various practical realizations of the antenna surface S meeting the desired spectral characteristics. As can be clearly seen, this process is quite complex. However, the analysis in Section 9.5.5 provides the initial mathematical results suggesting that this scheme can work if a physical solution exists (i.e., a solution in which the minimum attained in optimizing (9.61) is acceptable in the application at hand).

9.5.7 The Idea of the Antenna Generalized Transfer Function

The results achieved so far permit us to introduce a new concept of antenna transfer function that is more general than the ACGF itself. Indeed, by studying the structure of the interaction integrals (9.31), we may define the following dyadic function

$$\bar{\mathbf{H}}\left(\mathbf{k}, \mathbf{K}\right) := I\left(\mathbf{k} + \mathbf{K}\right) \bar{\boldsymbol{\Omega}}\left(\mathbf{K}\right). \tag{9.62}$$

In terms of this quantity, we can write the interaction integral (9.31) more compactly in the following form

$$\mathbf{J}\left(\mathbf{r}'\right) = \int_{\mathbb{R}^5} d^3k\, dp\, dq\, \bar{\mathbf{F}}\left(\mathbf{r}', \mathbf{k}\right) \cdot \bar{\mathbf{H}}\left(\mathbf{k}, \mathbf{K}\right) \cdot \mathbf{J}_s\left(\mathbf{K}\right). \tag{9.63}$$

Therefore, *the antenna generalized transfer function* $\bar{\mathbf{H}}\left(\mathbf{k}, \mathbf{K}\right)$ accepts two inputs, the spectral source antenna current $\mathbf{J}_s\left(\mathbf{K}\right)$ and the Fourier transform of the ACGF

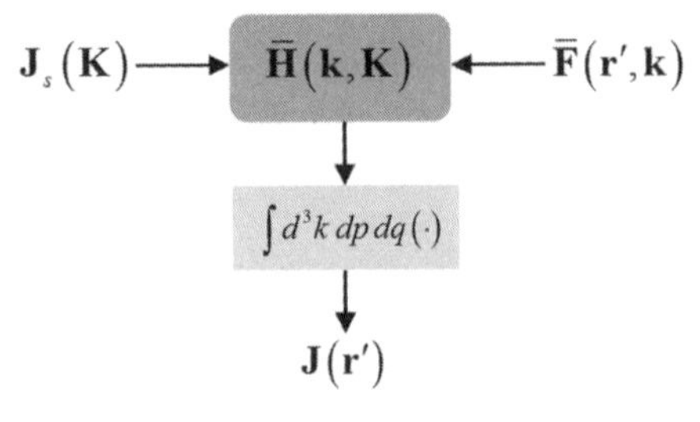

Figure 9.3 The concept of generalized antenna transfer function. After filtering the spectral-domain source antenna current and the receiving ACGF through the agency $\bar{\mathbf{H}}(\mathbf{k}, \mathbf{K})$, the results are integrated in order to form the received signal.

$\bar{\mathbf{F}}(\mathbf{r}', \mathbf{k})$, and produce (after summing over all of the spectral variables) the received signal $\mathbf{J}(\mathbf{r}')$. Figurer 9.3 illustrates the process of antenna-antenna interaction in terms of the generalized transfer function introduced above. In particular, one can trace how each Fourier competent $\mathbf{K}$ interacts with the receiving antenna Fourier mode $\mathbf{k}$ and contributes to the production of the received signal.[30] Therefore, the complete spectral structure of the complex process leading to the formation of the port signal of an antenna immersed in the near field of another antenna is now fully disclosed.

The two "inputs" $\mathbf{J}_s(\mathbf{K})$ and $\bar{\mathbf{F}}(\mathbf{r}', \mathbf{k})$, however, should be received with some qualification. In general, we expect that an 'input' to a system is independent of the system transfer function itself. This is not the case here for two reasons. First, the ACGF $\bar{\mathbf{F}}(\mathbf{r}', \mathbf{k})$ depends on the antenna surface S. Second, in case mutual coupling between the source and receiving antenna is taken into account, then the source current $\mathbf{J}_s(\mathbf{K})$ itself becomes function of the antenna surface S. However, the very same surface S is the sole determinant factor of the generalized transfer function $\bar{\mathbf{H}}(\mathbf{k}, \mathbf{K})$ itself. In spite of this, we believe that the concept introduced by the new transfer function $\bar{\mathbf{H}}$ remains important and even vital for the devolvement of a general formalism for the analysis and design of electromagnetic systems. The reason is that the function $\bar{\mathbf{H}}(\mathbf{k}, \mathbf{K})$ is simply the only *invariant* transfer function in the antenna-antenna interaction problem. Indeed, the fact that it depends mainly on the geometry of the antenna system,[31] as encapsulated in S, makes it indifferent to what actually exists in the electromagnetic environment surrounding the two

30 Cf. Section 9.5.8.

31 We are considering the infinite and homogeneous medium, characterized by the wavenumber k_0, inside which the entire system is immersed as a *fixed* background.

antennas. For example, in a complex system, both $\mathbf{J}_s(\mathbf{K})$ and $\bar{\mathbf{F}}(\mathbf{r}', \mathbf{k})$ will vary in general due to mutual coupling with other nearby objects (and also because of the nature of the spatial field structure injected into the system and circulating at a given stage of operation). To give the antenna S a system characterization that is immune to all such unpredictable changes, we have introduced the concept of the interaction function $\bar{\mathbf{H}}(\mathbf{k}, \mathbf{K})$ as the only truly invariant transfer function of the antenna system.

By focusing (in the process of antenna design and devolvement) on this transfer function, one can isolate the "second-order effects" expected to result from the actual operation of the device in a complex environment. For example, it is possible to start with a typical ACGF $\bar{\mathbf{F}}(\mathbf{r}', \mathbf{k})$ and use it in (9.63) in order to obtain a general feel of the overall characteristics $\bar{\mathbf{H}}(\mathbf{k}, \mathbf{K})$ required. Then, one can use the method described in Section 9.5.6 in order to obtain an approximation of the antenna shape S capable of realizing the needed spectral filtering characteristics. Afterwards, even when the ACGF $\bar{\mathbf{F}}(\mathbf{r}', \mathbf{k})$ and the source current deviate from the initial estimation, the overall performance of the antenna system, viewed from the spectral-domain perspective, is roughly the same. In this sense, the engineer can picture both $\mathbf{J}_s(\mathbf{K})$ and $\bar{\mathbf{F}}(\mathbf{r}', \mathbf{k})$ as genuine "input" functions to the generalized transfer function $\bar{\mathbf{H}}(\mathbf{k}, \mathbf{K})$, yet still both are capable of varying as the actual operation of the combined electromagnetic system requires.[32]

9.5.8 The Static Genesis

We are in a position now to put together all the results and conclusions arrived at so far in the analysis of antenna-antenna interactions within a unified theoretical framework, which will be called the *static genesis* of the received signal. By 'static' we emphasize that up to now the receiving mode antenna (Mode C) is fixed within the near-field zone of the source antenna. Mathematically, this is implied by the fact that a fixed global coordinate system was chosen for the mathematical description of the problem as illustrated in Figure 9.1. In Section 9.5.9, this restriction will be released. For now, however, focus will be on understanding the physical mechanism behind the production of the received signal in a fixed port.

The physical picture that appears the most natural in the description of the antenna-antenna interaction process is suggested by careful examination of the unified theme behind the various interaction integrals (9.31), (9.35), (9.63), (9.58),

32 We suggest further research along these lines of thinking in order to reach more concrete conclusions. This may probably be best attempted with simple systems.

which is the idea of *oscillator-oscillator coupling*. Indeed, the incident electromagnetic field is expanded (Fourier analyzed) into a sum of harmonic functions in space. The ACGF, which represents the system function of the antenna in the spatial domain, is also Fourier analyzed into its spatial harmonic modes. It turns out that the *spatial* interaction between the illumination field and the antenna manifests itself in exactly the same formal structure of the familiar interaction in time: That of field oscillator coupled with an antenna oscillator. If the regions s_n introduced in dividing the total surface S as in (9.33) are made very small, then each region may be considered an "atom." This terminology is not metaphorical because the interaction function I_n for this small region approaches that of a 2-dimensional sinc function. As is will know, such sinc function has a very narrow peak concentrated around its center, while it decays rapidly away from the peak. It follows that in this case each Fourier mode of the illumination field at $\mathbf{K}$ will interact *resonantly* with the ACGF Fourier mode at $-\mathbf{k}$. More precisely, most of the contributions to the interaction integrals will emerge from the spectral region centered around $\mathbf{k} + \mathbf{K} = 0$. Each other "atom" s_n will interact with the illumination field spatial harmonic modes in the same manner, and the total observed voltage will be simply the direct sum of the all the interactions. It is very clear then that the interaction of the illumination field and the antenna in space can be viewed in exactly the same (qualitative) manner of field-atom interaction in physics: That is, resonant interactions between field oscillators and antenna (current Green's function) oscillators, but in this case the whole process occurs in *space*.

The illumination field, being electromagnetic in character, has the additional peculiarity of being developable into a sum of propagating and nonpropagating modes. That implies that the picture outlined in the preceding paragraph is not complete, because a careful spectral analysis of the antenna near fields using the Weyl expansion revealed the existence of nonpropagating modes in the form of evanescent field. In order to assess the overall importance of this phenomenon, we will perform now a more precise analysis of the results obtained so far.

As mentioned above, if s_n is chosen very small, and moreover, say, with an approximately rectangular shape, then the function $I_n(\mathbf{k}, \mathbf{K})$ will look roughly like a 2-dimensional sinc function. However, for more general shapes and sizes of s_n, it is not always easy to have an idea about the degree of narrowness of the main peak. In this case, we can make use of the result (9.44), (9.37), and (9.38), which show that the spectral interaction kernel $I_n(\mathbf{k}, \mathbf{K})$ attains its maximum at $\mathbf{k}+\mathbf{K} = 0$ while decays away from this region. It is reasonable then to assume that the majority of the contribution of the field-ACGF interactions due to the "atom" s_n will come from the region around $\mathbf{k}+\mathbf{K} = 0$. However, a glance at (16.17) shows that the condition

$\mathbf{k} = -\mathbf{K}$ can be satisfied only for pure propagating illumination modes, i.e., when $p^2 + q^2 < 1$. The nonpropagating modes of the illumination field spectrum, i.e., the region $p^2 + q^2 > 1$, still interact with the Fourier modes of the ACGF, but their contributions are strongly attenuated by the spectral interaction kernel $I_n(\mathbf{k}, \mathbf{K})$.

Therefore, the picture of oscillator-oscillator interaction sketched above has to be qualified in the following manner. The total interaction of the illumination field with the receiving antenna can be always divided unambiguously into two distinct parts, resonant and nonresonant interactions, the former is the familiar oscillator-oscillator interaction (propagating mode interacts with Fourier modes of the ACGF), while the latter is an evanescent-oscillator interaction (evanescent mode interacting with a Fourier component of the ACGF). The ACGF will always manifests itself spectrally as a sum of spatial oscillator associated with each "atom" s_n. However, the illumination field is split into two parts, pure oscillator and nonpropagating (evanescent) modes. Since the essence of the near field of any source is the subtle manner in which such splitting into propagating and nonpropagating is enacted (Chapter 4), we conclude that the division of the interaction into the corresponding resonant and nonresonant parts above is of great physical significance.

Before explaining further this point, let us first write down the expressions of the divided interaction integrals. The general spatial bandwidth BW_n defined in (9.46) arises from two bandwidths, one associated with the $\mathbf{k}$-part of I_n, while the other is linked to the $\mathbf{K}$-dependence of the same function. More precisely, we have $|\mathbf{k}|^2 < BW_{n,\mathbf{k}}^2$ and $k_0^2 \left(p^2 + q^2\right) < BW_{n,\mathbf{K}}^2$, from which it follows that $|\mathbf{k}|^2 + k_0^2 \left(p^2 + q^2\right) < BW^2$ and $BW^2 = BW_{n,\mathbf{k}}^2 + BW_{n,\mathbf{K}}^2$. Now, the $\mathbf{K}$-bandwidth $BW_{n,\mathbf{K}}^2$ obviously satisfies $BW_{n,\mathbf{K}}^2 > k_0^2$ because attenuation in the $\mathbf{K}$-space starts only when $p^2 + q^2 > 1$. Therefore, we have $BW^2 > k_0^2$. The study now of the regions in the $\mathbf{k}$-space interacting resonantly and nonresonantly with the illumination field can be readily completed. We have

$$\begin{aligned} &\text{Resonant Interaction} \\ &\qquad p^2 + q^2 < 1\,, \;\; k_0^2 < |\mathbf{k}|^2 < BW_n^2, \\ &\text{Nonresonant Interaction} \\ &\qquad p^2 + q^2 > 1, \;\; 0 < |\mathbf{k}|^2 < BW_n^2 - k_0^2. \end{aligned} \tag{9.64}$$

We can now rewrite (9.63) in the following form

$$\mathbf{J}\left(\mathbf{r}'\right) = \bar{\mathbf{F}}\left(\mathbf{r}', \mathbf{k}\right) \hat{H}_{\mathrm{R}} \mathbf{J}_s\left(\mathbf{K}\right) + \bar{\mathbf{F}}\left(\mathbf{r}', \mathbf{k}\right) \hat{H}_{\mathrm{NR}} \mathbf{J}_s\left(\mathbf{K}\right). \tag{9.65}$$

The two operators $\hat{H}_{\mathrm{R}}$ and $\hat{H}_{\mathrm{NR}}$ are called the *resonant* and *nonresonant* operators, respectively, and are defined by their effects on a dyadic $\mathbf{k}$-function $\bar{\mathbf{Y}}\left(\mathbf{k}\right)$ and a vector $\mathbf{K}$-function $\mathbf{X}(\mathbf{K})$ in the following manner

$$\bar{\mathbf{Y}}(\mathbf{k})\,\hat{H}_{\mathrm{R}}\mathbf{X}(\mathbf{K}) := \int_{k_0^2<|\mathbf{k}|^2<BW^2} d^3k \int_{p^2+q^2<1} dpdq \\ \times \bar{\mathbf{Y}}(\mathbf{k})\cdot\bar{\mathbf{H}}(\mathbf{k},\mathbf{K})\cdot\mathbf{X}(\mathbf{K}), \tag{9.66}$$

$$\bar{\mathbf{Y}}(\mathbf{k})\,\hat{H}_{\mathrm{NR}}\mathbf{X}(\mathbf{K}) := \int_{0<|\mathbf{k}|^2<BW^2-k_0^2} d^3k \int_{p^2+q^2>1} dpdq \\ \times \bar{\mathbf{Y}}(\mathbf{k})\cdot\bar{\mathbf{H}}(\mathbf{k},\mathbf{K})\cdot\mathbf{X}(\mathbf{K}). \tag{9.67}$$

When BW goes to infinity, the expression (9.65) is exact. However, when BW is truncated to a finite value, then we obtain an approximation of the received signal $\mathbf{J}(\mathbf{r}')$ in the spirit of the discussion around the definition (9.46). The two integral operators $\hat{H}_{\mathrm{R}}$ and $\hat{H}_{\mathrm{NR}}$ are therefore dependent only on k_0 and the antenna shape S.[33]

We are in a position now to survey the results of our analysis by examining the general expression (9.65). We find that there are two fundamental insights revealed by the form of the interaction integral above regarding the problem of the static genesis:

1. The electromagnetic system is best understood at the fundamental level of inquiry we have been conducting so far as a *process* in the sense that the antenna is a *relational structure*. This structure is simply that transformational role enacted by the two operators $\hat{H}_{\mathrm{R}}$ and $\hat{H}_{\mathrm{NR}}$. In other words, the only *invariant* function of the antenna system is these two operators, while the two spectral quantities serving as "inputs," i.e., the source current $\mathbf{J}_s(\mathbf{K})$ and the ACGF $\bar{\mathbf{F}}(\mathbf{r}',\mathbf{k})$, may vary according to the electromagnetic environment and the operational details of the system at hand. Therefore, the source current and the ACGF are themselves not absolute characteristics of the antenna system, but instead manifest themselves in the interaction integral (9.65) as mere terms in the truly intrinsic *relations* encapsulated by the structure of the resonant and nonresonant operators $\hat{H}_{\mathrm{R}}$ and $\hat{H}_{\mathrm{NR}}$.

2. The actual value of the received signal $\mathbf{J}(\mathbf{r}')$ can be understood physically as the contributions of all of the ACGF oscillators $\bar{\mathbf{F}}(\mathbf{r}',\mathbf{k})$ after interacting resonantly with the propagating part of the illumination field spectrum, and nonresonantly with the evanescent part of this spectrum. Therefore, *an antenna interacts with another source in the latter's near-field zone in a manner dictated by the rather subtle way in which the illumination near field splits into propagating and nonpropagating parts*. This is because for a fixed global

33 Notice that BW is determined from $I_n(\mathbf{k},\mathbf{K})$, which are themselves functions of S.

coordinate system the spectral ACGF $\bar{\mathbf{F}}(\mathbf{r}', \mathbf{k})$ is also fixed, and it remains for the near-field illumination to vary from one configuration of sources to another, where by 'variation' we understand the above mentioned decomposition into purely propagating waves and evanescent modes.

It should be mentioned that although the nonresonant part of the interaction in (9.67) is significantly attenuated by the spectral interaction kernel $I_n(\mathbf{k}, \mathbf{K})$, we cannot in general ignore its contributions without further inquiry. Although we may find in many cases that it is possible, say as a first approximation, to consider only the resonant interaction with the purely propagating modes of the illumination field, we believe that for a deeper and more comprehensive understanding of the operational principles of devices designed within the framework of coupling in the near field zone, one has to retain the full expression (9.65) with both the resonant and nonresonant integral operators $\hat{H}_{\mathrm{R}}$ and $\hat{H}_{\mathrm{NR}}$ taken into account.

9.5.9 The Dynamic Genesis: Interaction Between Externally Applied Source and a Moving Receiving Antenna in the Near-Field Zone

9.5.9.1 Introduction

The general picture of antenna-antenna interaction within the framework of the static genesis is encapsulated in the equation (9.65), together with the resonant and nonresonant operators (9.66) and (9.67), respectively. However, although we have achieved a significant progress in our theoretical analysis of interaction by bracketing out the varying quantities and focusing instead mainly on the invariant structure, i.e., the operators $\hat{H}_{\mathrm{R}}$ and $\hat{H}_{\mathrm{NR}}$, our analysis is still incomplete because the last two operators themselves depend on the choice of the global coordinate system in Figure 9.1. To be sure, merely stating how the operators will change with a general rotation or translation of this global frame is superficial and does not shed any light on the problem from the fundamental viewpoint. A more careful approach was developed by the authors in Chapter 4 where two frames of reference were introduced, one is global while the other is local. The dynamic essence of the formation of the radiation field was identified afterwards as the complex manner in which the near field continuously split into propagating and nonpropagating parts while the perspective of the observer, represented here by the *local* coordinate system, changes continuously. For example, the concept of radial streamlines was proposed there as the most convenient tool in probing the structure of the near field, especially from the engineering point of view. However, this concept of radial streamlines cannot be directly applied to the antenna-antenna interaction problem in Figure 9.1. The reason is simply the following. In order to obtain a reasonable interaction mechanism, one must keep the mathematical representation of the ACGF of the receiving antenna fixed. Now in order to introduce a *second* coordinate system for the implementation of the idea of a local frame of reference associated with the fields radiated by the source antenna, the ACGF of the receiving antenna will *also* change with this local frame. Therefore, while the local frame is capturing how the source antenna's near field is splitting into propagating and nonpropagating parts, the ACGF is also changing, rendering the spectral resonance picture as oscillator-field interaction, incoherent since there is no longer a "fixed" ACGF oscillator interacting with varying spectrum of propagating and evanescent modes. We then achieve nothing using this approach.

This difficulty, however, can be overcome by noticing that the intention of introducing a local frame is merely to serve as a step toward the real object of investigation, which is the study of the *dynamic* aspects of the radiation field. In the analysis of Chapter 4, there was only a single antenna system under consideration, a fact that permitted us using a simple local frame comprised of a rotating coordinate system sharing the same origin with the global frame. Now in order to single out the dynamic aspects of interaction in the antenna-antenna problem of Figure 9.1, what is needed is not moving the global frame of the two-antenna system, but rather moving *one* antenna, say the receiving antenna, with respect to another. That is, we must introduce a *relative* motion of two antennas with respect to some fixed global frame. This was already suggested by Motions I-III in Figure 9.1. Our main aim in this section is to complete the study of the genesis of the received signal in antenna-antenna interactions by studying the dynamic aspects of the interaction understood here as motion of the receiving system with respect to the source. The present study may serve, beside other things, as a theoretical foundation for the design and devolvement of experimental techniques for probing the structure of the near fields of general antenna systems.

9.5.9.2 Study of General Euclidean Motion of the Receiving Antenna

We will consider the two geometrical transformations forming the classical Euclidean motion group, translation and rotation. Using the terminology of Figure 9.1, translation corresponds to Motion I while rotation to Motion II. Here, we keep the global coordinate system as the one fixed previously, i.e., with origin O inside the receiving antenna S. Let us start first with translation.

Suppose that the the antenna S is displaced into a new position $\mathbf{r}_a$, i.e., consider the transformation $S \rightarrow S_a$, where S_a is the surface positioned at $\mathbf{r}_a$. Due to this translation, a new current distribution $\mathbf{J}_a(\mathbf{r})$ will be induced on the antenna S_a. We are interested in the *changes* in the received signal at a specific location, say $\mathbf{r}'$. Let the ACGF of the antenna S_a be denoted by $\bar{\mathbf{F}}_a(\mathbf{r}',\mathbf{r})$. We can compute the received signal using the usual relation $\mathbf{J}_a(\mathbf{r}') = \int_{S_a} ds\, \bar{\mathbf{F}}_a(\mathbf{r}',\mathbf{r}) \cdot \mathbf{E}(\mathbf{r})$. However, since the source is externally applied, the new ACGF is simply the translated version of the antenna S, i.e., $\bar{\mathbf{F}}_a(\mathbf{r}',\mathbf{r}) = \bar{\mathbf{F}}(\mathbf{r}'-\mathbf{r}_a, \mathbf{r}-\mathbf{r}_a)$. Substituting this into the above integral and performing the transformation of variables $\mathbf{r}'' = \mathbf{r}-\mathbf{r}_a$ we obtain (after replacing $\mathbf{r}''$ by $\mathbf{r}$ at the end) the following formula $\mathbf{J}_a(\mathbf{r}') = \int_S ds\, \bar{\mathbf{F}}(\mathbf{r}'-\mathbf{r}_a,\mathbf{r}) \cdot \mathbf{E}(\mathbf{r}+\mathbf{r}_a)$. We proceed then in exactly the same manner in which (9.31) was derived from (9.30). Indeed, we obtain

$$\begin{aligned} \mathbf{J}_a(\mathbf{r}') = (2\pi)^{-3} & \int_{\mathbb{R}^3} d^3k \int_{\mathbb{R}^2} dpdq \\ & \times \bar{\mathbf{F}}(\mathbf{r}'-\mathbf{r}_a,\mathbf{k})\, I(\mathbf{k},\mathbf{K})\, e^{i\mathbf{K}\cdot\mathbf{r}_a} \cdot \mathbf{D}(\mathbf{K}). \end{aligned} \tag{9.68}$$

Comparing (9.68) with (9.31) and observing the definition (9.32), we find that the overall mechanism of interaction after translating the antenna S is rather similar to the untranslated case. The reason is that multiplication by a complex exponential does not affect the amplitude filtering characteristics of $I(\mathbf{k},\mathbf{K})$, and in particular, it leaves the spatial bandwidth intact. The illumination spectral field $\mathbf{D}(\mathbf{K})$ is the same, and also the spectral ACGF $\bar{\mathbf{F}}(\mathbf{r}'-\mathbf{r}_a,\mathbf{k})$ of the original configuration appears unchanged. Only the phase of $I(\mathbf{k},\mathbf{K})$ is modified by a linear phase shift $\mathbf{K}\cdot\mathbf{r}_a$. *However, it should be kept in mind that the above conclusion is valid only if none of the points of the translated antenna S lies within the source horizon.*

Next, consider a rotation of the receiving antenna S with respect to the global frame xyz by $\bar{\mathbf{R}}_a$. Again, if mutual coupling is neglected, then the ACGF of the rotated antenna S_a is given by $\bar{\mathbf{F}}_a(\mathbf{r}',\mathbf{r}) = \bar{\mathbf{F}}\left(\bar{\mathbf{R}}_a^{-1}\cdot\mathbf{r}', \bar{\mathbf{R}}_a^{-1}\cdot\mathbf{r}\right)$. The received signal is given by $\mathbf{J}_a(\mathbf{r}') = \int_{S_a} ds\, \bar{\mathbf{F}}\left(\bar{\mathbf{R}}_a^{-1}\cdot\mathbf{r}', \bar{\mathbf{R}}_a^{-1}\cdot\mathbf{r}\right) \cdot \mathbf{E}(\mathbf{r})$. Performing the transformation $\mathbf{r}'' = \bar{\mathbf{R}}_a^{-1}\cdot\mathbf{r}$ and changing back from $\mathbf{r}''$ to $\mathbf{r}$ at the end, we obtain $\mathbf{J}_a(\mathbf{r}') = \int_S ds\, \bar{\mathbf{F}}\left(\bar{\mathbf{R}}_a^{-1}\cdot\mathbf{r}',\mathbf{r}\right)\cdot\mathbf{E}\left(\bar{\mathbf{R}}_a\cdot\mathbf{r}\right)$. Next, we use the orthogonality relations of rotation matrices $\bar{\mathbf{R}}_a^T = \bar{\mathbf{R}}_a^{-1}$ and the identity $\mathbf{K}\cdot\left(\bar{\mathbf{R}}_a\cdot\mathbf{r}\right) = \left(\bar{\mathbf{R}}_a^T\cdot\mathbf{K}\right)\cdot\mathbf{r}$ in order to derive the following relation

$$\begin{aligned} \mathbf{J}_a(\mathbf{r}') = (2\pi)^{-3} & \int_{\mathbb{R}^3} d^3k \int_{\mathbb{R}^2} dpdq \\ & \times \bar{\mathbf{F}}\left(\bar{\mathbf{R}}_a\cdot{\mathbf{r}'}_a,\mathbf{k}\right) I\left(\mathbf{k},\bar{\mathbf{R}}_a^T\cdot\mathbf{K}\right) \cdot \mathbf{D}(\mathbf{K}). \end{aligned} \tag{9.69}$$

Again, the interaction mechanism is rather similar to original (before rotation) case. The reason is that simple rotation of the $\mathbf{K}$-argument of spectral interaction kernel $I(\mathbf{k},\mathbf{K})$ does not affect its amplitude filtering character, and in particular the bandwidth remains the same. We conclude then that *the picture of the static genesis* *contains in essence the most important features of interactions. By translating and/or rotating the receiving antenna system S, only translation and/or rotation of the spectral interaction kernel $I(\boldsymbol{k}, \boldsymbol{K})$ is needed in order to modify the conclusion of the static genesis,*[34] *while the rest of the interaction integral retains its original form intact. In particular, the spatial bandwidth of the antenna system is invariant with respect to the Euclidean motion group.*

9.5.9.3 Local Rotation of the Source Antenna: The Dynamic Aspects of the Near Field Structure

We come now to the most important analysis in this work, which is concerned with unconcealing the dynamic aspects of a given source antenna's radiated near field as revealed through its interaction with another antenna. This is not only important from the fundamental theoretical standpoint, but also we find such analysis crucial for the practical aspects as well. The reason is that in practice one can study one antenna only through the analysis of the way in which it interacts with another antenna or measuring system (say a special probe in the neat field).

The results (9.68) and (9.69) indicated that the static genesis framework can be used to relate the changes in the received signal due to motion of the receiving antenna to the mechanism of interaction conducted while the latter antenna was fixed. Since any motion in $\mathbb{R}^3$ can be analyzed into translation plus rotation, then we conclude that as long as the moving antenna does not enter the source horizon, the expressions (9.68) and (9.69) can be used to predict the interaction mechanism for the moving antenna. One particular compound motion is Motion III in Figure 9.1, which was defined as rotation of the receiving antenna such as its relative 3D orientation with respect to the source remains fixed. It is clear that simple translation (Motion I, translation along a circle centered around O') cannot implement this motion, but an additional special local rotation (Motion II, local rotation around O) has to be superimposed on the previous Motion I in order to correctly obtain Motion III. Such treatment is, first of all, quite complicated, and moreover not complete since we cannot treat the case when the moving receiving antenna enters the source horizon. A different approach to the problem is needed, which we propose as follows.

The key aspect in Motion III is that it is exactly equivalent to a rotation of the source antenna around O' while the receiving antenna is kept fixed, where the rotation is inverse to that originally intended when viewing the receiving antenna as the moving object. The advantages of rotating the *source* instead of

34 In the case of translation, this is true as long the antenna does not enter the source horizon.

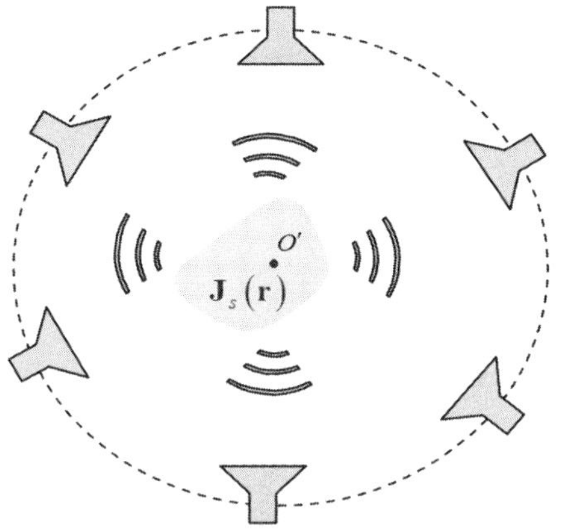

Figure 9.4 Study of the near-field structure of a source $\mathbf{J}_s(\mathbf{r})$ inclosed inside a region V_s via its interaction with another antenna rotating around O'. The received signal induced in the moving antenna varies according to the manner in which the source's near field are dynamically splitting into propagating and nonpropagating modes as the perspective of the observer varies through the motion of the receiving probe.

the receiving antenna are considerable. First, the global coordinate system and the relative position of the receiving antenna with respect to it are kept fixed, which facilitate our intention of studying the variation in the structure of the source antenna near field. Second, for the entire continuum of local rotations of the source around O', its antenna horizons are all identical and fixed as shown in Figure 9.1. The reason is that the global frame of reference is fixed, and the rotated source antenna remains all the time inside this horizon. Figure 9.4 illustrates this process. The motion of the probe around the source can be implemented by treating the probe as fixed while the source rotates locally around its center O'.

Now let us go back to the expression (9.30). The Fourier transform of the source will transform like $\mathbf{J}_s(\mathbf{k}) \rightarrow \mathbf{J}_s\left(\bar{\mathbf{R}}_a^T \cdot \mathbf{k}\right)$ as can be found from (9.21). However, the illumination field due to the source is always evaluated at the same locations $\mathbf{r} \in S$ because the global coordinate system is fixed. This implies that the spectral interaction kernel $I(\mathbf{k}, \mathbf{K})$ remains the same. Moreover, for the same reasons the spectral ACGF of the receiving antenna is unchanged by the local rotation of the source. We finally then arrive at our main result

$$\begin{aligned} \mathbf{J}\left(\mathbf{r}'; \bar{\mathbf{R}}_a\right) = (2\pi)^{-3} & \int_{\mathbb{R}^3} d^3k \int_{\mathbb{R}^2} dpdq \\ & \times \bar{\mathbf{F}}\left(\mathbf{r}', \mathbf{k}\right) \cdot \bar{\mathbf{H}}\left(\mathbf{k}, \mathbf{K}\right) \cdot \mathbf{J}_s\left(\bar{\mathbf{R}}_a^T \cdot \mathbf{K}\right). \end{aligned} \tag{9.70}$$

Notice that the spectral ACGF $\bar{\mathbf{F}}(\mathbf{r}',\mathbf{k})$ of the probe and its generalized transfer function $\bar{\mathbf{H}}(\mathbf{k},\mathbf{K})$ are fixed. Now, since the distance of probe to the source is constant, the main generator of change in the received signal can be best thought of as a change in the proportions of the resonant and nonresonant parts of the illumination field interaction with the receiving system in the spirit of the expression (9.65). Therefore, the result (9.70) suggests that the variation in the received signal of the rotating antenna (probe) in Figure 9.4 is due fully to the dynamic splitting of the source's near field into propagating and nonpropagating modes. This motivates our belief that a fair understanding of the (spectral) generalized transfer function of a well-designed probe may enable us to infer valuable conclusions about the near-field structure of a given source by carefully interpreting the variations in the received signal $\mathbf{J}\left(\mathbf{r}';\bar{\mathbf{R}}_a\right)$ for a wide range of specially chosen 3D rotations $\bar{\mathbf{R}}_a$. Notice that in contrast to (9.68), there is no restriction on the region of validity of (9.70).

9.5.9.4 The Effect of Mutual Coupling

Strictly speaking, the relations (9.68), (9.69), and (9.70) are not exact if mutual coupling is present. Indeed, the values of the source current $\mathbf{J}_s(\mathbf{r})$ and/or the receiving spectral ACGF $\bar{\mathbf{F}}(\mathbf{r}',\mathbf{k})$ will change with $\mathbf{r}_a$ and/or $\bar{\mathbf{R}}_a$. However, we will now show that although the quantitative prediction of the received signal has to take into account mutual coupling if necessary, the physical mechanism of interaction, and in particular the antenna system S's spatial bandwidth BW, remain unaffected.

In the presence of mutual coupling between the source and the receiving system, we must modify the current distributions according to the following general scheme

$$\mathbf{J}_s\left(\mathbf{r};\bar{\mathbf{R}}_a,\mathbf{r}_a\right)=\mathbf{J}_s\left(\bar{\mathbf{R}}_a\cdot\mathbf{r}-\mathbf{r}_a\right)+\delta\mathbf{J}_s\left(\mathbf{r};\bar{\mathbf{R}}_a,\mathbf{r}_a\right), \tag{9.71}$$

$$\begin{aligned}\bar{\mathbf{F}}\left(\mathbf{r}',\mathbf{r};\bar{\mathbf{R}}_a,\mathbf{r}_a\right)=&\;\bar{\mathbf{F}}\left(\bar{\mathbf{R}}_a\cdot\mathbf{r}'-\mathbf{r}_a,\bar{\mathbf{R}}_a\cdot\mathbf{r}-\mathbf{r}_a\right)\\&+\delta\bar{\mathbf{F}}\left(\bar{\mathbf{R}}_a\cdot\mathbf{r}'-\mathbf{r},\mathbf{r};\bar{\mathbf{R}}_a,\mathbf{r}_a\right).\end{aligned} \tag{9.72}$$

Formally speaking, these equations can be considered as a definition of the new functions $\delta\mathbf{J}$ and $\delta\bar{\mathbf{F}}$ arising from mutual coupling while $\mathbf{J}_s\left(\mathbf{r};\bar{\mathbf{R}}_a,\mathbf{r}_a\right)$ and $\bar{\mathbf{F}}\left(\mathbf{r}',\mathbf{r};\bar{\mathbf{R}}_a,\mathbf{r}_a\right)$ stand for the new total values of the source current and the spectral ACGF, respectively. In particular, we notice the special definitions $\mathbf{J}_s\left(\mathbf{r};\bar{\mathbf{I}},0\right):=\mathbf{J}_s(\mathbf{r})$ and $\bar{\mathbf{F}}\left(\mathbf{r}',\mathbf{r};\bar{\mathbf{I}},0\right):=\bar{\mathbf{F}}(\mathbf{r}',\mathbf{r})$, where $\bar{\mathbf{I}}$ is the unit dyad. In both cases, the origin of mutual coupling is the Euclidian motion of the source and/or receiving system according to the rotation $\bar{\mathbf{R}}_a$ and translation $\mathbf{r}_a$.

It is important to pay considerable attention to the way equations (9.71) and (9.72) have been written. First consider (9.71). If we rotate and translate the source antenna, then the new source current can be expressed as the sum of two terms. The first term on the LHS is the source current that would appear if there is *no* mutual coupling. As we saw in the derivation of equations like (9.68), (9.69), and (9.70), it is simply the rotation and translation of the spatial function (but the inverse operations for the corresponding Fourier transform). The new term $\delta\mathbf{J}_s$, however, does not arise in this simple way, but should be considered a totally new function of $\mathbf{r}$, $\bar{\mathbf{R}}_a$ and $\mathbf{r}_a$. It is beyond the scope of this chapter to determine how this functional variation can be expressed in a more concrete way at the very general level characterizing our investigation. Therefore, only the formal appearance of the mutual coupling term is retained in our equations. Now relation (9.72) can be treated in the same fashion as (9.71) with the difference that we should also transform the location of the received signal like $\mathbf{r}' \to \bar{\mathbf{R}}_a \cdot \mathbf{r}' - \mathbf{r}_a$ everywhere.

To simplify the presentation, let us introduce the following operator

$$\begin{aligned}&\bar{\mathbf{Y}}(\mathbf{k})\,\hat{I}(\mathbf{k},\mathbf{K})\,\mathbf{X}(\mathbf{K})\\&:=(2\pi)^{-3}\int_{\mathbb{R}^3}d^3k\int_{\mathbb{R}^2}dpdq\,\bar{\mathbf{Y}}(\mathbf{k})\cdot I(\mathbf{k},\mathbf{K})\,\bar{\mathbf{\Omega}}(\mathbf{K})\cdot\mathbf{X}(\mathbf{K}),\end{aligned} \tag{9.73}$$

valid for the dyadic function $\bar{\mathbf{Y}}(\mathbf{k})$ and the vector $\mathbf{X}(\mathbf{K})$. Now, in light of (9.71) and (9.72), we rotate and translate only the receiving antenna. Equations (9.68) and (9.69) will then be replaced by

$$\begin{aligned}&\mathbf{J}\left(\mathbf{r}';\bar{\mathbf{R}}_a,\mathbf{r}_a\right)=\\&\bar{\mathbf{F}}\left(\bar{\mathbf{R}}_a\cdot\mathbf{r}'-\mathbf{r}_a,\mathbf{k}\right)\hat{I}\left(\mathbf{k},\bar{\mathbf{R}}_a^T\cdot\mathbf{K}\right)e^{i\mathbf{K}\cdot\mathbf{r}_a}\bar{\mathbf{\Omega}}\cdot\mathbf{J}_s(\mathbf{K})\\&+\bar{\mathbf{F}}\left(\bar{\mathbf{R}}_a\cdot\mathbf{r}'-\mathbf{r}_a,\mathbf{k}\right)\hat{I}(\mathbf{k},\mathbf{K})\,\bar{\mathbf{\Omega}}\cdot\delta\mathbf{J}_s\left(\mathbf{K};\bar{\mathbf{R}}_a,\mathbf{r}_a\right)+\\&\delta\bar{\mathbf{F}}\left(\bar{\mathbf{R}}_a\cdot\mathbf{r}'-\mathbf{r}_a,\mathbf{k};\bar{\mathbf{R}}_a,\mathbf{r}_a\right)\hat{I}\left(\mathbf{k},\bar{\mathbf{R}}_a^T\cdot\mathbf{K}\right)e^{i\mathbf{K}\cdot\mathbf{r}_a}\bar{\mathbf{\Omega}}\cdot\mathbf{J}_s(\mathbf{K})\\&+\delta\bar{\mathbf{F}}\left(\bar{\mathbf{R}}_a\cdot\mathbf{r}'-\mathbf{r}_a,\mathbf{k};\bar{\mathbf{R}}_a,\mathbf{r}_a\right)\hat{I}(\mathbf{k},\mathbf{K})\,\bar{\mathbf{\Omega}}\cdot\delta\mathbf{J}_s\left(\mathbf{K};\bar{\mathbf{R}}_a,\mathbf{r}_a\right).\end{aligned} \tag{9.74}$$

The first term on the LHS is exactly the effect predicted in (9.68) and (9.69) under the assumption that no mutual coupling is present. The remaining three terms are new and present the effect of mutual coupling when both the source and the receiving antennas react back on each other because of the Euclidean motion of the receiving antenna. The second and third terms have roughly the same order of magnitude. They represent 1) the interaction between $\bar{\mathbf{F}}$ and $\delta\mathbf{J}_s$ on one hand, and 2) $\delta\bar{\mathbf{F}}$ and $\mathbf{J}_s$ on the other. The first type of interaction has no effect on the spectral interaction kernel $I(\mathbf{k},\mathbf{K})$. However, with type 2) of interaction, the $\mathbf{K}$-argument of $I(\mathbf{k},\mathbf{K})$ is rotated by $\bar{\mathbf{R}}_a^T$ and the whole function I multiplied by the phase factor $\exp(i\mathbf{K}\cdot\mathbf{r}_a)$. Finally, the fourth term in (9.74) represents a third type of interaction 3) between $\delta\bar{\mathbf{F}}$ and $\delta\mathbf{J}_s$. It has an order of magnitude yet smaller than types 1) and 2). Notice that both arguments of $I(\mathbf{k},\mathbf{K})$ are unaffected. In general, the price to paid for not transforming some or all of the spectral interaction kernel's arguments in the mutual coupling terms of (9.74) is that a new spectral source and/or spectral ACGF will appear, namely $\delta\bar{\mathbf{F}}$ or/and $\delta\mathbf{J}_s$, which depend on the rotation and translation in a complicated manner that cannot be disclosed at the current state of the ACGF formalism. Now the most important conclusion to be drawn from (9.74) is this: Regardless to the actual functional behavior of $\delta\bar{\mathbf{F}}$ or/and $\delta\mathbf{J}_s$, *the interaction mechanism with the presence of mutual coupling is still essentially the same as that of the static and dynamic genesis developed above*. The reason is that the only possible effects on the spectral interaction kernel $I(\mathbf{k},\mathbf{K})$ are rotation of the $\mathbf{K}$-argument and multiplication by $\exp(i\mathbf{K}\cdot\mathbf{r}_a)$. Such types of operations do not affect the "metrical" properties of the function $I(\mathbf{k},\mathbf{K})$ important for the spectral filtering process it performs in the physical mechanism of interaction as developed in Section 9.5.8. In particular, the spatial bandwidth BW of the antenna system remains the same.

In exactly the same manner, relation (9.70) will modified in the presence of mutual coupling to take the following form

$$\begin{aligned}\mathbf{J}\left(\mathbf{r}';\bar{\mathbf{R}}_a,\mathbf{r}_a\right) = {} & \bar{\mathbf{F}}\left(\mathbf{r}',\mathbf{k}\right)\hat{H}\left(\mathbf{k},\mathbf{K}\right)\mathbf{J}_s\left(\bar{\mathbf{R}}_a^T\cdot\mathbf{K}\right)\\ & +\bar{\mathbf{F}}\left(\mathbf{r}',\mathbf{k}\right)\hat{H}\left(\mathbf{k},\mathbf{K}\right)\delta\mathbf{J}_s\left(\mathbf{K};\bar{\mathbf{R}}_a^T\right)\\ & +\delta\bar{\mathbf{F}}\left(\mathbf{r}',\mathbf{k}\right)\hat{H}\left(\mathbf{k},\mathbf{K}\right)\mathbf{J}_s\left(\bar{\mathbf{R}}_a^T\cdot\mathbf{K}\right)\\ & +\delta\bar{\mathbf{F}}\left(\mathbf{r}',\mathbf{k}\right)\hat{H}\left(\mathbf{k},\mathbf{K}\right)\delta\mathbf{J}_s\left(\mathbf{K};\bar{\mathbf{R}}_a^T\right).\end{aligned} \tag{9.75}$$

Again, the first term on the LHS is exactly the no-mutual-coupling prediction (9.70). The next three terms are new and represent the effect of electromagnetic coupling between the probe and the source in Figure 9.4. Their interpretation, however, follows exactly the corresponding exposition around (9.74). In particular, we notice *the physical interaction mechanism associated with (9.70), the static and dynamic genesis, remains the same in the presence of mutual coupling.*

Part III

Nonlocal Metamaterials

Chapter 10

General Introduction and Motivation

10.1 INTRODUCTION

One of the major problems facing the electromagnetic engineer and researcher nowadays, a problem which probably distinguishes applied electromagnetics in this century from the preceding one, is the enormous growth in the spatial complexity of the environments inside which antennas and circuits are intended to work. For example, the increase in urban density and the steady inflation of the number of electronic devices used per unit square meter have led recently to interest in a more refined understanding of the full complexity of radiation and interactions in the near zone in the presence of complex media. Indeed, many of the design and analysis methods in applied electromagnetics tend to assume idealized environments, e.g., free space or ground plane with few other antennas or objects nearby. While these assumptions might be relevant for an initial analysis, they become increasingly difficult to sustain when there is an interest in novel phenomena and accurate representation of the operational details of new devices.

In the electromagnetic literature, the problem of radiation of point source in the presence of complex environments has been treated mostly within the conventional perspective of periodic structures. In such formulations, the environment is electromagnetically modeled using the effective-boundary conditions of macroscopic Maxwell's equations and the fields radiated by the source are computed by solving these equations for the unknown fields. These calculations tend to be difficult but once obtained, complete knowledge of the problem becomes at hand.

A different approach adopted in this part of the book attempts to treat the problem of radiation in complex environments from the perspective of *artificial media* or *metamaterials*. Although a complex environment surrounding an antenna presents tremendous difficulties when attempting to study theoretically the main structural features of the near field produced by the source distribution, a quite natural and simple approach for mathematically describing the environment is to use a *material* representation in which the overall effect of the medium reduces to the dielectric tensor $\varepsilon_{nm}(\mathbf{r}, \mathbf{r}'; t; t')$ connecting **D** and **E**. Since the electromagnetic field is a process occurring in space and time, and the surrounding (complex) environment consists, most interestingly, of strongly interacting *spatial* distributions of boundary conditions, we allow *both* space and time dependence to appear explicitly in the response function.

This approach to modeling the electrodynamics of complex media has been known in the physics literature, especially for dealing with the classical problems of optical scattering by crystals and source-free wave propagation in such periodic structures. The works [97] and [98] in particular systematized the focus on the spatial aspects of the material response of material media using the concept of *spatial dispersion*, which we take here as synonymous to *nonlocality*, although spatial dispersion as such has been well known much earlier, see for instance [139] and the references therein. Spatial dispersion, which means the dependence of the electric and magnetic material responses on the wavevector **k**, was treated microscopically in [140], [108]. Meanwhile, the topic of nonlocality, which does not figure prominently in classic electromagnetics books like [102] and [41], found its way into well-known monographs and textbooks such as [103], [105], [141], [142], in addition to the book [108] already mentioned. However, appreciation of the full span and potentials of nonlocal media does not seem to be widespread in electromagnetic teaching and research. In any case, with the recent interest in metamaterials, the topic of spatial dispersion seems to be coming back into the picture, see for example [143], [144]. Following these lines, an engineering-motivated approach to nonlocal media and spatial dispersionwill be proposed here, where the concept of nonlocality is taken as basic and the electrodynamics of artificial materials is investigated with focus on new physical phenomena, such as distortion-free negative group velocity propagation, which are not possible in principle if only temporal dispersion is considered.

10.2 SCOPE AND MOTIVATIONS

Traditionally, the research area classed under the label 'artificial materials,' or what has become popular nowadays as 'metamaterials,' is based on the idea of mimicking the way *natural* media respond to an applied electromagnetic field. The mechanism responsible for the electromagnetic character of the medium, for example optical properties, can be applied to repeat the whole process *artificially* in the sense that the atomic constituents of matter are individually manipulated and controlled in order to achieve a desired electromagnetic profile.

Conventional approaches to describe material responses rely on assuming that the external field induces multipole electric and magnetic moments in the medium, giving rise to polarization and magnetization density vectors.[1] This approach has its merits although it is theoretically problematic. The multipole method provides an extremely simple mathematical model that is adequate for a very wide range of applications. However, with the exploding progress in nanotechnology and experimental research, it is becoming increasingly important to employ a more general mathematical formalism that allows us to explore new dimensions in the material response that go beyond the traditional multipole description.

It is the opinion of the present authors that a significant proportion of future research in the field of artificial materials should be invested in studying the *spatial* degrees of freedom latent in the medium under interest, a space hitherto unexplored in depth with few notable exceptions [97], [98], [141], [108]. The purely spatial effects, for example spatial dispersion, have been often neglected because natural materials happen to have very small interatomic-spacing-to-operating-wavelength ratio, which implies that when a macroscopic field measurement is employed, all microscopic spatial information are washed away. However, there is nothing in principle prohibiting designing artificial media with arbitrary spatial response profile.[2] Tentative proposals for controlling wave propagation by manipulating both the temporal and spatial dispersion will be outlined in Chapters 12 and Chapter 13.

10.3 ENGINEERING NONLOCAL MEDIA

The engineering and design of new artificial media is the essence of the popular field of metamaterials. The idea is to manipulate the microscopic structure in order to produce tangible effects that can be recorded macroscopically by certain effective parameters like ϵ and μ. The main focus so far has been directed at manipulating the *temporal* dispersion of the medium.[3] However, with the steady improvement in technology, new spatial scales can be probed and manipulated, leading to interesting

1 Cf. Chapter 11.

2 The implementation of a particular solution of Maxwell's equations coupled with a suitable mechanical model is a *technological* problem, not a theoretical one. In this sense, the present treatment should be viewed as a theoretical contribution to the subject.

3 Roughly speaking, temporal dispersion is captured by the functional dependence of ϵ and μ on ω.

applications that were not possible before. One of these new phenomena is the nonlocal interaction between spatially separated parts of the materials, leading to what is called spatial dispersion.[4] In this case, the electromagnetic response of the medium fails to depend only on the position where we apply the external field, but depends also on the value of this field at other locations.[5]

It was observed that taking spatial dispersion into consideration may lead to qualitatively new phenomena not seen in conventional materials obeying classical optics (spatial dispersion is ignored). In particular, spatial dispersion can allow electromagnetic wave propagation with negative group velocity to occur, even when both the permittivity and permeability are positive [97]. Such interesting behavior was originally anticipated in connection with natural materials in crystal form, where spatial dispersion is manifest, for example, in the phenomena of exciton. Recently, the same original conclusions in [97] were reinstated [143], [144]. It is still possible, however, to put the problem in a wider context by referring not only to natural crystals, but also to any type of artificial materials. To demonstrate the philosophy of the engineering approach, consider Figure 10.1 where we take the medium function to be $\epsilon(\omega, \mathbf{k})$. The physics approach is illustrated in Figure 10.1(a) where the starting stage is assuming certain models for the natural material under consideration (usually crystal). Then, Taylor series expansion of some parameters in the model (the exciton model as in [139] or the permittivity function itself as in [97]) can be applied to estimate the medium function $\epsilon(\omega, \mathbf{k})$. The next step is to apply electromagnetic theory to study the resulting propagation. However, it is possible to invert this logic in the following way. In Figure 10.1(b), we start from certain wave propagation characteristics (e.g. negative group velocity, negative-refraction propagation, etc), and then derive the medium function, $\epsilon(\omega, \mathbf{k})$, such that Maxwell's theory will allow the desired wave propagation characteristic. This is done in Chapter 12 The future step is to find experimental methods to synthesize an artificial medium with this calculated function $\epsilon(\omega, \mathbf{k})$. The parallel devolvement for the case when the field is observed close to the source is briefly outlined in Chapter 13.

4 Following the literature, we use 'spatial dispersion' and 'nonlocality' synonymously.

5 Spatial dispersion manifests itself in the functional dependence of the medium parameters on the wave vector $\mathbf{k}$. Thus, when both temporal and spatial dispersion are present, we write the permittivity and the permeability functions as $\epsilon = \epsilon(\omega, \mathbf{k})$ and $\mu = \mu(\omega, \mathbf{k})$.

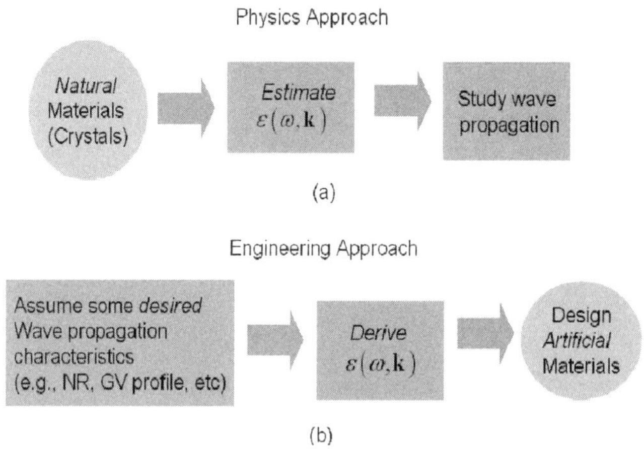

Figure 10.1 General philosophy of the study of electromagnetic wave propagation in dispersive media. (a) Physics approach. (b) Engineering approach.

10.4 THE PLAN AND GLOBAL STRUCTURE OF PART III

In Chapter 11, we review the Fourier-space formalism suitable for modeling the spatial effects of a given natural or artificial medium. The formalism is compatible with the traditional multipole approach but is conceptually easier to understand. This Fourier transform method introduced here is inspired by techniques developed in the physics community to attack plasma problems [97], [98]. There is a plethora of advantages in employing this particular point of view in this setting, the chief one being that the Fourier-space formalism is more general in its applicability to fluctuating fields with higher frequencies. Also, it naturally provides a complete characterization of the field in both space and time. Finally, being a spectral method, it allows for deeper understanding of localization phenomena and energy transfer and coupling mechanisms. One of the disadvantages is that it requires an additional mathematical background that is not usually part of the training of professional electromagnetic engineers. It also does not apply to static field problems. In general, the Fourier formalism does not conform with the conventional literature standards of notation and usage. The rest of the book can be read without the content of this chapter. It is included here for completeness.

The of core this Part III is deliberately divided into two Chapters 12 and 13 in order to highlight from the begining the need to separate between what we call "far-field metamaterials" and "near-field metamaterials." Indeed, the former has received the lion's share of the scientific community's attention during the last two or three decades, while the latter seems to be starting to emerge only very recently. One of the essential long-term goals of this book is to revitalize interest in exploring the latent potentials of artificially-engineered materials for electromagnetic applications through a clear understanding of how the spatial structure of the electromagnetic fields bifurcates into the two fundamentally different near and far zones. For this reason, Chapter 12 will concentrate on source-free propagation aspects in nonlocal media with special emphasis on the case of negative-refraction behavior. On the other hand, Chapter 13 will provide a short view of the generalization of the theory to include the presence of a source in a nonlocal media and how the inner structure of the fields close to the source can be understood this way. There is no claim here of completeness. The topic of nonlocal media is vast and still in its initial stages when it comes to the *engineering* aspects.

Chapter 11

Review of Spatial Electromagnetics (The Material Response Theory)

11.1 MAXWELL'S EQUATION

We start with the fundamental equations governing the Maxwellian fields **B** and **E** in matter-free regions. These are

$$\nabla \times \mathbf{E} = -\frac{\partial \mathbf{B}}{\partial t}, \tag{11.1}$$

$$\nabla \times \mathbf{B} = \mu_0 \mathbf{J} + \left(1/c^2\right) \frac{\partial \mathbf{E}}{\partial t}, \tag{11.2}$$

$$\nabla \cdot \mathbf{E} = \frac{\rho}{\varepsilon_0}, \tag{11.3}$$

$$\nabla \cdot \mathbf{B} = 0, \tag{11.4}$$

where c is the speed of light and $\varepsilon_0 = 8.854 \times 10^{-}12\ \mathrm{F/m}$ and $\mu_0 = 4\pi \times 10^{-}7\ \mathrm{H/m}$ are the permittivity and permeability of free space, respectively.

We notice that these equations are complete since they capture everything related to electromagnetic interactions. However, in order to solve Maxwell's equations *in the presence of matter*, one has to supply suitable decompositions of the source terms appearing in (11.2) and (11.3) in the following manner

$$\rho = \rho_{\text{ext}} + \rho_{\text{ind}} \tag{11.5}$$

and

$$\mathbf{J} = \mathbf{J}_{\text{ext}} + \mathbf{J}_{\text{ind}}, \tag{11.6}$$

where ρ_{ext} and $\mathbf{J}_{\text{ext}}$ are the imposed sources supplied externally. Matter will interact with the fields radiated by these sources and respond by generating induced sources ρ_{ind} and $\mathbf{J}_{\text{ind}}$. These induced sources cannot be deduced from Maxwell's equations themselves. They must be found upon constructing an appropriate mechanical model for matter in the radiation field.[1]

By observing the conservation of electric charge density $\rho(t, \mathbf{r})$, the equation of continuity for electromagnetism takes the following form

$$\frac{\partial \rho}{\partial t} + \nabla \cdot \mathbf{J} = 0. \tag{11.7}$$

Energy conservation is already built into the structure of Maxwell's equations in continuous media. Indeed, it is possible to directly derive the following relation

$$\frac{\partial}{\partial t}\left(\frac{1}{2}\varepsilon_0 \left|\mathbf{E}\right|^2 + \frac{1}{2}\left|\mathbf{B}\right|^2 \Big/ \mu_0\right) + \nabla \cdot \left(\frac{1}{\mu_0}\mathbf{E} \times \mathbf{B}\right) = -\mathbf{J} \cdot \mathbf{E}. \tag{11.8}$$

Let us supplement this equation with Lorentz force law

$$\mathbf{F} = q\mathbf{E} + \mathbf{v} \times \mathbf{B}. \tag{11.9}$$

One can carefully supply an interpretation for the various terms appearing on the RHS of (11.8) starting from the basic law of force (11.9). As it turns out, the time rate of the volume density of the work done by the electric current $\mathbf{J}$ on the electric field $\mathbf{E}$ is given by $-\mathbf{J} \cdot \mathbf{E}$. This provides us with an interpretation of the RHS of (11.8). Now, in order to interpret (11.8) as a continuity equation, we observe that, in vacuum, the quantities $\varepsilon_0 \left|\mathbf{E}\right|^2 \Big/ 2$ and $\left|\mathbf{B}\right|^2 \Big/ 2\mu_0$ can straightforwardly be interpreted as volume densities of electric and magnetic energies, respectively, stored in free space. It follows then that the last term, $\mathbf{E} \times \mathbf{B}/\mu_0$, can be easily interpreted to stand for the volume density of the power flow, or the electromagnetic flux.

1 It is here that the incompleteness of Maxwell's equations becomes critical. Strictly speaking, we need to augment the four Maxwell's equations listed above by additional laws prescribing how material particles are charged and how they behave in the presence of fields. Ultimately, this has to be done through quantum electrodynamics, although in the majority of applications a semi-classical approach is enough to derive results that are in conformity with experiments.

11.2 FOURIER TRANSFORM APPROACH TO THE GREEN'S FUNCTIONS

11.2.1 Maxwell's Equations in the Spectral Domain

As we are going to formulate the entire problem in terms of the Fourier transform, the usual spatio-temporal form of Maxwell's equations must be transformed into the spectral domain. In this section, we handle the problem of a source radiating in infinite isotropic and homogeneous medium. Maxwell's equations (11.1)–(11.4) can be written in the Fourier transform domain as

$$\mathbf{k} \times \mathbf{E}(\omega, \mathbf{k}) = \omega \mathbf{B}(\omega, \mathbf{k}), \tag{11.10}$$

$$i\mathbf{k} \times \mathbf{B}(\omega, \mathbf{k}) = -i\omega \mathbf{E}(\omega, \mathbf{k})/c^2 + \mu_0 \mathbf{J}(\omega, \mathbf{k}), \tag{11.11}$$

$$\mathbf{k} \cdot \mathbf{E}(\omega, \mathbf{k}) = -i\rho(\omega, \mathbf{k})/\varepsilon_0, \tag{11.12}$$

$$\mathbf{k} \cdot \mathbf{B}(\omega, \mathbf{k}) = 0. \tag{11.13}$$

The equation of continuity (11.7) can be also Fourier transformed into the form

$$\omega\rho(\omega, \mathbf{k}) = \mathbf{k} \cdot \mathbf{J}(\omega, \mathbf{k}). \tag{11.14}$$

The reader must note that these equations cannot be used to describe static fields, which may be tackled on their own by applying the Coulomb gauge. Therefore, throughout this chapter, we restrict ourselves to the case $\omega \neq 0$.

The program of performing calculations in electromagnetism using the Fourier transform method can be elucidated in the following manner

1. Express the magnetic field in terms of the electric field using Maxwell's equation (11.10)
$$\mathbf{B}(\omega, \mathbf{k}) = \mathbf{k} \times \mathbf{E}(\omega, \mathbf{k})/\omega. \tag{11.15}$$
2. Express the charge density in terms of the current density using the equation of continuity (11.14)
$$\rho(\omega, \mathbf{k}) = \frac{1}{\omega}\mathbf{k} \cdot \mathbf{J}(\omega, \mathbf{k}). \tag{11.16}$$
3. End up with a single equation in one unknown, $\mathbf{E}(\omega, \mathbf{k})$, and forcing term $\mathbf{J}(\omega, k)$; i.e., solve
$$\frac{\omega^2}{c^2}\mathbf{E}(\omega, \mathbf{k}) + \mathbf{k} \times [\mathbf{k} \times \mathbf{E}(\omega, \mathbf{k})] = -i\omega\mu_0 \mathbf{J}(\omega, \mathbf{k}). \tag{11.17}$$

Therefore, the program of solving Maxwell's equations reduces to solving a single algebraic equation in terms of the electric field $\mathbf{E}(\omega, \mathbf{k})$. All the other field and source components can be obtained from the solution of the electric field together with the given form of the source.

11.2.2 The Green's Function Tensor in the Spectral Domain

To obtain the Green's function in the Fourier domain, we first put (11.17) in a suitable form. We use the following identity

$$\mathbf{A} \times \mathbf{B} = \epsilon_{ijk} A_j B_k, \tag{11.18}$$

where ϵ_{ijk} is the permutation tensor.[2] Therefore, we have

$$\mathbf{k} \times \mathbf{E} = \epsilon_{ijk} k_j E_k. \tag{11.19}$$

Iterating, we obtain

$$\mathbf{k} \times \mathbf{k} \times \mathbf{E} = \epsilon_{ijk} k_j \epsilon_{kj'k'} k_{j'} E_{k'} = \epsilon_{ijk} \epsilon_{kj'k'} k_j k_{j'} E_{k'}. \tag{11.20}$$

We use the following basic identity

$$\begin{aligned} \epsilon_{abc}\epsilon_{ijk} = {} & \delta_{ai}\delta_{bj}\delta_{ck} + \delta_{ak}\delta_{bi}\delta_{cj} + \delta_{aj}\delta_{bk}\delta_{ci} \\ & - \delta_{bi}\delta_{aj}\delta_{ck} - \delta_{ak}\delta_{ai}\delta_{cj} - \delta_{bj}\delta_{ak}\delta_{ci}. \end{aligned} \tag{11.21}$$

Therefore, we have

$$\epsilon_{iab}\epsilon_{ijk} = \delta_{aj}\delta_{bk} - \delta_{ak}\delta_{bj}. \tag{11.22}$$

Using this identity in (11.17), we arrive at

$$\left[\left(\frac{\omega^2}{c^2} - k^2 \right) \delta_{nm} + k_n k_m \right] E_m(\omega, \mathbf{k}) = -i\omega\mu_0 J_n(\omega, \mathbf{k}). \tag{11.23}$$

The Green's function tensor is defined to satisfy the following equation

$$\left[\left(\frac{\omega^2}{c^2} - k^2 \right) \delta_{nm} + k_n k_m \right] G_{ml}(\omega, \mathbf{k}) = -i\omega\mu_0 \delta_{nl}(\omega, \mathbf{k}). \tag{11.24}$$

2 Throughout this chapter, the Einstein (repeated) summation index is used. That is, whenever an index is repeated in a given expression, summation is implied with respect to these indices.

Therefore, by inverting the matrix operator appearing in the equation above, the Green's function tensor is readily obtained in the following compact closed form

$$G_{nm}(\omega, \mathbf{k}) = \frac{-i\omega\mu_0}{\omega^2/c^2 - k^2}\left(\delta_{nm} - \frac{c^2}{\omega^2}k_n k_m\right) \tag{11.25}$$

Finally, we observe that it is possible to separate the field into two components, one transverse to the direction of the wave vector **k** (transverse mode), and another perpendicular to this direction, which we call longitudinal mode. The longitudinal mode is not involved in the radiation and is related to the near field. It contributes directly to the structure of the field surrounding the source.

11.3 REVIEW OF THE TRADITIONAL DESCRIPTION OF ELECTROMAGNETIC MATERIALS IN TERMS OF MULTIPOLE MOMENTS

The conventional old description of electromagnetic materials involves the introduction of two quantities to calculate the induced charge and current distributions. We review here the traditional view and show how it can be derived by a Fourier transform approach to the multipole expansion of the source.

The conventional idea is to assume that a given medium responds to both electric and magnetic fields by generating an induced *polarization* density **P** and *magnetization* **M**. However, this description is strictly valid when both the electric and magnetic responses can be unambiguously separated from each other. This is possible only when the fields are static; otherwise, it should be viewed as an approximation. Indeed, if rapid field fluctuations at the microscopic scale are taken into consideration, then the separation becomes ill-defined and even potentially problematic.

Let us see how **P** and **M** arise from the Fourier transform perspective. Consider an arbitrary charge and current distribution

$$\rho(t, \mathbf{k}) = \int d^3r e^{-i\mathbf{k}\cdot\mathbf{r}} \rho(t, \mathbf{r}), \tag{11.26}$$

$$\mathbf{J}(t, \mathbf{k}) = \int d^3r e^{-i\mathbf{k}\cdot\mathbf{r}} \mathbf{J}(t, \mathbf{r}). \tag{11.27}$$

Expand the exponential in Taylor series

$$e^{-i\mathbf{k}\cdot\mathbf{r}} = 1 - i\mathbf{k}\cdot\mathbf{r} + \frac{1}{2}(i\mathbf{k}\cdot\mathbf{r})^2 + \cdots. \tag{11.28}$$

Inserting (11.28) into (11.26), we obtain

$$\begin{aligned}
\rho(t,\mathbf{k}) &= \int d^3r\rho(t,\mathbf{r})\left[1 - i\mathbf{k}\cdot\mathbf{r} + \tfrac{1}{2}(i\mathbf{k}\cdot\mathbf{r})^2 + \cdots\right] \\
&= \int d^3r\rho(t,\mathbf{r}) - \int d^3ri\mathbf{k}\cdot\mathbf{r}\rho(t,\mathbf{r}) + \tfrac{1}{2}\int d^3r(i\mathbf{k}\cdot\mathbf{r})^2\rho(t,\mathbf{r}) + \cdots \\
&= -\int d^3rik_nr_n\rho(t,\mathbf{r}) - \tfrac{1}{2}\int d^3rk_nr_nk_mr_m\rho(t,\mathbf{r}) + \cdots \\
&= -ik_n\int d^3rr_n\rho(t,\mathbf{r}) - \tfrac{1}{2}k_nk_m\int d^3r\, r_nr_m\rho(t,\mathbf{r}) + \cdots \\
&= -i\mathbf{k}\cdot\mathbf{p}(t) - \tfrac{1}{2}k_nk_mq_{nm}(t) + \cdots,
\end{aligned} \tag{11.29}$$

where

$$p_n(t) = \int d^3rr_n\rho(t) \tag{11.30}$$

and

$$q_{nm}(t) = \int d^3r\, r_nr_m\rho(t) \tag{11.31}$$

are the dipole and quadrable moments, respectively. We also used the assumption that the charge distribution is neutral; $\int d^3r\rho(t,\mathbf{r}) = 0$. Similarly, by inserting (11.28) into (11.27), one obtains

$$\begin{aligned}
J_n(t,\mathbf{k}) &= \int d^3rJ_n(t,\mathbf{r})\left[1 - i\mathbf{k}\cdot\mathbf{r} + \tfrac{1}{2}(i\mathbf{k}\cdot\mathbf{r})^2 + \cdots\right] \\
&= \int d^3rJ_n(t,\mathbf{r}) - \int d^3r(i\mathbf{k}\cdot\mathbf{r})J_n(t,\mathbf{r}) + \tfrac{1}{2}\int d^3r(i\mathbf{k}\cdot\mathbf{r})^2J_n(t,\mathbf{r}) + \cdots \\
&= \underbrace{\int d^3rJ_n(t,\mathbf{r})}_{\mu_n(t)} - ik_m\underbrace{\int d^3r\, r_mJ_n(t,\mathbf{r})}_{\mu_{mn}(t)} - \tfrac{1}{2}k_lk_m\int d^3r\, r_lr_mJ_n(t,\mathbf{r}) + \cdots \\
&= \tfrac{\partial}{\partial t}p_n(t) - ik_m\tfrac{1}{2}\tfrac{\partial}{\partial t}q_{mn}(t) - i\epsilon_{mns}k_mm_s(t) + \cdots \\
&= \tfrac{\partial}{\partial t}p_n(t) - ik_m\tfrac{1}{2}\tfrac{\partial}{\partial t}q_{mn}(t) + i\epsilon_{nms}k_mm_s(t) + \cdots,
\end{aligned} \tag{11.32}$$

where equations (11.107) and (11.117) (see Appendix below) were utilized in obtaining the fourth equality, and the relation $\epsilon_{nms} = -\epsilon_{mns}$ is employed in the writing last equality. Also notice that the magnetization $\mathbf{m}$ is defined by (11.115).

By ignoring all quadrable and higher terms in (11.29) and (11.32), we find

$$\rho(t,\mathbf{k}) = -i\mathbf{k}\cdot\mathbf{p}(t), \tag{11.33}$$

$$\mathbf{J}(t,\mathbf{k}) = \frac{\partial}{\partial t}\mathbf{p}(t) + i\mathbf{k}\times\mathbf{m}(t). \tag{11.34}$$

Define the polarization and magnetization densities **P** and **M**, respectively, by the following relations

$$\mathbf{p}(t) = \int d^3r\mathbf{P}(t,\mathbf{r}) \tag{11.35}$$

and

$$\mathbf{m}(t) = \int d^3r\mathbf{M}(t,\mathbf{r}). \tag{11.36}$$

Inserting (11.33) and (11.34) into (11.26) and (11.27), it follows

$$\rho(t,\mathbf{k}) = -i\mathbf{k}\cdot\mathbf{P}(t,\mathbf{r}), \tag{11.37}$$

$$\mathbf{J}(t,\mathbf{k}) = \frac{\partial}{\partial t}\mathbf{P}(t,\mathbf{r}) + i\mathbf{k}\times\mathbf{M}(t,\mathbf{r}). \tag{11.38}$$

Therefore, by inverting the Fourier transforms (11.37) and (11.38), we obtain

$$\rho_{\text{ind}}(t,\mathbf{r}) = -\nabla\cdot\mathbf{P}(t,\mathbf{r}), \tag{11.39}$$

$$\mathbf{J}_{\text{ind}}(t,\mathbf{r}) = \frac{\partial}{\partial t}\mathbf{P}(t,\mathbf{r}) + \nabla\times\mathbf{M}(t,\mathbf{r}). \tag{11.40}$$

As can be seen now, this derivation ignores higher-order multipole without providing a clear-cut criterion for when and why this approximation is valid. Since we are attempting to construct a general theory for *both* near and far fields in the context of material response, it is important to employ a formulation that does not involve approximations that may not hold in certain media. Some other difficulties relate to the question of the convergence of the multipole expansion, which are seldom addressed in literature. Finally, there is the incompleteness issue in the expansion (11.28), which includes only terms with zero trace.

11.4 MATERIAL RESPONSE THROUGH THE FOURIER TRANSFORM APPROACH

We will now carefully introduce the equivalent representation of the electromagnetic material response in terms of the Fourier transform of the fields, not the actual field in space and time. There are several advantages in this approach. First, note that this approach does not apply to static fields, which are better addressed by the classical **P**-**M** approach. On other hand, certain complex electromagnetic effects, like spatial dispersion, magnetoelectric responses and optical activity, can be regarded as special cases of nonlocality.

It appears to the authors that operating directly on material systems with a formalism tailored especially to handle spatial dispersion is very advantageous. Besides its ability to deal with complex media exhibiting phenomena like magnetoelectric effects and optical activity, it can also provide a natural window to probe near-field interactions. Although we are still trying to mathematically identify the meaning of the near field, remember that one of the most immediate features that come to mind when thinking about fields in the near zone (close to the radiator or the scatterer) is that they tend to be localized, or, equivalently, contain short wavelength components that contribute significantly to the field structure. In this case, one is looking naturally for a mathematical device that characterizes electromagnetic wave phenomena in terms of the Fourier spatial modes, i.e., the **k**-components. Therefore, the formalism should look for information about the response of the system to particular wavevecotrs **k**. This is essentially the goal of integrating spatial dispersion into the theoretical description of material media.

Let us try to address in more detail some of the difficulties in the traditional approach to electromagnetic material response. By Fourier transforming (11.40) in time, we obtain

$$\mathbf{J}_{\text{ind}}(\omega,\mathbf{k}) = -i\omega\mathbf{P}(\omega,\mathbf{k}) + i\mathbf{k}\times\mathbf{M}(\omega,\mathbf{k})\,. \tag{11.41}$$

The problem here is that there exists no general *a priori* method to tell how the individual contributions of the quantities **P** and **M** divide between themselves in forming the total induced current. In this sense, one should view these two vectors as mere calculational tools, auxiliary devices used to compute the actually observed induced current $\mathbf{J}_{\text{ind}}$. In particular, there seems to be no harm in just setting the magnetization density **M** to zero and considering only a polarization density **P** contributing to the induced charge and current densities.

As we have just observed in Section 11.2.1, the program of calculating the fields through Maxwell's equations can be reduced to the solution of a single equation, namely (11.17), which contains a single unknown, the electric field vector $\mathbf{E}(\omega,\mathbf{k})$ itself. If the relation between the induced current density $\mathbf{J}_{\text{ind}}(\omega,\mathbf{k})$ and the electric field is known, then this relation, together with the master (11.17), can be used to completely solve the problem of light-matter interaction. It seems natural then to introduce a *single* material response tensor

$$(J_{\text{ind}})_m(\omega,\mathbf{k}) = \sigma_{mn}(\omega,\mathbf{k})\,E_n(\omega,\mathbf{k})\,, \tag{11.42}$$

where the matrix $\sigma_{mn}(\omega, \mathbf{k})$ is called the *conductivity tensor*. After solving for the electric field, all the remaining quantities, the magnetic field $\mathbf{B}(t, \mathbf{r})$ and the charge density $\rho(t, \mathbf{r})$, can be calculated from knowledge of the total current and the electric field.

One can replace the conductivity tensor by different equivalent representations that may turn out to be handy in some applications. In particular, we discuss here the polarization tensor $\alpha_{nm}(\omega, \mathbf{k})$ and the equivalent dielectric constant $\varepsilon_{nm}^{\mathrm{eq}}(\omega, \mathbf{k})$, defined by the following equations

$$\begin{aligned}\varepsilon_{nm}^{\mathrm{eq}}(\omega, \mathbf{k}) &= \delta_{nm} + \tfrac{i}{\omega\varepsilon_0}\sigma_{nm}(\omega, \mathbf{k}) \\ &= \delta_{nm} + \tfrac{1}{\omega^2\varepsilon_0}\alpha_{nm}(\omega, \mathbf{k}) \\ &= \delta_{nm} + \chi_{nm}(\omega, \mathbf{k}).\end{aligned} \tag{11.43}$$

The reader should notice that the equivalent dielectric function $\varepsilon_0\varepsilon_{nm}^{\mathrm{eq}}(\omega, \mathbf{k})$ is *not* the same as the conventional dielectric function defined in terms of the polarization and magnetization densities appearing in (11.40). In terms of the new dielectric function $\varepsilon_{nm}^{\mathrm{eq}}(\omega, \mathbf{k})$, we write

$$D_n(\omega, \mathbf{k}) = \sum_m \varepsilon_0\varepsilon_{nm}^{\mathrm{eq}}(\omega, \mathbf{k})\, E_m(\omega, \mathbf{k}). \tag{11.44}$$

It follows that in the Fourier transform approach to the material response, we effectively kill the magnetization vector $\mathbf{M}$ and collect all relevant physical processes into a single vector, the effective polarization density $\mathbf{P}$.

11.5 COMPARISON BETWEEN THE TRADITIONAL MULTIPOLE AND THE FOURIER TRANSFORM APPROACH TO THE MATERIAL RESPONSE

Within the multipole approach to the material response, two new fields are traditionally introduced, the electric induction $\mathbf{D}$ (the electric displacement vector), and the magnetic field strength $\mathbf{H}$. These are defined by the relations

$$\mathbf{D} := \varepsilon_0\mathbf{E} + \mathbf{P}, \tag{11.45}$$

$$\mathbf{H} := \frac{1}{\mu_0}\mathbf{B} - \mathbf{M}. \tag{11.46}$$

The electric susceptibility χ^e and the magnetic susceptibility χ^m are defined by the following equations

$$\mathbf{P} = \varepsilon_0 \chi^e \mathbf{E}, \tag{11.47}$$

$$\mathbf{M} = \frac{1}{\mu_0} \chi^m \mathbf{B}. \tag{11.48}$$

The effective dielectric constant, or electric permittivity ε, and the magnetic permeability μ, can now be defined in terms of the quantities above as

$$\mathbf{D}(\omega, \mathbf{k}) = \varepsilon \mathbf{E}(\omega, \mathbf{k}), \tag{11.49}$$

$$\mathbf{H}(\omega, \mathbf{k}) = \frac{1}{\mu} \mathbf{B}(\omega, \mathbf{k}). \tag{11.50}$$

We now proceed to derive the equivalence between this traditional approach and the Fourier formalism of Section 11.4. First, the current distribution is decomposed into two parts, one due to external (applied) sources, $\mathbf{J}_{\text{ext}}$, and the other, $\mathbf{J}_{\text{ind}}$ due to the interaction between the medium and the electromagnetic fields. We write

$$\mathbf{J}(t, \mathbf{r}) = \mathbf{J}_{\text{ext}}(t, \mathbf{r}) + \mathbf{J}_{\text{ind}}(t, \mathbf{r}). \tag{11.51}$$

The induced current is written using the conductivity tensor introduced in (11.42) and the result is substituted to the master equation (11.17). After simple re-arranging of terms, we find

$$\frac{\omega^2}{c^2} \mathbf{E}(\omega, \mathbf{k}) + \mathbf{k} \times \mathbf{k} \times \mathbf{E}(\omega, \mathbf{k}) + i\omega\mu_0 \bar{\bar{\sigma}}(\omega, \mathbf{k}) \cdot \mathbf{E}(\omega, \mathbf{k}) = -i\omega\mu_0 \mathbf{J}_{\text{ext}}(\omega, \mathbf{k}). \tag{11.52}$$

Now let us calculate by means of the ϵ-μ method. In this case, the induced current is written in terms of both the polarization and magnetization current densities $\mathbf{P}$ and $\mathbf{M}$ as shown in (11.40). Using (11.47) and (11.48) in (11.41), we find

$$\mathbf{J}_{\text{ind}}(\omega, \mathbf{k}) = -i\omega\varepsilon_0 \chi^e \mathbf{E}(\omega, \mathbf{k}) + i\mathbf{k} \times \frac{\chi^m}{\mu_0} \mathbf{B}(\omega, \mathbf{k}). \tag{11.53}$$

But from Maxwell's equations in the Fourier domain, specifically (11.10), we know that

$$i\mathbf{k} \times \mathbf{B}(\omega, \mathbf{k}) = \frac{i}{\omega} \mathbf{k} \times \mathbf{k} \times \mathbf{E}(\omega, \mathbf{k}). \tag{11.54}$$

The induced current in (11.53) becomes then

$$\mathbf{J}_{\text{ind}}(\omega, \mathbf{k}) = -i\omega\varepsilon_0 \chi^e \mathbf{E}(\omega, \mathbf{k}) + i\frac{\chi^m}{\omega\mu_0} \mathbf{k} \times \mathbf{k} \times \mathbf{E}(\omega, \mathbf{k}). \tag{11.55}$$

Combining (11.51) and (11.55) and substituting the result into (11.17), we arrive after some rearranging to

$$\begin{aligned}&\tfrac{\omega^2}{c^2}\mathbf{E}(\omega,\mathbf{k})+\mathbf{k}\times\mathbf{k}\times\mathbf{E}(\omega,\mathbf{k})\\&\qquad+\tfrac{\omega^2}{c^2}\chi^e\mathbf{E}(\omega,\mathbf{k})-\chi^m\mathbf{k}\times\mathbf{k}\times\mathbf{E}(\omega,\mathbf{k})=-i\omega\mu_0\mathbf{J}_{\mathrm{ext}}(\omega,\mathbf{k})\end{aligned}\tag{11.56}$$

By comparing (11.52) and (11.56), we conclude that we must have

$$i\omega\mu_0\bar{\bar{\sigma}}(\omega,\mathbf{k})\cdot\mathbf{E}(\omega,\mathbf{k})=\frac{\omega^2}{c^2}\chi^e\mathbf{E}(\omega,\mathbf{k})-\chi^m\mathbf{k}\times\mathbf{k}\times\mathbf{E}(\omega,\mathbf{k}).\tag{11.57}$$

In tensor form, (11.57) becomes

$$i\omega\mu_0\sigma_{nl}(\omega,\mathbf{k})E_l(\omega,\mathbf{k})=\frac{\omega^2}{c^2}\chi^e E_n(\omega,\mathbf{k})-\chi^m\left[k_nk_l-k^2\delta_{nl}\right]E_l(\omega,\mathbf{k}).\tag{11.58}$$

Since the equality holds for arbitrary E_l, we obtain

$$\sigma_{nl}(\omega,\mathbf{k})=\frac{1}{i\omega\mu_0}\left\{\frac{\omega^2}{c^2}\chi^e\delta_{nl}-\chi^m\left[k_nk_l-k^2\delta_{nl}\right]\right\}.\tag{11.59}$$

From (11.43), we reach to

$$\begin{aligned}\varepsilon_{nl}^{\mathrm{eq}}(\omega,\mathbf{k})&=\delta_{nl}+\tfrac{i}{\omega\varepsilon_0}\tfrac{1}{i\omega\mu_0}\left\{\tfrac{\omega^2}{c^2}\chi^e\delta_{nl}-\chi^m\left[k_nk_l-k^2\delta_{nl}\right]\right\}\\&=\delta_{nl}+\chi^e\delta_{nl}-\tfrac{c^2}{\omega^2}\chi^m\left(k_nk_l-k^2\delta_{nl}\right)\\&=(1+\chi^e)\delta_{nl}-\tfrac{c^2}{\omega^2}\chi^m\left(k_nk_l-k^2\delta_{nl}\right).\end{aligned}\tag{11.60}$$

Finally, we use the definitions (11.45), (11.46), (11.47), (11.48) to write

$$\varepsilon_{nm}^{\mathrm{eq}}(\omega,\mathbf{k})=\left(\frac{\varepsilon}{\varepsilon_0}\right)\delta_{nm}-\frac{c^2}{\omega^2}\left(1-\frac{\mu_0}{\mu}\right)\left(k_nk_m-k^2\delta_{nm}\right).\tag{11.61}$$

This is the main equation we are looking for. It shows that a medium which is magnetic in the ε-μ approach translates into spatial dispersion in the Fourier approach. It follows also that the two dielectric functions are the same only if there is no spatial dispersion.

11.6 GENERAL PROPERTIES OF THE MATERIAL RESPONSE TENSORS

The requirement that the electromagnetic fields should by themselves satisfy Maxwell's equations cannot fully specify how the very same fields will behave in a material environment. Such behavior is dictated by a more complex structure consisting of the mechanical response of the microscopic constituents coupled with the electromagnetic fields. In this section, we survey and present rigorously the most important *non*-electromagnetic restrictions imposed on the material tensor. Such restrictions can be conveniently gathered under the heading 'General properties of the Material Tensor' since they involve quite broad characteristics that are wider than the particular dynamical laws encapsulated by the Maxwell's equations.

Our main equations will be the relation between the electric flux density vector and the electric field in both the spatio-temporal and spectral domain. These are, respectively,

$$D_n(\omega, \mathbf{k}) = \sum_m \varepsilon_0 \varepsilon^{\mathrm{eq}}_{nm}(\omega, \mathbf{k}) E_m(\omega, \mathbf{k}), \tag{11.62}$$

$$D_n(t, \mathbf{r}) = \varepsilon_0 \int dt' \int d^3 r' \sum_m \varepsilon^{\mathrm{eq}}_{nm}(t - t', \mathbf{r} - \mathbf{r}') E_m(t', \mathbf{r}'). \tag{11.63}$$

These equations describe electromagnetic processes in homogeneous, isotropic or anisotropic media. It is important to keep in mind that within the Fourier-space formalism the equivalent dielectric tensor is *inherently* a tensor; even when the medium under consideration is isotropic, the dielectric function is still generally a tensor. Also, the reader may notice from (13.2) that the field induced at particular time t and location $\mathbf{r}$ depends generally on the applied field at *different* times and locations. We say that the medium exhibit "memory" in both the temporal and spatial sense. The spatial sense of the this memory, which is going to be the main concern for us here, is called *nonlocality*.[3]

11.6.1 The Reality of the Fields

Since the fields appearing in (13.2) are all real, the properties of the Fourier transform dictate that the negative and positive frequencies appearing in the spectrum of

3 Whenever there is no risk of confusion, we drop the superscript 'eq' from $\varepsilon^{\mathrm{eq}}_{nm}(\omega, \mathbf{k})$ and refer to the equivalent dielectric function as merely the *dielectric tensor*.

the fields are both essentially equivalent to each other. Formally, we have

$$\varepsilon_{nm}^{*}(\omega,\mathbf{k}) = \varepsilon_{nm}(-\omega,-\mathbf{k}). \tag{11.64}$$

We express in this requirement a relation that any material tensor describing the responses of the medium to real quantities must satisfy.

11.6.2 Dissipative and Non-Dissipative Processes

The material tensorial response is the Fourier transform of a real quantity and hence generally complex. The real part and the imaginary part of this tensor are usually interpreted as those responsible for dispersion and losses (dissipation), respectively. In this section, we provide the mathematical evidence in support of this interpretation.

We start by decomposing an arbitrary response tensor into hermitian and antihermitian parts

$$\varepsilon_{nm}(\omega,\mathbf{k}) = \varepsilon_{nm}^{\mathrm{H}}(\omega,\mathbf{k}) + \varepsilon_{nm}^{\mathrm{A}}(\omega,\mathbf{k}), \tag{11.65}$$

where

$$\varepsilon_{nm}^{\mathrm{H}}(\omega,\mathbf{k}) = \frac{1}{2}\left[\varepsilon_{nm}(\omega,\mathbf{k}) + \varepsilon_{mn}^{*}(\omega,\mathbf{k})\right], \tag{11.66}$$

$$\varepsilon_{nm}^{\mathrm{A}}(\omega,\mathbf{k}) = \frac{1}{2}\left[\varepsilon_{nm}(\omega,\mathbf{k}) - \varepsilon_{mn}^{*}(\omega,\mathbf{k})\right]. \tag{11.67}$$

It is obvious that the two parts satisfy

$$\varepsilon_{nm}^{\mathrm{H}*}(\omega,\mathbf{k}) = \varepsilon_{mn}^{\mathrm{H}*}(\omega,\mathbf{k}), \tag{11.68}$$

$$\varepsilon_{nm}^{\mathrm{A}*}(\omega,\mathbf{k}) = -\varepsilon_{mn}^{\mathrm{A}*}(\omega,\mathbf{k}). \tag{11.69}$$

We now recall our interpretation in Section 11.1 of the term $-\mathbf{J}\cdot\mathbf{E}$ as the density of the rate of energy transfer by the current $\mathbf{J}$ into the electric field $\mathbf{E}$. The current can be decomposed into external and induced parts as $\mathbf{J} = \mathbf{J}_{\mathrm{ex}}+\mathbf{J}_{\mathrm{ind}}$. Thus, the total work done by the *medium* on the electric field is obtained by integrating $-\mathbf{J}_{\mathrm{ind}}\cdot\mathbf{E}$ in both time and space as

$$-\int dt\int d^3r\mathbf{J}_{\mathrm{ind}}(t,\mathbf{r})\cdot\mathbf{E}(t,\mathbf{r}) = -\int\frac{d\omega d^3k}{(2\pi)^4}\mathbf{J}_{\mathrm{ind}}(\omega,\mathbf{k})\cdot\mathbf{E}^{*}(\omega,\mathbf{k}), \tag{11.70}$$

where the power theorem of Fourier transforms was used in writing the equality. We now have

$$\begin{aligned}&\int \tfrac{d\omega d^3k}{(2\pi)^4}\mathbf{J}_{\text{ind}}(\omega,\mathbf{k})\cdot\mathbf{E}^*(\omega,\mathbf{k})\\&\quad=\int \tfrac{d\omega d^3k}{(2\pi)^4}\tfrac{1}{2}\left[\mathbf{J}^*_{\text{ind}}(\omega,\mathbf{k})\cdot\mathbf{E}(\omega,\mathbf{k})+\mathbf{J}_{\text{ind}}(\omega,\mathbf{k})\cdot\mathbf{E}^*(\omega,\mathbf{k})\right].\end{aligned}\tag{11.71}$$

In deriving this, the integral was first divided into its negative and positive frequency parts, and then a transformation of variables was applied to the negative frequencies integral. Finally, the symmetry condition (reality condition) given in (11.64) was applied. Employing (11.42) in (11.71), we can write

$$\begin{aligned}&-\int \frac{d\omega d^3k}{(2\pi)^4}\mathbf{J}_{\text{ind}}(\omega,\mathbf{k})\cdot\mathbf{E}^*(\omega,\mathbf{k})\\&\quad=\int \frac{d\omega d^3k}{(2\pi)^4}\frac{1}{2}[\sigma_{nm}(\omega,\mathbf{k})E_m(\omega,\mathbf{k})E_n^*(\omega,\mathbf{k})\\&\qquad+\sigma^*_{mn}(\omega,\mathbf{k})E_n^*(\omega,\mathbf{k})E_m(\omega,\mathbf{k})]\\&\quad=\int \frac{d\omega d^3k}{(2\pi)^4}\sigma^H_{nm}(\omega,\mathbf{k})\,E_m(\omega,\mathbf{k})\,E_n^*(\omega,\mathbf{k}).\end{aligned}\tag{11.72}$$

Therefore, it is the hermitian part of the conductivity tensor which contributes to the dissipation of energy by the medium. Equivalently, by considering the relation between the conductivity and the equivalent dielectric tensor (11.43), we find that it is the antihermitian part of the dielectric tensor that contributes to energy dissipation by the medium.

11.6.3 Onsager Relations

Since any material responses tensor is ultimately based on a mechanical model, of which the dynamical equations must satisfy certain space-time symmetry transformations, there exist quite general restrictions on the mathematical form of a physically realizable material tensor. In order to give the reader some idea about such requirement, we list the classical dynamical equation for the particle motion, namely the Lorentz force (11.9). By writing the force as $\mathbf{F} = d\mathbf{p}/dt$, where $\mathbf{p}$ is the linear momentum, it is an easy matter to verify that the equation of motion is invariant under the transformations

$$t\to -t,\quad \mathbf{p}\to -\mathbf{p},\quad \mathbf{B}\to -\mathbf{B}\tag{11.73}$$

The same conclusion can be obtained if the Lorentz force law is replaced by the Schrodinger equation.

Notice that a time-reversal corresponds to the substitution $\omega \rightarrow -\omega$ in the Fourier domain. The reversal of the sign of the momentum corresponds to reversing the sign of the wavevector. Finally, the reversal of the sign of the magnetic field is shown explicitly in the following standard form of the Onsager relations[4]

$$\varepsilon_{nm}^{\mathrm{eq}}(\omega,-\mathbf{k})\big|_{-\mathbf{B}} = \varepsilon_{mn}^{\mathrm{eq}}(\omega,\mathbf{k})\big|_{\mathbf{B}}\,. \tag{11.74}$$

The Onsager relations places severe restrictions on the physically allowable form of the material response. We discuss below particular examples of isotropic spatially dispersive media.

Let us focus on materials that don't respond to the magnetic field. In this case, the Onsager relations reduce to the situation in which the tensorial response is required to be invariant under the transformation

$$\mathbf{k} \longrightarrow -\mathbf{k}, \quad n \longleftrightarrow m. \tag{11.75}$$

First, notice that in the Fourier transform approach, even when the medium is isotropic, the response is still described by a tensorial quantity, c.f. (11.61). For isotropic media that are spatially dispersive, we can analyze the situation by pure matrix-theoretic arguments. Indeed, the only available vector in this case is k_m, while the only available tensors are δ_{nm} and ϵ_{nml}. It can be shown that the Onsager relations leads to the result that we can construct only three independent second-rank tensors. A popular choice in the condensed-matter physics literature is the following

$$\varepsilon_{nm}^{\mathrm{eq}}(\omega,\mathbf{k}) = \varepsilon^{L}(\omega,k)\,\kappa_n\kappa_m + \varepsilon^{T}(\omega,k)\,(\delta_{nm}-\kappa_n\kappa_m) + i\varepsilon^{R}(\omega,k)\,\epsilon_{nml}\kappa_l, \tag{11.76}$$

where

$$\kappa_m = k_m/k, \quad k = |\mathbf{k}|\,. \tag{11.77}$$

Here, the quantities $\varepsilon^{L}(\omega,k), \varepsilon^{T}(\omega,k), \varepsilon^{R}(\omega,k)$ are the longitudinal, transverse, and rotational permittivities, respectively. The rotatory parts can be ignored in media that don't exhibit optical activity. Notice that for media in which both the longitudinal and transverse parts happen to be equal to each other, the equivalent dielectric tensor reduces to the scalar case.

11.6.4 The Kramers–Kronig Relations

The fact that the dielectric tensor is a response function imposes a restriction on the relationship between the real and imaginary part. This restriction is due to causality

4 The symmetry relations (11.64) are used to simplify the final form.

and can be rigorously derived by standard techniques in the theory of complex functions.[5] Kramers-Kronig relations say that the real and imaginary parts of the Fourier transform of a function that is causal (i.e., a function with inverse Fourier transform identically zero for a time interval in the form $-\infty < t < t_0$) satisfy

$$\varepsilon_{nm}^{\mathrm{eq,H}}(\omega,\mathbf{k}) - \delta_{nm} = \frac{i}{\pi}\wp\int_{-\infty}^{\infty} d\omega' \frac{\varepsilon_{nm}^{\mathrm{eq,A}}(\omega',\mathbf{k}) - \delta_{nm}}{\omega - \omega'}, \tag{11.78}$$

$$\varepsilon_{nm}^{\mathrm{eq,A}}(\omega,\mathbf{k}) = \frac{i}{\pi}\wp\int_{-\infty}^{\infty} d\omega' \frac{\varepsilon_{nm}^{\mathrm{eq,H}}(\omega',\mathbf{k})}{\omega - \omega'}, \tag{11.79}$$

where $\wp$ symbolizes the Cauchy principal value.[6] Equations (11.78) and (11.79) show that if dissipation is known, then dispersion can be uniquely determined (and vice versa) by applying the Hilbert transform operator to the available data.

One can see that when spatial dispersion is present, then in the case of non-dissipative medium, i.e., a medium with negligible losses which, as can be seen from Section 11.6.2, corresponds to $\varepsilon_{nm}^{\mathrm{eq,A}}(\omega,\mathbf{k}) = 0$, the dispersion behavior dictated by $\varepsilon_{nm}^{\mathrm{eq,H}}(\omega,\mathbf{k})$ is restricted to only the class of functions of ω which have zero Hilbert transform. It can be shown that such functions take the basic form $1/(\omega - \omega_m)$ with constant ω_m. This explains partially why such basic form pops out very frequently in practice. Moreover, they also demonstrate the power of Kramers-Kronig relations in being able to severely restrict the allowable functional form of dispersion in lossless media.

The general lesson we learn from taking causality into consideration when thinking about designing artificial media is that once losses are neglected for the entire frequency range $-\infty < \omega < \infty$ (or the medium is designed to have small losses globally), the *global* form of dispersion is no more a free degree of freedom but, instead, takes a particular form. However, in practice we seldom achieve or require particular specifications of the losses and/or dispersion to hold for the entire frequency range. Noticing that the Hilbert transform relations in (11.78) and (11.79) are global operators, i.e., they involve integration over the entire frequency range in order to know the value at a single frequency (nonlocal or memory-dependent in

5 The causality restriction translates formally to the following setting. Imagine that the medium is excited by an applied electric field **E**. The material responses, for example through (11.47), will appear in the form of a forced (induced) quantity, here the polarization density **P**. If the applied field was zero for time $t < 0$, then causality implies that there must be no induced polarization in this time interval.

6 These equations represent a Hilbert transform relation between the hermitian and antihermitian parts, which play the role of real and imaginary parts, respectively, in the case of matrices (linear operators).

frequency), we just need to restrict ourselves to a finite frequency and wavenumber range upon which the desired losses and dispersion characteristics are required to apply. By this route, Kramers–Kronig relations cannot impose a serious restriction on the design and analysis of artificial media.

11.7 ADVANCED PROPERTIES OF THE MATERIAL TENSOR

In this Section, we look at the material tensor through the point of view of complex analysis. The motivation for such study is that certain characteristics of signals excited in media, like short-term disturbances and damped waveforms are best understood analytically if viewed using the mathematical device of Laplace transform instead of the familiar Fourier transform, the latter being ideally suited for the task of analyzing the steady-state behavior of a given system. As will be shown below, there are general restrictions on the mathematical form of the response functions when viewed in the complex plane. Knowledge of such global restrictions is vital in the theory and practice of metamaterials.

11.7.1 Stability Restrictions

From the physical point of view, a passive medium cannot generate energy and hence all propagating signals must be damped or decaying when the distance goes to infinity.[7] Mathematically, this translates to the requirement that all poles are located in the left-half plane (LHP). We call the following the statement of the stability condition of material media

All poles of the material tensor must be located in the LHP. (11.80)

To see why this should be the case, just (Laplace) invert a spectral component in the form $1/(\omega - \omega_0 + i\gamma/2)$ and notice the sign of the resulting exponential factor. For signals to exponentially decay, the algebraic sign of the factor γ must be positive.

11.7.2 Causality Restrictions

Although we have already looked at causality in the study of Kramers -Kronig relations, we want to understand here this topic at a deeper level. Consider the

7 Notice that for a range that is bounded, both growing and decaying signals are possible. For example, consider a multilayered medium. In one intermediate layer both growing and decaying waves are permitted.

Fourier transform of a causal function $f(t)$ given by

$$f(\omega) = \int_0^\infty dt\, f(t)\, e^{i\omega t}. \tag{11.81}$$

Let us study the asymptotic behavior of this function when $t \to \infty$. We first notice that when $\mathrm{Im}\{\omega\} > 0$, the integral in (11.81) has a finite value since the integrand approaches zero as t grows to infinity. Moreover, on repeatedly differentiating this integral, we conclude also that all derivatives of $f(\omega)$ are finite. Therefore, the function $f(\omega)$ is analytic in the upper half complex plane (UHP). We have then

A causal function is analytic in the UHP. (11.82)

An immediate corollary is that

A causal function has no poles or branch points in the UHP. (11.83)

This principle forms the mathematical background behind the derivation of Kramers-Kronig relations.

11.7.3 Landau Condition

The Laplace transform of a signal is defined as

$$F(s) := \int_0^\infty dt\, f(t)\, e^{-st}. \tag{11.84}$$

Therefore, the s-plane and the complex ω-plane are related by $s = i\omega$, which means that 'upper' and 'lower' in the language of one transform translates into 'right' and 'left', respectively, in the language of the other transform. The inverse Laplace transform is given by the equation

$$f(t) = \frac{1}{2\pi i} \int_{\Gamma - i\infty}^{\Gamma + i\infty} ds\, F(s)\, e^{st}, \tag{11.85}$$

where Γ specifies how the integration contour should be chosen. Landau condition states that

The contour in (11.85) is to the left of all singularites in the s-plane. (11.86)

Therefore, the integration contour must be above all singularities in the complex ω-plane. It can be shown then that the resulting function does not depend on the particular path provided it satisfies the Landau condition.

11.8 WAVE PROPAGATION

11.8.1 Dispersion Relations

By wave modes or wave propagation we understand electromagnetic disturbances that can propagate in a source-free medium. In our case, the medium response is described by the nonlocal model characteristic of the Fourier approach.

Equation (11.52) is the *inhomogeneous* wave equation in our medium. From definition (11.42), the induced current expressed in terms of the vector potential (temporal gauge) is given by

$$J_{\text{ind},\,m}(\omega, \mathbf{k}) = \alpha_{mn}(\omega, \mathbf{k})\, A_n(\omega, \mathbf{k})\,. \tag{11.87}$$

In tensor form, we can write then (11.52) as

$$\Xi_{nm}(\omega, \mathbf{k})\, A_m(\omega, \mathbf{k}) = -\frac{\mu_0 c^2}{\omega^2} J_{\text{ext},\,n}(\omega, \mathbf{k})\,, \tag{11.88}$$

where

$$\Xi_{nm}(\omega, \mathbf{k}) = \frac{c^2}{\omega^2}\left(k_n k_m - k^2 \delta_{nm}\right) + \chi_{nm}(\omega, \mathbf{k})\,. \tag{11.89}$$

If the source term in (11.88) is set to zero, we obtain the *homogeneous* wave equation describing the propagation of waves in a source-free environment, i.e., the eigenmodes. However, as we found in Section 11.6.2, the antihermitian part of the tensor $\Xi_{nm}(\omega, \mathbf{k})$ is responsible of dissipation or energy generation in the medium. Such term must be omitted from the final homogeneous equation describing pure wave propagation. The desired equation of motion is therefore given by

$$\Xi^{\text{H}}_{nm}(\omega, \mathbf{k})\, A_m(\omega, \mathbf{k}) = 0, \tag{11.90}$$

where $\Xi^{\text{H}}_{nm}(\omega, \mathbf{k})$ describes the hermitian part of the tensor $\Xi_{nm}(\omega, \mathbf{k})$. The reader should notice that there is a thermodynamic hypothesis implicit in the derivation of this fundamental equation. That is, dissipation is treated as equivalent to source, and so the antihermitian part is removed even when it describes only a passive medium. Such hypothesis, equivalence of source and sink, is an additional postulate that cannot be derived from Maxwell's equations and should be supplied by an external theory, in this case thermodynamics of continuous media.

Notice that (11.90) is a matrix equation. From linear algebra, the necessary and sufficient condition for the existence of a nontrivial solution is that the determinant of the hermitian matrix $\Xi^{\text{H}}_{nm}(\omega, \mathbf{k})$ is identically zero. That is, the dispersion relation is given by

$$\det\left[\Xi^{\mathrm{H}}_{nm}\left(\omega,\mathbf{k}\right)\right]=0. \tag{11.91}$$

In general, this dispersion relation has potentially many solutions, each is called a *mode* or *branch*. We write the solution of the lth mode of the dispersion equation (11.91) as

$$\omega=\omega_l\left(\mathbf{k}\right). \tag{11.92}$$

For each mode, there corresponds a vector A_n satisfying equation (11.90). Such vector is called the *polarization* of the wave mode. within the scheme of Fourier-space electromagnetics, there exists a detailed theory of how to obtain and classify polarizations in various types of media, which is based on direct application of results from tensor calculus. However, we omit such details for the limitations of space.

11.8.2 The Green's Function

The solution of the inhomogeneous wave equation (11.88) can be formally written as

$$A_n\left(\omega,\mathbf{k}\right)=-\frac{\mu_0c^2}{\omega^2}G_{nm}\left(\omega,\mathbf{k}\right)J_{\mathrm{ext},\,m}\left(\omega,\mathbf{k}\right), \tag{11.93}$$

where $G_{nm}\left(\omega,\mathbf{k}\right)$ is the Green's function dyad in the spectral domain. From matrix theory, an expression of this dyad can be immediately written as

$$G_{nm}\left(\omega,\mathbf{k}\right)=\frac{\mathrm{cof}_{nm}\left[\Xi_{n'm'}\left(\omega,\mathbf{k}\right)\right]}{\det\left[\Xi_{nm}\left(\omega,\mathbf{k}\right)\right]}, \tag{11.94}$$

where cof_{nm} is the cofactor matrix. In deriving this result, only the hermitian part of the operator $\Xi_{nm}\left(\omega,\mathbf{k}\right)$ is used, and therefore the Green's dyad as it stands here is hermitian. However, when inverting the Fourier transform for the purpose of calculating the fields in the spatio-temporal domain, a singularity in the spectral domain is encountered around $\omega=\omega_l\left(\mathbf{k}\right)$. The traditional solution of this problem is to carefully enforce suitable causality conditions. Technically, the determinant is expanded in the following form

$$\det\left[\Xi_{nm}\left(\omega,\mathbf{k}\right)\right]\approx\frac{\partial}{\partial\omega}\det\left[\Xi_{nm}\left(\omega,\mathbf{k}\right)\right]\Bigg|_{\omega=\omega_l\left(\mathbf{k}\right)}\left\{\omega-\omega_l\left(\mathbf{k}\right)+i0\right\}, \tag{11.95}$$

where the expansion illustrated here is taken around the lth mode pole. By formally inverting the Fourier transform using the Dirac delta function, we obtain the following expression for the antihermitian part of the Green's function

$$D^{\mathrm{A}}_{nm}(\omega,\mathbf{k}) = \sum_{l} -i\pi\omega_l(\mathbf{k})\,\mathbf{e}^{*}_{l,m}(\mathbf{k})\,\mathbf{e}_{l,n}(\mathbf{k})\,F_l\left[\Xi_{nm}(\omega,\mathbf{k})\right]\delta\left(\omega-\omega_l(\mathbf{k})\right), \tag{11.96}$$

where the *scalar* quantity $F_l[\Xi_{nm}]$ depends only on $\mathbf{k}$ and can be directly determined by the dispersion profile of the medium, but its explicit expression is not of direct concern to us here. The unit vectors $\mathbf{e}_{l,n}(\mathbf{k})$ describe the polarization of the lth mode.[8] This derivation shows that although the antihermitian part was not originally taken into consideration in writing the expression of the Green's dyad (11.94), causality considerations *forces* us to introduce an antihermitian part. As we will show in Section 11.9, it is precisely this antihermitian part that contributes to the radiated field.[9]

It is interesting to observe again the role played by thermodynamics in the solution of Maxwell's equations. Indeed, the ultimate origin of the causality consideration introduced above can be tracked back to the thermodynamic requirement that energy decays away from the source and toward the sink. Maxwell's equations themselves are blind to the direction of power flow; they can support both (temporally) forward and backward waves. However, thermodynamics appears to fix the sign of the pole contribution around the real ω-axis and hence effectively imposes a particular form on the solution of the field equations. The reader can better appreciate the subtlety of this fact by recalling that the very concepts of source and sink are thermodynamic in nature and cannot be based ultimately on Maxwell's equations. An impulsive excitation, say an ideal Dirac delta function, can be mathematically introduced to the theory in a straightforward manner, e.g., using generalized function theory. However, the choice of the *sign* of the imaginary part of the pole associated with source or sink depends on energetics and dissipation, a topic that is best described macroscopically by classical thermodynamics. Since the ultimate origin of the antihermitian part of the Green's function, as shown above in (11.96), is causality, and the particular form of this depends in turn on thermodynamic consideration, and knowing that it is this part of the Green's dyad that is responsible of radiation (see Section 11.9), we can claim that the ultimate answer

8 Notice that for the case of transverse modes, the degeneracy of the eigenvalue problems requires a special treatment. Indeed, in this case one has to resort to the use of polarization *matrices*. We omit such details here.

9 The hermitian part contributes to the non-propagating field (near field) surrounding the source.

to the question of why an antenna can radiate appears to be purely thermodynamic in nature.

11.9 LOCALIZATION OF ELECTROMAGNETIC ENERGY RADIATED BY ANTENNAS EMBEDDED IN COMPLEX MEDIA

In this section, we provide some applications for the general Fourier approach in the characterization of the material responses to the electromagnetic fields as sketched above. We perform an explicit calculation of the electromagnetic energy radiated by an arbitrary antenna in a medium described by a nonlocal response tensor. We will show that the Fourier approach described in this chapter may help understanding the structure of the near-field surrounding the antenna, and therefore the possibility of localizing energy in complex artificial media.

The method relies on calculating the total energy of the radiated field using the Fourier integral. We start from the statement of energy conservation as stated in (11.8). The current $\mathbf{J}$ appearing on the RHS is replaced by the current distribution on the antenna, which is taken as an externally-controlled source $\mathbf{J}_{\text{ext}}$. As discussed in Section 11.1, the energy (work) density transferred to the surrounding field by this current is given by $-\mathbf{J}_{\text{ext}} \cdot \mathbf{E}$. The trick in performing general calculation is to introduce a new quantity $U_l(\mathbf{k})$, which is defined as *the density in the* $\mathbf{k}$*-space of the energy added to the surrounding field by the antenna when radiating through the* l*th mode*. That is, by energy conservation, the net time-averaged total energy transferred to the field through all modes is given by the following equation

$$\int_{-T/2}^{T/2} dt \int d^3r \mathbf{J}_{\text{ext}}(t, \mathbf{r}) \cdot \mathbf{E}(t, \mathbf{r}) = \sum_l \int \frac{d^3k}{(2\pi)^3} U_l(\mathbf{k}). \tag{11.97}$$

Expressing the electric field in terms of the vector potential in the temporal gauge as $\mathbf{E}(\omega, \mathbf{k}) = i\omega \mathbf{A}(\omega, \mathbf{k})$, using the Green's function of (11.93), and employing the Pareseval (power) theorem of Fourier analysis, we write the LHS as

$$\begin{aligned}
\int_{-T/2}^{T/2} dt \int d^3r \mathbf{J}_{\text{ext}}(t, \mathbf{r}) \cdot \mathbf{E}(t, \mathbf{r}) &= -\int \tfrac{d\omega d^3k}{(2\pi)^4} \mathbf{J}^*_{\text{ext}}(\omega, \mathbf{k}) \cdot \mathbf{E}(\omega, \mathbf{k}) \\
&= \int \tfrac{d\omega d^3k}{(2\pi)^4} \tfrac{i}{\varepsilon_0 \omega} J^*_{\text{ext},m}(\omega, \mathbf{k}) G_{mn}(\omega, \mathbf{k}) J_{\text{ext},n}(\omega, \mathbf{k}) \\
&= \int \tfrac{d\omega d^3k}{(2\pi)^4} \tfrac{i}{\varepsilon_0 \omega} J^*_{\text{ext},m}(\omega, \mathbf{k}) \left[G^{\text{A}}_{mn}(\omega, \mathbf{k}) + G^{\text{H}}_{mn}(\omega, \mathbf{k})\right] J_{\text{ext},n}(\omega, \mathbf{k})
\end{aligned} \tag{11.98}$$

Due to the presence of the factor i in the integrand, together with the fact that the integral must be real, it follows from the basic properties of hermitian and antihermitian functions (operators) that only the antihermitian part of the Green's dyad contributes to the radiation field. Now, by inserting into (11.98) the value given in (11.96), we finally arrive at an expression of the energy density in the following form

$$U_l(\mathbf{k}) = F_l\left[\Xi_{nm}(\omega,\mathbf{k})\right]\left|\mathbf{e}_l^*(\mathbf{k})\cdot\mathbf{J}_{\text{ext}}(\omega_l(\mathbf{k}),\mathbf{k})\right|^2, \tag{11.99}$$

where F_l is a positive real function with particular details not of direct concern to us here.[10] We will propose an interpretation for the physical meaning of $U_l(\mathbf{k})$. Consider the inverse Fourier transform

$$u_{F,l}(\mathbf{r}) = \Im^{-1}\left\{\sqrt{U_l(\mathbf{k})}\right\}. \tag{11.100}$$

Next, we use the Parseval theorem to write

$$\int d^3r\, u_{F,l}(\mathbf{r})\left(u_{F,l}(\mathbf{r})\right)^* = \int \frac{d^3k}{(2\pi)^3}\sqrt{U_l(\mathbf{k})}\sqrt{U_l(\mathbf{k})}. \tag{11.101}$$

Since the RHS is by definition the total energy radiated by the antenna in the lth mode, it follows that the integrand of the LHS, namely $u_l(\mathbf{r}) := \left|u_{F,l}(\mathbf{r})\right|^2$ can be interpreted as the *spatial* distribution of the energy density radiated by the antenna through the lth mode. We have

$$\begin{aligned} u_l(\mathbf{r}) = \textstyle\int\int d^3k d^3k' \sqrt{F_l\left[\Xi_{nm}(\omega,\mathbf{k})\right]F_l\left[\Xi_{nm}(\omega,\mathbf{k}')\right]^*} e^{i(\mathbf{k}-\mathbf{k}')\cdot\mathbf{r}} \\ \times\left|\mathbf{e}_l^*(\mathbf{k})\cdot\mathbf{J}_{\text{ext}}(\omega_l(\mathbf{k}),\mathbf{k})\,\mathbf{e}_l^*(\mathbf{k}')\cdot\mathbf{J}_{\text{ext}}(\omega_l(\mathbf{k}'),\mathbf{k}')\right|. \end{aligned} \tag{11.102}$$

This new quantitative measure contains information about the spatial structure of the time-averaged energy surrounding a radiator specified by its externally enforced current distribution $\mathbf{J}_{\text{ext}}(t,\mathbf{r})$. For example, it can be directly used in studying the localization of the radiated energy surrounding an antenna located inside an artificial medium described by the dispersion profile $\chi_{nm}(\omega,\mathbf{k})$. In other words, information about the conversion of the radiation field into near-field (non-propagating Fourier modes) can be investigated by suitable computation of the measure introduced above.

10 Although this is by no means obvious, one way of re-arranging terms and factors in the derivation of (11.99) can lead to a simple interpretation of the factor F_l in terms of the ratios of various stored energies.

Another important fact suggested by the expression (11.99) is the following. Although we are interested here in the localization of energy around the antenna, i.e., the *time-averaged* spatial distribution of energy, we can see from the first argument of $\mathbf{J}_{\text{ext}}$ that this distribution does depend on the temporal characteristics of the current distribution of the antenna.[11] In other words, one can engineer the localization of electromagnetic energy around an antenna not just by manipulating the spatial distribution of the current on the antenna, but also by the independent degree of freedom of the time excitation. In principle, the two arguments of the *external* source function $\mathbf{J}_{\text{ext}}$ are obviously independent of each other. The art of antenna design can be viewed as the method of controlling the functional dependence of the current on its two spatial and temporal arguments.

11.10 APPENDIX: MAGNETIC MOMENTS IN TERMS OF ELECTRIC MOMENTS

11.10.1 The Magnetic Moment Term

Multiply the equation of continuity (11.7) by r_l and integrate over all space to get

$$\int d^3r \frac{\partial \rho\left(t,\mathbf{r}\right)}{\partial t} r_l \;=-\int d^3r\, r_l \nabla\cdot\mathbf{J}\left(t,\mathbf{r}\right). \tag{11.103}$$

Consider first the LHS of (11.103). By employing the definition of the dipole moment (11.30), we write immediately

$$\int d^3r \frac{\partial \rho\left(t,\mathbf{r}\right)}{\partial t} r_l = \frac{\partial}{\partial t}\int d^3r \rho\left(t,\mathbf{r}\right) r_l = \frac{\partial}{\partial t} p_l\left(t,\mathbf{r}\right). \tag{11.104}$$

Now we consider the RHS of (11.103). Write the divergence as $\nabla\cdot\mathbf{J}\left(t,\mathbf{r}\right) = \left(\partial/\partial r_s\right) J_s\left(t,\mathbf{r}\right)$ and integrate by parts through the variable r_s to obtain

$$\begin{aligned}
\int d^3r\, r_l \nabla\cdot\mathbf{J}\left(t,\mathbf{r}\right) &= \int d^3r\, r_l \tfrac{\partial}{\partial r_s} J_s\left(t,\mathbf{r}\right)\\
&= \int d^2r \int dr_s\, r_l \tfrac{\partial}{\partial r_s} J_s\left(t,\mathbf{r}\right)\\
&= \int d^2r \left[J_s\left(t,\mathbf{r}\right) r_l\big|_{r_s=-\infty}^{r_s=+\infty} - \int dr_s \tfrac{\partial r_l}{\partial r_s} J_s\left(t,\mathbf{r}\right)\right]\\
&= \int d^2r \left[J_s\left(t,\mathbf{r}\right) r_l\big|_{r_s=-\infty}^{r_s=+\infty} - \int dr_s\, \delta_s^l J_s\left(t,\mathbf{r}\right)\right]\\
&= \int d^2r \left[J_s\left(t,\mathbf{r}\right) r_l\big|_{r_s=-\infty}^{r_s=+\infty} - \int dr_s\, J_l\left(t,\mathbf{r}\right)\right].
\end{aligned} \tag{11.105}$$

11 At first look, this dependence may appear hidden in the dispersion relation that relates each temporal frequency ω_l to the wavevector $\mathbf{k}$.

Using the assumption that the surface current $J_s(t, \mathbf{r})$ vanishes on the surface of the integration volume, we obtain

$$\int d^3r\, r_l \nabla \cdot \mathbf{J}(t, \mathbf{r}) = -\int d^3r\, J_l(t, \mathbf{r}) := -\mu_l(t, \mathbf{r}). \tag{11.106}$$

From (11.103), (11.104), and (11.106), we finally arrive at

$$\mu_l(t, \mathbf{r}) = \frac{\partial}{\partial t} p_l(t, \mathbf{r}). \tag{11.107}$$

11.10.2 The Magnetic Quadrable Term

Multiply the equation of continuity (11.7) by $r_l r_m$ and integrate over all space to get

$$\int d^3r \frac{\partial \rho(t, \mathbf{r})}{\partial t} r_n r_m = -\int d^3r\, r_n r_m \nabla \cdot \mathbf{J}(t, \mathbf{r}). \tag{11.108}$$

Consider first the LHS of (11.108). By employing the definition of the electric quadrable moment (11.31), we write immediately

$$\int d^3r \frac{\partial \rho(t, \mathbf{r})}{\partial t} r_n r_m = \frac{\partial}{\partial t}\int d^3r \rho(t, \mathbf{r})\, r_n r_m = \frac{\partial}{\partial t} q_{nm}(t). \tag{11.109}$$

Now let us take the RHS of (11.108). We first decompose the magnetic moment into the sum of symmetric and anti-symmetric parts as follows

$$x_n J_m = \frac{1}{2}(x_n J_m + x_m J_n) + \frac{1}{2}(x_n J_m - x_m J_n). \tag{11.110}$$

Now, write again the divergence as $\nabla \cdot \mathbf{J}(t, \mathbf{r}) = (\partial / \partial r_s) J_s(t, \mathbf{r})$ and integrate by parts through the variable r_s to obtain

$$\begin{aligned}
\int d^3r\, r_n r_m \nabla \cdot \mathbf{J}(t, \mathbf{r}) &= \int d^3r\, r_n r_m \tfrac{\partial}{\partial r_s} J_s(t, \mathbf{r}) \\
&= \int d^2r \int dr_s\, r_n r_m \tfrac{\partial}{\partial r_s} J_s(t, \mathbf{r}) \\
&= \int d^2r \left[J_s(t, \mathbf{r})\, r_n r_m \Big|_{r_s=-\infty}^{r_s=+\infty} - \int dr_s\, \tfrac{\partial}{\partial r_s}(r_n r_m)\, J_s(t, \mathbf{r}) \right] \\
&= \int d^2r \left[J_s(t, \mathbf{r})\, r_n r_m \Big|_{r_s=-\infty}^{r_s=+\infty} - \int dr_s\, (r_n \delta_s^n + r_m \delta_s^m)\, J_s(t, \mathbf{r}) \right] \\
&= \int d^2r \left[J_s(t, \mathbf{r})\, r_n r_m \Big|_{r_s=-\infty}^{r_s=+\infty} - \int dr_s\, \{ r_n J_s(t, \mathbf{r}) + r_m J_s(t, \mathbf{r}) \} \right]
\end{aligned} \tag{11.111}$$

Using again the assumption that the surface current $J_s(t, \mathbf{r})$ vanishes on the surface of the integration volume, we obtain

$$\int d^3r\, r_n r_m \nabla \cdot \mathbf{J}(t, \mathbf{r}) = -\int d^3r\, [r_n J_s(t, \mathbf{r}) + r_m J_s(t, \mathbf{r})]. \tag{11.112}$$

From (11.108), (11.109), and (11.112), we reach

$$\int d^3r\, [r_n J_s(t, \mathbf{r}) + r_m J_s(t, \mathbf{r})] = \frac{\partial}{\partial t} q_{ns}(t). \tag{11.113}$$

The antisymmetrical part in (11.110) can be written readily in the form $1/2\epsilon_{lmn}\mu_{mn}(t)$, where magnetic quadrable moment μ_{mn} is defined as

$$\mu_{nm}(t) = \int d^3r\, r_m\, J_n(t, \mathbf{r}). \tag{11.114}$$

Therefore, one can express the *axial* vector as

$$\mathbf{m}(t) = \frac{1}{2} \int d^3r\, \mathbf{r} \times \mathbf{J}(t, \mathbf{r}). \tag{11.115}$$

It follows then

$$\begin{aligned}
\epsilon_{lsn} m_n(t) &= \tfrac{1}{2} \int d^3r \epsilon_{lsn}\, (\mathbf{r} \times \mathbf{J}(t, \mathbf{r}))_n \\
&= \tfrac{1}{2} \int d^3r \epsilon_{lsn}\, \epsilon_{ns'n'} r_{s'} J_{n'}(t, \mathbf{r}) \\
&= \tfrac{1}{2} \int d^3r\, (\delta_{ss'} \delta_{nn'} - \delta_{sn'} \delta_{ns'})\, r_{s'} J_{n'}(t, \mathbf{r}) \\
&= \tfrac{1}{2} \int d^3r\, [r_s J_n(t, \mathbf{r}) - r_n J_s(t, \mathbf{r})]
\end{aligned} \tag{11.116}$$

where the the definition of the cross product (11.18) was used in the second equality, and the identity (11.22) was employed for the third equality. Thus, from (11.109), (11.110), (11.113), and (11.116) we finally arrive to

$$\mu_{ln}(t) = \frac{1}{2}\frac{\partial}{\partial t}q_{ln}(t) + \epsilon_{lns}m_s(t). \tag{11.117}$$

Chapter 12

The Far-Field Theory of Nonlocal Metamaterials

In Chapter 12, we develop a general theoretical scheme for the *engineering* approach to electromagnetic wave propagation in dispersive materials. Our investigation is carried out through two stages. First, we focus on the special case where the group velocity is negative, which may lead (if the medium is lossless) to negative refraction. In the second stage, we go beyond the first-order approximation of the group velocity by deriving the second-, and third-order corrections of the power flow due to the spatial dispersion profile.

12.1 LINEAR PHENOMENOLOGICAL MODEL FOR THE MEDIUM RESPONSE

In this section, we review the basic theory of electromagnetic wave propagation in a homogeneous, and dispersive medium described by the dielectric tensor $\bar{\bar{\epsilon}}(\omega, \mathbf{k})$ and the permeability $\bar{\bar{\mu}}(\omega, \mathbf{k})$.

The general relation between the electric displacement $\mathbf{D}$ and the electric field $\mathbf{E}$ is given by [98]

$$\mathbf{D}(\mathbf{r}, t) = \int dt' \int d^3r' \bar{\bar{\varepsilon}}(\mathbf{r} - \mathbf{r}', t - t') \cdot \mathbf{E}(\mathbf{r}', t'), \tag{12.1}$$

where it has been assumed that the medium is time-invariant and spatially homogeneous. The Fourier transform of the field is defined as

$$\mathbf{D}(\omega, k) = \int dt \int d^3r \mathbf{D}(\mathbf{r}, t) e^{j\mathbf{k}\cdot\mathbf{r}} e^{-j\omega t}, \tag{12.2}$$

which when applied into (12.1) will lead to

$$\mathbf{D}(\omega, \mathbf{k}) = \bar{\bar{\varepsilon}}(\omega, \mathbf{k}) \cdot \mathbf{E}(\omega, \mathbf{k}). \tag{12.3}$$

Applying the same logic to the magnetic field, we also find

$$\mathbf{B}(\omega, \mathbf{k}) = \bar{\bar{\mu}}(\omega, \mathbf{k}) \cdot \mathbf{H}(\omega, \mathbf{k}), \tag{12.4}$$

where we have

$$\bar{\bar{\varepsilon}}(\omega, \mathbf{k}) = \int d\tau \int d^3R\, \bar{\bar{\varepsilon}}(\mathbf{R}, \tau) e^{j\mathbf{k}\cdot\mathbf{R}} e^{-j\omega\tau}, \tag{12.5}$$

$$\bar{\bar{\mu}}(\omega, \mathbf{k}) = \int d\tau \int d^3R\, \bar{\bar{\mu}}(\mathbf{R}, \tau) e^{j\mathbf{k}\cdot\mathbf{R}} e^{-j\omega\tau}. \tag{12.6}$$

For a source-free region, Maxwell's equations are given in the following form

$$\nabla \times \mathbf{E} = -\frac{\partial \mathbf{B}}{\partial t}, \;\; \nabla \times \mathbf{H} = \frac{\partial \mathbf{D}}{\partial t}, \;\; \nabla \cdot \mathbf{B} = 0, \;\; \nabla \cdot \mathbf{D} = 0. \tag{12.7}$$

Assume that a plane monochromatic wave is excited and propagated with fields given by

$$\mathbf{E}(\mathbf{r}, t) = \mathbf{E}_0(\omega, \mathbf{k}) e^{-j\mathbf{k}\cdot\mathbf{r}}, \;\; \mathbf{H}(\mathbf{r}, t) = \mathbf{H}_0(\omega, \mathbf{k}) e^{-j\mathbf{k}\cdot\mathbf{r}}. \tag{12.8}$$

Substituting these fields into the two curl Maxwell's equation (12.7), taking the curl of both sides, and using (12.3) and (12.4) with scalar (isotropic) medium functions, we obtain

$$\mathbf{k} \times \mathbf{k} \times \mathbf{E}_0(\omega, \mathbf{k}) - \left(\omega^2/c^2\right) \varepsilon(\omega, \mathbf{k}) \mu(\omega, \mathbf{k}) \mathbf{E}_0(\omega, \mathbf{k}) = 0, \tag{12.9}$$

and from the remaining two divergence Maxwell's equation (12.7) we find

$$\varepsilon(\omega, \mathbf{k}) \mathbf{k} \cdot \mathbf{E}_0(\omega, \mathbf{k}) = 0, \tag{12.10}$$

$$\mu(\omega, \mathbf{k})\, \mathbf{k} \cdot \mathbf{H}_0(\omega, \mathbf{k}) = 0. \tag{12.11}$$

To separate between the transverse (T) and longitudinal (L) waves we assume in the next part that the medium is nonmagnetic ($\mu = 1$) [**?**]. Let us assume that

$$\epsilon(\omega, \mathbf{k}) \neq 0. \tag{12.12}$$

Then, from (12.10) we obtain $\mathbf{k} \cdot \mathbf{E}_0(\omega, \mathbf{k}) = 0$. This condition when applied to (12.9) immediately gives the dispersion relation for the transverse waves

$$\mathbf{k} \cdot \mathbf{k} = (\omega/c)^2\, n^2(\omega, \mathbf{k}), \tag{12.13}$$

where we have *defined* the index of refraction as

$$n(\omega, \mathbf{k})^2 := \varepsilon(\omega, \mathbf{k})\, \mu(\omega, \mathbf{k}). \tag{12.14}$$

The longitudinal modes can be obtained by setting $\epsilon(\omega, \mathbf{k}) = 0$. Therefore, from (12.10) we see that $\mathbf{k} \cdot \mathbf{E}_0(\omega, \mathbf{k}) \neq 0$. That is, contrary to the transverse wave, the wave vector here is not orthogonal to the field amplitude.

Generally speaking, the dispersion relations for the L and T modes are different and should be distinguished from each other by using the appropriate indices *l* and *t*, respectively, whenever possible. However, in this chapter the main focus will be on transverse waves so these subscripts will be omitted for the simplicity of notation. Notice that when spatial dispersion is ignored, $\varepsilon(\omega, \mathbf{k}) = \varepsilon(\omega)$. Hence, for the longitudinal modes the equation $\varepsilon(\omega, \mathbf{k}) = 0$ can be satisfied only at discrete frequencies. In other words, the group velocity $\partial\omega_l/\partial\mathbf{k}$ is zero and no energy flow can be associated with the longitudinal modes (one exception for this is some forms of plasmas [98]). Now, when spatial dispersion is considered, relation (12.12) is not only satisfied at a continuous range of frequencies, but may give non-zero group velocity, contributing to the power flow in the medium.

12.2 NEGATIVE GROUP VELOCITY MEDIA

Let us start with a very general refraction of index given by $n = n(\omega, \mathbf{k})$. The resulting dispersion relation for the transverse mode propagating in infinite, homogeneous, and isotropic medium is given by (12.13). The group velocity is defined as [102], [41]

$$\mathbf{v}_g = \frac{\partial\omega}{\partial\mathbf{k}} = \nabla_{\mathbf{k}}\omega = \hat{x}\frac{\partial\omega}{\partial k_x} + \hat{y}\frac{\partial\omega}{\partial k_y} + \hat{z}\frac{\partial\omega}{\partial k_z}. \tag{12.15}$$

Our goal now is to derive an equation connecting the spatial and temporal dispersion such that the resulting medium supports negative group velocity (NGV) propagation.

Differentiate both sides of (12.13) with respect to k_α, where $\alpha = x, y, z$, we get

$$\frac{k_\alpha}{k} = \frac{\partial n(\omega, \mathbf{k})}{\partial k_\alpha}\frac{\omega}{c} + \frac{\partial \omega}{\partial k_\alpha}\frac{n(\omega, \mathbf{k})}{c}. \tag{12.16}$$

Using the following chain rule

$$\frac{\partial n(\omega, \mathbf{k})}{\partial k_\alpha} = \frac{\partial n(\omega, \mathbf{k})}{\partial \mathbf{k}} \cdot \frac{\partial \mathbf{k}}{\partial k_\alpha} + \frac{\partial n(\omega, \mathbf{k})}{\partial \omega}\frac{\partial \omega}{\partial k_\alpha}, \tag{12.17}$$

equation (12.16) can be solved for $\partial \omega / \partial k_\alpha$ to give

$$v_{g\alpha} = \frac{\partial \omega}{\partial k_\alpha} = \frac{\frac{k_\alpha}{k} - \frac{\omega}{c}\frac{\partial n}{\partial \mathbf{k}} \cdot \frac{\partial \mathbf{k}}{\partial k_\alpha}}{\frac{n}{c} + \frac{\omega}{c}\frac{\partial n}{\partial \omega}}. \tag{12.18}$$

Let us now calculate the dot product between $\mathbf{v}_g$ and $\mathbf{k}$. We write

$$\mathbf{v}_g \cdot \mathbf{k} = \sum_\alpha v_{g\alpha} k_\alpha = \frac{1}{\frac{n}{c} + \frac{\omega}{c}\frac{\partial n}{\partial \omega}} \sum_\alpha \left[\frac{k_\alpha}{k} - \frac{\omega}{c}\frac{\partial n}{\partial \mathbf{k}} \cdot \frac{\partial \mathbf{k}}{\partial k_\alpha}\right] k_\alpha, \tag{12.19}$$

and after multiplying the numerator and the denominator by k

$$\mathbf{v}_g \cdot \mathbf{k} = \frac{k}{\frac{n}{c} + \frac{\omega}{c}\frac{\partial n}{\partial \omega}} \sum_\alpha \left[\frac{k_\alpha^2}{k^2} - \frac{\omega}{c}\frac{\partial n}{\partial \mathbf{k}} \cdot \frac{\partial \mathbf{k}}{\partial k_\alpha}\frac{k_\alpha}{k}\right]. \tag{12.20}$$

Notice that

$$\sum_\alpha \frac{k_\alpha^2}{k^2} = 1 \tag{12.21}$$

and

$$-\sum_\alpha \frac{\omega}{c}\frac{\partial n}{\partial \mathbf{k}} \cdot \frac{\partial \mathbf{k}}{\partial k_\alpha}\frac{k_\alpha}{k} = -\frac{\omega}{c}\frac{\partial n}{\partial \mathbf{k}} \cdot \frac{\mathbf{k}}{k} = -\frac{\omega}{c}\frac{\partial n}{\partial k}, \tag{12.22}$$

where the relation $\partial \mathbf{k} / \partial k_\alpha = \hat{u}_\alpha$ has been used. Therefore, (12.20) reduces to

$$\mathbf{v}_g \cdot \mathbf{k} = k\frac{1 - \frac{\omega}{c}\frac{\partial n}{\partial k}}{\frac{n}{c} + \frac{\omega}{c}\frac{\partial n}{\partial \omega}}. \tag{12.23}$$

It can be shown by the same procedure that $|\mathbf{v}_g|\,|\mathbf{k}| = |\mathbf{v}_g \cdot \mathbf{k}|$. Thus, the angle cosine $\cos\theta = \mathbf{v}_g \cdot \mathbf{k}/|\mathbf{v}_g|\,|\mathbf{k}|$ is either 1 or -1. This is expected since we assumed the medium to be homogeneous and isotropic.[1] Therefore, we define the *negative* group velocity as the case when the angle between $\mathbf{v}_g$ and $\mathbf{k}$ is $180°$. By defining $\gamma := |\mathbf{v}_g|$ and assuming $k > 0$, (12.23) can give the following result for NGV

$$\frac{\omega}{c}\frac{\partial n(\omega, \mathbf{k})}{\partial k} - \frac{\gamma}{c}\left(1 + \omega\frac{\partial}{\partial\omega}\right) n(\omega, \mathbf{k}) = 1. \tag{12.24}$$

12.3 THE PHYSICAL MEANING OF NEGATIVE GROUP VELOCITY

Negative refraction (NR) must occur at the interface separating the conventional from the meta-material if the Poynting vector **S** and the wave vector **k** in the metamaterial are oriented opposite to each other. That is, if we have

$$\mathbf{S} \cdot \mathbf{k} < 0. \tag{12.25}$$

We will refer to (12.25) here as the main *sufficient* condition for obtaining NR in our metamaterial. The question now is whether the requirement

$$\mathbf{v}_g \cdot \mathbf{k} < 0 \tag{12.26}$$

is equivalent to condition (12.25). This is identical to asking whether the Poynting vector **S** and the group velocity $\mathbf{v}_g$ are oriented in the same direction. The answer is that in general they are not [101], [97]. The two vectors $\mathbf{v}_g$ and **S** become parallel if the medium is lossless or has small dissipation. In this case, it is possible to write [101], [97], [41]

$$\mathbf{S} = W\mathbf{v}_g, \tag{12.27}$$

where W is the total energy density stored in the medium. Since at thermodynamic equilibrium $W > 0$ [102], it follows that S and $\mathbf{v}_g$ are parallel. For lossy media, the angle between these two vectors may vary considerably depending on the material; no *a priori* conclusion can be stated without examining the specific dispersion and losses profile.

It is common in the literature of metamaterials nowadays, following the original work of Veselago [145], to associate negative-refraction media with the handedness as being left-handed (LH), in contrast the normal right-handedness (RH) of

1 In other words, the dependence of the refraction index n on the wave vector $\mathbf{k}$ can be written idnetically as either $n(\mathbf{k})$ or $n(k)$.

conventional materials. However, it has been noticed long before Veselago's work that negative refraction is a more general phenomenon that should be addressed in terms of group velocities, not the algebraic signs of the medium parameters [118], [119]. In particular, it was predicted that negative refraction may occur even when both ϵ and μ are positive [97], a situation consistent with (12.24), which gives the exact details of how to choose the temporal and spatial dispersion of the medium such that the resulting waves propagate with NGV. If, furthermore, the medium has low dissipation, equation (12.25) is satisfied and the medium will support NR.

From (12.23), we write

$$\mathbf{v}_g = \frac{1 - \frac{\omega}{c}\frac{\partial n}{\partial k}}{\frac{n}{c} + \frac{\omega}{c}\frac{\partial n}{\partial \omega}} \frac{|\mathbf{k}|}{k} \tag{12.28}$$

and

$$\mathbf{v}_p = \frac{\omega}{k}\frac{|\mathbf{k}|}{k} = \frac{c}{|n|}\hat{a}_{\mathbf{k}}. \tag{12.29}$$

The correct interpretation of the group velocity is that it is the speed of propagation of the smoothly varying wave packet's *envelope* of relatively small bandwidth (first-order approximation). This velocity is the same as the energy velocity in lossless media but in general lossy materials this is not correct [101], [113].

Assume that a certain direction in space is chosen as a reference. Thus, with respect to this direction each of $\mathbf{v}_g$ and $\mathbf{v}_p$ can be either positive or negative. Will now show that the condition (12.25) represents a more general definition of a broader type of metamaterials by isolating and identifying the following four distinct cases, depending on the algebraic signs of the group velocity and the wave vector.

Case I ($v_p > 0, v_g > 0,\ n > 0.$) This is the conventional medium. Here the wave envelope and phase propagate away from the source and positive refraction occurs all the time.

Case II ($v_p < 0, v_g > 0,\ n < 0.$) This is the so-called Veselago medium. Here the wave envelope propagates away from the source while phase propagates towards the source. Negative refraction occurs in this case.

Case III ($v_p > 0, v_g < 0,\ n > 0.$) This represents the main interest of this chapter. Here the wave envelope propagates toward the source while the phase propagates away from the source. However, although n is positive, negative refraction may occur if the medium has small dissipation and a carefully chosen profile of the spatial dispersion is implemented.

Case IV ($v_p < 0, v_g < 0$, $n < 0$.) Here, both the wave envelope and phase propagate toward the source. In this case negative refraction will appear for small losses but this cannot be achieved using only temporal dispersion.

To get a better understanding of the four cases listed above we need to resort to important distinction between normal and anomalous dispersion.[2] We will prove now the previous statements. The first and the second cases are self-evident and no further illustrations are needed here. For the third case, assume first that the medium has small losses so we can apply (12.27) and write

$$\mathbf{S} \cdot \hat{a}_{\mathbf{k}} = W \frac{c - \omega \partial n / \partial k}{n + \omega \partial n / \partial \omega}. \tag{12.30}$$

Consider first a medium showing only temporal dispersion ($\partial n / \partial k = 0$). Since $n > 0$, then the only way for v_g to become negative is to have $\partial n / \partial \omega < 0$. This is, however, the region of anomalous dispersion, which corresponds usually to high losses. This means that negative refraction is *not* guaranteed in this case. We must stress here that a metamaterial in which the group velocity is negative is still meaningful even when there is no negative refraction. We need to refer to $\mathbf{v}_g$ as only the velocity in which a wave packet propagates without appreciable distortion [113]. Such media has been already demonstrated experimentally more than three decades ago where the group velocity was reportedly measured with supraliminal negative values in carefully designed media having anomalous dispersion [146], [147], [148].

When considering spatial dispersion, the dot product in (12.30) can be made negative by solutions of equation (12.24) as we will show in later sections. In this case, no assumption about the sign of $\partial n / \partial \omega < 0$ is necessary and condition (12.26) can be satisfied in low dissipation media, leading to negative refraction. Thus, spatial dispersion is the crucial factor in permitting negative refraction in such kind of metamaterials (Case III).

Finally, Case IV will be treated briefly here. Consider first the scenario when the spatial dispersion is neglected. Here, since n is already negative, (12.28) may suggest that obtaining NGV in a negative phase velocity medium is possible without operating in the region of anomalous dispersion. However, in Appendix 12.5 we show that causality considerations do *not* allow this. If the losses are small, then it is

2 Normal dispersion is characterized by a medium function, say *n* for example, which is monotonically increasing. Hence, $\partial n / \partial \omega > 0$. Anomalous dispersion is then defined as the opposite case when $\partial n / \partial \omega < 0$. In general, we know from experiments that anomalous dispersion is correlated with lossy media [116], [33], [113]. For a rigorous proof that anomalous dispersion is a necessary condition for the medium to be lossy see [102].

impossible to have both NGV and also $v_p < 0$. Experimental data in [149], [150] are consistent with this conclusion as it shows that v_g and v_p becomes simultaneously negative in the region of anomalous dispersion. Therefore, in Case IV it is not guaranteed to observe NR even though $n < 0$. The situation again will change when spatial dispersion is considered where careful choice of the dispersion profile may lead to NGV in the normal dispersion region, leading therefore to NR.

12.4 EXACT SOLUTION FOR THE DISPERSION ENGINEERING EQUATION

12.4.1 Geometric Interpretation

Before proceeding into the exact analytical solution of (12.24), it will be very insightful to provide a geometrical interpretation of this solution. One of the applications of such interpretation will be setting the stage for necessary and sufficient conditions required for the existence and the uniqueness of the solution. Those are very important as they will illuminate the physically feasible conditions in which spatial and temporal dispersion can conspire with each other to produce NR propagation with a pre-specified group velocity profile.

The relation $\omega = \omega(k)$ is simply the dispersion law of the medium. We may say that this equation determines a family of curves in the plane upon which the general solution $n = n(\omega, k)$ will be constructed. Notice that this function is a surface in the ω-k-n 3-dimensional space. Therefore, as shown in Figure 12.1, one can consider the family of curves $\omega = \omega(k)$ as base curves upon which the solution surface would be found.

12.4.2 Development of the Exact Solution

Let us consider the dispersion relation $\omega = \omega(k)$ as an implicit parametrization of the solution in terms of *k*. Then it is possible to write

$$\frac{d}{dk} n\left(\omega\left(k\right), k\right) = \frac{\partial n}{\partial k} + \frac{\partial n}{\partial \omega} \frac{d\omega}{dk}. \tag{12.31}$$

From the defintion of group velocity and the dispersion relation we have

$$\frac{d\omega}{dk} = -\gamma\left(\omega, k\right). \tag{12.32}$$

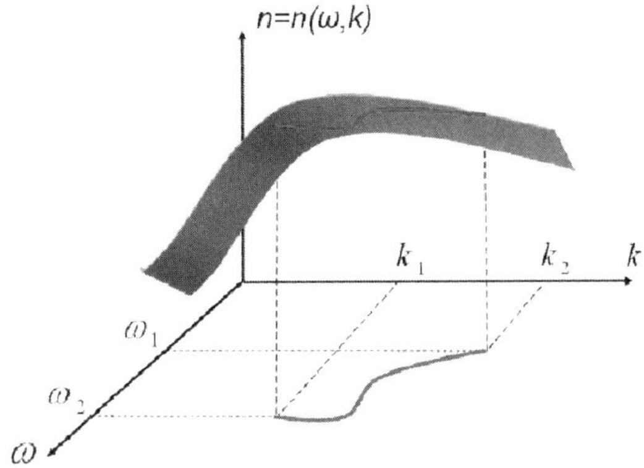

Figure 12.1 Geometric interpretation for a solution to the dispersion engineering equation (12.24). The solution $n = n(\omega, k)$ is a surface in the ω-k-n 3-dimensional space.

Thus, from (12.31) and (12.32) we obtain

$$\begin{aligned}\frac{\omega}{c}\frac{dn}{dk} - \frac{\gamma}{c}n &= \frac{\omega}{c}\left\{\frac{\partial n}{\partial k} + \frac{\partial n}{\partial \omega}\frac{d\omega}{dk}\right\} - \frac{\gamma}{c}n \\ &= \frac{\omega}{c}\left\{\frac{\partial n}{\partial k} - \gamma\frac{\partial n}{\partial \omega}\right\} - \frac{\gamma}{c}n \\ &= \frac{\omega}{c}\frac{\partial n}{\partial k} - \frac{\omega\gamma}{c}\frac{\partial n}{\partial \omega} - \frac{\gamma}{c}n\end{aligned} \tag{12.33}$$

And from (12.31) we readily get the following *ordinary* differential equation (ODE)

$$\frac{\omega\left(k\right)}{c}\frac{dn\left(\omega,k\right)}{dk} - \frac{\gamma\left(\omega,k\right)}{c}n\left(\omega,k\right) = 1. \tag{12.34}$$

Therefore, the solution to the original partial differential equation (12.24) can be thought of as solving the ODE (12.34) along the path (curve) described by the ODE (12.32). Notice that γ is in general an arbitrary positive function of both ω and k. Therefore, although the problem has been reduced into two ODEs, still no general solution is available analytically.

12.4.3 Solution for k-Dependent Group Velocity

Consider the boundary-value problem consisting of the PDE (12.24) together with

$$\frac{\partial\gamma}{\partial\omega} = 0,\ n\left(\omega, k = k_1\right) = \phi\left(\omega\right)\ \ \forall\ \omega_1 < \omega < \omega_2, \omega_1 > 0, k_1 > 0, \tag{12.35}$$

where $k_1 < k_2$ and $\omega_1 < \omega_2$ are positive real numbers. Assuming that γ is function of *k* only, it is possible to directly integrating equation (12.33) to obtain

$$\omega(k) = -\int dk\gamma + a. \tag{12.36}$$

Substituting (12.36) into (12.34) we find

$$\left(-\int dk\gamma + a\right)\frac{dn}{dk} - \gamma n = c. \tag{12.37}$$

Let us assume for the moment that $\omega \neq 0$ for any *k*. Then we can write

$$\frac{dn}{dk} - \frac{\gamma}{-\int dk\gamma + a} n = \frac{c}{-\int dk\gamma + a}. \tag{12.38}$$

This equation admits the following exact solution

$$n(k) = e^{-F}\left(c\int \frac{dk\, e^F}{-\int dk\gamma(k) + a} + b\right), \quad F = \int \frac{dk\gamma(k)}{\int dk\gamma(k) - a}, \tag{12.39}$$

where *a* and *b* or constants to be determined later. Since we are solving the ODE (12.34) along the trajectory specified by (12.32), then *b* is not independent of *a*, and we may write in general $b = f(a)$, where the function *f* is to be fixed by enforcing the boundary condition imposed on the function $n = n(\omega, k)$.

12.4.4 Solution for Constant Group Velocity

Let us evaluate the general solution for the case when γ is constant; i.e., we want to impose the condition that the group is velocity is constant but negative. In this case, (12.36) gives

$$\omega(k) = -\gamma k + a. \tag{12.40}$$

In Figure 12.2 we show the geometric structure of this case. The linear segments shown between the two lines $k = k_1$ and $k = k_2$ represent the permissible characteristic curves. Substituting (12.40) it into the general solution we obtain after evaluating the integrals

$$n = \frac{ck + f(a)}{-\gamma k + a}. \tag{12.41}$$

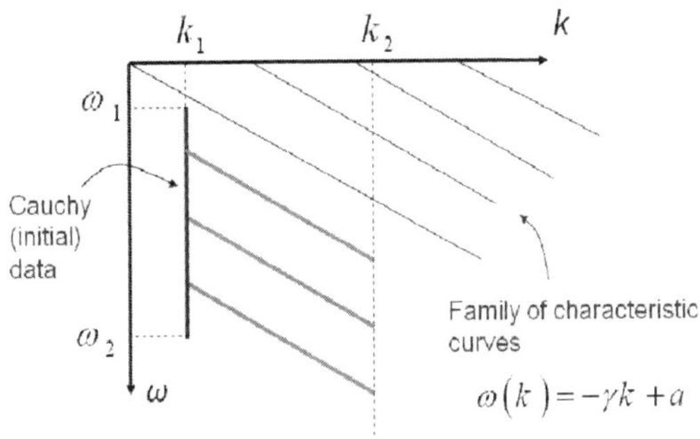

Figure 12.2 Geometric interpretation for solution of the dispersion eningeering equation (12.24). The solution $n = n(\omega, k)$ is a surface in the ω-k-n 3-dimensional space.

Substituting $a = \omega + \gamma k$ we get

$$n(\omega, k) = \frac{ck + f(\omega + \gamma k)}{\omega} \tag{12.42}$$

To find the function f, we apply the boundary condition $n(\omega, k = k_1) = \phi(\omega)$ to get $f(\omega + \gamma k_1) = \omega\phi(\omega) - ck_1$. Using the transformation $x = \omega + \gamma k_1$ we determine the function f to be

$$f(x) = (x - \gamma k_1)\,\phi(x - \gamma k_1) - ck_1. \tag{12.43}$$

The final solution can be written then as

$$n(\omega, k) = \frac{c(k - k_1)}{\omega} + \frac{1}{\omega}\left[\omega + \gamma(k - k_1)\right]\phi(\omega + \gamma(k - k_1)). \tag{12.44}$$

The importance of the general expression given in (12.44) is evident. Dispersion engineering in this case amounts to choosing the right spatial dispersion profile, starting at initial data consisting of the temporal dispersion at a specific value of the wavenumber k, such that the resulting wave propagation exhibits a constant NGV. Therefore, while the the desired anti-parallel nature of $\mathbf{S}$ and $\mathbf{k}$ is obtained, the group velocity does not vary with frequency, leading to minimal distortion in signal transmission for communication applications for example.

For a medium with small losses, Kramers-Kronig relations implies

$$\frac{\partial n}{\partial \omega} = \frac{(k - k_1)(c - \gamma\phi)}{-\omega^2} + \frac{\gamma (k - k_1)}{\omega} \frac{\partial \phi}{\partial \omega} > 0. \tag{12.45}$$

Let us choose an initial data in the normal dispersion regime (low losses) such that $\partial \phi / \partial \omega = A\omega^2, \; A > 0$. In this case

$$\frac{\partial n}{\partial \omega} = \frac{(k - k_1)(c - \gamma\phi)}{-\omega^2} + \gamma (k - k_1) A\omega, \tag{12.46}$$

which can be to satisfied, for example, at sufficiently high frequencies. Another possibility would be to choose $\phi(\omega + \gamma(k - k_1)) > c/\gamma > 0$.

12.4.5 Zero-Temporal Dispersion

The consideration of spatial dispersion will lead to a new picture for the special case when temporal dispersion is ignored. We start by the following simple theorem: Assuming constant $\gamma > 0$ (constant negative group velocity), there is no possibility to achieve NR when the temporal dispersion is zero. To prove this, we notice that ti follows immediately from (12.44) that for constant NGV and $\partial n / \partial \omega = 0$ we get $\phi(\omega + \gamma k) = -c/\gamma$, in which it follows that the corresponding spatial dispersion profile is simply $n(k) = -c/\gamma$. Thus, the refraction index is also independent of *k* and negative. However, we proved before that in order to have NR we must have very small losses. In this case a negative index of refraction can be achieved only with double negative material. To achieve negative refraction in such a medium we *must* have temporal dispersion [145]. Therefore, we conclude that there is no physical solution corresponding to *n* above.

Moreover, from (12.28) it follows that in the case of zero-temporal dispersion the group velocity takes the form

$$v_g = \frac{c}{n} - \frac{\omega}{n} \frac{\partial n}{\partial k}. \tag{12.47}$$

We notice two important things here. First, although *n* does not depend on frequency, the group velocity will have a linear dependence on frequency for nonzero spatial dispersion. Thus, we cannot say that having a refraction index that does not depend on frequency means a group velocity that is constant. This statement is true only if spatial dispersion is ignored.

Second, from (12.47) it is clear that one can achieve NGV if we choose $\partial n/\partial k$ large enough. In particular, if *n* is positive, we just need to satisfy $\partial n/\partial k > c/\omega$. Therefore, if losses are small, it is possible to achieve NR in media that has no temporal dispersion.

12.5 APPENDIX: PROOF FOR CASE IV

From (12.28) and (12.29) we can write

$$\mathbf{v}_g = \frac{n}{n + \omega\partial n/\partial\omega}\mathbf{v}_p. \tag{12.48}$$

For negative phase velocity media we have $n = -\sqrt{\mu\varepsilon}$. Therefore, it is possible to write

$$\begin{aligned} n + \omega\partial n/\partial\omega &= n + \omega\partial\left(-\sqrt{\varepsilon\mu}\right)/\partial\omega \\ &= -\sqrt{\varepsilon\mu} - \tfrac{\omega}{2\sqrt{\varepsilon\mu}}\left[\varepsilon\tfrac{\partial\mu}{\partial\omega} + \mu\tfrac{\partial\varepsilon}{\partial\omega}\right] \\ &= -\sqrt{\varepsilon\mu}\left\{1 + \tfrac{\omega}{2\varepsilon\mu}\left[\varepsilon\tfrac{\partial\mu}{\partial\omega} + \mu\tfrac{\partial\varepsilon}{\partial\omega}\right]\right\}. \end{aligned} \tag{12.49}$$

From Kramers-Kronig relations we know that in a medium with small losses $\partial/\partial\omega\left(\omega\varepsilon\right) > 0$ and $\partial/\partial\omega\left(\omega\mu\right) > 0$ [102]. This in turns yields $\partial\varepsilon/\partial\omega > -\varepsilon/\omega$ and $\partial\mu/\partial\omega > -\mu/\omega$. Noticing that both ϵ and μ are negative, the previous two inequalities when combined together will give $\mu\partial\varepsilon/\partial\omega + \varepsilon\partial\mu/\partial\omega < -2\varepsilon\mu/\omega$, or

$$1 + \frac{\omega}{2\varepsilon\mu}\left[\varepsilon\frac{\partial\mu}{\partial\omega} + \mu\frac{\partial\varepsilon}{\partial\omega}\right] < 0. \tag{12.50}$$

Therefore, (12.48), (12.49), and (12.50) lead to group and phase velocities with signs opposite of each other.

Chapter 13

The Near-Field Theory of Nonlocal Metamaterials

13.1 INTRODUCTION

The purpose of this chapter is to continue and expand the proposals in Chapter 12, which were mainly concerned with source-free prorogation, to the more difficult and richer problem of radiation in the near zone of sources embedded in complex media exhibiting nonlocality. The chapter is brief and does not present in full length the technical details of the results attained therein. The general theory of nonlocal electromagnetic theory, including further examples and details, will be treated by one of the authors in a future monograph.

The main motivations for undertaking this study are the following:

1. By modeling complex environments surrounding conventional antenna systems, such as mobile and MIMO arrays, using nonlocal material response function, it is possible to obtain new insights into the fundamental changes in the performance of these conventional systems brought up by operating them in non-standard or non-idealized surroundings.
2. Investigations of nonlocal media can be considered a subject of interest in itself since nonlocal metamaterials constitute a major extension of the now traditional concept of metamaterials, the latter being typically understood within the classic picture of temporal dispersion.
3. It is well known that nano-scale structures cannot be modeled using macroscopic Maxwell's equations for all purposes. The most general level of the electrodynamics of nanostrucutres involves a nonlocal kernel in the material

response that is essential at the microscopic level. Although averaging these microscopic information can lead sometimes to classic material response with no effective spatial dispersion, for completeness and rigorous understanding of the physics of nano-scale problems, nonlocality remains essential.

4. New engineering applications can emerge from considering the novel physical phenomena exhibited by nonlocal media. Indeed, in this chapter we point out several of such potential applications related to energy localization, retrieval, harvesting.

It will be shown in Section 13.3 below that the essence of the near-field theory of nonlocal media is the derivation of a new dyadic Green's function for point sources embedded in such media. This will be achieved by working in the Fourier domain and it involves a significant generalization of the classic Weyl expansion. The new Weyl expansion is very different from the conventional one since new poles in the complex domains have to be included in the complex plane counter integration encountered while performing the inverse Fourier transformation needed to obtain the Green's function in the spatial domain. We will not be able to go into all the details of the derivation but some of their immediate physical consequences and potential novel applications will be provided thereafter.

13.2 BASIC MODEL FOR ELECTROMAGNETIC NONLOCALITY IN MATERIAL MEDIA

In this chapter, we follow the Fourier transform approach to the electrodynamics of nonlocal media presented in Chapter 11, which is the most natural mathematical formalism suitable for the study of spatial problems. The approach is widely used in the physical literature, especially the literature related to the optics of crystals. In general, all electric and magnetic responses are treated in a single dielectric tensor instead of the conventional division into electric and magnetic polarizations. In agreement with this proposed revision, which is essentially contained in [97], [98], [103], we write the relation between the electric field **E** and displacement vector **D** as

$$D_n\left(\omega,\mathbf{k}\right)=\sum_m \varepsilon_{nm}^{\mathrm{eq}}\left(\omega,\mathbf{k}\right)E_m\left(\omega,\mathbf{k}\right), \tag{13.1}$$

$$D_n\left(t,\mathbf{r}\right)=\int dt'\int d^3r'\sum_m \varepsilon_{nm}^{\mathrm{eq}}\left(t-t',\mathbf{r}-\mathbf{r}'\right)E_m\left(t',\mathbf{r}'\right), \tag{13.2}$$

where $\varepsilon_{nm}^{\mathrm{eq}}$ is the equivalent dielectric tensor of the Fourier space approach. The relation between this tensor and the conventional dielectric and magnetic susceptibilities can be found in Chapter 11.

It can be shown from first principles that the most general form of the dielectric function is given by [97], [105]

$$\begin{aligned}\varepsilon_{nm}^{\mathrm{eq}}(\omega,\mathbf{k}) &= \varepsilon^{L}(\omega,k)\,\kappa_n\kappa_m \\ &+\varepsilon^{T}(\omega,k)\left(\delta_{nm}-\kappa_n\kappa_m\right)+i\varepsilon^{R}(\omega,k)\,\epsilon_{nml}\kappa_l,\end{aligned} \tag{13.3}$$

where $\kappa_m = k_m/k, \quad k = |\mathbf{k}|$. Here, the quantities $\varepsilon^L(\omega,k), \varepsilon^T(\omega,k), \varepsilon^R(\omega,k)$ are the longitudinal, transverse, and rotational permittivities, respectively. The rotatory parts can be ignored in media that don't exhibit optical activity.

We postulate additional simplifying hypotheses about this environment by assuming that the effective medium is 1) infinite, 2) isotropic, 3) homogeneous in space and 4) shift-invariant in time, 5) lossless, 6) nonmagnetic, and 7) optically inactive. Here, we also make the further assumption that 8) the longitudinal and transverse dielectric functions are equal. In this case, we can write [140], [105]

$$\mathbf{D}(\mathbf{r}) = \int_{\mathbb{R}^3} d^3r\varepsilon\left(\mathbf{r}-\mathbf{r}',\omega\right)\mathbf{E}\left(\mathbf{r}\right). \tag{13.4}$$

The effective medium[1] modeling the surrounding complex environment reduces then to a general *scalar* response function, namely $\varepsilon(\mathbf{r}-\mathbf{r}',\omega)$.

Although it is occasionally mentioned in textbooks that spatial dispersion is possible only with *tensor* dielectric function, the special assumptions stated above insure that a scalar function is enough. Note that for this condition to be true, the equality of the longitudinal and transverse responses is essential. This will require a high symmetry and in practice one may conjuncture that most media don't possess such symmetry. However, as shown above, the idealized medium described in this chapter is physically possible and hence realizable. Moreover, it is the most simple nonlocal medium imaginable, and therefore it is essential that an initial theory of near-field nonlocal electromagnetics be developed for such basic and fundamental environments. As will be seen in the following parts, even with such simplified nonlocal media, completely new physical phenomena come in to the picture.

By the term *nonlocal medium* we refer to a material described by response functions similar to (13.1). As we noticed previously, it follows from this definition that the material exhibits a memory-like behavior in the sense that the response to a

1 Starting from equation (13.4), we drop the superscript 'eq' from dielectric function symbols.

field excreted at a particular location appears to depend on the field values at *other* locations. We will show below that this phenomenon is quite general and does not refer just to a particular physical process occurring in natural crystals.

First, notice that we arrive at the definition (13.1) through a Fourier transform approach to the electromagnetic fields. Moreover, we are able to derive a relation connecting the traditional multipole approach and the Fourier approach. It is then found that nonlocality or spatial dispersion arises very naturally in accounting for nonmagnetic media. However, it is in the nature of the Fourier approach itself to introduce the spatial spectral variable $\mathbf{k}$ into the description of the material medium, and hence one can view nonlocality as a characteristics of the formalism itself, rather than a particular label given to an exotic physical process, for example exciotons-polaritons in crystal optics.

Let us start by providing a global qualitative look at the response of material media in classical and quantum physics. This view will serve as standard theoretical background upon which we measure our understanding of how to design artificial media.

Imagine that the material is composed of a system of uncoupled (hence, independent) oscillators. Each oscillator can interact with the applied electromagnetic fields by producing a dipole moment $\mathbf{p}$. From the basic picture of Lorentz models, we can express the functional dependence of this induced dipole moment on the temporal frequency ω by the broad Lorentzian form $\zeta/\left(\omega^2 - \omega_0^2\right)$, where ω_0 is a constant called the eigenfrequency or the resonance frequency, and ζ the oscillator strength. In general, each independent oscillator will resonate with the applied field according to its own eigenfrequency and strength, and the medium's overall response will be taken as the sum of all individual resonances. In this view, it is useful to think of each oscillator as representing an 'atom', even when its actual physical dimensions are much larger than real atoms. The essential idea in the art of artificial material design is taking this conceptual framework into its extreme by assuming that one can manipulate each atom individually in order to control and tailor the resulting material responses. The assumption that the atoms are *un*coupled will be translated to the fact that the resulting eigenfrequency ω_0 and oscillator strength ζ don't depend on wavelength, or equivalently on $\mathbf{k}$. For natural materials observed and studied through macroscopic electromagnetics, the atomic separation, for example in periodic structures like crystals, denoted here by a, is very small compared with the operating wavelength, i.e., we have $a/\lambda \ll 1$. In this case, all atoms appear to be in perfect phase synchronization and no significant coupling mechanism takes place.[2]

The situation is dramatically different in periodic structures, like photonic crystals and frequency selective surfaces, where, in this case, the operating wavelength can become appreciable compared with the characteristic spatial scale of the separation between the atoms (or unit cells), and hence interesting electromagnetic behavior can arise, like stopbands, localization, etc. However, it is still possible to describe all these complex structures by employing an effective dielectric function that is *nonlocal*. Such function can contain the full information of the symmetry group of the periodic structure. Therefore, Maxwell's equations, written in terms of these equivalent response functions, can be appropriated to describe the electromagnetism of the medium without explicitly enforcing a set of boundary conditions.[3] Aside from the economic advantage of such formulation, allowing the effective dielectric function to become nonlocal has the merit of bringing the full power of the conceptual framework of effective medium theory right to the fore even though the artificial medium under consideration may not satisfy the natural condition of infinitesimally small atomic constituents.

13.3 DERIVATION OF THE DYADIC GREEN'S FUNCTION OF A SPATIALLY RESPONSIVE (NONLOCAL) MEDIUM

In order to study the structure of the near field associated with a source embedded within this medium, we need to compute the dyadic Green's function of the problem. This we endeavor to achieve using the Fourier transform method. In this case, the relation (13.4) reduces to $\mathbf{D}(\mathbf{k}) = \varepsilon(k,\omega)\,\mathbf{E}(\mathbf{k})$. Here, $\varepsilon(k,\omega) := \mathcal{F}\{\varepsilon(\mathbf{r}-\mathbf{r}',\omega)\}$, where $\mathcal{F}$ is the Fourier transform operator applied to the spatial variable $\mathbf{r}-\mathbf{r}'$ and $k := |\mathbf{k}|$.

Let a point source $\mathbf{J}(\mathbf{r}) = \hat{\alpha}\delta(\mathbf{r}-\mathbf{r}')$ be located at $\mathbf{r}'$. Working in the spatial Fourier space, we can derive from Maxwell's equations and the definition of the dyadic Green's function the following algebraic (tensor) relation $-\mathbf{k}\times\mathbf{k}\times\bar{\mathbf{G}}(\mathbf{k},\mathbf{r}') - \omega^2\varepsilon(k,\omega)\,\bar{\mathbf{G}}(\mathbf{k},\mathbf{r}') = \bar{\mathbf{I}}e^{-i\mathbf{k}\cdot\mathbf{r}'}$ [103],[108]. Solving for $\bar{\mathbf{G}}(\mathbf{k},\mathbf{r}')$ and then applying the inverse Fourier transform operator, we find

$$\bar{\mathbf{G}}(\mathbf{r},\mathbf{r}') = \frac{1}{(2\pi)^3}\int_{\mathbb{R}^3} d^3k\,\frac{\left[\bar{\mathbf{I}}\omega^2\varepsilon(k,\omega) - \mathbf{k}\mathbf{k}\right]e^{i\mathbf{k}\cdot(\mathbf{r}-\mathbf{r}')}}{\omega^2\varepsilon(k,\omega)\left[k^2-\omega^2\varepsilon(k,\omega)\right]}. \tag{13.5}$$

2 The fundamental pre-condition for this to be true is that the fields are averaged on a spatial scale much larger than this natural characteristic spatial scale, i.e., the atomic separation a.

3 For example, see [142].

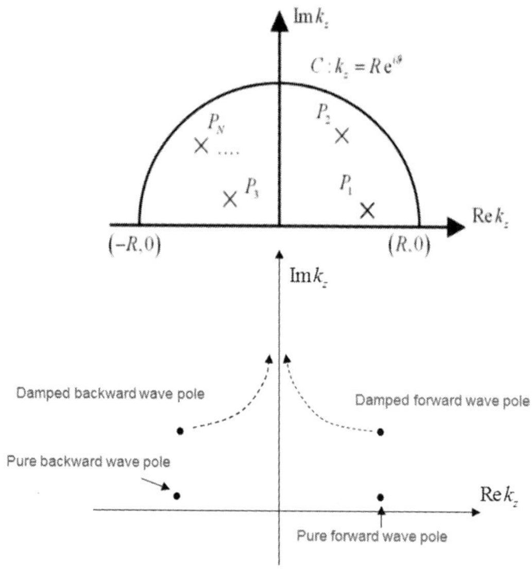

Figure 13.1 Distribution of complex poles introduced by the nonlocality of the medium.

In order to apply the theory of the near field developed in Chapter 4, we introduce a cartesian coordinate system xyz and attempt a spectral expansion of the Green's function along the z-axis as in the classical Weyl expansion. Therefore, we must evaluate first the k_z integral in (13.5) and express the result as a spectral expansion in terms of the transverse variables k_x and k_y. Notice that since the medium is assumed to be completely symmetric, the evaluation of (13.5) in the manner just described does not depend on the orientation of the chosen (local) coordinate system. Our hope then is to investigate whether the integral can be split into propagating and nonpropagating parts and subsequently break the rotational symmetry in order to get an insight into the structure of the near field as was done successfully with antennas radiating in free space.[4]

However, in order to proceed further, a specific but still general enough restriction on the dielectric function $\varepsilon(k, \omega)$ must be made. We will assume that the spatial response of the environment is nonresonant. In this case, $\varepsilon(k, \omega)$ can be considered analytic in **k**. Furthermore, since the medium is completely symmetric, we must also have $\varepsilon(k, \omega) = \varepsilon(-k, \omega)$.[5] Therefore, only *even* powers will emerge in the Taylor series expansion

$$\varepsilon(k, \omega) = \sum_{n=0}^{\infty} \varepsilon_n(\omega) k^{2n} = \varepsilon_0 + \varepsilon_1 k^2 + \varepsilon_2 k^4 + \cdots. \tag{13.6}$$

The truncation of the series by N terms will be called the N*th-order class of nonresonant environment*, and this is the most general situation to be encountered in practice.[6] It turns out that the finite series corresponding to the Nth-order medium can be treated in exact manner using the methods of contour integration in the complex k_z plane. Indeed, in this case, the integrand in (13.5) when viewed as a complex function of k_z will possess *pole* singularities given by the solution of the two equations

$$\varepsilon\left(k_x^2 + k_y^2 + k_z^2, \omega\right) = 0, \tag{13.7}$$

$$k_z^2 = \omega^2 \varepsilon\left(k_x^2 + k_y^2 + k_z^2, \omega\right) - k_x^2 - k_y^2, \tag{13.8}$$

where the unknown is k_z. Here, (13.7) corresponds to longitudinal poles while the (13.8) characterizes transverse poles.

4 Cf. Chapter 4

5 Cf. Chapter 11.

6 We are excluding pure quantum elementary excitations like excitons [97], [139], which exhibit resonant spatial dispersion effects. Such special scenarios can be treated on case-to-case basis. For details about the microscopic quantum origin of spatial dispersion in a model independent way, see [141], [142].

The two relations above are *the fundamental near field engineering equations* since their solution will completely determine the structure of the field near to the source in spatially responsive environments. We expect that the design and synthesis of final physical prototypes of specially-engineered nonlocal media for near applications will require adequate understanding of the mathematical behavior of (13.7) and (13.8).

For an Nth-order class, (13.7) and (13.8) are polynomial equations in k of order $2N$. Therefore, they are effectively also polynomial equations in k_z with the same order when k_x and k_y are treated as constants. Therefore, by factoring the denominator of the integrand of (13.5), we can express the latter as a rational function multiplied by $\exp(i\mathbf{k} \cdot (\mathbf{r} - \mathbf{r}'))$. Furthermore, we notice that for Nth-order medium with $N > 0$, we have

$$\lim_{|k_z|\to\infty} \frac{\bar{\mathbf{I}}\omega^2\varepsilon(\mathbf{k},\omega) - \mathbf{kk}}{\varepsilon(k,\omega)\left[k^2 - \omega^2\varepsilon(k,\omega)\right]} = 0. \tag{13.9}$$

Therefore, all the conditions of Jordan's lemma are met and we can evaluate the k_z-integral in (13.5) using the residue theorem [67]. The residues are those corresponding to the roots of (13.7) and (13.8). The Green's function of the nonlocal medium can then be expanded in the following form

$$\bar{\mathbf{G}}(\mathbf{r},\mathbf{r}') = \sum_{n=1}^{2N}\left[\bar{\mathbf{G}}_n^T(\mathbf{r},\mathbf{r}') + \bar{\mathbf{G}}_n^L(\mathbf{r},\mathbf{r}')\right], \tag{13.10}$$

where

$$\bar{\mathbf{G}}_n^{T,L}(\mathbf{r},\mathbf{r}') = \frac{1}{\omega^2}\int_{R^2} dk_x dk_y \times \frac{\left[\bar{\mathbf{I}}\omega^2\varepsilon(k,\omega) - \mathbf{kk}\right]e^{-ik_x(x-x')-ik_x(y-y')}e^{-ik_{T,L,n}|z-z'|}}{\prod\limits_{\substack{m=1\\m\neq n}}^{N}\left(k_{T,L,n}^2 - k_{T,m}^2 - k_\rho^2\right)\prod\limits_{\substack{m=1\\m\neq n}}^{N}\left(k_{T,L,n}^2 - k_{L,m}^2 - k_\rho^2\right)\sqrt{k_{T,L,n}^2 - k_\rho^2}}. \tag{13.11}$$

which were obtained by direct application of the residue calculus [67].

The expression (13.11) provides the required generalization of the classic Weyl expansion when the point source is embedded into a nonlocal medium. As can be seen from at a glance, this is considerably more complicated that the case when there is only temporal dispersion. In fact, the derivation above implies that whatever is the complexity of a given temporal dispersion, no essential change in the Weyl expansion will follow. However, only with *spatial* dispersion can completely new phenomena arise. Brief proof and examples of how this happens will be given in the remaining parts of this chapter.

13.4 RUDIMENTS OF A THEORY FOR COMPLEX POLES ENGINEERING IN NONLOCAL MEDIA

From the fundamental theorem of algebra, each of these two equations in k will have $2N$ (possibly repeated) complex roots, say k_n^T, k_n^L, $n = 1, 2, ..., 2N$, for transverse and longitudinal poles, respectively. There exists two rules governing the distribution of these roots in the complex k_z plane, see Figure 13.1. The first is that since only even powers in k appear in (13.7) and (13.8), then for each root $k_n^{T,L}$, $-k_n^{T,L}$ is also a root.

The second rule results from our assumption that the medium is lossless, which makes all the coefficients of k^{2n} real. In this case, for each complex roots, the conjugate is also a root. In general, among each set of $2N$ roots, N will correspond to the expression of the Green's function for $z > 0$ while the remaining poles (obtained by negating the first set) are associated with the (symmetric) expression for $z < 0$. Therefore, the number of *independent* poles is N transverse poles and N longitudinal poles, with a possible reduction in number when some of these poles are complex.

We will not present the calculations of the k_z integral but mention mainly that the expression provides effectively a spectral expansion of the Green's functions in terms of modes indexed by k_x and k_y, where k_z for the nth transverse/longitudinal pole is given by the relation

$$k_{z,n}^{T,L} = \pm\sqrt{\left(k_n^{T,L}\right)^2 - k_x^2 - k_y^2}. \tag{13.12}$$

The choice of the sign is made in each case to satisfy the radiation condition at infinity.

Detailed discussion of the physical interpretation of the results will be given elsewhere. Here, we mainly present some of the most interesting features. We start by the observation that if the poles are all real, then the structure of the near field is similar to the situation studied for free space Part I. The situation becomes dramatically different when one of the poles is in the second (forth) quadrant, with very small imaginary part. This pole type will give rise to *backward wave propagation*. For small k_x and k_y, it will effectively couple to the far field zone, giving rise to a negative-group-velocity wave, in perfect agreement with the conclusion arrived to in Chapter 12 using a source-free analysis.

For complex roots with non-negligible imaginary parts, the two rules governing the distribution of the poles will force additional poles to appear in the other quadrants of the complex k_z plane. The field associated with the four poles is not pure evanescent, but has the form of *attenuating propagating mode*, where propagation includes *both* forward *and* backward phase shift. If all the poles are of this type, which is possible because equations (13.7) and (13.8) have even order, then *no* pure propagating modes in the antenna radiation are possible and we predict then that the field is fully localized by the medium. Figure 13.1(bottom) summarizes the physical interpretation of the nonlocal medium Green's function poles.

13.5 APPLICATIONS FOR NEAR-FIELD ENGINEERING, METAMATERIALS, AND NANOELECTROMAGNETICS

We mentioned in the introduction that the first motivation for understanding the physics of radiation and propagation in nonlocal media is the need to model complex environments surrounding conventional systems such as mobile devices. Indeed, when the medium is very dense, it is possible as a first approximation to simplify the fields by spatial averaging on an effective wavelength larger than the typical inter-element spacings between the constituents of the complex environment. In this case, the first important model of the new electromagnetic features introduced by this surrounding is the inclusion of explicit dependence on $\mathbf{k}$ in the dielectric tensor of the effective medium.

One may also work in the converse direction. Specifically, it is possible to envision the scenario where the electromagnetic engineer already starts with an artificial medium described by a nonlocal dielectric function in an attempt to use the new properties of this environment to modify and shape the radiation characteristics of an antenna inserted into his/her medium, see Figure 13.2.

The theory developed in this chapter provides some indications about how this program can be realized. Indeed, by engineering the poles of the nonlocal MTM to lie deep in the complex plane, we showed than radiation energy can become totally confined within a finite shell around the source. The thickness of this shell depends directly on how far one can push the complex poles away from the real axis in Figure 13.1. This potential localization can be used in circuits and systems to store electromagnetic energy in the medium surrounding them, possibly recovering this energy in later time or sending it to another location.

Another potential engineering potential of using nonlocal MTMs for near-field engineering is depicted in Figure 13.3 and is based on the expansion of the medium Green's function to several terms each corresponding to longitudinal/transverse pole as shown in (13.10). Each pole will correspond to a "channel" described by its corresponding Green's function $\bar{\mathbf{G}}_n^{T,L}(\mathbf{r},\mathbf{r}')$. The total radiated

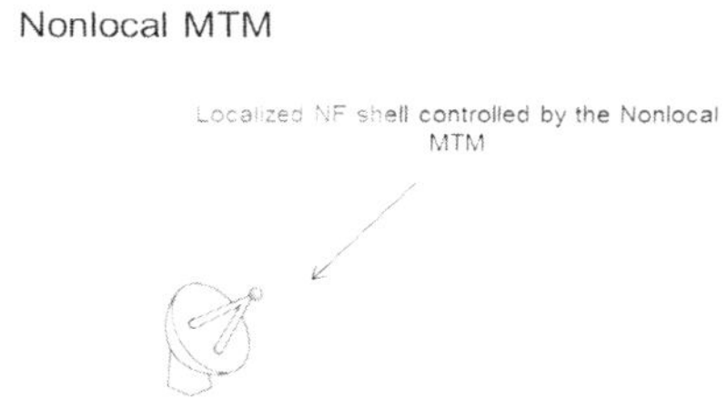

Figure 13.2 A nonlocal MTM is used to engineer the NF of an antenna embedded inside the medium.

field in the presence of the nonlocal MTM is the sum of all these channels or filters. Therefore, in a way formally resembling far-field antenna arrays, *a nonlocal MTM with properly designed poles can act like a near-field array though there is only one antenna present*, giving rise to what we call *virtual near-field arrays*, a phenomenon physically impossible without nonlocality. This feature can be used in shaping the near field. Indeed, since the MTM designer can change with the locations of the complex poles, the relative form and strength of each "channel" $\bar{\mathbf{G}}_n^{T,L}(\mathbf{r},\mathbf{r}')$ can be changed in turn in order to effect a desired form in a way similar to pattern shaping of far-field antenna arrays radiating in free space.

Finally, we present brief remarks on the emerging field of nanoelectromagnetics. Since nano-antennas and nano-scatterers appear to obey a form of Maxwell's equations essentially nonlocal in nature [142], knowledge of how macroscopic systems behave in the presence of complex environments modeled as nonlocal effective media (a direct example is periodic structures) can help researchers further understand the hitherto unexplored consequences of replacing traditional Maxwell's equations by the new nonlocal ones valid at the microscopic scale. Conversely, research in nanoelectromagnetics can influence microwave antenna design by exploiting the formal similarly in the mathematical structures of the electrodynamics at the nanoscale and nonlocal macroscopic electromagnetics such as the one developed in this chapter.

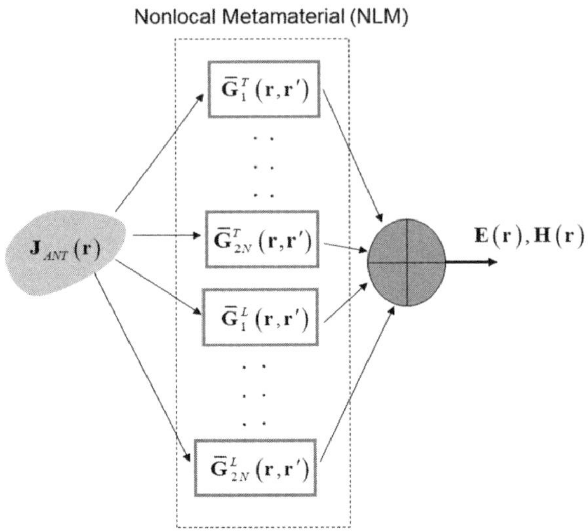

Figure 13.3 Virtual near-field array created by the additional complex poles introduced by nonlocality.

13.6 CONCLUSION

The chapter proposed a program for the study of electromagnetic radiation in the near zone of sources embedded in complex media modeled as nonlocal materials. The basic method first obtaining the Green's function of this medium (the electromagnetic fields radiated by point source) and then understanding the physical content of the solution. Working in the spectral domain, the nonlocal medium Green's function has been obtained by generalizing the Weyl expansion to nonlocal medium. It was found that nonlocality introduces qualitatively new features in the solution not seen before in temporal dispersive media. For instance, the spatial structure of the antenna near field can be controlled by a properly designed surrounding nonlocal medium. It is found that backward propagation in the near field zone can exist in the near-zone of nonlocal media. Furthermore, it is also possible to achieve full localization of the field, which is significant for energy applications. This opens the door for novel potentials for controlling and engineering the radiation fields by carefully designing special complex surrounding media. Applications of nonlocal metamaterials for near field engineering were described and some connections with nanoelectromagnetics pointed out.

Part IV

Applications

Chapter 14

Basic Applications of the Antenna Current Green's Function

14.1 INTRODUCTION

In both theoretical and applied electromagnetics, it has always been the objective to search for intuitive methods for visualizing and handling the analysis of propagation and radiation problems. Such methods, provided they exist in a mature form, permit the systematic extension of techniques originally developed for simple systems in order to deal successfully with much larger and considerably more complex structures. The availability of advanced methodologies, however, rests entirely on a correct and rigorous mathematical understanding of the nature of the physical problem at hand; afterwards one may decide, following typical patterns of progressive learning through trial and error, on which, *methodology* best reflects the most productive approach to the description and analysis of *general* radiating structures.

The Antenna Current Green's Function (ACGF) formalism has been recently proposed (in a brief outline) for the analysis of general electromagnetic systems [5], with the theoretical, conceptual, and physical foundations developed extensively in Chapters 8 and 9. In this chapter, we focus on the *empirical* side of the ACGF formalism, outlining a general scheme for its application to the description of electromagnetic interactions in arbitrary antenna arrays working in the *receiving* mode.

The chapter is organized into three major sections. The first part, Section 14.2, deals with theoretical issues pertinent to the formulation of the ACGF of receiving antenna systems, but with focus on the applied requirements of such systems.

We first present in Section 14.2.1 the background needed to follow subsequent derivations. In Section 14.2.2, we develop the concrete form of the receiving ACGF when the excitation and observation of the received signals are enacted by means of wire systems. In Section 14.2.3, the formalism is generalized to deal with array problems. We follow this in Section 14.2.4 by an in-depth analysis of the mechanism of interaction between the geometry of an arbitrarily-shaped antenna surface and the general polarization of an incoming plane wave. A detailed system-theoretic model is singled out and illustrated for the process of attaining a better grasp of the fundamentals of the interaction picture. Finally, we provide in Section 14.5 a simplified abstract antenna system designed to present in a clear form the main ingredients of the physics of the antenna problem in the receiving mode.

In the second part of the chapter, Section 14.3, we take a closer look at some of the issues that are likely to be relevant to practical applications, but while still remaining conceptually in the discursive mode of general analysis. Some elementary numerical issues relevant to the estimation of the ACGF are discussed in a very broad manner in connection with discretization error. A simple model is advanced to illustrate the connection between the error in computing the exact ACGF and the actually observed received signals predicted using the approximation of this ACGF.

The third part, Section 14.4, we develop briefly the ACGF for all-wire antenna systems. The goal is the demonstration of some of the general theoretical discussions presented in the previous two sections within a numerical setting. The key concept is the application of the Singularity Expansion Method (SEM) to the ACGF of the system. Indeed, we perform a spectral expansion of the ACGF of the wire-antenna system and show that this ACGF can be approximated (to a high accuracy) by a very simple analytical model based on the SEM. The interaction with illuminating homogeneous (plane) and inhomogeneous waves is then evaluated analytically and the resulting expressions are verified by direct comparison with full-wave analysis. Finally, a discussion of the physical meaning of the simple analytical models is presented in which the interaction of the illumination field is illustrated as a resonant/nonresonant interaction schema.

Although the boundary condition assumed throughout this Part is the simple perfect electric (PEC) condition, the entire analysis can be applied *verbatim* to any macroscopic boundary condition, as will be shown by the authors elsewhere. However, treatment of the general case in the present work will only increase the complexity of the discussion and obscure the main insights to be developed below, which are not restricted to any particular boundary condition. Therefore, we will explicitly work here only with PEC structures. On the other hand, given some boundary condition, the main thrust of our analysis will be focus on *general* results derived for *arbitrary* antennas obeying the adopted major electromagnetic boundary condition. No special antenna examples will be investigated till Section 14.4. Throughout this chapter and as everywhere in the book, an implicit time-harmonic excitation $\exp(-i\omega t)$ is assumed and suppressed everywhere.

14.2 THE ANTENNA CURRENT GREEN'S FUNCTION FORMALISM FOR RECEIVING ANTENNA SYSTEMS

14.2.1 Rudiments of the General Formulation

In this section, we rehearse the ACGF formalism developed in Chapter 8 in a simplified form suitable for the applications to follow. We start with a general antenna system comprised of an arbitrary-shaped PEC object. The number of connected and disconnected surfaces supporting the PEC boundary condition is kept arbitrary, and so the formalism applies for both single antennas and antenna arrays. The system is immersed in infinite, homogeneous free space with permittivity ϵ_0 and permeability μ_0. Let the total PEC surface of the antenna system be S. We then distinguish a proper subset of this surface, $U \subset S$, such that an input excitation electric field $\mathbf{E}_t^{\text{ex}}(\mathbf{r}) := \hat{n} \times \mathbf{E}^{\text{ex}}(\mathbf{r})$, where $\hat{n}$ is normal to the surface S,[1] is applied, i.e., we have $\mathbf{E}_t^{\text{ex}}(\mathbf{r}) \neq 0$ only for $\mathbf{r} \in U$. Due to the disturbance of the electromagnetic boundary condition of the problem caused by the presence of the excitation field $\mathbf{E}_t^{\text{ex}}(\mathbf{r})$ within the excitation region U, the total antenna system will respond by producing a current distribution on the surface S. The relation between the input $\mathbf{E}_t^{\text{ex}}(\mathbf{r})$ and the output $\mathbf{J}(\mathbf{r})$ can be written in terms of a suitable electromagnetic *linear* operator $\mathcal{L}$ appearing in the equation

$$\mathbf{J}(\mathbf{r}) = \mathcal{L}\mathbf{E}_t^{\text{ex}}(\mathbf{r}). \tag{14.1}$$

We are not concerned with the exact details of the operator $\mathcal{L}$. The *computational* problem can be solved using integral or differential equation solvers, but the *physical* problem always remains the simple one encapsulated by the equation (14.1): The input vector field tangential to S, the excitation electric field, will produce via the system operator $\mathcal{L}$ another vector tangential field, the antenna current distribution, which is also tangential to S. This suggests the idea of a system transfer function, which we introduce now under the rubric of the antenna

1 Strictly speaking, we should write $\hat{n}(\mathbf{r})$ instead of $\hat{n}$, but we omit this for simplicity.

current Green's function (ACGF). Part of the novelty of this ACGF concept is that it describes a system existing only on a 2-dimensional surface, namely S.

Now suppose that the antenna system is excited by a point source located at $\mathbf{r}'$, i.e., let $\mathbf{E}_t^{\text{ex}}(\mathbf{r}) = \hat{\alpha}\delta(\mathbf{r}-\mathbf{r}')$, where $\hat{\alpha}$ is a unit vector tangential to S at $\mathbf{r}'$ and δ is the Dirac delta function. The antenna current Green's function $\bar{\mathbf{F}}(\mathbf{r},\mathbf{r}')$ is defined by the relation

$$\bar{\mathbf{F}}(\mathbf{r},\mathbf{r}')\cdot\hat{\alpha} = \mathcal{L}\hat{\alpha}\delta(\mathbf{r}-\mathbf{r}'), \tag{14.2}$$

which is valid for *arbitrary* vector $\hat{\alpha}$ *tangential* to the antenna system surface S. It follows that the ACGF is properly a *dyadic* function (a tensor), or probably most easily put, a *matrix*.[2] Consequently, it was proved in Chapter 8 that

$$\mathbf{J}(\mathbf{r}) = \int_U ds'\bar{\mathbf{F}}(\mathbf{r},\mathbf{r}')\cdot\mathbf{E}_t^{\text{ex}}(\mathbf{r}'). \tag{14.3}$$

The relation (14.3) presents an intuitive picture of the excitation of an antenna system. In other words, it says that the total current excited on the overall antenna surface S is given by a linear sum of the contributions originating from all point excitations located within the energizing port area U, each scaled by a proper weight specified by the ACGF $\bar{\mathbf{F}}(\mathbf{r},\mathbf{r}')$. The latter function is fixed for a given antenna system, while the geometry of the port (i.e., the geometry of the surface U) and the particular details of the excitation, i.e., the functional form of $\mathbf{E}^{\text{ex}}(\mathbf{r}')$, are left open. Since the data $\bar{\mathbf{F}}(\mathbf{r},\mathbf{r}')$ need to be computed only once and then stored for subsequent processing, the exact expression (14.3) represents the most complete picture of the operation of antenna systems in the transmitting mode.

In the receiving mode, however, one expects *another* ACGF to be constructed and used in formulas resembling (14.3). However, it can be proved using Lorentz reciprocity theorem that the receiving mode ACGF, say $\bar{\mathbf{L}}(\mathbf{r},\mathbf{r}')$, which is defined by $\mathbf{J}_{\text{rx}}(\mathbf{r}) = \int_S ds'\,\bar{\mathbf{L}}(\mathbf{r},\mathbf{r}')\cdot\mathbf{E}_t^{\text{inc}}(\mathbf{r}')$, is related to the transmitting mode ACGF $\bar{\mathbf{F}}(\mathbf{r},\mathbf{r}')$ by the simple relation $\bar{\mathbf{L}}(\mathbf{r},\mathbf{r}') = \bar{\mathbf{F}}^T(\mathbf{r}',\mathbf{r})$, where T denotes the transpose operation [13]. Therefore, the receiving mode induced current $\mathbf{J}_{\text{rx}}(\mathbf{r}')$ due to illumination by the incident field $\mathbf{E}_t^{\text{inc}}(\mathbf{r})$ is determined by the relation

$$\mathbf{J}_{\text{rx}}(\mathbf{r}') = \int_S ds\bar{\mathbf{F}}^T(\mathbf{r},\mathbf{r}')\cdot\mathbf{E}_t^{\text{inc}}(\mathbf{r}). \tag{14.4}$$

Notice that here the integration must be performed on the *entire* antenna surface S. The received current can be computed for any $\mathbf{r}'\in U$; however, in this case the

2 It was found in Chapter 8 that the nine components of the 3D cartesian ACGF tensor are not all independent even without applying reciprocity considerations. However, this fact has no bearing on what follows in this chapter.

transmitting mode ACGF $\bar{\mathbf{F}}(\mathbf{r}, \mathbf{r}')$ has to be computed for *all* the corresponding values of $\mathbf{r}' \in U$ in order to use (14.4) for predicting $\mathbf{J}_{\mathrm{rx}}(\mathbf{r}')$.

14.2.2 Formulation of the ACGF for Surface-Wire Antenna Systems

The formulation developed above assumes that the entire antenna system is represented geometrically by a 2D surface. To be sure, this is the most general situation encountered in actual physical systems. However, in realistic applications of the formalism to practical problems, it is desirable to have a simplified version of the theory reflecting certain reduced but important models. In many devices, energy is customarily injected or removed by one-dimensionally extended structures, forming the physical ports of the antenna, e.g., coaxial lines, thin microstrip lines, linear probes, etc. In such situations, the cross sectional area of the physical port is usually small compared to the total surface area of the antenna body.

This motivates the consideration of an antenna system comprised of a surface S joined (not necessarily physically touching) by a generally curved wire W. The central axis of W consists of the curve $\mathcal{C}$.[3] We also consider a radius function $a = a(\tau)$, where τ is the parameter of the curve $\mathcal{C}$, such that at the location $c(\tau)$ on $\mathcal{C}$ the radius of the wire W is given by $a(\tau)$. Assuming thin-wire approximation, which is valid when $\max_{\tau}\{\pi a(\tau)^2\}/area\,(S) \ll 1$ holds, the degrees of freedom associated with the current on the wire reduces to the one-dimensional data specified by vectors tangent to $\mathcal{C}$ as will be illustrated shortly. The surface S forms the main body of the radiating structure while the curve $\mathcal{C}$, together with the radius function a, stand for the reduced geometrical model of the antenna's port.[4] Therefore, the antenna system ANT can be defined simply as the set-theoretic union of S and $\mathcal{C}$, i.e., $ANT := S \cup W$. For brevity, we will describe the construction of the wire's surface W from the data $\mathcal{C}$ and a by the operator form $W = clnd(a, \mathcal{C})$, where $clnd$ stands for the operator forming the cylindrical surface with axis $\mathcal{C}$ and radius a.

After choosing a global coordinate system to describe the surface-wire antenna system, it is possible to construct a function that assigns to each point on the curve $\mathbf{r} \in \mathcal{C}$ the tangent to the curve at the point; this tangent is given by the unit vector $\hat{\alpha}^0(\mathbf{r}) = dc(\tau)/d\tau$, where τ is the parametric length of $\mathcal{C}$ and the derivative is evaluated at τ_0 such that $\mathbf{r} = c(\tau_0)$. We can then find a vector tangential to W parallel to the tangential to $\mathcal{C}$. From now on, we refer to the former as $\hat{\alpha}^0(\mathbf{r})$,

3 A *curve* is defined in differential geometry as a smooth map $c : [a, b] \to \mathbb{R}^3$, where $b > a \geq 0$ [?]. In this way, the curve $\mathcal{C}$ is simply the set $\mathcal{C} = \{x \in \mathbb{R}^3 | \, x = c(\tau) \text{ for some } \tau \in [a, b]\}$.

4 The still interesting case of all-wire antenna systems will be analyzed numerically in Section 14.4.

where $\mathbf{r} \in W$. This tangent vector $\hat{\alpha}^0(\mathbf{r})$ will be taken as the representation of the polarization of the surface current on the corresponding wire $W = clnd\,(a, C)$.[5]

On the other hand, it is also possible to construct at each point on the surface $\mathbf{r} \in S$ two orthonormal unit vectors $\hat{\alpha}^1(\mathbf{r})$ and $\hat{\alpha}^2(\mathbf{r})$ tangential to S at that point. The procedure is more complicated than the case with curves but is pretty much standard and can be found in any textbook on differential geometry, for example [59], [61], [58].

Based on this simple geometrical language, we find that the ACGF of the transmitting mode of the antenna system ANT can be written in the form

$$\begin{aligned}\bar{\mathbf{F}}_{ANT}(\mathbf{r}, \mathbf{r}') = \hat{\alpha}^0(\mathbf{r})\,\hat{\alpha}^0(\mathbf{r}')\,F_{00}(\mathbf{r}, \mathbf{r}') \\ +\hat{\alpha}^1(\mathbf{r})\,\hat{\alpha}^0(\mathbf{r}')\,F_{10}(\mathbf{r}, \mathbf{r}') + \hat{\alpha}^2(\mathbf{r})\,\hat{\alpha}^0(\mathbf{r}')\,F_{20}(\mathbf{r}, \mathbf{r}').\end{aligned} \tag{14.5}$$

Here $\hat{\alpha}^n(\mathbf{r})\,\hat{\alpha}^m(\mathbf{r})$ stands for the familiar tensor product.[6] The transfer function $F_{00}(\mathbf{r}, \mathbf{r}')$ presents the net effect of a point source, located at $\mathbf{r}'$ on the wire W, observed at location $\mathbf{r}$ also on the wire W, connecting therefore an excitation polarized along $\hat{\alpha}^0(\mathbf{r}')$ with an induced current polarized along $\hat{\alpha}^0(\mathbf{r})$.[7] On the other hand, the transfer functions $F_{10}(\mathbf{r}, \mathbf{r}')$ and $F_{20}(\mathbf{r}, \mathbf{r}')$ provide information about the effect of a point source, also located at $\mathbf{r}'$ on the wire W and polarized along $\hat{\alpha}^0(\mathbf{r}')$, observed at a point $\mathbf{r}$ but this time located on the surface S, and with polarization specified by $\hat{\alpha}^1(\mathbf{r})$ and $\hat{\alpha}^2(\mathbf{r})$.[8]

The receiving mode ACGF $\bar{\mathbf{L}}_{ANT}(\mathbf{r}', \mathbf{r})$ is analogously given by

$$\begin{aligned}\bar{\mathbf{L}}_{ANT}(\mathbf{r}', \mathbf{r}) = \hat{\alpha}^0(\mathbf{r}')\,\hat{\alpha}^0(\mathbf{r})\,F_{00}(\mathbf{r}', \mathbf{r}) \\ +\hat{\alpha}^0(\mathbf{r}')\,\hat{\alpha}^1(\mathbf{r})\,F_{01}(\mathbf{r}', \mathbf{r}) + \hat{\alpha}^0(\mathbf{r}')\,\hat{\alpha}^2(\mathbf{r})\,F_{02}(\mathbf{r}', \mathbf{r}).\end{aligned} \tag{14.7}$$

5 These definitions are meaningful only for *thin* wires. In particular, the points on W and C belonging to the same cross section are identified in our notation (this 'identification' can be made rigorously using the concept of equivalent class). Otherwise, there will be no advantage in ignoring other possible directions of the current polarization, say the azimuthal components.

6 Readers unfamiliar with the operation of tensor product in differential geometry can simply consider $\hat{\alpha}^n(\mathbf{r})\,\hat{\alpha}^m(\mathbf{r})$ as an entity *defined* by the relation

$$[\hat{\alpha}^n(\mathbf{r})\,\hat{\alpha}^m(\mathbf{r})] \cdot \hat{\alpha}^l(\mathbf{r}) := \hat{\alpha}^n(\mathbf{r})\left[\hat{\alpha}^m(\mathbf{r}) \cdot \hat{\alpha}^l(\mathbf{r})\right], \tag{14.6}$$

valid for arbitrary vectors $\hat{\alpha}^{n,m,l}$.

7 In other terms, $\mathbf{r} \notin clnd\,(a, C)$ or $\mathbf{r}' \notin clnd\,(a, C) \rightarrow F_{00}(\mathbf{r}, \mathbf{r}') = 0$.

8 In other terms, $\mathbf{r} \notin S$ or $\mathbf{r}' \notin clnd\,(a, C) \rightarrow F_{10}(\mathbf{r}, \mathbf{r}') = F_{20}(\mathbf{r}, \mathbf{r}') = 0$.

The crucial statement of reciprocity $\bar{\mathbf{L}}_{ANT}(\mathbf{r}', \mathbf{r}) = \bar{\mathbf{F}}^T_{ANT}(\mathbf{r}, \mathbf{r}')$ [13] consequently implies

$$\begin{gathered} F_{00}(\mathbf{r}, \mathbf{r}') = F_{00}(\mathbf{r}', \mathbf{r}), F_{10}(\mathbf{r}, \mathbf{r}') = F_{01}(\mathbf{r}', \mathbf{r}), \\ F_{20}(\mathbf{r}, \mathbf{r}') = F_{02}(\mathbf{r}', \mathbf{r}). \end{gathered} \tag{14.8}$$

By putting (14.7), (14.8), and (14.4) together, we arrive at the main result

$$\mathbf{J}_{\text{rx}}(\mathbf{r}') = \hat{\alpha}^0(\mathbf{r}') \int_{ANT} ds\, \mathbf{u}(\mathbf{r}', \mathbf{r}) \cdot \mathbf{E}_t^{\text{inc}}(\mathbf{r}), \tag{14.9}$$

where

$$\begin{aligned} \mathbf{u}(\mathbf{r}', \mathbf{r}) := \hat{\alpha}^0(\mathbf{r}) F_{00}(\mathbf{r}, \mathbf{r}') + \hat{\alpha}^1(\mathbf{r}) F_{10}(\mathbf{r}, \mathbf{r}') \\ + \hat{\alpha}^2(\mathbf{r}) F_{20}(\mathbf{r}, \mathbf{r}'). \end{aligned} \tag{14.10}$$

Therefore, we have proved the following simple but worth mentioning result: ***The ACGF of the surface-wire antenna system, where the physical port is located in the wire part, has the structure of a one-dimensional tensor***. That is, the ACGF of the surface-curve system ANT is not properly a full (3×3) matrix as in the general case of the tensor $\bar{\mathbf{F}}$ appearing in (14.3), but rather a 1×3 row tensor. Furthermore, by inspecting the expressions (14.9) and (14.10), we find that the receiving port signal can be completely recovered from the *transmitting* mode data $\mathbf{u}(\mathbf{r}', \mathbf{r})$ by a simple inner-product-and-integration operation applied to the illuminating field tangential component $\mathbf{E}_t^{\text{inc}}(\mathbf{r})$. We again recall that this field is arbitrary and so (14.9) is valid for *both* far- and near- field electromagnetic interaction with the receiving array. Some numerical predictions using (14.9) are presented in Section 14.4 for far-field illumination.

14.2.3 Extension to Array Configurations

The mathematical treatment of array problems follows from the general relations (14.3) and (14.4). Due to their importance in practical applications, we pause here to present the relevant expressions and discuss their physical significance and interpretation.

Assume we are given an array of N elements, each consisting of a radiating surface S_n and a feed-wire system W_n. The total surface S and feed systems W are expressed in terms of the individual components as

$$S = \bigcup_{n=1}^{N} S_n, W = \bigcup_{n=1}^{N} W_n. \tag{14.11}$$

By substituting (14.11) to (14.9), the received current at the mth port is given by

$$\mathbf{J}_{\mathrm{rx}}\left(\mathbf{r}_m\right) = \hat{\alpha}_m^0\left(\mathbf{r}_m\right)\sum_{n=1}^{N}\int_{S_n+W_n} ds\,\mathbf{u}_n\left(\mathbf{r}_m,\mathbf{r}\right)\cdot\mathbf{E}_t^{\mathrm{inc}}\left(\mathbf{r}\right), \tag{14.12}$$

where

$$\begin{aligned}\mathbf{u}_n\left(\mathbf{r}_m,\mathbf{r}\right) := {} & \hat{\alpha}_n^0\left(\mathbf{r}\right)F_{00}^n\left(\mathbf{r}_m,\mathbf{r}\right) \\ & +\hat{\alpha}_n^1\left(\mathbf{r}\right)F_{01}^n\left(\mathbf{r}_m,\mathbf{r}\right)+\hat{\alpha}_n^2\left(\mathbf{r}\right)F_{02}^n\left(\mathbf{r}_m,\mathbf{r}\right).\end{aligned} \tag{14.13}$$

Here, $\mathbf{r}_m$ denotes the position of the receiving port of the mth antenna, i.e., $\mathbf{r}_m \in W_m$. We can easily see that $\mathbf{r} \notin W_n$ and $\mathbf{r} \notin S_n$ entail $F_{00}^n\left(\mathbf{r}_m,\mathbf{r}\right) = 0$ and $F_{01}^n\left(\mathbf{r}_m,\mathbf{r}\right) = F_{02}^n\left(\mathbf{r}_m,\mathbf{r}\right) = 0$, respectively.

Expression (14.12) allows us to interpret the interaction of a receiving array with an arbitrary incident field as the exact sum of N channels or subsystems, each characterized by a Green's function of the form $\hat{\alpha}_m^0\left(\mathbf{r}_m\right)\mathbf{u}_n\left(\mathbf{r}_m,\mathbf{r}\right), n = 1,...,N$, specifying how an excitation at location $\mathbf{r}$ is transformed into the mth receiving port signal observed at $\mathbf{r}_m$.

Further inspection of the form (14.13) suggests that each of these subsystem Green's functions can in turn be split into the sum of two terms. The first is $\hat{\alpha}_m^0\left(\mathbf{r}_m\right)\hat{\alpha}_n^0\left(\mathbf{r}\right)F_{00}^n\left(\mathbf{r}_m,\mathbf{r}\right)$, which provides information about the interaction between the feed system W_n of the nth antenna with the currently observed mth feed $\mathbf{r}_m \in W_m$. The second subsystem Green's function is the one corresponding to $\hat{\alpha}_m^0\left(\mathbf{r}_m\right)\left[\hat{\alpha}_n^1\left(\mathbf{r}\right)F_{01}^n\left(\mathbf{r}_m,\mathbf{r}\right)+\hat{\alpha}_n^2\left(\mathbf{r}\right)F_{02}^n\left(\mathbf{r}_m,\mathbf{r}\right)\right]$, and supplies information about how two orthogonal excitations at the surface S_n of the nth antenna interact with the receiving port W_m at $\mathbf{r}_m$.

For various applications, it is of interest to spell out the ACGF of the transmitting mode. From reciprocity, we can immediately write,

$$\bar{\mathbf{F}}\left(\mathbf{r},\mathbf{r}_m\right) = \sum_{n=1}^{N}\bar{\mathbf{F}}_n\left(\mathbf{r},\mathbf{r}_m\right), \tag{14.14}$$

where

$$\begin{aligned}\bar{\mathbf{F}}_n\left(\mathbf{r},\mathbf{r}_m\right) = \Big[& \hat{\alpha}_n^0\left(\mathbf{r}\right)\hat{\alpha}_m^0\left(\mathbf{r}_m\right)F_{00}^n\left(\mathbf{r},\mathbf{r}_m\right) \\ & +\hat{\alpha}_n^1\left(\mathbf{r}\right)\hat{\alpha}_m^0\left(\mathbf{r}_m\right)F_{10}^n\left(\mathbf{r},\mathbf{r}_m\right) \\ & +\hat{\alpha}_n^2\left(\mathbf{r}\right)\hat{\alpha}_m^0\left(\mathbf{r}_m\right)F_{20}^n\left(\mathbf{r},\mathbf{r}_m\right)\Big].\end{aligned} \tag{14.15}$$

The excitation of the antenna array in the transmitting mode proceeds according to an externally applied field of the form

$$\mathbf{E}_t^{\mathrm{ex}}(\mathbf{r}) = \sum_{m=1}^{N} \hat{\alpha}_m^0(\mathbf{r}_m)\, b_m \delta(\mathbf{r} - \mathbf{r}_m), \tag{14.16}$$

where each of the N ports is energized by a field with strength and phase given by the complex number b_n. From (14.3), (14.14), and (14.15), we arrive at

$$\mathbf{J}(\mathbf{r}) = \sum_{m=1}^{N}\sum_{n=1}^{N} b_m \bar{\mathbf{F}}_n(\mathbf{r}, \mathbf{r}_m) \cdot \hat{\alpha}_m^0(\mathbf{r}_m). \tag{14.17}$$

This represents the radiating current of the array due to a complex vector of excitation $[b_n]_{n=1}^{N}$ as in (14.16). Some numerical examples illustrating this general process will be given in Section 14.4.

It is worth mentioning that in order to compute the receiving mode signal in terms of transmitting mode data, it is *not* enough to have access to the *total* current (14.17). Instead, one must acquire a set of N independent ACGFs, each obtained by setting $b_m = 1$ and $b_n = 0$ for $n \neq m$; then by repeating this procedure for $m = 1, ..., N$, the N $\hat{\alpha}_m^0(\mathbf{r}_m)$-components of the ACGFs in (14.15) can be determined, which is the information needed to predict the received signal via (14.12).

14.2.4 Interaction with Plane Waves: An Excursus on Polarization

The ACGF formalism permits a decoupling of the illuminating field from the physical body of the receiving antenna, where the latter acts as a continuously-distributed linear system. We will now show that this fundamental insight allows us to isolate the purely geometrical factors, which are involved in the interaction of an arbitrarily polarized incident field with the generally curved shape of the antenna's physical body surface, from the purely electromagnetic response of the antenna. In other words, it is possible to locate the various components in the polarization-port interaction within the receiving antenna formula (14.4) by carefully studying the manner in which the ACGF varies *as a 2D tensor* defined on the surface S.

In order to better clarify the nature of the contribution made here, the reader may consider the simple example of linear wire antenna working in the receiving mode. In this case, the purely geometric considerations of how the polarization of the incident field interacts with the shape of the antenna is well understood.

For simplicity, assume that the plane wave is linearly polarized. In this case, polarization matching can be defined as the inner product of the direction of the electric field vector and the unit vector oriented alone the wire. However, when the antenna surface is arbitrarily curved, there seems to be no general way to theoretically describe this purely geometric interaction between the polarization of the illumination field and the antenna surface, which is now a function of the general position on this surface.[9] The following discussion provides, to our knowledge, the first general analysis of this geometric interaction, which turned out to depend essentially on the concept of an exact transfer function in space, i.e., the ACGF defined on the antenna system surface S.

For a direct demonstration of this point, we compute the geometrical factors needed in the evaluation of the total interaction of an incident field, possessing a randomly-oriented polarization vector, with a fixed-shape single antenna element embedded in array configurations. As will be seen shortly, the computation of the expected value of the relevant statistical variables in the field-antenna interaction scheme is performed only once for a given shape, while the electromagnetic data may vary according to the array geometry, excitation mechanism, or modifications in the electromagnetic material properties. Such interesting general conclusion appears to be one of the gains obtained from developing a rigorous system-function formalism for antenna systems using the new concept of the ACGF.

Consider an incident plane wave given by the form $\mathbf{E}^{\mathrm{inc}}(\mathbf{r}) = \mathbf{E}_0(\mathbf{k}_0)\exp(i\mathbf{k}_0 \cdot \mathbf{r})$. Since we are assuming a PEC boundary condition, the magnetic field of the incident wave does not interact with the antenna. Using (14.4), we write the signal induced in the physical port U as

$$\mathbf{J}_{\mathrm{rx}}(\mathbf{r}') = \int_S ds\, \bar{\mathbf{F}}^T(\mathbf{r}, \mathbf{r}') \cdot \mathbf{E}_0(\mathbf{k}_0) \exp(i\mathbf{k}_0 \cdot \mathbf{r}). \tag{14.18}$$

Since the amplitude vector $\mathbf{E}_0(\mathbf{k}_0)$ does not depend on $\mathbf{r}$, it is not difficult to see that (14.18) entails that the induced signal is given in terms of the *spatial Fourier transform* of the vector $\bar{\mathbf{F}}^T(\mathbf{r}', \mathbf{r}) \cdot \mathbf{E}_0(\mathbf{k}_0)$ evaluated at the "spatial frequency" $\mathbf{k}_0$. It still needed, however, to decouple the dependence on the polarization vector $\mathbf{E}_0(\mathbf{k}_0)$. In order to achieve this, we first expand the receiving mode ACGF in terms of a suitable set of locally tangential orthonormal vectors $\hat{\alpha}^n(\mathbf{r}')$ and $\hat{\alpha}^m(\mathbf{r})$ as follows

$$\bar{\mathbf{F}}^T(\mathbf{r}, \mathbf{r}') = \sum_{n,m} \hat{\alpha}^n(\mathbf{r}')\, \hat{\alpha}^m(\mathbf{r})\, F_{mn}(\mathbf{r}, \mathbf{r}'), \tag{14.19}$$

9 Because, as is known in elementary differential geometry, curvature varies from one point to another on general 2D surface. For example, see [60].

where the indices m and n run through 1 to 2. The four functions $F_{mn}(\mathbf{r},\mathbf{r}')$ are those due to the *transmitting* mode (reciprocity utilized). Substituting (14.19) into (14.18), we obtain after using (14.6)

$$\mathbf{J}_{\text{rx}}(\mathbf{r}') = \sum_{n,m} \hat{\alpha}^n(\mathbf{r}') \int_S ds \left[\hat{\alpha}^m(\mathbf{r}) \cdot \mathbf{E}_0(\mathbf{k}_0)\right] \times F_{mn}(\mathbf{r},\mathbf{r}') \exp(i\mathbf{k}_0 \cdot \mathbf{r}). \tag{14.20}$$

Therefore, although the factor $\hat{\alpha}^n(\mathbf{r}')$ has been moved outside the Fourier integral, the integrand still contains the factor $\hat{\alpha}^m(\mathbf{r}) \cdot \mathbf{E}_0(\mathbf{k}_0)$, which ties the polarization vector $\mathbf{E}_0(\mathbf{k}_0)$ with the inner *geometrical* structure of the ACGF as represented in this case by $\hat{\alpha}^m(\mathbf{r})$. We suggest using the product theorem of the Fourier transform to re-write (14.20) in the following illuminating form

$$\mathbf{J}_{\text{rx}}(\mathbf{r}') = \sum_{m,n} \hat{\alpha}^n(\mathbf{r}') F_{mn}(\mathbf{k},\mathbf{r}') * P^m_{\mathbf{k}_0}(\mathbf{k})\big|_{\mathbf{k}=\mathbf{k}_0}, \tag{14.21}$$

where

$$F_{mn}(\mathbf{k},\mathbf{r}') := \int_S ds\, F_{mn}(\mathbf{r},\mathbf{r}') \exp(i\mathbf{k} \cdot \mathbf{r}), \tag{14.22}$$

$$P^m_{\mathbf{k}_0}(\mathbf{k}) := \mathbf{E}_0(\mathbf{k}_0) \cdot \hat{\alpha}^m(\mathbf{k}), \tag{14.23}$$

$$\hat{\alpha}^m(\mathbf{k}) := \int_S ds\, \hat{\alpha}^m(\mathbf{r}) \exp(i\mathbf{k} \cdot \mathbf{r}). \tag{14.24}$$

The star '$*$' denotes the convolution operation.[10] Equation (14.21) states that the received signal can be understood as the sum of convolutions of the Fourier transform of the ACGF components with the Fourier transforms of the polarization interactions factors defined in (14.23), where the convolution is evaluated at the wavevector of the incident field $\mathbf{k}_0$.

The advantages of the new form (14.21) can be elucidated as follows. From (14.19), we can see that the ACGF contains two sets of data, the first, a purely geometric one, is reflected by the orthonormal vectors $\hat{\alpha}^n(\mathbf{r}')$ and $\hat{\alpha}^m(\mathbf{r})$, while the second is purely electromagnetic and consists of the response functions

10 The product theorem states that $\mathcal{F}\{fg\} = \mathcal{F}\{f\} * \mathcal{F}\{g\}$, where $\mathcal{F}$ stands for the Fourier transform operator. A sufficient condition for the applicability of the product theorem is that both f and g are square integrable [68]. This, however, follows immediately in our case from the fact the ACGF, which is a continuous function, is defined on the *bounded* support S.

$F_{mn}(\mathbf{r}, \mathbf{r}')$.[11] From (14.18), it is possible to see that the received signal can be best understood as related to the Fourier transform of the ACGF. However, by the expression (14.21), we now see that this Fourier transform is actually given as the *convolution* of two spectral functions (Fourier transforms). The first spectral function is given by (14.22) and is a function independent of the details of the illumination field. The second spectral function (14.23) is given simply as the inner product of the plane wave amplitude $\mathbf{E}(\mathbf{k}_0)$ and the Fourier transforms of the ACGF's excitation position vectors $\hat{\alpha}^m(\mathbf{k})$, the latter being a purely geometrical function. Now, ***while both*** $F_{mn}(\mathbf{k}, \mathbf{r}')$ **and** $\hat{\alpha}^m(\mathbf{k})$ ***are fixed for a given electromagnetic problem, i.e., by being not dependent on the variations of the illumination field, the quantity*** $P^m_{k_0}(\mathbf{k})$ ***varies according to*** $\mathbf{E}(\mathbf{k}_0)$ ***in a simple way as specified in*** (14.23).

To better appreciate the significance of the decoupling claimed in (14.21) between the geometrical and electromagnetic responses, let us introduce a plane wave incidence problem where the polarization of the wave is random. More precisely, we assume that for a given $\mathbf{k}_0$, the direction of the electric field vector, which is contained in a plane perpendicular to $\mathbf{k}_0$, is specified by the angle ψ, the rotation being around $\mathbf{k}_0$ with respect to some reference. That is, we assume linear polarization in the form $\mathbf{E}_0(\mathbf{k}_0, \psi) = E_0(\mathbf{k}_0, \psi)\,\hat{e}(\mathbf{k}_0, \psi)$, where $\hat{e}$ is a unit (real) vector. Other types of polarization, circular and elliptic, can also be treated but the logic of the demonstration to follow remains unaltered. Henceforth, the probability density function (pdf) of the polarization $\mathbf{E}_0(\mathbf{k}_0, \psi)$ will be denoted by $f_{\mathbf{k}_0}(\psi)$.[12] The received signal becomes a function of the polarization in the form $\mathbf{J}_{\rm rx}(\mathbf{r}', \mathbf{k}_0; \psi)$, i.e., we treat $\mathbf{k}_0$ as a deterministic entity while the polarization specified by ψ is a random variable with a known statistical distribution. We would like to compute the expected value of the received signal in (14.18), which is given by

$$\langle \mathbf{J}_{\rm rx}(\mathbf{r}', \mathbf{k}_0; \psi)\rangle_\psi := \int_0^{2\pi} d\psi \mathbf{J}_{\rm rx}(\mathbf{r}', \mathbf{k}_0; \psi) f_{\mathbf{k}_0}(\psi)\,. \tag{14.25}$$

11 Notice, however, that the geometric and electromagnetic structures of the ACGF are not totally independent of each other. One way to see this is by developing a special tensor transformational calculus, which demonstrates how the geometrical and electromagnetic information get entangled with each other when transforming from one local coordinate system on S to another. But for a *fixed* set of local coordinate systems, it is possible to make a boundary line between the geometric and electromagnetic information as mentioned above. Since we are not going to change the coordinate system in the analysis of this chapter, the foregoing considerations are enough.

12 For the generalization to other types of polarization, a *joint* pdf of the familiar polarization parameters can be introduced.

Using the general form (14.21), we can reach the following result

$$\mathbf{J}_{\mathrm{rx}}(\mathbf{r}') = \sum_{m,n} \hat{\alpha}^n(\mathbf{r}') F_{mn}(\mathbf{k},\mathbf{r}') * \left\langle P^m_{\mathbf{k}_0,\psi}(\mathbf{k})\right\rangle_\psi \Big|_{\mathbf{k}=\mathbf{k}_0}, \tag{14.26}$$

where the *characteristic polarization* **k**-*function* is defined as

$$\left\langle P^m_{\mathbf{k}_0,\psi}(\mathbf{k})\right\rangle_\psi := \hat{\alpha}^m(\mathbf{k}) \cdot \int_0^{2\pi} d\psi \mathbf{E}_0(\mathbf{k}_0,\psi) f_{\mathbf{k}_0}(\psi). \tag{14.27}$$

In deriving (14.26), we have interchanged the s- and ψ- integrals in (14.20), an operation permitted because the two integrals are finite [?]. Next, the transition from the resulting equation to (14.26) proceeds according to the same line of reasoning used in moving from (14.20) to (14.21), i.e., via invoking the product theorem of the Fourier transform.

Our main result is the expression (14.26), which presents a significant advance in the understating of how antennas interact with plane waves with random polarization. In words, it says that the purely geometrical aspects of polarization-antenna interactions can be completely decoupled from the purely electromagnetic response. Indeed, the geometrical polarization-antenna interaction is given by the spectral **k**-function $\left\langle P^m_{\mathbf{k}_0,\psi}(\mathbf{k})\right\rangle_\psi$, while the purely electromagnetic response of the antenna system, independent of the incident wave, is that due to the other spectral **k**-function $F_{mn}(\mathbf{k},\mathbf{r}')$. Remembering that convolution operations are intimately connected with linear systems acting as a filter, it is possible to appreciate the value of the insight provided by the derivation of (14.21): The average received signal of an arbitrarily-shaped antenna interacting with a plane wave possessing random polarization is expressible as a linear sum of **k**-shifted basic response functions, in this case $\left\langle P^m_{\mathbf{k}_0,\psi}(\mathbf{k}_0-\mathbf{k})\right\rangle_\psi$, each weighted by the antenna (purely electromagnetic) response functions $F_{mn}(\mathbf{k},\mathbf{r}')$. For each statistical distribution of the incident plane waves, there is one and only one such characteristic polarization function $\left\langle P^m_{\mathbf{k}_0,\psi}(\mathbf{k})\right\rangle_\psi$, which is a function of only the geometrical shape of the antenna. On the other hand, the electromagnetic response $F_{mn}(\mathbf{k},\mathbf{r}')$ may change by several mechanisms. For example, by changing the feed location, the function $F_{mn}(\mathbf{k},\mathbf{r}')$ will change (notice the already explicit dependence on $\mathbf{r}'$). Also, by placing another object near the antenna, close enough to produce an electromagnetic coupling, the response functions $F_{mn}(\mathbf{k},\mathbf{r}')$ will change even for the same port location $\mathbf{r}'$. However, *none* of these mechanisms will alter the characteristic polarization **k**-function (14.27).

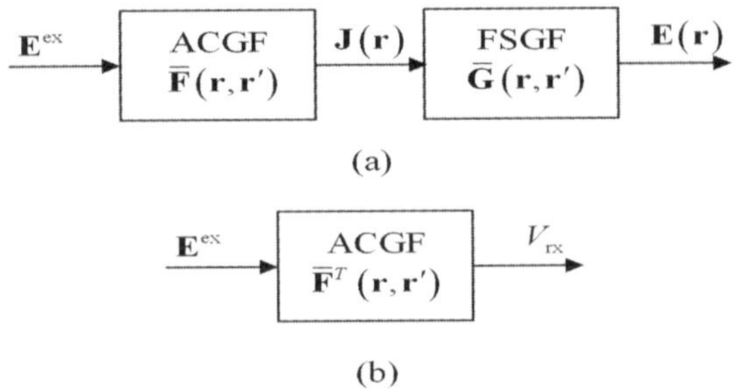

Figure 14.1 System-theoretic formulation of the interaction of arbitrarily-shaped receiving antenna with arbitrarily-polarized plane-wave illumination. The system model combines the 1) statistical, 2) geometrical, and 3) pure electromagnetic aspects of general antenna array systems.

For example, imagine an array composed of identical radiating and receiving elements. Therefore, since the geometry of each element is the same, effectively one characteristic polarization function $\left\langle P^{m}_{\mathbf{k}_0,\psi}(\mathbf{k})\right\rangle_{\psi}$ will be used for the entire array. If the distances between the elements are sufficiently small, the relative position and orientation of each antenna within the array will change the purely electromagnetic responses encapsulated in the functions $F_{mn}(\mathbf{k},\mathbf{r}')$ because of the phenomenon of mutual coupling (electromagnetic interaction between nearby elements). However, this will not change $\left\langle P^{m}_{\mathbf{k}_0,\psi}(\mathbf{k})\right\rangle_{\psi}$ as mentioned above since it depends only on the *local* geometry of antenna element and the illumination field. Therefore, with the expression (14.26) we have achieved the interesting insight that ***the purely geometrical interaction aspects between randomly polarized incident plane waves impinging on an arbitrarily curved antenna surface, on one hand, and the purely electromagnetic response of the antenna, on the other hand, can be completely decoupled in such a way that the former acts like a linear*** **k*-filter on the latter*.**

This interpretation is illustrated in Figure 14.1. In the top figure, we show a block diagram representation of the interaction of an antenna with randomly polarized plane waves. The "input" to the system is the spectral response function $F_{mn}(\mathbf{k},\mathbf{r}')$. The **k**-filter's "impulse response" is the characteristic polarization **k**-function $\left\langle P^{m}_{\mathbf{k}_0,\psi}(\mathbf{k})\right\rangle_{\psi}$. The proper output of this system is the mnth component $J^{mn}_{\text{rx}}(\mathbf{k},\mathbf{r}')$ of the induced current.[13] Now, this system view is actually a "meta-system" description. The reason is that the "input" $F_{mn}(\mathbf{k},\mathbf{r}')$ can be changed

13 See (14.26).

in two ways (as mentioned above), mainly by changing the feed location and/or placing a nearby object capable of electromagnetically coupling with our antenna under consideration, i.e., in effect both methods entail a change in the purely electromagnetic system as such. The engineer interested in studying the performance of a system involving the same antenna shape design, but excited by different feed positions or placed within a complex environment, can utilize the metasystem-theoretic description in Figure 14.1 because the polarization-geometry interaction aspects are totally buried in the **k**-filter's $\left\langle P^{m}_{\mathbf{k}_0,\psi}\left(\mathbf{k}\right)\right\rangle_{\psi}$, which is invariant with respect to the above mentioned factors.

On the other hand, it is possible to analyze the problem in a direct system level, rather than the previous meta-systemic picture, by considering the interpretation given in the bottom of Figure 14.1. Here, the feed position and the electromagnetic environment are fixed, but the polarization aspects are varied. The electromagnetic response $F_{mn}\left(\mathbf{k},\mathbf{r}'\right)$ plays then the role of the **k**-filter impulse response, and the output signal can be understood in the familiar interpretation of the convolution as the superposition of properly scaled shifted versions $F_{mn}\left(\mathbf{k}_0-\mathbf{k},\mathbf{r}'\right)$.

In both interpretations of Figure 14.1, the meta-systemic and the systemic, we should observe a peculiarity in the antenna problem. The proper **k**-filter impulse response in the top figure involves the spectral value $\mathbf{k}_0$; also, in the bottom figure, the input **k**-function involves $\mathbf{k}_0$; this dependence manifests itself through the appearance of $\mathbf{k}_0$ as subscript in the characteristic polarization function $P^{m}_{\mathbf{k}_0,\psi}$. That does not reduce the significance of the convolution interpretation outlined above, but suggests that care should be paid to this technical point when working with expressions like (14.21) and (14.26). Indeed, the convolution operation is applied only to the normal (main) functional **k**-arguments appearing in the functional expressions of Figure 14.1. In other words, the *subscript* $\mathbf{k}_0$ does not enter into the convolution but acts merely as an index.

14.3 SOME EMPIRICAL CONSIDERATIONS IN THE ACGF FORMALISM

14.3.1 Some General Remarks

This work can be considered as a generalization of the specific approaches and formulations usually presented in the literature of applied electromagnetics. The ACGF formalism adapted for analyzing receiving mode antenna arrays has something in common with the conventional approach, but also it represents a significant advance. The thing we find in common with existing literature is the fact that in (14.4) we compute the receiving mode signal at a *localized* region, in this case U. That is, we are still interested in a terminal-like quantity as is the case with the circuit paradigm in applied electromagnetics. However, in light of recent advances in our understanding of the field-theoretic aspects of general electromagnetic systems, we highlight the fact that expression (14.4) presents a decoupling of the field illumination from the actual physical body of the antenna system. In other words, reciprocity is not employed in connection with two fixed sets of sources and fields in order to compute the system circuit Y-matrix [28], but rather applied in a peculiar manner to the system function $\bar{\mathbf{F}}(\mathbf{r}, \mathbf{r}')$ itself. In this way, the expression given in (14.4) is valid for *arbitrary* field illumination, in particular, it is still valid when the exciting field is that belonging to the *near* field of another object close to the antenna system under consideration.

Besides its fundamental theoretical importance, the use of the ACGF, computed only once in the transmitting mode, has some pragmatic advantages when compared with other conventional methods found in literature. These include reduced memory requirements, especially for statistical analysis of fixed arrays; the ability to identify and isolate the contributions of the various physical parts of the antenna systems to the total received signals; and the utility of the method in providing information going beyond the circuit Y-matrix, such as the manner in which diverse field variations determine the overall system performance, especially in connection with mutual coupling. This is mainly the outcome of the ability to use (14.4) as an *exact* relation for *arbitrary* field illumination.

14.3.2 On the Approximation of the Receiving Antenna System via the ACGF Formalism

It will be demonstrated now that in practical antenna configurations, which use port excitation more complex than the simple point-source excitation type, it is still possible to gain a considerable insight into the operation of the antenna in the receiving mode using transmitting mode data. This follows conceptually from the geometric fact that in principle one may approximate any complex port by a superposition of suitable point-like ports. This will require considering the issue of *averaging* field quantities at the physical port in order to obtain simple scalar circuit quantities like voltage and current. The remaining parts of this Section will address this topic at a very general level. Numerical examples for special antennas can be found in Section 14.4.

When the excitation field is a simple delta source $\mathbf{E}^{\mathrm{ex}}(\mathbf{r}) = \hat{\alpha}(\mathbf{r}_0)\,\delta(\mathbf{r}-\mathbf{r}_0)$, equation (14.3) tells us that the corresponding current distribution obtained in this case is given simply by $\mathbf{J}_{\mathbf{r}_0}(\mathbf{r}) = \bar{\mathbf{F}}(\mathbf{r},\mathbf{r}_0)\cdot\hat{\alpha}(\mathbf{r}_0)$. It is important to keep in mind all the time that this current distribution $\mathbf{J}_{\mathbf{r}_0}(\mathbf{r})$ is *not* a Green's function, but it is *numerically* equal to the Green's function that connects a point excitation at $\mathbf{r}_0$ polarized along $\hat{\alpha}(\mathbf{r}_0)$ with the field vector induced at location $\mathbf{r}$. Now consider the situation in which a field excitation $\mathbf{E}^{\mathrm{ex}}(\mathbf{r})$ that is *not* composed of a discrete sum of Dirac delta functions is applied within the physical port area U. Denote the current distribution obtained by means of (14.3) by $\mathbf{J}_U(\mathbf{r})$. It should be obvious that this current is not even numerically equal to any Green's function whatsoever; in particular, $\mathbf{J}_U(\mathbf{r})$ does not represent a proper transfer function connecting *arbitrary* excitation in U with the induced current at $\mathbf{r}$. The nature of $\mathbf{J}_U(\mathbf{r})$ in general depends strongly on the nature of $\mathbf{E}^{\mathrm{ex}}(\mathbf{r})$. If a new excitation is applied, also within U, there is no general recipe for expressing the new induced current in terms of the old current when the latter is obtained with respect to the old excitation field. This observation explains some of the differences between the ACGF formalism and other methods existing in literature based on direct applications of the reciprocity theorem. The method of this chapter is essentially a Green's function strategy (and reciprocity was applied to it only in a later stage via a special theorem [13],) so our attention must be directed toward antenna responses to *point* sources, not arbitrary input fields.

However, we notice that the quantity $\bar{\mathbf{F}}(\mathbf{r},\mathbf{r}')$ is actually a *family* of current distributions functions (functions of $\mathbf{r}$) indexed by the source position $\mathbf{r}'$. There exists to date no theoretical method allowing us to relate different members of this family indexed by different source variables. In this situation, the only method available is brute-force numerical computation in which we store all the functions $\bar{\mathbf{F}}(\mathbf{r},\mathbf{r}')\,,\mathbf{r}'\in U$, with sufficiently dense sampling of the port region U. Notice that this need to be done only once since the *exact* receiving mode relation (14.4) is valid now for arbitrary illumination field.

Although this can be achieved in principle, it might be computationally expensive in many cases, and one would like to consider a simpler approach that does not involve computing a family of ACGFs but rather a single computation. In many common full-wave electromagnetic solvers, realistic port models are integrated within the main code in order to provide users with the ability to compute the system performance and make direct comparison with existing experimental setups. This situation is more urgent in general electromagnetic systems, where several antennas are connected with each other, beside other circuits, through waveguide structures. The excitation field is then that of the waveguide propagating mode, and hence not a simple point source. The ACGF formalism can still of course deal with such situations using (14.3), provided we know the eigenmodes of the excitation waveguide, and then employing (14.4) to predict the receiving antenna port signal; but we would like to directly investigate the empirical aspects as they arise in practice.

In order to give a general view applicable to any antenna system, We will employ the ACGF formalism to derive the main relation connecting a non-impulsive field excitation with the actually measured currents in physical experiments. The antenna system to be considered in this derivation will be the reduced surface-wire system of Section 14.2.2. The reasons for this choice is the simplicity of the resulting derivation and also the direct relevance to the physical set ups normally encountered in practice.

Let us first start with the simplest form of an excitation field deviating from the ideal Dirac delta field. Consider the following *U-characteristic field* defined by

$$\mathbf{E}_U^{\mathrm{c}}(\mathbf{r}) := \hat{\alpha}^0(\mathbf{r}) \begin{cases} 1, & \mathbf{r} \in U, \\ 0, & \text{otherwise.} \end{cases} \tag{14.28}$$

Here, U refers to a segment of the wire W used to model the antenna wire part (which supports the physical port) that is exposed to excitation.[14] It is the segment of $W = \mathit{clnd}(a, \mathcal{C})$ on which a *constant* excitation field, polarized along the vector $\hat{\alpha}^0(\mathbf{r})$ parallel to the tangent to $\mathcal{C}$, is applied to the antenna system. An arbitrary nonuniform excitation, e.g., an eigenfunction of the waveguide feed line impinging on the region U of the feeding line $\mathcal{C}$, can be approximated by a superposition of functions all having the form of (14.28).[15]

We will assume that the antenna is excited by some U-characteristic field, i.e., $\mathbf{E}^{\mathrm{ex}}(\mathbf{r}) = \mathbf{E}_U^{\mathrm{c}}(\mathbf{r})$. Naturally, one starts with the smallest possible U that a mesh

14 That is, $U = \mathit{clnd}(a, C_0)$, where $C_0 \subseteq C$ is a segment of the curve $\mathcal{C}$.

15 More precisely, let the total region of nonuniform excitation be expanded as the sum

$$U = \bigcup_{n=1}^{N} U_n$$

of N mutually disjoint segments U_n. The excitation field is then approximated as

$$\mathbf{E}^{\mathrm{ex}}(\mathbf{r}) = \sum_{n=1}^{N} \beta_n \mathbf{E}_{U_n}^{\mathrm{c}}(\mathbf{r}),$$

where β_n represents the average value of the actual excitation in the segment U_n. The smaller the segments, the better the approximation. However, notice that in this case the approximate excitation is discontinuous (but remains *piecewise* continuous). This has no bearing on our analysis and results. A more sophisticated version of our argument can be developed in such a way that the approximating functions are themselves continuous. We omit this construction for space limitation.

descretization can achieve for a given operating frequency. A small subdivision is assumed to correspond to a "point" excitation, and hence the claim that the obtained current distribution is numerically equal to the proper ACGF of the problem. However, as we pointed above, this is strictly speaking not true. The quantity obtained by exciting the antenna system using the field $\mathbf{E}_U^{\mathrm{c}}(\mathbf{r})$ can be computed from (14.3) and is given by

$$\mathbf{J}_U(\mathbf{r}) = \int_U ds' \, \bar{\mathbf{F}}(\mathbf{r}, \mathbf{r}') \cdot \hat{\alpha}^0(\mathbf{r}'). \tag{14.29}$$

Here, $\mathbf{J}_U(\mathbf{r})$ stands for the current at **r** due to a U-characteristic excitation. We will now suppose that this current distribution, *erroneously* identified with the ACGF, is used in (14.4) in order to compute the receiving mode induced current in U, which we denote by $J_{\mathrm{rx}}(U)$. More specifically, we use (14.29) and (14.4) to compute the quantity

$$J_{\mathrm{rx}}(U) := \gamma \int_S ds \int_U ds' \hat{\alpha}^0(\mathbf{r}') \cdot \bar{\mathbf{F}}^T(\mathbf{r}, \mathbf{r}') \cdot \mathbf{E}^{\mathrm{inc}}(\mathbf{r}), \tag{14.30}$$

where γ is a constant inserted here for units consistency but its value will be fixed in few moments.

The fact that the regions U and S are both compact allows us using a well-known result in real analysis in order to interchange the order of integrations in (14.30) [72]. Performing this operation and then using again (14.4) to redefine the s-integral, we arrive at

$$J_{\mathrm{rx}}(U) = \gamma \int_U ds' \hat{\alpha}^0(\mathbf{r}') \cdot \mathbf{J}_{\mathrm{rx}}(\mathbf{r}'). \tag{14.31}$$

This is our main result. In other words, (14.31) says that ***the induced current, computed by*** **simply** ***treating the current distribution generated in the transiting mode due to a*** U***-characteristic excitation as the proper ACGF, is proportional to the average of the correct receiving mode induced current within the region*** U. Now we notice that all physical measurement devices essentially involve an averaging over a small but finite spatial region. We can then conveniently choose $\gamma = 1/area(U)$. In practice, such a choice can always be made by a proper calibration of the device used to measure the receiving mode induced current.

Relation (14.31), although can be rendered intuitive by constructing a suitable toy model like the one presented in Appendix 14.5, is actually nontrivial because it involves interchanging the order of two repeated integrations, an operation that is not always permissible.[16] In the opinion of the authors, the result (14.31) represents the essential statement of the *empirical* content of the ACGF formalism. It connects the numerically computed estimation, $\mathbf{J}_{\text{rx}}(\mathbf{r}')$ on the RHS, with the physically measured value in a real set up, $J_{\text{rx}}(U)$ of the LHS. Indeed, the error in the theoretical estimation of the true ACGF of the problem, which cannot be obtained analytically in most cases, is "balanced" by the fact that measuring devices produce quantitative values only for averages. It appears then that the size of the discretization in the numerical strategy used to invert the electromagnetic operator $\mathcal{L}$ in (14.2), roughly the size of U in (14.28), is delimited by the spatial resolution of the measuring device.

14.4 APPLICATION OF THE ACGF FORMALISM TO THE ANALYSIS OF LINEAR WIRE ANTENNA SYSTEMS

In this section, we apply the general ACGF formalism to the analysis of electromagnetic field interactions with receiving wire antenna systems. The numerical examples provided below serve to illustrate the basic ideas, rather of being exhaustive. Indeed, it seems there is a wide range of traditional problems, e.g., mutual coupling, that can be studied in a rigorous fashion using the idea of the ACGF as an exact transfer function in space. Such further studies will be taken up elsewhere.

14.4.1 General Formulation

Assume a configuration of N linear thin-wire antennas. The antennas are distributed in space at locations described by the position vectors of the delta-source excitation $\mathbf{r}_m, m = 1, 2, ..., M \leq N$. Each antenna is oriented alone a direction described by the rotation matrix R_n applied to $\hat{z}$ chosen as an arbitrary initial direction. At the terminal of each antenna, a load with complex impedance Z_{L_n} is connected.

Since we are working with thin-wire antennas, both the delta source polarization and the direction of the induced current on the antenna are known a priori: The direction of the antenna element itself. Fixing the polarity of the source and current,

16 For example, in the case of infinite radiators, say a structure no more complicated than a PEC infinite plane sheet, the s-integral in (14.30) becomes infinite. It is not clear a priori that in general this integral will be uniformly convergent for all values of $\mathbf{r}'$. Consequently, the derivation of (14.31), provided it is possible, becomes more involved, and may require special treatment. In this book, however, we focus our attention on physically realizable antennas, and to those correspond only *compact* regions S.

we use the unit vector $\hat{\alpha}_n := \mathrm{R}_n \hat{z}$ to describe the orientation of the nth antenna. The source index m should then be mapped to its corresponding antenna, and to accomplish this we use the device $\sigma : m \to \sigma_m$, where in this notion σ_m is the index of the antenna excited by the mth source. We then write

$$\bar{\mathbf{F}}(\mathbf{r}, \mathbf{r}_m) = \sum_{n=1}^{N} \hat{\alpha}_n \hat{\alpha}_{\sigma_m} F_{nm}(\mathbf{r}, \mathbf{r}_m). \tag{14.32}$$

The function F_{nm} gives the current response at the nth antenna due to a delta-source excitation located at the port of the σ_mth antenna.[17] Substituting (14.32) to (14.14), it is evident that

$$\bar{\mathbf{F}}(\mathbf{r}) = \sum_{n=1}^{N} \sum_{m=1}^{M} \hat{\alpha}_n \hat{\alpha}_{\sigma_m} F_{nm}(\mathbf{r}, \mathbf{r}_m), \tag{14.33}$$

which completely characterizes the N-element antenna array system in the transmitting mode when all M sources are active at their respective ports.

We are interested in the interaction of the array described by the ACGF (14.33) with an incident electric field given by $\mathbf{E}(\mathbf{r})$. Applying the inverse reciprocity theorem to (14.32) and using (14.4), we obtain

$$\mathbf{J}(\mathbf{r}_m) = \hat{\alpha}_{\sigma_m} \sum_{n=1}^{N} \int_{S_n} ds\, \hat{\alpha}_n F_{nm}(\mathbf{r}, \mathbf{r}_m) \cdot \mathbf{E}(\mathbf{r}), \tag{14.34}$$

where S_n is the surface of the nth wire antenna. Equation (14.34) gives an exact expression for the received mode current at the mth port in response to an arbitrary near- or far- field $\mathbf{E}(\mathbf{r})$.

A direct numerical verification of the relation (14.34) is given as follows. We use the Method of Moment (MoM) to compute the ACGF of the transmitting mode by exciting the system by an approximation of a delta source. The antenna system is a dipole comprised of two identical thin wires excited by the delta source at an infinitesimal gap. The inverse reciprocity theorem derived in Chapter 9 is used to relate the ACGF of the receiving mode to this data and the induced open circuit (Norton) current source I_{oc} as predicted by (14.34) is computed by numerical integration. The receiving antenna's port is connected to a load impedance of 50Ω

17 Notice that we describe the polarization of both the source field and the response current using the same set of unit vectors because of the peculiarity of thin-wire linear antennas. In the most general case, two different sets of vectors must be used.

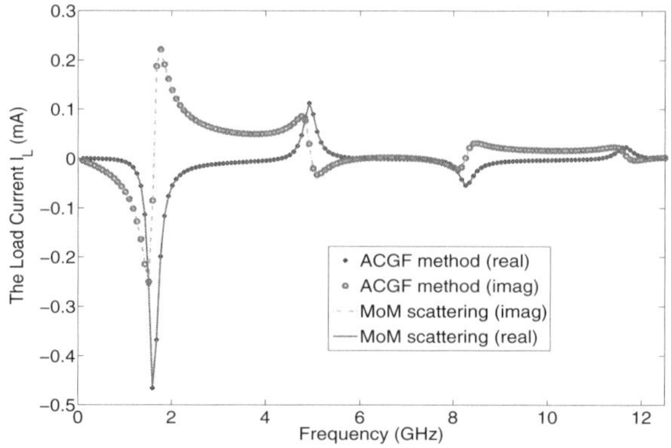

Figure 14.2 Interaction of a linear wire antenna of length 90 mm with an incident plane wave at $\theta = 5^o$. The wire has a radius of 3 mm and the feed location is at the center. The receiving mode antenna has a load impedance of 50 Ω.

and the load current is computed by means of the relation

$$I_L(f) = \frac{Z_{Th}(f)}{Z_L + Z_{Th}(f)} I_{oc}(f), \tag{14.35}$$

where $Z_{Th}(f)$ is the input impedance of the antenna at frequency f computed in the transmitting mode. The results are compared with a full-wave scattering solution (MoM) in Figures 14.2 and 14.3 for plane waves of angles of incidence given by $\theta = 5, 75^o$, respectively. Excellent agreement is found in all angles of incidence but only those two cases are presented here for brevity.

14.4.2 Spectral Analysis of the ACGF of Linear Wire Antennas

We will approach the spectral structure of the ACGF of wire antenna systems not from the standpoint of the Fourier integral, but rather from the much simpler perspective of finite sum of complex exponential. The disadvantage is merely theoretical, which is that the spectral representation is approximate, not exact. The advantages are, however, considerable. We will achieve a drastic reduction in the computational complexity of the resulting expressions. Indeed, we now show that the interaction of the linear wire antenna with homogeneous and inhomogeneous

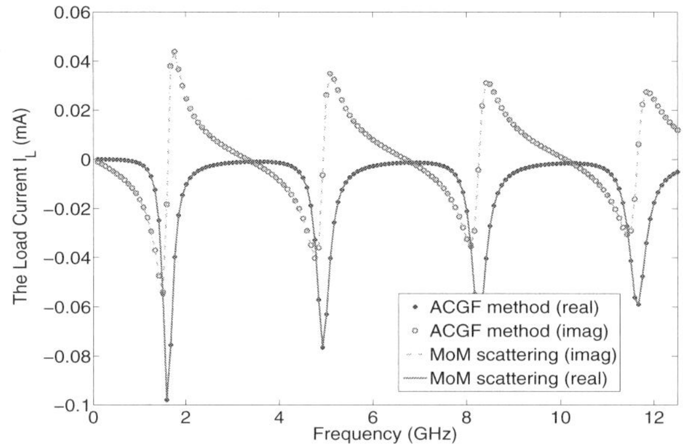

Figure 14.3 The same problem as in Figure 14.2 but with incident wave at $\theta = 75^o$.

waves can be approximated by simple analytic expressions that are as accurate as full-wave numerical solution.

Using Proney's or the matrix pencil methods, we write

$$F_{nm}(\mathbf{r}, \mathbf{r}_m) \simeq \begin{cases} \sum_{l=1}^{P_{nm}} a_l^{nm} e^{\kappa_l^{nm} s}, & \mathbf{r} \in S_n \\ 0, & \text{, otherwise.} \end{cases} \tag{14.36}$$

Here, a_l^{nm} and k_l^{nm} are complex numbers determined by the algorithm. In this expression, P_{nm} is the number of terms in the spectral expansion of the ACGF F_{nm} and is a parameter of the method. Recall that s in (14.36) is a local length parameter of the nth wire as introduced in Section 14.2.2.

In order to verify the analytical expression (14.36), we obtained an approximation of the 0.75λ dipole antenna described in the caption of Figure 14.2 using the classical Prony's algorithm. The comparison with the numerically obtained ACGF is given in Figure 14.4. As can be seen, excellent agreement is observed throughout the entire span of the antenna system with only three complex exponentials.

In the previous example, each wire was individually modeled analytically by finding its Prony's approximation. Since the feed location was chosen at the center, the models for the two wires are identical (by symmetry). In order to demonstrate an asymmetrical case, a different antenna system is presented in which the dipole antenna is comprised of two unequal wires with final results shown in Figure 14.5. Notice that, as anticipated in the theoretical part of this part, the ACGF is strongly

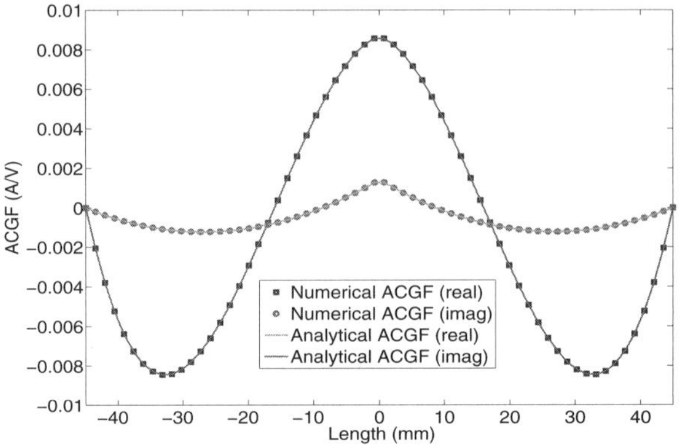

Figure 14.4 Comparison between the analytical approximation of the ACGF using (14.36) and the numerical ACGF obtained using MoM. The antenna system is the same problem as in Figure 14.2 but with frequency fixed at 5 GHz. The analytical approximation (14.36) uses three complex exponentials (for each wire) with amplitudes $a_1 = -0.0030 + j0.0141$, $a_2 = 5.4979 - j0.6061$, $a_3 = 3.1206 + j1.8557$; the corresponding exponents are $\kappa_1 = 136.05 - j31.817$, $\kappa_2 = -5.2724 - j82.0551$, $\kappa_3 = 9.7095 + j81.6741$.

dependent on the location of the source. In this case, each wire was approximated by its own Prony's model (details can be found in the caption of Figure 14.5).

14.4.3 Interaction with Homogeneous and Inhomogeneous Waves

Consider an incident wave of the form $\mathbf{E}(\mathbf{r}) = \mathbf{E}_0(\mathbf{k}_0)\exp(i\mathbf{k}_0 \cdot \mathbf{r})$ in which $\mathbf{k_0}$ can be either pure real or pure imaginary vector (homogeneous and inhomogeneous waves, respectively). By substituting (14.36) into (14.34), we obtain

$$\mathbf{J}(\mathbf{r}_m) \simeq \hat{\alpha}_{\sigma_m} \sum_{n=1}^{N} \mathbf{E}_0(\mathbf{k}_0) \cdot \hat{\alpha}_n \times \sum_{l=1}^{P_{nm}} a_l^{nm} \frac{e^{(\kappa_l^{nm} + i\mathbf{k}_0 \cdot \hat{\alpha}_n)s_n^e} - e^{(\kappa_l^{nm} + i\mathbf{k}_0 \cdot \hat{\alpha}_n)s_n^i}}{\kappa_l^{nm} + i\mathbf{k}_0 \cdot \hat{\alpha}_n}. \tag{14.37}$$

Here, s_n^i and s_n^e are the length parametric values of the initial and end points, respectively, of the nth linear wire antenna.

In order to simplify the analysis, we consider a wire antenna oriented along the z-axis with length $2L$. Its port is located at the wire center (placed at the origin),

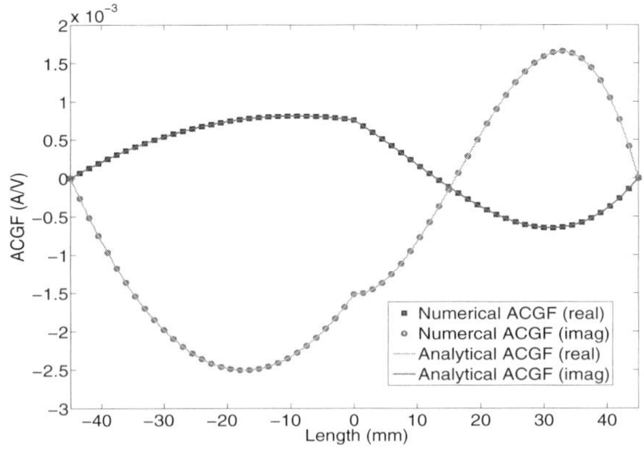

Figure 14.5 Comparison between the analytical approximation of the ACGF using (14.36) and the numerical ACGF obtained using MoM. The antenna system is a dipole antenna with the upper segment of length 0.75λ and the lower with 0.375λ at frequency 5 GHz. The analytical approximation (14.36) uses four complex exponentials. For the 0.75λ wire, the amplitudes are $a_1 = -84.883 - j3.799$, $a_2 = -2834.4 + j9.2359$, $a_3 = 3167.3 + j154.07$, $a_4 = -247.21 + j2915.1$; the corresponding exponents are $\kappa_1 = -2.5156 + j3.8101$, $\kappa_2 = 3.7737 + j2.4996$, $\kappa_3 = -3.7863 - j2.5557$, $\kappa_4 = 2.5285 - j3.7541$. For the 0.375λ segment, a model of three complex exponentials is found with amplitudes $a_1 = -21336.0 + j4424.9$, $a_2 = 14609.0 + j16118.0$, $a_3 = 6725.7 - j20541.0$; the corresponding exponents are $\kappa_1 = 0.1768 + j0.2668$, $\kappa_2 = 0.1424 - j0.2867$, $\kappa_3 = -0.3192 + j0.0199$.

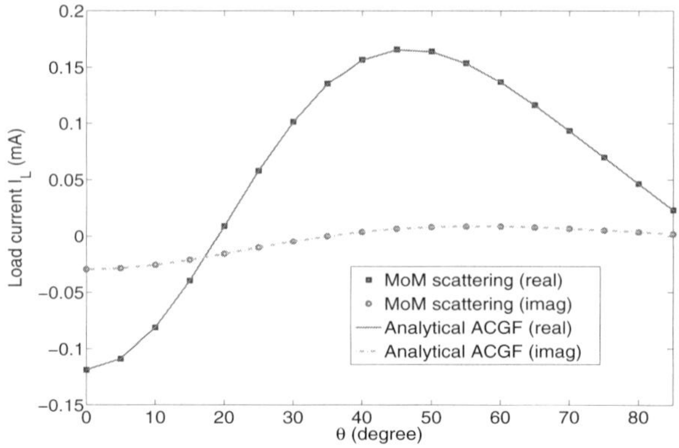

Figure 14.6 The same problem as in Figure 14.2 but with frequency fixed at 5 GHz. The angle of incidence θ is varied and the prediction of the analytical formula (14.37) is compared with full-wave MoM solution.

where a load impedance Z_L is connected. Using (14.37), the interaction with a plane wave produces a received voltage given by

$$V_L = (\mathbf{E}_0 \cdot \hat{z}) \frac{Z_L Z_{Th}}{Z_{Th} + Z_L} \sum_{l=1}^{P} a_l I\left(\kappa_l, \mathbf{k}_0 \cdot \hat{z}\right), \tag{14.38}$$

where we define the *spectral interaction function* $I(\kappa, k)$ by

$$I\left(\kappa, k\right) := \frac{e^{(\kappa + ik)L} - e^{-(\kappa + ik)L}}{\kappa + k}. \tag{14.39}$$

Therefore, in principle, an arbitrary-size wire antenna system can be effectively replaced with a simple analytical expression capable of predicting the interaction with arbitrary plane waves with a level of accuracy matching full-wave scattering results.

The analytical expression (14.38) is verified in Figure 14.6 by working with the 0.75λ dipole system. Again, very good agreement is observed demonstrating the ability of the obtained simple analytical expression to predict the correct received signals for all angles of incidence on the basis of the ACGF formalism developed earlier.

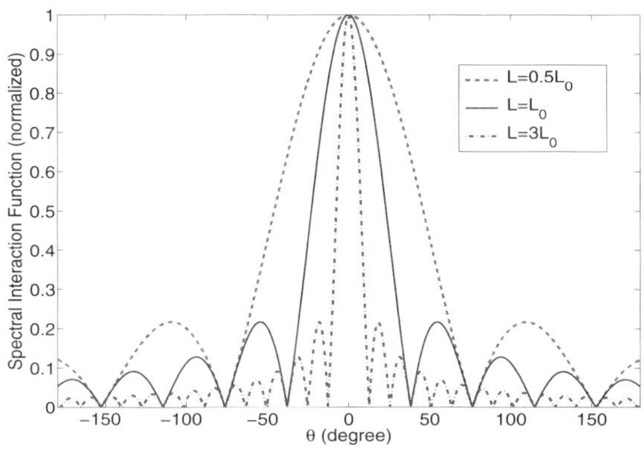

Figure 14.7 Study of the spectral interaction function (14.39) for varying half-dipole length L. The results are normalized with respect to $1/2L$. In the figure, $L_0 = 45\,\mathrm{mm}$.

14.4.4 General Remarks on the Results

We conclude by discussing the physical meaning of the results obtained so far. The spectral interaction function defined by (14.39) appears as the basic building blocks through which the received signal due to arbitrary illumination homogeneous or inhomogeneous waves can be computed. Indeed, as can be seen from relation (14.38), the induced signal is simply a weighted sum of the values of the interactions $I(\kappa_l, k)$ evaluated for a given plane wave wavenumber k (along the wire antenna) and a given κ_l. In order to have some idea about the general behavior of the solution, we show in Figure 14.7 the spectral interaction function $I(\kappa_l, k)$ evaluated at $\kappa_l = 0$ for several wire lengthes L. It is clear that this function behaves as a filter (in the spectral variable k) centered at $k = 0$ with a bandwidth determined by the extension of the antenna system. For larger systems (larger L), the spectral resolution significantly increases. This example confirms the conclusions arrived at but on a general theatrical ground in Chapter 9.

Notice that for the particular example worked out above, the pertinent spectral variable k is simply $k_0 \cos\theta$. Therefore, we have $k \in [-k_0, k_0]$. Now Figure 14.7 clearly shows that the spectral interaction function $I(\kappa_l, k)$ is centered at κ_l (with a spatial k-bandwidth of roughly $2\pi/L$). Therefore, an incident plane wave will interact *resonantly* with the receiving system only if $\mathrm{Im}\kappa_l \in [-k_0, k_0]$. For this reason, we will call the numbers κ_l the complex poles of the ACGF, while a_l serves as the strengths or amplitudes of these poles. If the condition $\mathrm{Im}\kappa_l \in [-k_0, k_0]$ is satisfied, then the ACGF pole at κ_l will contribute resonantly to the received signal with a relative strength (weighting factor) determined by a_l.

It is of interest to notice that most practical wire antenna systems give rise to complex poles, not pure imaginary.[18] This can be seen from a glance at the numerical examples given in the captions of Figures 14.4 and 14.5. At the frequency of choice, the resonance interaction band is $[-104.7\mathrm{m}^{-1}, 104.7\mathrm{m}^{-1}]$. The three ACGF poles of Figure 14.4 fall within this resonance range, although they also possess real parts that cause the response to deviate from the theoretical peak level. This suggests that full resonance can be obtained in the case of inhomogeneous waves, i.e., interaction with the near field of another source.

14.5 TOY MODEL: A DISCRETE ANTENNA SYSTEM

In order to illustrate the intuitive character of the foregoing formulation of the fundamental problem of how antennas work in the transmitting and receiving mode, and the nature of their interrelationships, we develop in this part a fictitious toy model of a discrete antenna composed of a finite number of points. Each point can be excited by an externally applied electric field and respond by producing a current distribution. Moreover, distant points can respond to an excitation not involving any of them. Assume further that each point is 'internally identical' to all other points, i.e., there is no *absolute* identity distinguishing one point from the other. Only the *relative* position of two points with respect to another set of points is considered as a legitimate way of marking or baptizing one point as "different" from the other. This imaginary situation can be easily attained if we assume that the discrete antenna is comprised of geometrical points imbedded in empty infinite, isotropic, and homogeneous space, say the standard Euclidean space $\mathbb{R}^3$.

For simplicity, let us first assume that the problem is scalar (acoustic). Our first scalar discrete antenna configuration is shown in Figure 14.8(top), which consists of four points labeled by the numerals from 1 to 4. Point 1 acts in this model as the antenna port. In the top figure, the transmitting mode is illustrated, with various weights F serving as transfer functions from Point 1 to Points 2-4. Let the excitation field be given by a unit-amplitude delta function $E^{\mathrm{ex}} = \delta(\mathbf{r} - \mathbf{r}_1)$ located at the position $\mathbf{r}_1$. The "current distribution" on this antenna is given by

$$\begin{aligned} J(\mathbf{r}) = {} & F(\mathbf{r}_1, \mathbf{r}_1)\,\delta(\mathbf{r} - \mathbf{r}_1) + F(\mathbf{r}_2, \mathbf{r}_1)\,\delta(\mathbf{r} - \mathbf{r}_2) \\ & + F(\mathbf{r}_3, \mathbf{r}_1)\,\delta(\mathbf{r} - \mathbf{r}_3) + F(\mathbf{r}_4, \mathbf{r}_1)\,\delta(\mathbf{r} - \mathbf{r}_4). \end{aligned} \tag{14.40}$$

18 The reason is that the ACGF has to decay into exactly zero at the terminals of the wire system.

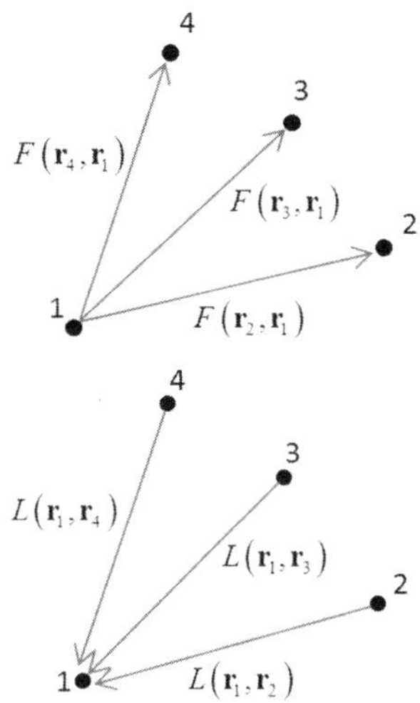

Figure 14.8 Discrete antenna system with Point 1 acting as a transmit-receive port. Top: Transmitting Mode. Bottom: Receiving mode. In both cases, the self-interaction terms $F(\mathbf{r}_1, \mathbf{r}_1)$ and $L(\mathbf{r}_1, \mathbf{r}_1)$ are not shown explicitly.

This form has a simple and intuitive physical meaning. The antenna system in the transmitting mode can be considered as a set of four subsystems, each being given by the pair of points $(1, n)$, where $n = 1, 2, 3, 4$. The subsystem $(1, n)$ has an "input terminal" represented by Point 1 while the "output terminal" is Point n. It possesses a transfer function specified by the complex number $F(\mathbf{r}_1, \mathbf{r}_n)$, which is the transmitting mode ACGF. The actual antenna current is, therefore, simply the linear superposition of the outputs of all these subsystems. The antenna can then be viewed as an ensemble of linear subsystems orchestrated by a common excitation point (in this example, Point 1).

Let us move now to the bottom of Figure 14.8. Here, the antenna, being operated in the receiving mode, is excited by an arbitrary illuminating field, which becomes after interacting with discrete locations

$$\begin{aligned} E^{\text{inc}}(\mathbf{r}) = e_1\delta(\mathbf{r} - \mathbf{r}_1) + e_2\delta(\mathbf{r} - \mathbf{r}_2) \\ + e_3\delta(\mathbf{r} - \mathbf{r}_3) + e_4\delta(\mathbf{r} - \mathbf{r}_4), \end{aligned} \tag{14.41}$$

where the quantities $e_n, n = 1, 2, 3, 4$ are complex constants. The received current can be expressed as

$$J(\mathbf{r}) = [L(\mathbf{r}_1, \mathbf{r}_1) e_1 + F(\mathbf{r}_1, \mathbf{r}_2) e_2 \\ + F(\mathbf{r}_1, \mathbf{r}_3) e_3 + F(\mathbf{r}_1, \mathbf{r}_4) e_4] \, \delta(\mathbf{r} - \mathbf{r}_1) . \tag{14.42}$$

Now, it is not difficult to see that again the discrete antenna system can be viewed as an ensemble of subsystems, each represented by the pair $(n, 1)$, and is associated with transfer functions $L(\mathbf{r}_1, \mathbf{r}_n), n = 1, 2, 3, 4$, respectively.

We can appreciate the difference between the transmitting and receiving modes of the antenna at two levels. The first level is straightforward. Here, we observe that in the receiving mode, the antenna interacts with a field that generally illuminates the *entire* antenna's body.[19] This can be seen at once by comparing (14.41) with (14.42). In the former case, only the localized excitation field applied at Point 1 interacts with the antenna. In the latter case, Point 1 receives contributions from field-antenna interactions at *all* points, including Point 1 itself, which serves here as the receiving port. Now, all other points in the situation depicted in Figure 14.8(bottom) also receive in terns contributions from the excited points. For example, it is possible to write down equations similar to (14.42) for each of the other Points 2-4. However, we restricted ourselves to Point 1 in the receiving mode because here we work within the paradigm of circuit theory, in which only terminal-like quantities are taken into consideration.[20] It can be seen then that the physics of the receiving mode, if understood as the manner in which ensembles of linear subsystems superimpose themselves upon each other in order to give rise to the total contribution at terminal-like quantities, is not identical to the physics of the transmitting mode, the reason being that the receiving mode signal depends on the particular circumstances of *how* the incident field interacts with the entire body of the antenna system.

The second level is a more profound one. There is an essential difference in the antenna system, with respect to its two principal modes of operation, that is usually overlooked. We first start by observing the *formal* similarities between the top and the bottom of Figure 14.8. Indeed, for both modes of operation, the same paths between Point n and Point 1 are traversed twice in opposite directions. This fact, that two interactions appear to occur along the same line, is probably the unconscious motivation for researchers to look for reciprocity theorems assuring

19 Unless the incident field is a localized beam, which is typical in optics, not in microwave applications.

20 See Chapter 8 for extended discussion of the role of circuit theory in the analysis of general electromagnetic systems.

us that somehow the net interaction between two pairs of points is the same if we merely swap the terms of one pair. In the case of electromagnetic systems obeying certain conditions (e.g., see [28]), it is possible indeed to prove this intuitive guess. We ask then the following question, Is this reciprocity relation obvious a priori? Our answer is that it is not obvious, and the fact that it can be proved for many practical electromagnetic systems should not deceive us into thinking that there is something intuitively clear about reciprocity. Actually, there is a *fundamental* difference between the two modes of operation illustrated in Figure 14.8. We explain this difference as follows.

By assumption all parts of the discrete antenna system in Figure 14.8 are endowed by *internally identical* electromagnetic structure. That amounts to the supposition that it is not possible to distinguish between any two points if focus is concentrated on some inner structure responsible for generating the differential relation between the respective two points (if such relation exists).[21] Now, in general two transfer functions $F(\mathbf{r}_1, \mathbf{r}_n)$ and $F(\mathbf{r}_1, \mathbf{r}_m)$ are different for $n \neq m$. One may wonder what is the reason for this difference if the points are by definition internally identical. The answer is simply that two points may differ in their *relative* position with respect to other parts of the system considered as a frame of reference. Let us take a concrete example. In Figure 14.8(top), we may ask ourselves about the possible reason that may explain a difference in the numerical values of $F(\mathbf{r}_1, \mathbf{r}_2)$ and $F(\mathbf{r}_1, \mathbf{r}_3)$. The answer is that Point 2 *is* different from Point 3 when compared to each other with respect to their position relative to Point 4. The subsystem comprised of Points 1 and 2 is then different from the subsystem comprised of Points 1 and 3, a difference to be explained by reference to an external part of the system serving as a frame of reference, in this case Point 4.

Now let us focus on one pair of points, say the one composed of Points 1 and 2. In the transmitting mode, an excitation field is placed at Point 1 while the induced current is observed at Point 2. In the receiving mode, the excitation is at

21 There is nothing extraordinary about such assumption. It is actually the standard situation encountered in classical macroscopic electrodynamics. For example, two perfectly conducting spheres are internally identical electromagnetically because the PEC boundary condition of each is not disturbed by the existence of the other close to it. Note that this situation is not applicable in the case of quantum systems in particular, and nanoscale electrodynamics in general. In the latter case, the mere presence of nearby object may lead to direct coupling capable of changing the electromagnetic properties of the subsystem under consideration. Finally, we mention that this footnote does not suggest that the points in Figure 14.8 are infinitesimal PEC spheres, since in such case the model looses its depth. These points are *geometrical* points, i.e., infinitesimal primitive part-structures; and therefore they are *not* infinitesimal *physical* points, e.g., atoms or fundamental particles surrounded by void. In other words, the part-points of the discrete antenna are "formal atoms," not material atoms.

Point 2 while the observation is enacted at Point 1. By an argument that should now be familiar to the reader, Points 1 and 2 are different from each other (for example, compare their relative positions to Point 4). Consequently, the physical problem of placing the excitation at Point 1 is different in general from the corresponding problem consisting of placing the excitation this time at Point 2. Indeed, the electromagnetic environment in the latter case is different from the former, where by 'electromagnetic environment' we mean the system of *other* points existing near the point under consideration. It follows that associated with the pair of Points 1 and 2 are *two* subsystems, one to be represented by the *ordered* pair $(1, 2)$, while the other is denoted by $(2, 1)$; the former being a subsystem in the transmitting mode, while the other the generally different subsystem in the receiving mode.

Our conclusion is that the applicability of reciprocity-type theorems to a wide range of electromagnetic problems, in such a way leading to the fact that any two subsystems obtained in the transmitting and receiving modes and involving the same pair of parts are equivalent, is indeed a remarkable finding which is by no means logically obvious. Notice that the statement of reciprocity indicates that switching the positions of the source and observation points does not change the transfer functions connecting the two points; *and this turns out to be true regardless of the electromagnetic environments surrounding these two points, which in general are quite diverse and nonuniform.*[22]

The reader at this point may reflect on why in the electromagnetic *radiation* problem we find that the characteristic Green's function is indeed reciprocal. The explanation is not difficult to find. Consider the scalar problem and its Green's function $g(\mathbf{r}, \mathbf{r}') = \exp(ik|\mathbf{r} - \mathbf{r}'|)/|\mathbf{r} - \mathbf{r}'|$. Here, we find that the configuration space of the problem consists of the entire Euclidean space $\mathbb{R}^3$. Indeed, the "parts" of this systems are the set of all geometrical points $\mathbf{r} \in \mathbb{R}^3$. The space $\mathbb{R}^3$ is topologically and geometrically homogeneous: All points look exactly alike. Therefore, given two pairs of radiation problems, one described by $(\mathbf{r}, \mathbf{r}')$ and the other by $(\mathbf{r}', \mathbf{r})$, and such that $\mathbf{r}' \neq \mathbf{r}$, we find that it is impossible to distinguish between the two situations since there exists no other "special electromagnetic environment" closer to one point than the other. In this case, the statement of reciprocity is logically and conceptually obvious.

We finally mention that there is nothing essentially new in the generalization of the analysis above when the discrete antenna toy model is assumed to be vectorial instead of scalar. The corresponding equations only get more complicated since the scalar transfer functions need now to be made dyadic functions. The transpose

22 The reader may notice that our definition of 'electromagnetic environment' above has nothing to do with effective material response characterizations, such as dispersive, anisotropic, chiral, etc.

operation in the relation between $\bar{\mathbf{F}}$ and $\bar{\mathbf{L}}$ is intuitively clear in light of (14.6) since a vector that appears as part of the source in one configuration becomes part of the observation in the other configuration, and vice versa with the other vector. The passage to the limit of continuous antenna system can be made formally using the calculus of generalized functions and all summations in equations like (14.41) and (14.42) become integrals like (14.3). In our simple discrete antenna systems, the number of parts was four while in the passage to the limit the actual antenna system acquires the size of the cardinal of the continuum.

14.6 RECEIVING ANTENNA SYSTEM CIRCUIT MODEL

We review the receiving antenna circuit model that can be used in connection with the fundamental expression (14.4). It is worth adding that none of the numerical or experimental models are exact, and hence they cannot be proved from the ACGF formalism itself without appending the theory with further assumptions, for example, new axioms for circuit theory. This, however, is not usually the practice in applied electromagnetics, and hence there is no formalized theory to work with. In all cases, the empirical validation of the circuit and MoM models has been accumulating since the past sixty years and the evidence is overwhelming that such computational schemes can be successfully employed to tackle most of the practical problems to be encountered in engineering applications. The ACGF formalism provides a rigorous exact framework for understanding existing numerical and experimental antenna methods and for extending them in the near and far future.

The equivalent circuit of a receiving antenna consists of Z_{ANT}, the input impedance measured across the physical port terminals in the *transmitting* mode. The load impedance Z_L is connected in series at the terminals of the antenna through a transmission line. The induced voltage v_{oc} is the equivalent *open*-circuit voltage at the antenna terminals due to an external far-field excitation and is equal to the voltage that would develop if the load impedance was removed.[23] In terms of the ACGF formalism, it can be found by an elementary circuit-theoretic argument that $|v_{\mathrm{oc}}| = Z_{ANT}\,(2\pi a)\,|\mathbf{J}_{\mathrm{rx}}|$, where here $\mathbf{J}_{\mathrm{rx}}$ and the radius function a are computed at the location of the physical port.

23 When considering an array system of N receive antenna elements, this concept of induced voltage can be generalized using network theory [29].

Chapter 15

Antenna Current Green's Function as a Method to Compute Near-Field Interactions

15.1 INTRODUCTION

There is a growing interest within the applied electromagnetic community in systems involving parts and subsystems interacting at close distance. Examples are compact antenna arrays, wireless devices working in dense and crowded electromagnetically changing environments such as MIMO and DoA applications, miniaturized circuits and radiators, near-field communication, and wireless energy transfer. A common denominator in all these applications is the existence of the problem of illumination by *near* fields. In theory, the near field is much more complex than the far fields or waveguide modes. Although it is not completely arbitrary but appears to enjoy a very specific mathematical structure of its own, e.g., see Part I, the understanding near-zone problems remains challenging compared with the other limit cases (examples of the latter include far zone and waveguide-based excitations).

Field illumination by a *generic* source at close distance involves not only propagating modes with wavelength roughly determined by the formula $\lambda = f/c$, but also *short*-wavelength components or *evanescent* modes that represent the rapid variation in the near field [35]. Note that for a specially prepared discretization mesh, it is possible to compute the response to any near field using full-wave analysis methods such as MoM [23]. However, for each new type of near field, e.g., different composition of short-wavelength components, it is necessary to change the mesh in order to obtain an accurate solution to the problem [38], [39]. Recently, some methods to compute the response of antennas to generic near fields were proposed but for special antennas [42], [44]. These methods, though useful, don't address the problem at the most general level, besides being already contained in the method presented here as a special case.

Investigation of the question whether there is a *general* method to compute the response of the antenna to such generic excitation led the authors to the development of the antenna current Green's function (ACGF) formalism. The Chapters 8 and 9 contain the bulk of the theoretical foundations of the approach to the ACGF as an exact transfer function in space. Some applications to mutual coupling in arrays are reported in [15] and Chapter 16. The numerical and experimental examples found in Chapter 14 and [15], however, included only far field illumination. This chapter provides the first numerical study of the near-field interaction problem using the antenna current Green's function method.

The chapter is organized as follows. Section 15.2 provides a motivation for the study of the problem of EM devices interactions with near-field illumination by pointing out how this case arises in practice. Section 15.3 provides the theoretical background needed to understand this chapter in a self-sufficient manner. It reviews the concept of the ACGF in light of the specific needs of the current investigation, explains how the general theoretical presentation in Chapter 8 can be adapted for near field applications. Some of the main themes discussed there will be put into concrete demonstration in Section 15.4, which provides a detailed example comprised of wire antenna solved using the method of moment. We compute the ACGF and discuss its validity and physical significance within the specific context in which it appears. Finally, we end up with conclusion and summary.

15.2 SOME PRELIMINARY MOTIVATIONS

Before starting the technical analysis itself, let us pause for a while to reflect on how the problem of near-field illumination of EM devices arises in practice. There are several possibilities. First, we may think of our device D as directly excited by another antenna S acting like a source producing near fields sufficiently strong to force D to respond by generating a signal at its output terminal. Figure 15.1 illustrates this case, which we call the S-Rx scenario.

In literature, such situation is dealt with in the following manner. Both Device D and Source S are treated as a *single* system in the full-wave solution of the problem in order to take into account the extent of mutual coupling between the

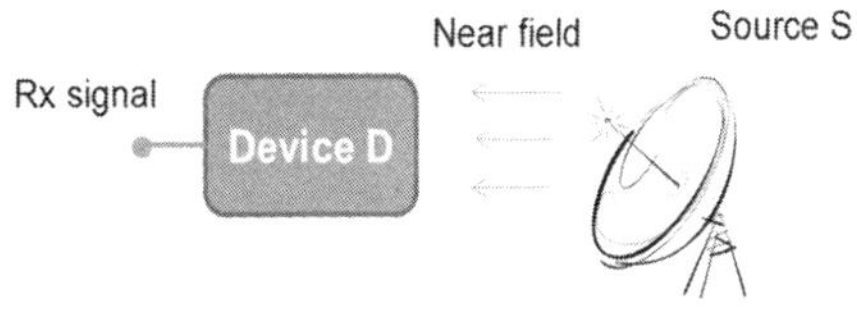

Figure 15.1 The S-Rx scenario. Interaction between a Source S close to Device D (typically an antenna system). The field impinging on Device D is a near field rich with short-wavelength components. The device responds to the near field by generating a received (Rx) signal.

two [23], [38], [39]. However, in practice one needs to study how D is working as an *independent* unit that can be embedded into various environments, where a source like S is simply one potential element of this environment among others. In other words, our main focus is the system D *itself*, while sources are relegated to the background. Therefore, the first natural step is to ignore the effect of the source S on the electromagnetic responsitivity of D and deal with the latter as a system interacting mainly with illumination fields. Here, the field is a near field and the device will behave according to this type of excitation. For example, one can assume that the device is electromagnetically shielded in a proper way or designed to minimize its interaction with nearby objects. After developing some basic understanding of the physics of how an independent D responds to arbitrary near field produced by S, we can move subsequently to the next stage where the effect of mutual coupling is included in the problem. This will most likely manifest itself as a *perturbation* on the characteristics studied in the previous step. Detailed investigations of mutual coupling will be tackled by the authors in a separate publication, but see Chapter 16.

Let us move now to a more common problem encountered in real-life situations. Figure 15.2 depicts what we choose to call the S-O-Rx scenario. Here, the source antenna S is located in the far zone of the device D. The far field generated by S will first interact with an object O close to the device D. Next, the field scattered by O will impinge on D as a *near* field. Therefore, although the entire structure D+O is located in the far zone of the source S, then, provided we again focus attention on the device of interest D, the latter is eventually excited by near fields, not far fields. In most of the standard literature, the idealized scheme in which the receiving system D exists alone in free space (or above a ground plane) is assumed. In electromagnetically dense environments, such as highly populated urban areas,

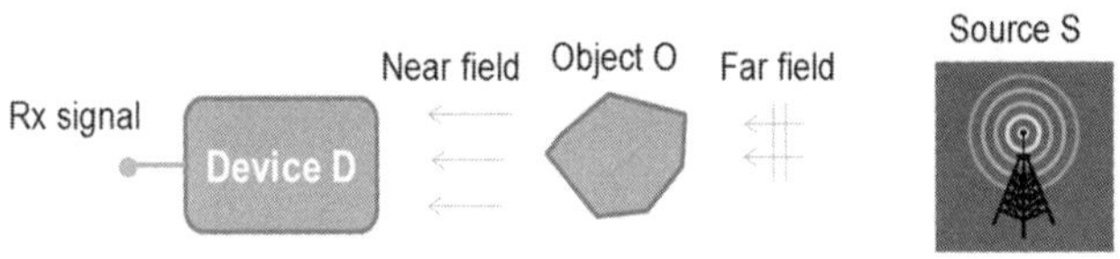

Figure 15.2 The S-O-Rx scenario. Interaction between a Source S far from a Device D (typically an antenna system) is mediated by a Object O placed in the vicinity of the device. The far field produced by the distant source will be converted upon interaction with Object O to a near field (scattered field) impinging in turn on Device D. The device responds to the near field by generating a received (Rx) signal.

this is no longer satisfactory. Even more, in nanoscale problems there is always a strong energy exchange with the surrounding objects unless extreme care is taken to eliminate this interaction [43]. Therefore, we believe that there is a need to reexamine the electromagnetic problem as a whole from this perspective of a device interacting solely with some near field.

Interest in describing the antenna using the ACGF method derives from the well-known advantages of classical system theory in electrical engineering. Indeed, by finding a characteristic function (transfer or Green's function) of the device, we achieve

1. Deeper understanding of the physics of the device. This is because the Green's function as a transfer function contains the theoretically and physically *complete* information needed to compute the output for *generic* input regardless to the environment and without the need to re-meshing.
2. Reduced computational efforts. This is because one may use the same device operating within very different systems and environments. Without re-performing a full-wave analysis of the device + new enviroment/system every time, we can directly compute the output by just specifying what is the illumination field at hand.
3. Although sometimes it is not easy to know the illumination field itself without further calculations, we can still use the Green's function productively in theoretical research, test, and devolvement, for example, by performing *statistical* analysis to predict the expected received signal in such-and-such scenarios or working conditions, e.g., DoA and MIMO applications often involve such kind of analysis.

4. Knowing the antenna current Green's function and employing it for near-field excitation calculations is a key step in finding new methods to compute and compensate for mutual coupling. This line of research is treated elsewhere, e.g, see Chapter 15.

To our best knowledge, such study has not been systematically considered in the computational literature. The authors believe that it is convenient at the present time to spend considerable efforts on understanding the physical and computational aspects of near-field interactions in electromagnetic systems, both simple and complex. Although the topic of the near field is notoriously difficult, we hope that a group effort within the larger community will eventually produce significant progress. This chapter is part of an initial contribution in this specific direction.

15.3 CONCEPTUAL AND COMPUTATIONAL ASPECTS OF THE ANTENNA CURRENT GREEN'S FUNCTION METHOD FOR NEAR-FIELD INTERACTIONS

15.3.1 The ACGF Approximation Techniques

Existence of the ACGF was established in Chapter 8 within the framework of distribution theory. The method of the proof was to actually construct the function using sequences of trial approximations and then showing that some subsequence does converge to the exact value of the antenna received signal. Therefore, it is possible to exploit the existence proof itself for the study of the concrete implementation of the ACGF in numerical contexts. This topic will be briefly illuminated here.

The distributional ACGF $\bar{\mathbf{F}}(\mathbf{r}', \mathbf{r})$ is replaced by a sequence of regular dyadic functions $\bar{\mathbf{F}}_n(\mathbf{r}', \mathbf{r}), n = 1, 2, ..., \infty$. While these functions approximate the exact ACGF of the problem, they differ essentially from $\bar{\mathbf{F}}(\mathbf{r}', \mathbf{r})$ in being *ordinary* functions, rather than a distribution. In fact, each function $\bar{\mathbf{F}}_n(\mathbf{r}', \mathbf{r})$ can be represented as the current on the antenna generated in response to a special vector surface excitation function $f_S(\mathbf{r}', \mathbf{r})$. This special excitation belongs to a very wide range of localized fields all approximating in a certain sense a generalization of the concept of the Dirac delta function familiar in system theory. For example, any localized smooth pulse can work for this excitation.[1] Therefore, the ACGF of the antenna problem can be approximated by simply exciting the antenna by a sequence of surface Dirac functions and choosing a suitable approximation level to work with.

1 For the rigorous description of the set of allowed delta sequences, see Chapter 8.

The Method of Moment (MoM), Finite Element Method (FEM), Finite Difference Time Domain Method (FDTD), can all be used to perform this computation. For example, to confirm the method, FEM was used in [15] (with comparison with measurement) while the MoM was utilized in Chapter 14 in conjunction with the singularity expansion method.[2] It is possible, moreover, to rely on special measurement methods to obtain the ACGF. For instance, by exciting the antenna by an extremely concentrated field and measuring the induced current distribution, it might be possible to bypass the need to employ very dense mesh in the numerical solution.

15.3.2 The ACGF and Traditional Full-Wave Solvers

Since electromagnetic fields are either far or near fields, an arbitrarily complex field is most probably a near field. Commercial EM codes usually deal with two types of excitation, wave ports and far-field illumination. Both types of excitations necessitate generating a sufficiently fine mesh with a degree of resolution that can be quickly estimated from the operating frequency. The reason is that both far-field and wave port excitations involve *propagating* modes, which have a wavelength roughly around v/f, where v is the speed of propagation in the medium of interest and f the frequency. The situation, however, is very different with near-field excitations. In this case, not only wavelengthes between ∞ and λ_0 are available, but also *short* wavelength components $0 < \lambda < \lambda_0$ corresponding to the *non*propagating modes. Those rapidly decaying field modes make the *a priori* prediction of the proper resolution of the full-wave numerical solution mesh very difficult for *generic* field excitation. In other words, only if the nature of the field impinging on the antenna is known *in advance* can we specify the suitable mesh of the problem. We believe that in order to characterize a device in a manner that is *independent of the nature of the illumination field*, one must have something new in addition to the traditional numerical solver: The *transfer function* of the system, or, equivalently, the *antenna current Green's function*.

In our opinion, the reason why this new concept was not fully pursued during the last few decades relates to the nature of the research problems that have been deemed important by the practicing community. It has been widely believed that far fields interaction with antennas is the dominant type of interactions. This is certainly correct in many communication and radar applications operating in typical environments, where the idealization of properly isolated and shielded

2 However, we pose the possibility that other special numerical methods need to be developed in order to perform this computation in a more accurate and systematic fashion.

device can always be made. However, currently things look different compared with before. Indeed, urban environments are electromagnetically dense. Devices are always embedded in complex environments. Typical practical scenarios now involve multiple antennas and circuits constantly interacting with each other. Moreover, in nanoscale structures and metametrials, the smallness of the devices and unit cells forces interactions with the surroundings to become very critical (as an example, consider near-field nano-optics and subwavelength imaging [43]). In all such cases, two key terms come to mind: near field and mutual coupling. Interaction at short distances typically involve near fields.

Finally, we mention that for general near-field illumination, the MoM matrix cannot be adequate even for a single frequency. The reason is that no fixed accurate mesh can be found for *arbitrary* field excitation. In practice, only special inputs are typically taken into account, mainly plane-wave illumination and wave ports. The problem of how to study the antenna response to *generic* field form has not been investigated in a systematic fashion within the applied EM community.

15.4 NUMERICAL ANALYSIS OF NEAR-FIELD INTERACTIONS IN LINEAR WIRE ANTENNAS

As an a concrete example demonstrating the ACGF method for near-field excitation, we consider a linear wire antenna system illuminated by an external near field $\mathbf{E}^{\mathrm{ex}}(\mathbf{r})$. In order to numerically solve the problem, a thin-wire electric field integral equation (EFIE) for the current distribution on the antenna is solved using the Method of Moment [23], [38]. The operator equation is written as

$$\hat{n} \times \mathbf{E}^{\mathrm{ex}}(\mathbf{r}) = \hat{n} \times \mathcal{L}\mathbf{I}(\mathbf{r}), \tag{15.1}$$

where $\mathbf{I}$ is the current on the wire with the latter's outward normal vector pointing along $\hat{n}$. The electromagnetic operator $\mathcal{L}$ is the one associated with the EFIE.

For simplicity, we work with a linear (triangular) basis function. The code is verified by comparison with the solver WIPL-D, which uses higher-order basis functions to model the current on wire segments [39].

The MoM expansion of the current is given by

$$I(z) = \sum_{l=1}^{N} I_l f_l(z). \tag{15.2}$$

Here we assume that the antenna is oriented along the z-axis and that the origin coincides with the middle point of the wire. The complex numbers I_n give the unknown current values at the mesh locations z_l, for $l = 1, 2, ..., N$, while $f_l(z)$ are the basis functions (for explicit expressions in the triangular case, see for example [38]). Using this numerical model, we intend to approximate the ACGF of the wire numerically as

$$F(z, z') = \sum_{l=1}^{N} I_l(z') f_l(z). \tag{15.3}$$

Note that the MoM current values I_l become functions of the location of excitation z'.

Let the tangential component of the external field be denoted by $E^{\text{ex}}(z)$. From the fundamental definition of the ACGF, the current induced on the antenna can be given by

$$I(z') = \sum_{l=1}^{N} I_l(z') \int_{S_l} dz\, f_l(z) E^{\text{ex}}(z), \tag{15.4}$$

where the integration is on the segment S_l on which the lth basis function $f_l(z)$ is defined. The integration is performed numerically using Gauss-Legendre technique with five points on each segment. Note that in order to use the Tx mode ACGF (15.3) to write (15.4), we have invoked the inverse reciprocity theorem.

The expression (15.4) gives the main formula of the ACGF approach when the MoM is used as the computational medium through which we implement the exact theory in Chapter 8. Its accuracy depends on how good the current values $I_n(z')$ are in modeling the response to a delta function at $z = z'$. In order to assess the method, we need also to know how good the obtained ACGF is in modeling the response of the antenna to rapid fluctuations or short-wavelength components present in the illuminating near field.

A convenient way allowing both goals to be achieved at once is to choose a delta sequence [71] (a sequence of ordinary functions approximating the idealized Dirac delta source) that consisting of decaying exponentials. One possible such choice is the sequence of functions

$$E^n(z) = (1/2)\, n \exp(-n|z|), \tag{15.5}$$

for $n = 1, 2, ...$ The main advantage in using the special sequence (15.5) is that it has exactly the same form of the *evanescent* modes radiated by a point source at the origin. As is well known in electromagnetic near-field theory, the major difference between far fields and generic fields (for example, near fields, directed beams, scattered fields) is that the latter contain rich and complex mixture evanescent modes in addition to the typical propagating modes [35]. Mathematically, a generic near field can be written as

$$\mathbf{E}\left(\mathbf{r};\bar{\mathbf{R}}\right)=\mathbf{E}_{\mathrm{pr}}\left(\mathbf{r};\bar{\mathbf{R}}\right)+\mathbf{E}_{\mathrm{ev}}\left(\mathbf{r};\bar{\mathbf{R}}\right), \tag{15.6}$$

where $\mathbf{E}_{\mathrm{pr}}$ and $\mathbf{E}_{\mathrm{ev}}$ are the propagating and evanescent parts, respectively. Here, $\bar{\mathbf{R}}$ is a 3D rotation matrix specifying orientation of the local coordinate frame used in deciding the direction of the axis along which the splitting into propagating and evanescent modes is enacted (Chapter 4).

Consequently, since it is this type of evanescent modes $\mathbf{E}_{\mathrm{ev}}\left(\mathbf{r};\bar{\mathbf{R}}\right)$ that represents the rapid part of generic near field variations, knowledge of the accuracy of the ACGF method in terms of exponentially-decaying excitations like (15.5) automatically provides information about how good the ACGF thus obtained is in predicting the response of the antenna system to this level of short-wavelength details in the near field.

We may now use (15.4) to illustrate the applicability of this argument in actual examples. Let each ACGF approximation be written as

$$F^{n}\left(z,z'\right):=\mathcal{L}^{-1}E^{n}\left(z\right)=\sum_{l=1}^{N}I_{l}^{n}\left(z'\right)f_{l}\left(z\right). \tag{15.7}$$

for $n=1,2,\ldots$. It can be shown that all the requirements of the special delta sequence (15.5) stated in Chapter 8 are satisfied and therefore, according to the convergence theory of the ACGF, it is always possible to obtain production of the near field response in antennas using a sufficiently small exponential illumination field.[3] Therefore, we conclude that the corresponding current values $I^{n}\left(z'\right)$ obtained by (15.4) will converge to the exact solution as was shown in Chapter 8.

Consider a linear wire antenna with length 0.25λ and radius 0.001λ. We choose four members of the exponential delta sequence (15.5). Figure 15.3 provides the shape of the illumination field along the wire extension. The ACGF approximations corresponding to each exponential field are shown in Figure 15.4. In general, as n in the excitation fields of (15.5) becomes large, the sequence of pulses approaches the idealized delta function. Since the electric-field integral equations of this particular problem is *not* continuous (the EFIE operator is unbounded [70]), the convergence of the delta sequence functions (15.5) to the Dirac does not necessarily imply that the corresponding current distribution will converge. However, the analysis in Chapter 8 proved that convergence to the correct current *does* happen for

3 The details of the proof are lengthy but straightforward and can be readily obtained by mimicking the argument of Chapter 8.

all operators obeying the reciprocity theorem, which include in particular the EFIE operator of the antenna problem (15.1). Thus, from the purely numerical viewpoint, one can work with any approximation level of the exact ACGF by choosing the proper value n. In general, the higher the accuracy of the approximating ACGF, the larger is n.

To verify the prediction of the current distribution using the ACGF with a direct approach, we present in Figure 15.5 comparison with the scattering problem (15.1) solved directly using the MoM for two choices of n. In the MOM scattering code, we generate a nonuniform mesh appropriate to the shape of the exponential delta sequence member for each choice of n in the input field (15.5). For the computation of the ACGF in the transmitting mode, we use a single mesh sufficiently dense to provide accurate prediction for all the input excitation near field cases considered. Excellent agreement can be observed. Although this agreement in the convergence results with increasing n was proved mathematically in Chapter 8 and verified numerically for far-field illumination in Chapter 9, they are confirmed here numerically for non-plane wave, i.e., inhomogeneous or evanescent fields, for the first time. Note that since a generic electromagnetic field can always be written as a proper mixture of propagating and evanescent modes Chapter 4, the results here provide evidence that the ACGF can be used to compute the response to any excitation field. This is because the basic building blocks of this excitation, namely propagating and evanescent modes, are dealt with successfully using the ACGF method.

Although the method of moment was used in this work to investigate the near-field ACGF approach, we mention that since the ACGF has been established on a rigorous basis going back to Maxwell's equations Chapter 8, it could be advantageous to develop special numerical methods or measurement procedures to obtain the Green's function in ways that go beyond conventional numerical methods.

In order to go beyond linear wire antennas, for example to consider more complex structures such as multilayered microstrip and dielectric resonator antennas, it is required to develop a general EM solver that can deal with the special excitations (the delta sequence) applied to the antenna in order to obtain its ACGF. Most commercial EM solvers provide only plane wave and wave port excitations, or sometimes special waves like Gaussian beams. In studying such problems, more factors come into the picture, for instance how the geometrical shape of the antenna interacts with spectral content of the non-standard NF applied to that part under consideration. Such detailed study, including questions of convergence, were dealt

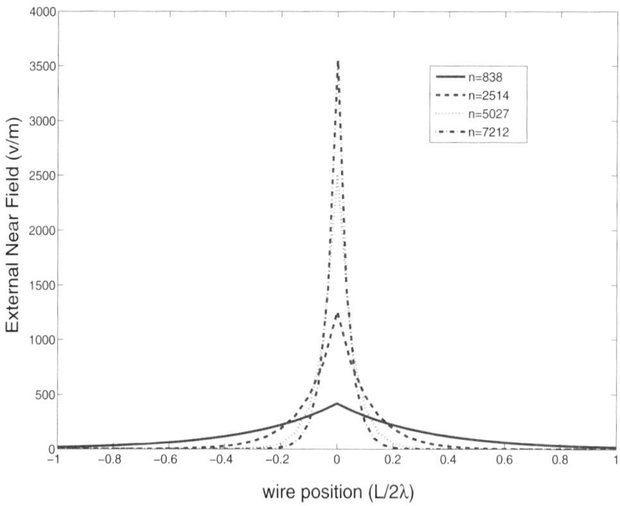

Figure 15.3 A sequence of exponentially localized near electric field excitations (as in (15.5)) approximating the Dirac delta functions applied around the center of a 0.25λ linear thin-wire antenna with radius 0.001λ.

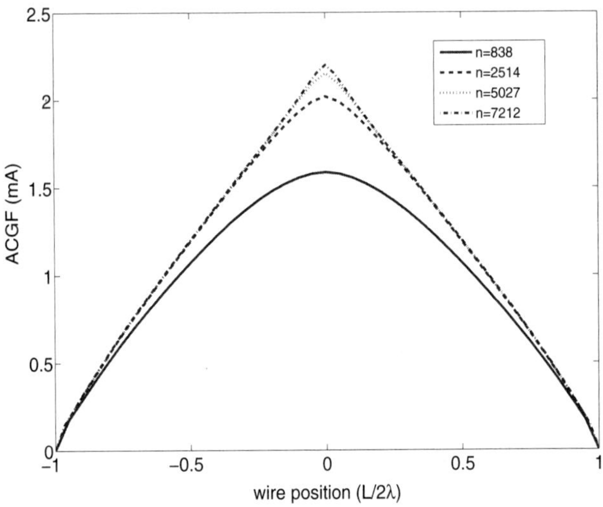

Figure 15.4 The approximations of the ACGFs of the antenna computed using the MoM in response to the excitations of in Figure 15.3.

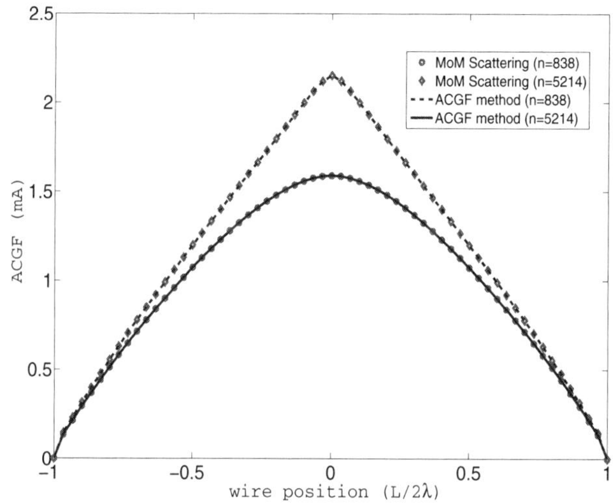

Figure 15.5 Comparison between the ACGF and direct MoM scattering code for the antenna described in the caption of Figure 15.3.

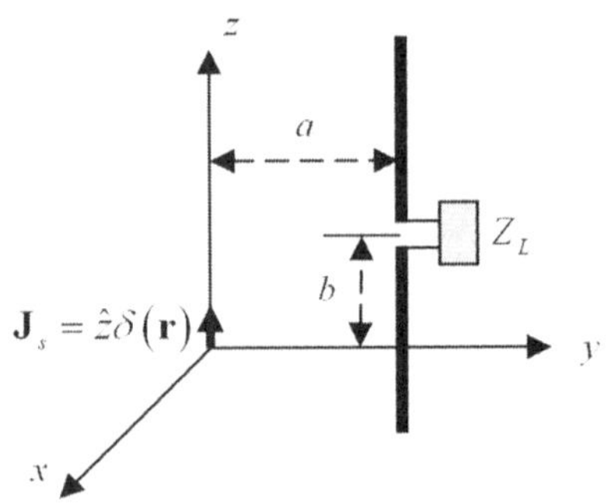

Figure 15.6 Interaction of a linear wire antenna of length $2L$ with the near field produced by a point source.

with theoretically in Chapter 8, and numerically for linear wire antennas here. The numerical study of 2D and 3D structures will be taken up in future publications.

15.4.1 Interaction with Point Sources

We will utilize the analytical expression (14.37) to compute the received signal induced in a linear wire antenna due to interaction with the near field of a point source $\mathbf{J}_s$ as shown in Figure 15.6. For simplicity, we consider only a source polarized in the z-direction. The wire antenna is shown to be in the yz plane.

As was suggested in Chapter 4, the salient feature of the near field is its dynamic splitting into propagating and nonpropagating (evanescent, inhomogeneous) parts. For a receiving antenna oriented along the z-direction as in Figure 15.6, only the z-component of the source will interact with the wire. Let us denote the propagating and nonpropagating parts of this component by $E_z^{\rm pr}$ and $E_z^{\rm ev}$, respectively. The field radiated by the source current $\mathbf{J}_s$ can be put in the following form (see Chapter 4 for details)

$$E_z^{\rm pr} = ik_0\left(A_{\rm pr}^0 - A_{\rm pr}^2\right), E_z^{\rm ev} = ik_0\left(A_{\rm ev}^0 + A_{\rm ev}^2\right), \tag{15.8}$$

where

$$A_{\rm pr}^n := \int_0^1 du\, u^n e^{ik_0|z|u} J_0\left[k_0\rho\sqrt{1-u^2}\right], \tag{15.9}$$

$$A_{\mathrm{ev}}^{n} := \int_0^\infty du\, u^n e^{-k_0|z|u} J_0\left[k_0\rho\sqrt{1+u^2}\right]. \tag{15.10}$$

Here, $\rho = \sqrt{x^2+y^2}$.

The received load voltage V_L will be given by $V_L = V_L^{\mathrm{pr}} + V_L^{\mathrm{ev}}$, where

$$V_L^{\mathrm{pr}} = \frac{ik_0 Z_L Z_{Th}}{Z_{Th}+Z_L} \sum_{l=1}^{L_{nm}} a_l \left[B_{\mathrm{pr}}^0(\kappa_l) - B_{\mathrm{pr}}^2(\kappa_l)\right], \tag{15.11}$$

$$V_L^{\mathrm{ev}} = \frac{ik_0 Z_L Z_{Th}}{Z_{Th}+Z_L} \sum_{l=1}^{L_{nm}} a_l \left[B_{\mathrm{ev}}^0(\kappa_l) + B_{\mathrm{ev}}^2(\kappa_l)\right]. \tag{15.12}$$

Here we define

$$B_{\mathrm{pr}}^n(\kappa_l) := \int_0^1 du\, u^n J_0\left[k_0\rho\sqrt{1-u^2}\right] I(\kappa_l, ik_0 u), \tag{15.13}$$

$$B_{\mathrm{ev}}^n(\kappa_l) := \int_0^\infty du\, u^n J_0\left[k_0\rho\sqrt{1+u^2}\right] I(\kappa_l, -k_0 u). \tag{15.14}$$

Here we define the spectral interaction function as

$$I(\kappa, k) := \frac{e^{(\kappa+k)(L+b)} - e^{(\kappa+k)(-L+b)}}{\kappa+k}. \tag{15.15}$$

Note that in this particular calculation, $\rho = a$.[4] The expressions (15.12) and (15.11) are quite general. Their derivation depended crucially on the powerful Green's function theorem (14.3) and the spectral expansion (14.36). They show that for arbitrary wire antennas, the response to an arbitrary near field generated by a z-directed point source can be fully understood in terms of the behavior of *only four* different spectral integrals (15.13) and (15.14).[5]

15.5 CONCLUSION

This chapter presented an investigation of the recently proposed antenna current Green's function (ACGF) formalism for applications concerning the analysis of antenna arrays working in the receiving mode. A rigorous formulation was presented

4 Cf. Figure 15.6.

5 Further types of spectral integrals will be needed for sources polarized in directions other than $\hat{z}$. However, in this chapter, we focus on the simple example given in Figure 15.6.

for the commonly encountered practical situation when an antenna is fed by a linear structure or a waveguide with a cross-sectional area small compared to the overall surface area of the effectively radiating structure. For achieving the highest level of generality in the method, the formulation assumed an arbitrarily-curved radiating surface and also a generally curved wire extension. It was found then that the ACGF tensor takes a very simple form in this case. Exact formulas for predicting the signals induced in the receiving mode using transmitting mode data, and valid for arbitrary illumination field, were then derived rigorously for the first time using the idea of the current Green's function.

Several in-depth discussions were presented to probe the structure of field-antenna interaction as exemplified in the above mentioned general receiving array system. We then demonstrated how the method can be applied in the general case when the excitation of the structure is enacted by means of fields more complicated than the point-source model used in defining the ACGF of the system. It was found that the predicted value of the received signal, which would result if an approximation of the ACGF is attempted using a numerical method, will be equal to the spatial average of the correct value of the induced current, e.g., those obtained through measurements. This result can help motivating further interplay between theory and measurement in antenna practice since it provides a theoretical view of measurement based on an exact Green's function method, not the numerical approximation of full-wave analysis, and hence can never be exact.

The chapter ended by a numerical study of wire antenna systems. The basic formulas are developed for the special thin-wire array problem and the main results were verified by direct computations using method of moment. A new turn in the discussion was taken by performing a spectral analysis on the ACGF of the wire-antenna system. The analysis, using Prony's algorithm, obtained an analytical approximate of the ACGF form and was used successfully in predicting the interaction with plane-wave illumination in simple analytical form. The physical meaning of the results was explicated in which the important difference between resonant and nonresonant interactions with antenna systems was introduced and numerically illustrated for the case of wire antennas.

The value of the analysis presented here is that it can be considered the first time in literature in which systematic and rigorous models of generalized antenna systems are presented using an exact transfer function in space. One of the advantages obtained through this step is the ability to separate the input of the system, the general illumination field, from the description of the antenna system itself. The latter is further analyzed into two components, one termed the purely geometric level while the other is the purely electromagnetic response. The pure geometric level depends only on the shape of the generally curved antenna surface. The pure electromagnetic response depends on the feed mechanism, the nature of the materials composing the antenna system, and the electromagnetic environment. Such detailed insight into the various levels of electromagnetic field interactions with general antenna systems is possible only when an exact concept of spatial Green's function is developed in detail, with attention paid to the differential geometry of the antenna surface. We expect that the basic ideas presented in this chapter can help establish a deeper understanding of how antenna (and other electromagnetic) systems in the receiving mode work under very general operational conditions. Indeed, the neat separation of the interaction scheme into the three distinct levels mentioned above has direct relevance to engineering studies and development since it allows identifying the relevant degrees of freedom in one particular mode of operation while fixing the others.

Chapter 16

Electromagnetic Mutual Coupling in General Antenna Systems

16.1 GENERAL INTRODUCTION

16.1.1 Broad View and Outline

This chapter aims at providing a broad new view on the topic of mutual coupling in general electromagnetic systems, with emphasis here put on antenna arrays. The topic of mutual coupling is not new, and has been recognized as an important research area since the early beginnings of antenna theory. The basic problem encountered here is how the behavior of one electromagnetic device changes when placed next to another device or nearby object. Although there is a universal agreement on this deceptively simple definition, there exists a wide range of different answers to the immediate question about how to describe the device's behavior of interest in the mutual coupling problem as defined above. Unfortunately, there has been a tendency throughout the last six or seven decades to reduce mutual coupling to coupling through the ports of the antenna system under consideration. While port coupling is indeed part of the general process of electromagnetic mutual coupling, it by no means exhausts the problem. One of the main objectives of this chapter is to illuminate mutual coupling through a more comprehensive scheme. For example, we will learn how to monitor interactions by looking through the near field, rather than mutual impedance or the far field. Since future devices are being reduced in size and are operating in dense and crowded environments (just think of the high average number of mobile phone devices existing in each square meter nowadays), it is expected that characterization of coupling through the near field will become increasingly important in the near future.

However, the need to study mutual coupling through the near field faces very serious obstacles. First of all, we don't have a complete understanding of the structure of the near field comparable with the case of the far field. If an antenna system is being illuminated by a near field instead of plane wave, the physics of coupling and interactions are in general quite different. The main difficulty is that it is not known in advance what wavelength components exist in the near field impinging on the system. In contrast, with plane-wave illumination only one wavelength is possible for a given frequency f, that given by the relation $\lambda = c/f$, where c is the speed of light in the space into which the antenna is radiating and receiving. Consequently, the task of analyzing the problem of mutual coupling in the presence of complex illumination field becomes involved since in general antenna arrays behave differently with different wavelength field components acting as excitations. This difficulty will be solved in two steps. Initially, we will utilize the near-field theory developed by the authors to provide suggestions about how best should the near-field structure be analyzed in order to obtain a physically meaningful picture. The next step is to utilize the antenna current Green's function (ACGF) in order to compute the response of the device to generic excitations. The two-step approach will then be combined and illustrated for antenna arrays.

The second difficulty concerns the problem of how to compute mutual coupling for large-and-complex antenna systems. This is the problem that has dominated research interest in the last period. It concerns the well-known prohibitive cost of solving Maxwell's equations for large number of antenna arrays. We will provide a two-step approach to deal with this problem. In the first step, we employ the infinitesimal dipole model (IDM) method to model arbitrary antenna arrays based on measured near field data. This bypasses the need to solve Maxwell's equations on a computer. The obtained model is analytical and can be used to predict the fields radiated by the original array everywhere in the exterior region. In order to incorporate the effect of mutual coupling, a physics-based approach, inspired by multiple scattering in solids, will be introduced. This approach, here applied in a simple and intuitive fashion, will be further developed later in a more rigorous manner using the ACGF formalism. In fact, we will show that using perturbation theory, one can compute mutual coupling in arbitrary antenna arrays with a given error tolerance without the need to solve Maxwell's equations for the entire coupled system.

The key idea of our approach to mutual coupling will be isolating and properly defining the single most important quantity describing mutual coupling in the widest general way possible. This quantity turns out to be the *mutual coupling Green's function*, which acts like a transfer function for mutual coupling in generic electromagnetic systems. Indeed, such function can predict those popular and standard mutual coupling effects at the ports of the arrays plus the impact on the radiation pattern. But it *also* can estimate the impact on the near field brought by coupling and interaction with nearby objects. However, the most important feature of the mutual coupling Green's function is the insight it provides into the physical aspects of energy exchange between antennas and their environments. By analyzing the structure of the coupling transfer function, we hope to be able to understand in depth how two antennas couple and react against each other. Examples will be provided in which new pictures are generated by our method explicating the dynamic structure of the near field between two antennas (wires) and how energy is localized in the space lying between them. This new analysis can find applications in various problems that depend crucially on interactions at short distance, for example detection of underground objects, near field communications, near field radar and MIMO, and many others.

16.1.2 Motivations and Context

The main motivations for this research come from the unique present historical juncture of electromagnetic and electronic research: Some of the most challenging technological problems facing the electromagnetic engineer in the future will require a deeper understanding of the physical and computational aspects connected with the problem of interaction and energy exchange in general antenna systems. In fact, steady expansion of the electromagnetic technological infrastructure is now facing the following two limiting problems:

- The increasing complexity and/or density of the working environment.
- The need to reduce the size of electromagnetic and electronic systems.

Limitations of space push toward smaller spatial extents, a situation which inevitably raises difficulties regarding how various parts of the device mutually interact with each other. On the other hand, the need to run devices in dense and complex environments brings in the unavoidable task of characterizing how interactions with nearby objects may modify the ideal performance of the device. The two problems above, the internal and external ones, are fundamentally the same when viewed from a proper general theoretical perspective: *electromagnetic mutual coupling between two or more objects.*

Following this view, it is expected that by rigorously formulating the problem using exact techniques, one hopes to develop a deeper understanding of the scientific dimensions of the issue at hand, an understanding capable of nurturing fruitful solutions to concrete problems. This chapter aims at achieving this double goal: to sketch out a general theoretical formulation of mutual coupling using methods recently developed by the authors; and to devise special computational and statistical methodologies convenient to common practical array and system designs.

There are numerous advantages for developing a general theory of electromagnetic mutual coupling in applied electromagnetics. We mention below only few of what we think might be important

1. A theory of mutual coupling must involve a theory of electromagnetic energy exchange between radiators or between radiators and their environment. Such theory is still not fully developed in the classical regime of macroscopic electromagnetic, especially with respect to the fundamental role played by evanescent modes in light of recent progress in near field optics and engineering. In fact, the topic of energy exchange is well developed for light-matter interactions, which typically involves quantum-mechanical effects. For the case of *classic* antennas and structures where the wavelength is comparable with the spatial dimensions along which the electromagnetic boundary condition is placed, to our knowledge no general theory of energy exchange exists up to date. Note that such theory has to deal with such poorly understood concepts like reactive, localized, and stored energies, which are still controversial. We anticipate that a theory of mutual coupling will help clarify many of the subtle aspects involved with electromagnetic energy exchange in general, particularly within the prospects of the potential wealth of numerical and physical data that such a theory of antenna's coupling may generate.

2. A deeper physical understanding of mutual coupling is an *a priori* condition for being able to devise satisfactory methods and techniques for mutual coupling compensation in general antenna arrays. Since the size of the array can be considerably reduced by decreasing the average inter-element spacings, correcting deviation from ideal behavior caused by such reduction of size is one of the most effective methods to achieve the long-term goal of system miniaturization.

3. It is conceivable that a theory of mutual coupling will allow new types of applications to arise. As experience has shown during the last several decades of research in applied electromagnetics, strong interactions can lead to qualitatively new electromagnetic behavior. For example, strong coupling between

narrowband systems can make the resulting (combined) system wideband. An in-depth analysis of the structure of electromagnetic interactions, therefore, might be a good method for generating new behavior exploitable for future applications.

4. Energy, here electromagnetic energy, can be manipulated and controlled only by working with strong mutual coupling systems. The reason is that radiation tends to weaken considerably at long distances due to the well-known $1/R^2$ spherical wave spreading loss. This implies that energy harvesting, control, and transportation, will likely be implemented through a setup that involves *near*-field interactions with *strong* mutual coupling. In order to understand how to design such systems, a general theory of mutual coupling is necessary.

16.1.3 The Overall Structure of The Theory of Electromagnetic Mutual Coupling

In Figure 16.1, we present the various investigations to be discussed in this chapter about electromagnetic mutual coupling together with the multiple possible connections existing among them.

The Antenna Current Green's function (ACGF) will be used to define mutual coupling rigorously by introducing the concept of mutual coupling Green's function (Section 16.2.1). Here, we isolate that part of total ACGF that is responsible of mutual coupling alone and find that it provides a complete and exact description of mutual coupling effects, including input port parameters, far field pattern, and the entire near field. For studying the behavior of antenna arrays in the receiving mode, it is possible to use this mutual coupling ACGF to predict the received port voltages for arbitrary field illumination. This is not possible with knowledge of only the S-matrix, which highlights one of the important applications of the new mutual coupling ACGF.

The dipole model approach to mutual coupling (Section 16.4) will be introduced in order to provide a simple and efficient method capable of dealing with arbitrary antenna arrays with or without strong mutual coupling. The method accepts a set of measured near field data and construct a model for the source using optimization methods. Mutual coupling here is *not* defined from the limited viewpoint of port coupling, but rather through the near field itself. An initial theory of mutual coupling is then presented in which we suggest that coupling occurs because of a multiple scattering effect in the space between the two antennas. A technique to compensate for mutual coupling effects in the dipole model of a single antenna is then introduced and verified by working with near field data.

Inspired by the intuitive approach to mutual coupling via infinitesimal dipole models, the basic mutual coupling ACGF is now shown to be computable in terms of a perturbation series that involves only the *isolated* elements' ACGFs and the *forward* interaction operator (Section 16.5). The forward interaction operator is in general small since it usually involves the perturbation effects coming from the nearest elements. Hence, it is possible to compute low-order corrections due to mutual coupling for very large arrays using the perturbation series without the need to invert the full electromagnetic operator of the problem. The approach also will confirm our assumption in the dipole model approach that mutual coupling is a multiple scattering effect.

The new theory of the near field will then be combined with the formulation of mutual coupling in terms of the ACGF method to provide a deeper analysis of interactions between two antennas. We show that certain modifications of existing Method of Moment (MoM) codes introduce a direct method to measure the degree of localization of energy in the space between the antennas. In particular, we keep intact the entire numerical apparatus concerned with mesh descretization, choice and implementations of basis functions, but split the Green's functions into two parts in a physically meaningful and illuminating way. The method consists of septation the interaction operator into propagating and evanescent part then computing the ratio of the nonpropagating part to the total interaction. New pictures illustrating interesting features in energy coupling between wire antennas will be given there. In particular, we provide an experiment in which one wire is used as a probe to study the structure of the field radiated by another wire.

As can be seen from Figure 16.1, the various sections of this chapter converge into the block enclosed by dashed line, which represents an integration of the overall achievement in one unified theory of electromagnetic energy exchange. *This final theory has not been attained yet*, but we hope that the multiple investigations conducted in this chapter and the remaining parts of the book will help bringing us closer to attaining this ambitious objective.

16.2 WHAT IS MUTUAL COUPLING IN APPLIED ELECTROMAGNETICS?

The term 'mutual coupling,' although widely used in the literature of applied electromagnetics, is seldom clearly defined in a precise and general way. Instead, it is usually left for the specific context inside which the term happens to be currently introduced and used to decide the particular definition of the problem

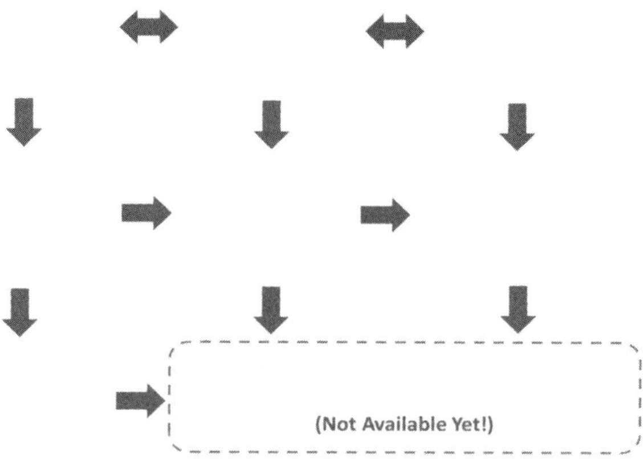

Figure 16.1 Road map for the structure of the mutual coupling theory presented in this chapter.

at hand. The lack of systematic investigations of the semantic aspects of mutual interactions is partially responsible, we believe, for the occasionally controversial claims made about what has been actually achieved regrading the analysis and mitigation of problems caused by the onset of mutual coupling effects in the system under consideration. Indeed, even though excellent separate proposals for methods successfully handling mutual coupling degradation in special common problems is not lacking, an overall view of the subject combined with more general approach to both analysis and computation of energy exchange between arbitrary systems is highly needed now. Moreover, we believe that some of the existing methods cannot actually deal with more general problems. The absence of a *general* theory of mutual coupling is one of the reasons behind this immature belief in the unrestricted applicability of many popular mutual compensation methods. We will not attempt here anything like a literature survey or comprehensive comparative studies, but proceed immediately to construct a general definition of mutual coupling using new theoretical contributions made recently (see chapters on the near field and the antenna current Green's function formalism) that address the topics of near fields, localized energy, and the antenna current Green's function. Based on the new formulation of mutual coupling introduced here for the first time, we propose some methodological techniques for handling the subject on a computational and statistical basis.

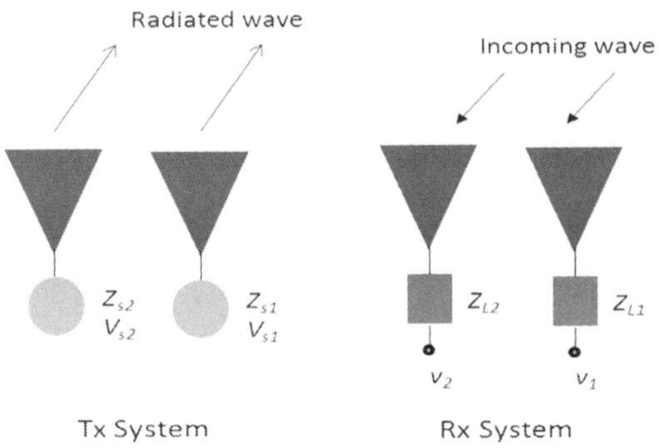

Figure 16.2 Simplest mutual coupling scenario involving a two-element antenna system. In the transmitting (Tx) system at the left, a source with internal voltage V_s and impedance Z_s is connected to the element and used to energize the system. In the receiving (Rx) system at the right, an incoming wave (usually plane wave) is used to excite the two-element array. The terminals of the antennas are connected to a load impedances $Z_{L1,2}$ and the resultant received voltages $V_{1,2}$ are observed.

Before attempting to define the problem of mutual coupling using the concept of the antenna current Green's function, it is appropriate to ponder for a while the existing understanding of the problem currently prevailing in the specialized literature. This presentation will help the reader to better appreciate the advantage of the rigorous and exact general formulation proposed later, and also will improve our ability to relate the theoretical framework to practice in terms of actual computations and/or measurements.

Figure 16.2 illustrates a minimal mutual coupling problem consisting of a two-element antenna array system operating in both the transmitting and receive mode, represented by the left and half graphs, respectively. The common understanding is that a separate antenna element is designed and developed as an *isolated* object, i.e., and antenna operating in idealized environment, typically infinite free space or half space. Subsequently, the device thus obtained is operated not in its original ideal design environment, but in the real world where other objects, possibly identical antenna elements, come into close proximity within its neighbourhood. Due to the peculiarity of electromagnetic devices, if the separation between the objects is not large with respect to the operating wavelength, then strong mutual interactions may exist and affect the performance of the device under consideration. In that case, the element in the array will not exhibit the original behavior, and, therefore, the actual array or complex system will deliver a *modified* output. *The deviation of the array's output from that anticipated based on simple addition (sum or superposition) of the idealized (isolated) performances of the constitutive elements is exactly what is called* ***mutual coupling effect***. Let us explain this with few common examples.

There are three main features in the electromagnetic performance of all antenna arrays:

1. Terminal port voltages and/or currents.
2. Near fields.
3. Far fields.

Most of the attention of commercial, industrial, and scientific research activities seem to be focused on the first and the third features, i.e., the terminal (port) characteristics and the far field. The near field is seldom directly addressed in research, and industry has been slow in appreciating its potentials. However, later on the near field will receive prominent attention, so we focus for now on the remaining two factors.

Consider first terminal voltages. The entire circuit connected to the ports of the receiving array can be replaced by their equivalent impedance Z_L, which in general varies from one port to another. When the antenna array operates without mutual coupling, each element will act as if the other objects, e.g., the antenna nearest to it, does not exist. The ideal received voltages will be denoted by V_1 and V_2 as in Figure 16.2 (right). Due to mutual coupling, the idealized voltages will be changed to V_1' and V_2'. We define the mutual coupling effect by the quantities

$$\delta V_1 := V_1' - V_1, \ \delta V_2 := V_2' - V_2. \tag{16.1}$$

In other words, for the quantification of deviation of the observed voltages due to mutual coupling, we compare V_1' and V_2' with the idealized (no mutual coupling) voltages V_1 and V_2.

This definition of mutual coupling is the most common one found in literature (the definition of mutual coupling for the far field is similar and will be given below). The majority of the so-called mutual coupling compensation techniques are designed to address a situation defined in these terms. There are several shortcomings, however, in this approach. We mention the most important:

- The definition is not general. It applies only to those systems where the terminal voltages are the major object of interest.

- The definition of the mutual coupling effects δV depends on the nature of the illumination field. In the case of (16.1), a plane wave excitation is assumed. As will be demonstrated later, this is far from exhausting the problem. For example, in dense and crowded environments, near-field illuminations tend to dominate.
- The precise nature of the role played by the load impedance is not clearly separated from pure electromagnetic effects. In Figure 18.4, mutual coupling is intuitively understood as a process emerging from some sort of electromagnet coupling between the two antennas. The variable load impedance, on the other hand, is an *external* factor relevant only to the remaining parts of the system. The input Thevenin impedance of each antenna is added to the load impedance in order to compute the port current, but this former impedance itself varies with mutual coupling, leading to variations in the power delivered to the load emerging from changes in the circuit power divider ration. In our opinion, such pure circuit effects should be distinguished from pure electromagnetic effects.

The general definition of mutual coupling to be presented later will take into account such difficulties.

For the transmitting antenna system, the most important performance measure is the far field. Let the far fields radiated by idealized single elements be given by $\mathbf{E}_1(\theta,\phi)$ and $\mathbf{E}_2(\theta,\phi)$. By superposition, the expected total far field is given by $\mathbf{E}_1(\theta,\phi)+\mathbf{E}_2(\theta,\phi)$. Due to mutual coupling, the actual far field will be given by $\mathbf{E}'(\theta,\phi)$. The far-field mutual coupling effect $\delta\mathbf{E}(\theta,\phi)$ is then defined as

$$\delta\mathbf{E}\left(\theta,\phi\right)=\mathbf{E}'\left(\theta,\phi\right)-\mathbf{E}_1\left(\theta,\phi\right)-\mathbf{E}_2\left(\theta,\phi\right). \tag{16.2}$$

In other words, *the signature of the far-field mutual coupling effect is the failure of simple superposition of idealized antenna element to account for the actually observed radiation pattern*. The function $\delta\mathbf{E}\left(\theta,\phi\right)$ measures this degree of deviation. The near-field mutual coupling effect is defined in exactly the same manner.

For matching consideration, it is useful to extend the above definitions to deal directly with the input impedance. Let the Thevenin impedance of a radiating antenna element in isolation be given by Z_T. Then the same impedance with mutual coupling will be modified to Z_T'. The input impedance mutual coupling effect δZ_T is defined as

$$\delta Z_{T_1}:=Z_{T_1}'-Z_{T_1},\ \delta Z_{T_2}:=Z_{T_2}'-Z_{T_2}. \tag{16.3}$$

That is, mutual coupling will modify the input impedance seen at its excitation port when radiating in the presence of another nearby object.

16.2.1 A Rigorous and Exact Formulation of the Problem of Mutual Coupling

The classic problem of mutual coupling can be described in very simple terms as follows. Suppose you are given two electromagnetic systems A and B. Assume a frequency-domain approach where a time-harmonic dependence $\exp(-\omega t)$ is presupposed and suppressed everywhere. It can be shown that the full performance of the two systems individually is fully accounted for using the antenna current Green's functions $\bar{\mathbf{F}}_A(\mathbf{r}, \mathbf{r}')$ and $\bar{\mathbf{F}}_B(\mathbf{r}, \mathbf{r}')$ for the systems A and B, respectively [13]. These functions can be used to predict the exact response of the systems under consideration for both the transmitting and receiving modes Chapters 8 and 9. Indeed, the received signals in antennas A and B can be written in the following proper superposition form

$$\begin{aligned} \mathbf{J}_A(\mathbf{r}) &= \int_{S_A} ds'\, \bar{\mathbf{F}}_A(\mathbf{r}, \mathbf{r}') \cdot \mathbf{E}^{\text{in}}(\mathbf{r}'), \\ \mathbf{J}_B(\mathbf{r}) &= \int_{S_B} ds'\, \bar{\mathbf{F}}_B(\mathbf{r}, \mathbf{r}') \cdot \mathbf{E}^{\text{in}}(\mathbf{r}'). \end{aligned} \tag{16.4}$$

Here, S_A and S_B refers to the surfaces of antennas A and B. It is important to notice that in this context each of the two transfer functions $\bar{\mathbf{F}}_A(\mathbf{r}, \mathbf{r}')$ and $\bar{\mathbf{F}}_B(\mathbf{r}, \mathbf{r}')$ above is computed for an *isolated* system, i.e., when the system under consideration works in infinite, isotropic, local, and homogeneous space, say free space.

Next, when each of the two systems is brought into close proximity to the other, the possibility of significant coupling between the devices cannot be ruled out. In this case, the actual transfer function is no longer equal to the one computed for isolated environment. In general, we write the new antenna current Green's functions as $\bar{\mathbf{F}}'_A(\mathbf{r}, \mathbf{r}')$ and $\bar{\mathbf{F}}'_B(\mathbf{r}, \mathbf{r}')$.

Finally, define *the mutual coupling transfer functions* $\delta\bar{\mathbf{F}}_A(\mathbf{r}, \mathbf{r}')$ and $\delta\bar{\mathbf{F}}_B(\mathbf{r}, \mathbf{r}')$ by the relations

$$\begin{aligned} \delta\bar{\mathbf{F}}_A(\mathbf{r}, \mathbf{r}') &:= \bar{\mathbf{F}}'_A(\mathbf{r}, \mathbf{r}') - \bar{\mathbf{F}}_A(\mathbf{r}, \mathbf{r}'), \\ \delta\bar{\mathbf{F}}_B(\mathbf{r}, \mathbf{r}') &:= \bar{\mathbf{F}}'_B(\mathbf{r}, \mathbf{r}') - \bar{\mathbf{F}}_B(\mathbf{r}, \mathbf{r}'). \end{aligned} \tag{16.5}$$

It is this deviation of the modified ACGF from its value when considered in isolated environment what we define precisely as the effect of electromagnetic mutual coupling. As one may notice, the theoretical tool of the antenna current Green's function allows us to define mutual coupling in general electromagnetic systems using rigorous exact terms.

The functions $\delta\bar{\mathbf{F}}_A(\mathbf{r}, \mathbf{r}')$ and $\delta\bar{\mathbf{F}}_B(\mathbf{r}, \mathbf{r}')$ can be thought of as the mathematical representation of mutual coupling in terms of exact transfer function in space. In other words, given any illumination field impinging on the A + B system, we can compute separately the contribution to the receiving mode signal arising solely from the effect of mutual coupling on each subsystem A and B. Indeed, we have the following expressions for the perturbation on the total signals

$$\begin{aligned} \delta \mathbf{J}_A(\mathbf{r}) &= \int_{S_A} ds' \, \delta \bar{\mathbf{F}}_A(\mathbf{r}, \mathbf{r}') \cdot \mathbf{E}^{\text{in}}(\mathbf{r}'), \\ \delta \mathbf{J}_B(\mathbf{r}) &= \int_{S_B} ds' \, \delta \bar{\mathbf{F}}_B(\mathbf{r}, \mathbf{r}') \cdot \mathbf{E}^{\text{in}}(\mathbf{r}'). \end{aligned} \tag{16.6}$$

Here, similar to (16.5), we can write $\mathbf{J}'_A(\mathbf{r}) = \delta \mathbf{J}_A(\mathbf{r}) + \mathbf{J}_A(\mathbf{r})$ and $\mathbf{J}'_B(\mathbf{r}) = \delta \mathbf{J}_B(\mathbf{r}) + \mathbf{J}_B(\mathbf{r})$ for the signals received by the respective antennas at location $\mathbf{r}$ due to general illumination field $\mathbf{E}^{\text{in}}(\mathbf{r}')$. Since the expressions (16.6) are exact and general, the complete solution of the problem of mutual coupling reduces to finding the perturbation transfer functions $\delta \bar{\mathbf{F}}_A(\mathbf{r}, \mathbf{r}')$ and $\delta \bar{\mathbf{F}}_B(\mathbf{r}, \mathbf{r}')$.

16.3 INTERLUDE: APPLICATIONS OF THE ANTENNA CURRENT GREEN'S FUNCTION MUTUAL COUPLING FORMULATION

In the next few sections, we will sketch out briefly the main features of some major applications of the above rigorous and exact definition of electromagnetic mutual coupling using the antenna current Green's function. Before going into details, we provide a general philosophical anticipation of these applications serving as a rational motivation for their perusal.

The key difference between the popular definitions of mutual coupling presented in Section 16.2 and the rigorous and exact formulation of Section 16.2.1 is that in the former all of the definitions (16.1), (16.2), (16.3) appear as special cases of the latter. Indeed, the common denominator in the definitions of mutual coupling in terms of far- and near- fields, input impedance, mutual impedance, etc, is that all such formulations deal with *outputs* of the antenna problem. On the other hand, the definition (16.5) differs essentially in being worked out in terms of the exact general *transfer function* of the problem throughout which *any output whatsoever* can be computed. Stated differently, by finding the perturbations in the ACGF caused by mutual coupling, the most basic interaction problem is essentially solved. (This can be inferred from (16.6)). It is such *system viewpoint* – facilitated by the concept of exact transfer function in space – what will form the crux of the next three applications.

16.4 THE INFINITESIMAL DIPOLE MODEL (IDM) APPROACH TO MUTUAL COUPLING

16.4.1 Introduction

This part of the chapter is concerned with the analysis of the near field produced by array elements with arbitrary degree of couplings. The use of such arrays has been a cornerstone in wireless and radar technology since its early invention in the previous century. The function of the array depends on the configuration and the design of the elements, and for retaining full control of the radiated fields one usually runs into the important situation when strong electromagnetic coupling between the elements is present. In this case, the near-field structure is no longer the superposition of the fields radiated by isolated elements. Instead, the near field acquires a very complex structure, and the exact manner in which the far field is generated from this structure is not well understood yet. In this part, we extend the equivalent dipole method to deal with strongly interacting array elements. The method is chosen for its computational simplicity and conceptual clarity.

16.4.2 Mutual Coupling and Computational Electromagnetics

With the increasing sophistication and complexity in present-day electronic and electromagnetic devices and systems, the demand for fast and efficient methods for performing the analysis, design, and devolvement of a wide variety of tasks is still a crucial issue in applied science and engineering. In particular, the *computational* demand becomes a prohibitive limiting factor in the analysis and design of large-and-complex antenna arrays for many important applications such as radars and smart antennas. In such cases, it is usually required to perform repeated computations of the radiated field and the echo scattered by nearby objects or distant targets in order to facilitate signal processing schemes in conjunction with a particular application in mind. Since accurate characterization of such systems is usually feasible mainly in the full-wave analysis mode, there is a need to consider alternative methods that can give the same accuracy but without the need to solve the expensive entire boundary-value problem every time a modification in the array under consideration is introduced.

The method proposed in this part is suited to these and other applications in the sense that it aims to replace the complex and expensive numerical problem by a more tractable, semi-analytical approach in a way that avoids re-solving the entire problem for every array configuration. The approach is based on extending the method originally presented in [21] for single antennas but to handle this time the analysis of strongly-coupled antenna array elements. The *equivalent dipole model method* is based on replacing a complicated radiation problem by a simple model consisting of few infinitesimal dipoles radiating in free space or semi-half free space. The crux of this idea is to avoid solving the actual boundary-value problem and search instead for equivalent source representations for the problem at hand. The price, of course, is that the obtained dipole model is not unique. However, it can still be regarded acceptable in the sense that the model is valid only in the region *exterior* to some domain enclosing the antenna. The obtained dipoles re-produce the actual field in this exterior region with high accuracy and can be used to analyze antenna-antenna and antenna-scatterer interactions, therefore reducing considerably the computational cost.

In the formulation [21], the dipole model was employed to study successfully antenna-antenna interactions when the separation between the elements was large enough (weak coupling). However, the original method suffers from its inability to deal with strong mutual coupling effects for general arrays. In this section, we develop a new idea to systematically account for antenna-antenna interactions whose mutual coupling is significant, with our attention focused on the operational details of implementing the method for general arrays. When mutual coupling is strong, the equivalent current distribution obtained for the single element using the method available now in literature [21] fails to predict correctly the near-field behavior when this element is placed in the actual array environment. However, a hypothesis pertinent to the nature of mutual coupling in antenna arrays is developed in this part to modify the dipole model in order to predict correctly the new field in the strong mutual coupling scenario [11]. The method is based on modeling mutual coupling, i.e., the electromagnetic interaction between radiating elements, as a *multiple scattering effect* taking place between the antenna element and the nearby right and left elements. For simplicity, we limit our investigations in this section to linear array topology. It turns out that such hypothesis predicts correctly, within the original method's error, the electromagnetic interaction. Moreover, only one dipole model, taking into effect the first-neighborhood interactions, can be used to predict the correct near field for arbitrary-size arrays.

16.4.3 The Infinitesimal Dipole Method and the Near Field

The main emphasis of the infinitesimal dipole method is on developing efficient methods suitable for the description and analysis of the antenna near fields in various configurations of weakly or strongly interacting array elements. After developing the main method, we end in Section 16.4.9, where we provide an in-depth analysis of the antenna arrays near fields. It has been proposed recently by the authors (see also chapter on near field) that the decomposition of the radiated fields into propagating and nonpropagating (evanescent) fields constitutes the single most important feature in the dynamic description of the processes forming electromagnetic radiation as discussed in Chapter 4. Indeed, variations of the manner in which the near-field splits into propagating and nonpropagating parts as the observer varies his perspective (orientation of the observation frame) while moving around the antenna give rise to many important features in the antenna radiation pattern. Moreover, the coupling between antenna elements can be viewed as a field effect in which the respective near-field shells of two elements placed in the vicinity of each other tend to mutually interpenetrate and hence couple directly through their fields themselves. Traditionally, the only near-field measure studied in details has been the input impedance, predominantly for matching considerations. However, it has been shown by the authors that this approach is limited when consideration focuses instead on the electromagnetic fields themselves.[1]

For these motivations, there is a need for a fresh re-examination of the fundamental issues pertaining to the principles of operations of antenna arrays viewed mainly from the near-field perspective. The reasons for this can be stated as follows

1. First, the radiation pattern of the array is itself the outcome of a complex genetic process in which the near field continuously differentiates and changes in form giving rise to a determinate shape for the radiation at the far zone. Understanding the structure of the near fields then allows us to better grasp fresh potentials for controlling the radiated fields by modifying the current distribution on the array.
2. Second, electromagnetic interactions between array elements, especially elements placed very close to each other, is a direct coupling phenomenon between near fields shells as stated above. The physics of the near field then dictates the manner in which electromagnetic coupling will affect the performance of the array under consideration.
3. Third, for future applications, which tend to increase dramatically in complexity, such as metamaterial environments (artificial media) and nanostructures, it is important to get a hold on the fine details of the spatial structure

1 Cf. Chapter 4.

of the near field in order to theoretically pre-plan new devices and guide the research and devolvement process.

The spectral analysis of the near field, outlined in Section 16.4.9 of the present part, is one such perspective on the spatial structure capable of shedding light on the three issues above. Although the utilization of the evanescent fields has found its way in near-field nano-optics, especially for imaging applications, very little research in this direction has been conducted for antennas working in the microwave regime. For these reasons, Section 16.4.9 develops this important applications for the equivalent dipole model in order to provide the working community with reliable theoretical tools for the analysis and design of antenna arrays.

16.4.4 The Basic Method of Equivalent Dipole Source Models

We provide here the minimum necessary background to follow the main arguments of this section but the reader may refer to [21] and [22] for further information about the technical aspect of the implementation. Some of the most basic details relating to the mathematical description of the dipole model used in this part are collected in the Appendices.

The electric field radiated by a single dipole with moment $\mathbf{p}_m$ and location $\mathbf{r}_m$ is given by the well-known formula

$$\mathbf{E}(\mathbf{r}) = \frac{\mu_0 \omega^2}{4\pi} \bar{\mathbf{G}}(\mathbf{r}) \cdot \mathbf{p}_m. \tag{16.7}$$

The scalar Green's function is defined as

$$g(\mathbf{r}) = \frac{e^{ikr}}{r}, \tag{16.8}$$

while the dyadic Green's function is given in terms of the scalar solution using the following well-known relation

$$\bar{\mathbf{G}}(\mathbf{r}) = \left(\bar{\mathbf{I}} + \frac{1}{k^2} \nabla\nabla \right) g(\mathbf{r} - \mathbf{r}_m), \tag{16.9}$$

where $k = \omega/c = 2\pi/\lambda$ is the wavenumber in free space, c and λ are the speed of light and the wavelength, respectively, both in free space, and $\bar{\mathbf{I}}$ is the unit dyad.

The basic idea of the dipole model method is to generate the actual field of an arbitrary radiating structure by an expansion in the form [133],[134],[21]

$$\mathbf{E}_{\mathrm{DM}}(\mathbf{r}) = \sum_{m=1}^{M} \frac{\mu_0 \omega^2}{4\pi} \bar{\mathbf{G}}(\mathbf{r}) \cdot \mathbf{p}_m, \tag{16.10}$$

where $\mathbf{E}_{\mathrm{DM}}$ stands for the Dipole Model (DM) fields and μ_0 is the magnetic permeability of free space. Viewed in light of (16.10), the dipole model is an *analytical expansion* of a given arbitrary field in terms of *infinitesimal* dipoles (point sources) with the appropriate moments, locations, and orientations. When considered as such, there is no way to predict the number of dipoles in advance. Clearly, if M is large, the proposed approach loses its strength as analytical method and becomes on par with classical full-wave numerical solution of Maxwell's equations (subject to the appropriate boundary condition), or other deceptively similar methods [135]. However, it was found in previous research that the number of dipoles is actually small, typically in the range of five to ten dipoles for many antennas of practical interest [134],[21]. This vital observation will be exploited in this work by trying to see if such small size dipole model can be recruited in studying complicated problems. The Method of Moment (MoM) [23] will be used in this part to obtain the near field training data set needed to find the dipole model and in combining an existing model with the numerical solver to study more complicated configurations.

Figure 16.3 illustrates the general process. The expression of the error function $F = ||\mathbf{E}_{\mathrm{DM}} - \mathbf{E}_{\mathrm{A}}||$, a highly non-trivial choice in optimization, can be found in Appendix 16.6.1. The objective function is a nonlinear measure which is difficult to optimize. Notice that each dipole requires seven optimization parameters (Appendix 16.6.1). Therefore, for, say, 10 dipoles, the dimensionality of the search landscape is 70, necessitating a powerful optimization algorithm to deal with the problem. In this part, we employ the Quantum Particle Swarm Optimization (QPSO) algorithm developed in [22] for electromagnetic applications in order to search for the best dipole model. The applicability of the algorithm for a wide range of practical problems was already demonstrated in [21].

The training data set is ideally obtained by a high-resolution accurate near-field measurement setup. Once the data is available, the dipole model can be found using the QPSO algorithm (or any other similar powerful search method) and an analytical model for the antenna can then be utilized to predict the field anywhere outside the forbidden region (see below). Therefore, there is in principle no need to numerically solve Maxwell's equations. The dipole method is an ideal choice when

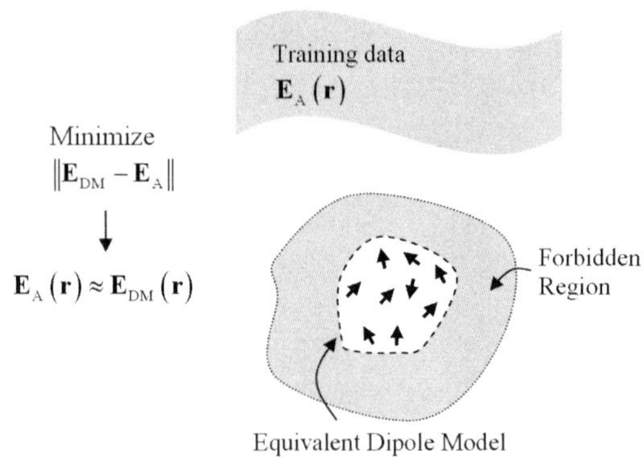

Figure 16.3 General schematics for the procedure of obtaining an equivalent dipole model for an arbitrary radiating structure. Only a training set of near-field data, not necessarily on a closed surface, are needed to guide the QPSO algorithm in the search for accurate dipole models. A suitable error function $||\mathbf{E}_{\mathrm{DM}} - \mathbf{E}_{\mathrm{A}}||$ is introduced and used as an objective function in the QPSO algorithm. The outcome of the search is a dipole model that can re-produce the training set and predict the near and far fields at other locations.

the geometry of the radiating structure is not well known (e.g., imaging problems) or very complicated to deal with numerically. In this part, which is concerned with developing the proof of concept, we employ the MoM to compute the actual fields. Moreover, we show that the dipole model has a considerable advantage when combined with MoM to deal with large array problems.

It is important to notice that the dipole model, being a current distribution that does not satisfy the actual boundary condition of the problem at hand, is not unique. That is, there exists a region directly surrounding the antenna in which the dipole model field deviates significantly from the actual field. The most extreme case is when the field is observed at the (discrete) locations of the dipoles, where the singularity will produce infinite (non-physical) field strength. The forbidden region is explicitly indicated in Figure 16.3 to sort out this important restriction on the equivalent dipole model: such models can *not* be used directly to study what happens *at* the antenna itself. However, they can used to accurately describe the field *everywhere* in the *exterior* region.

16.4.5 Analysis of Linear Arrays of Patch Antennas

In this section, we provide a general investigation of a concrete antenna array. The configuration is linear. The chosen antenna element is a wideband PEC patch excited by L-shaped probe. This antenna is intermediate in complexity between linear metallic wires and dielectric antennas. Notice that the basic method works well when applied to a wide range of antennas, for example dielectric resonator antennas [21]. However, no study of the effect of mutual coupling between array elements on the infinitesimal dipole model was reported before and hence we endeavor to achieve this in this work.

Figure 16.4 illustrates the details of the antenna geometry. A dipole model was obtained for this antenna radiating as a single element at the frequency 4.5 GHz. Next, we investigate the performance of the antenna within an array environment consisting of a linear configuration of identical elements as shown in Figure 16.5. All elements are located above an infinite ground plane with uniform separation given by d. Image theory is used to model the ground plane, replacing effectively the free space Green's function by the half-space Green's function, and reducing therefore the computational demand of the optimization problem. (The reader should notice that the situation with *finite* ground plane cannot be modeled exactly using this method because of the complexity of the Green's function of such ground plane. We expect then that our model may not be very successful with small ground planes in which edge diffraction effects are important. The problem can be solved,

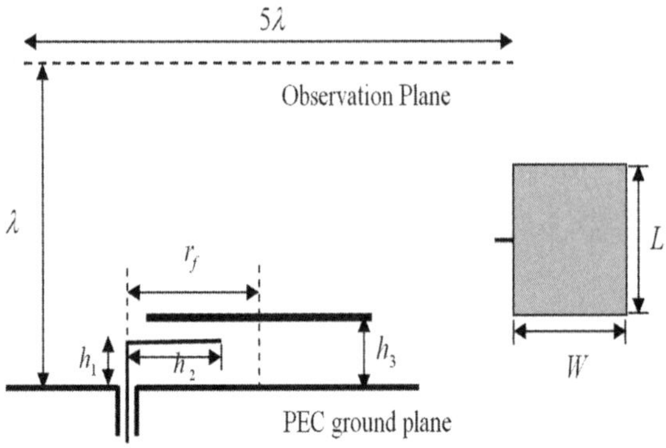

Figure 16.4 Conducting patch excited by L-Shaped coaxial probe. $h_1 = 5\,\text{mm}$, $h_2 = 11.14\,\text{mm}$, $h_3 = 8$ mm, $r_f = 13.84\,\text{mm}$, $W/2$ = 12.84 mm, and $L/2$ = 15.44 mm.

however, by including the effect of the edge in the actual near field data used to search for the best dipole model. The only potential difficulty is that a larger number of dipoles may be needed in order for the optimization algorithm to converge into an accurate model.)

The array at the top of Figure 16.5 is the actual configuration. At the bottom of the same figure we show an equivalent dipole model. Each dashed box has a size roughly the same as the physical volume occupied by the actual antenna but need not be exactly identical. (It has been found by the authors' experience that allowing some variations in the initialization of the box inside which the QPSO algorithm will search for the dipoles is potentially advantageous in obtaining good convergence results.) It is very important to notice that this dipole model was not obtained from a training set data due to the actual *array* at the top. The procedure is detailed as follows

1. Obtain a dipole model for the *single* antenna radiating in free environment.

2. Position the dipole model at the location of each antenna in the array configuration. The field radiated by this model is obtained analytically in a straightforward manner by shift operations (displacement in space) essentially equivalent to using the array factor of linear configurations.

3. To obtain the effect of the arbitrary complex excitation, a scaling of only the dipole moments is needed. This is based on the fact that Maxwell's equations

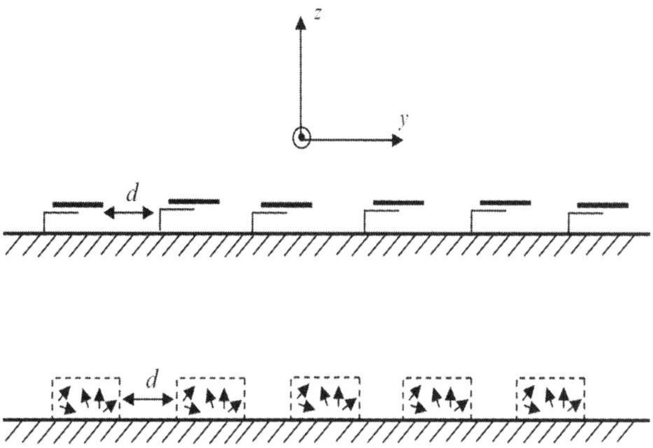

Figure 16.5 (Top Array) A linear array of patch antennas with uniform separation between the elements given by d. The radiating elements are located above an infinite ground plane. (Bottom Array) The equivalent dipole model configuration for the same array.

are linear and once a solution (in this case the dipole model) was obtained, there is no need to rerun the simulation/measurement to obtain the solution for another (proportional) excitation. That is, let the dipole model moments due to voltage excitation V be given by $p_u(V)$. Assume that a dipole model moment $p_u(V_0)$ was obtained for voltage V_0. The new moments due to a complex voltage V are then given by

$$p_u(V) = p_u(V_0)V, \tag{16.11}$$

where u is one of x, y, and z.

Therefore, what was accomplished here is first obtaining the dipole model for an *isolated* radiating element, and then superimposing several *identical* dipole models in order to mimic the original array. The models are 'identical' in the sense that they differ from each other only with respect to a scaling factor. All the models have in common the same *relative* distribution of infinitesimal dipoles (relative locations, orientations, moments) within their allocated volume positions.

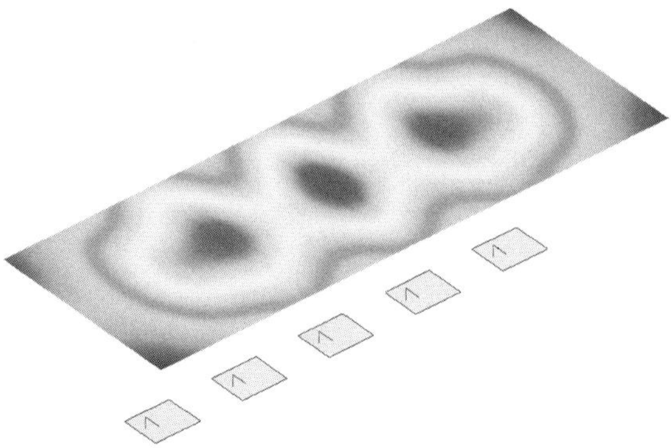

Figure 16.6 A set of near field (magnitude of the total electric field) for the comparison between the array dipole model and the actual configuration solved by MoM. The distance of the set is λ from the PEC ground plane. The dimensions of the rectangular is $9\lambda \times \lambda$. The number of observation points $N_{\text{ops}} = 10,000$. The array elements are excited with voltage sources (from left to right) 1V, 16V, $1 + j1$V, $2j$V, and $-3j$V. The inter-element spacing is 1.1λ.

16.4.6 Weak Mutual Coupling

We first rehearse the situation when the mutual coupling between the elements is weak. In this case, one can replace each antenna by the dipole model that was obtained for a single antenna and superimpose as many dipole models as we wish. Figure 16.6 illustrates an example of 5 PEC patch antennas with uniform separation of 1.1λ.

For the problem shown in Figure 16.6, the obtained global error (see Appendix 16.6.2 for definition) is 5.34%, which is very close to the original global error obtained by the QPSO algorithm for a single (non-interacting) antenna. From the experience of the authors, a global error in the range 3-8% is considered acceptable [21]. The reader should notice that there is nothing special about the particular near-field data of Figure 16.6. Indeed, arbitrary different sets, but with sufficient resolution, can be chosen as long as the observation plate does not get very close to the antennas (more on this below).

In Figure 16.7, we present a comparison between the fields computed by the MoM and those predicted by the dipole model (DM). We show these particular components since they represent the maximum strength of the electric and magnetic fields and reflect the variation across the *length* of the near field rectangle shown

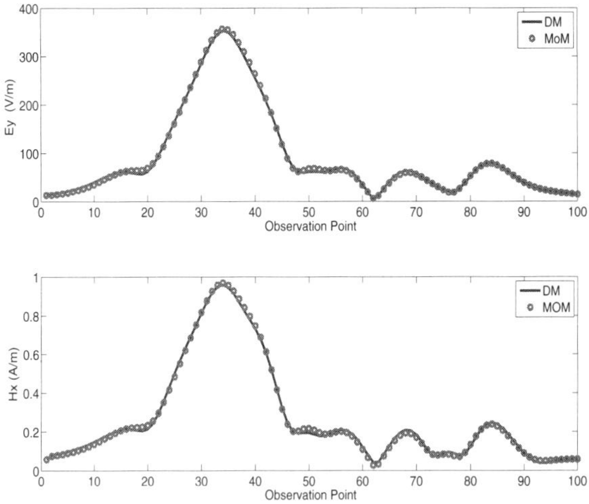

Figure 16.7 Comparison between the Dipole Model (DM) and the Method of Moment (MoM) for two components of the field due to the 5-element linear array of Figure 16.6. The field is plotted across a line passing midway across the length of the near field rectangular shown in Figure 16.6.

in Figure 16.6. This excellent agreement was obtained for a vector of complex excitations (see caption of Figure 16.7 for the exact numerical values) chosen arbitrarily. The MoM solution was computed due to these excitations while (16.11) was used to obtain the modified dipole model array factor.

In Figure 16.8, we study the performance of the dipole model when the near-field observation rectangle is brought closer to the array under test. As can be seen, the dipole model is still capable of predicting the MoM fields even at distance from the ground plane as close as 0.4λ (roughly 0.28λ from the PEC patch itself). On the other hand, at distance 0.2λ (the observation rectangle is almost touching the PEC patches), the error becomes prohibitively large (around 45%). However, it is instructive to examine several electric and magnetic field components corresponding to this worst-case error. Figure 16.9 illustrates the comparison for E_x, E_y, H_x, and H_y. We notice two points in the log scale plot there. First, although the global error is high, the dipole model can still capture very roughly the field behavior. Second, where it comes to the quantitative measure of the difference between the dipole field and the actual field, it appears that most of the error arises

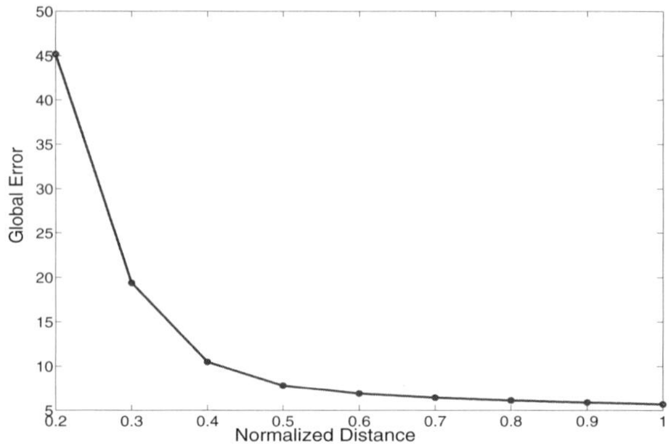

Figure 16.8 The variation in the global error (in percentage) when the near-field observation rectangle in Figure 16.6 is brought closer to the array. The horizontal axis shows the distance measured from the ground plane normalized to the operating wavelength $\lambda = 66.67\mathrm{mm}$. At the distance 0.2λ the near-field rectangle is almost touching the antenna array.

from non-physical singularities introduced by the dipoles, which don't exist in the actually smooth current computed by the MoM. Indeed, away from the non-physical singularities, it seems that this worst-case dipole model can still provide a reliable estimation, at least on the qualitative level, for the actual field produced by the array. A quick glance at the expression of the global error introduced in (16.65) can convince the reader that this is indeed the case. That is, when the observation point is very close to a dipole location, the fields radiated by that dipole becomes very large, theoretically approaching infinity. However, since the MoM field is finite, the ratio forming the error measure will tend to infinity, regardless to the agreement or disagreement at *other* observation points.

16.4.7 Arrays with Strong Mutual Coupling

We now systematically study the effect of decreasing the distance between the elements in the patch antenna linear array on the global error. Figure 16.10 demonstrates this parametric study. It is clear that when the separation shrinks to a certain critical distance, the mutual coupling starts to significantly increase the global error. For this particular array, this happens, loosely speaking, around $d = 0.5\lambda$, indicating the onset of considerable electromagnetic coupling between the elements starting

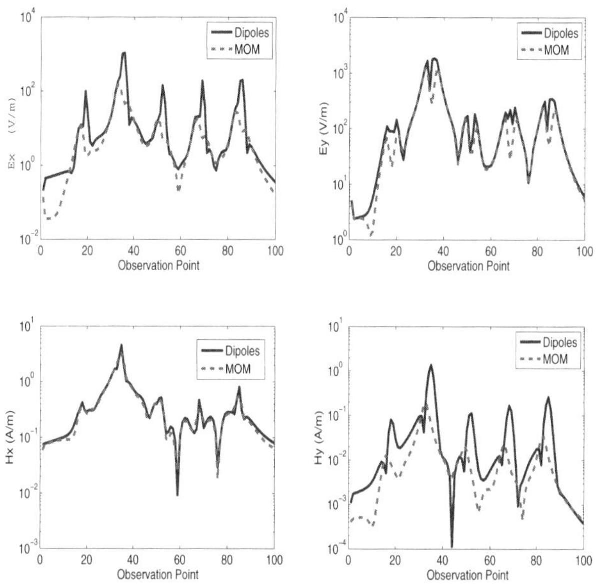

Figure 16.9 Comparison between the Dipole Model (DM) and the Method of Moment (MoM) when the near-field rectangle is 0.2λ from the ground plane, i.e., for the case corresponding to the global error of roughly 45% in Figure 16.8.

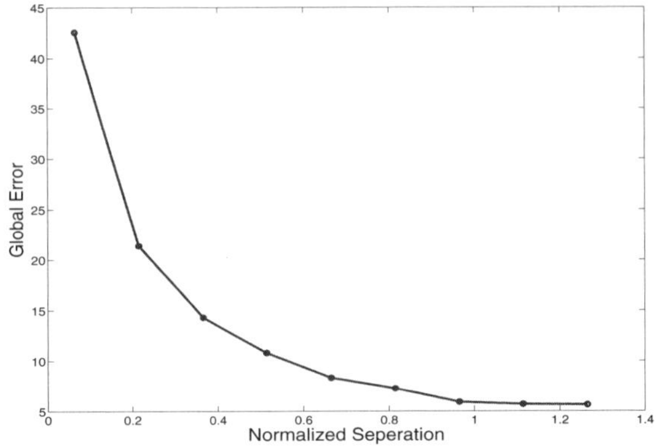

Figure 16.10 The global error (in percentage) versus the separation (normalized to the wavelength $\lambda = 66.67\mathrm{mm}$) between the PEC patch antennas in the array of Figure 16.6.

from this separation. In the current section, we will take the case of $d = 0.21\lambda$ as a representative example of strong mutual coupling.

It is obvious that the dipole model that was obtained with only an isolated radiating element cannot be directly used to predict the actual field when several interacting elements are radiating simultaneously within the same environment. This means that the current distribution obtained by the QPSO algorithm is invalidated by the presence of nearby *interacting* elements. Indeed, the results of the previous section show that when the array elements are not interacting with each other, the same dipole model can be used for all the elements provided we account correctly for the array factor. The question now is whether it is possible to account for the interaction itself using the dipole model without going into the optimization problem of searching for a dipole model for the *entire* array considered as a *single* large antenna. The latter task is practically not feasible for the following reasons

1. The number of dipoles, and hence the dimensionality of the optimization problem, becomes very large. For example, suppose we have a 10-element array and we use 10 dipoles for each antenna but all are dispersed within a search box encompassing the entire array physical domain. In this case, the number of optimization parameters is $100 \times 7 = 700$, an extremely large size to work with in a global optimization problem.

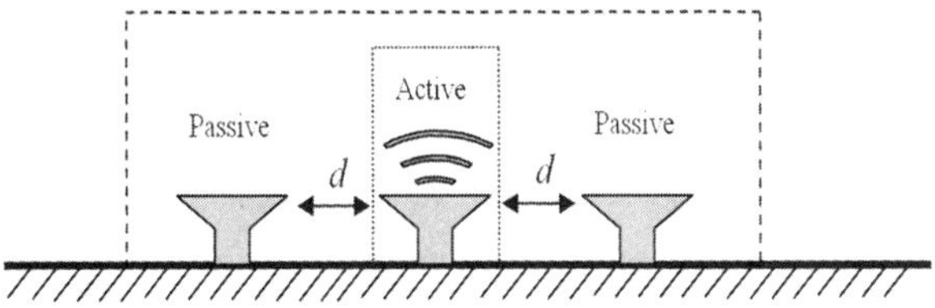

Figure 16.11 The basic building block for a modified equivalent dipole model that takes into account the interaction between closely packed array elements. The antenna in the middle is excited by a voltage source while the two antennas to the right and left are passive, i.e., the voltage source is removed but the source impedance is left in the numerical model or experiment.

2. Notice that, to make things even worse, since the effective antenna size of the entire array is large, a large number of training data points is required to capture the physics of the near field due to the array under test. This increases tremendously the computational demand on the objective function which, when combined with the very large size of the optimization landscape, will make the use of the QPSO algorithm or any other global optimization method practically useless.
3. Even if all the previous computational demands were met, the dipole model we obtain through such procedure is valid only for a fixed-size array. By Adding or removing few elements, the dipole model will fail to predict correctly the actual fields.

In the next section, we present a simple method that appears to mitigate the previous difficulties.

16.4.8 A Method to Account for Strong Mutual Coupling Using Equivalent Dipole Models

16.4.8.1 Introduction

Consider an antenna radiating in an ideal environment consisting of multiple identical antennas arranged in a linear array configuration (cf. Figure 16.5). We assume that the separation d between the elements is chosen such that strong mutual coupling will modify the near field generated by the array.

What is the origin of mutual coupling in antenna arrays? Of course, the answer to this question is quite complex and depends crucially on the type, function, design, and environment associated with the array under test. However, our inquiry is about the origin and nature of electromagnetic interactions in general. Indeed, there may be many manifestations for such mutual coupling, like surface modes propagation in dielectric substrates or just space-to-space direct coupling of energy from one element to another, but it might be reasonable to visualize all interactions in the following unified manner inspired by the study of scattering in solids.

16.4.8.2 Elementary Theory of Electromagnetic Interactions

Consider two antennas A and B. Both antennas are excited and hence generate an electromagnetic field in the surrounding region. If the two antennas are close to each other, the field due to antenna A, call it $\mathbf{E}_A$, will hit antenna B and scatters back, now as $\mathbf{E}_B$, to the location of antenna A. This scattered field $\mathbf{E}_B$ will in turn scatter by antenna A back to antenna B, and the process goes on indefinitely. The solution of the new boundary-condition problem in which both antennas exist as scattering object will give rise to what is usually refereed to in literature as the multiple scattering effect [120]. The field due to such situation is the self-consistent field that is simply the formal sum of the infinite series of small back-and-forth scattering events developing between antenna-objects A and B. For arbitrary antenna structure, there is no analytical solution of such multiple scattering problems and one has to resort to brute force full-wave numerical solution of Maxwell's equations with the right boundary condition. However, it is the objective of the work presented in this part to avoid solving actual boundary-condition problems and rely instead on the QPSO algorithm to search for equivalent dipole models. We accomplish this by postulating two hypothesis:

1. **Hypothesis I**: The electromagnetic interaction between radiating elements in antenna arrays is solely due to the multiple scattering effect between active (energized) element and passive nearby elements.
2. **Hypothesis II**: The interaction is, to the first-order approximation, due to the nearest neighbor elements.

Accoutring to Hypothesis I, although in the actual array all elements are energized (active), we assume that the effect of an antenna B on an antenna A is mainly due to the multiple scattering of the fields generated by antenna A by antenna B *treated as a passive object*. That is, we ignore the effect of antenna B being actually active in the array and assume that its effect on the near field is due to the multiple scattering series formed between the objects A and B. Next, according to Hypothesis II, we assume that only the nearest elements to the energized antenna contribute to the mutual coupling. In the linear array configuration, this second assumption stipulates only two passive elements to be considered in the interaction picture, one to the right and the other to the left of the active element. The power of this Hypothesis is that it allows us to obtain a basic building block suitable for analyzing antenna arrays with *arbitrary* size without going into the computationally demanding process of running the QPSO algorithm again for each new size.

In Figure 16.11, we demonstrate what we mean by the phrase 'basic building block.' There we see an active antenna surrounded by two passive antennas to the right and the left. The separation d is assumed to be small enough to allow for multiple scattering effects between the array elements to become significant. Now, either high-resolution near field measurement setup or an accurate numerical full-wave solution can be used to obtain a suitable near field data set. The QPSO algorithm is then employed to find an equivalent dipole model for this new structure. It is important to stress the following fact here: The dipoles in the optimization code are allowed to move only within the *dotted* box in Figure 16.11 enclosing the *active* antenna only. The interpretation of the new dipole model obtained through this search is that the current distribution (dipoles' locations, orientations, and moments) contains the effect of nearby elements on the current distribution of the active element only. Therefore, we can use this new (modified) current distribution to build arbitrary-size arrays by applying the basic procedure developed in Section 16.4.5. The *dashed* box in Figure 16.11 represents then the basic structure accounting for mutual coupling within the limit of the approximations assumed in the two basic hypothesis above.

16.4.8.3 Numerical Validation of the Proposed Method

This procedure has been carried out and an accurate equivalent dipole model was obtained for the configuration of Figure 16.11 with the patch antennas of Figure 16.4 as the radiating element and separation $d = 0.21\lambda$. The MoM was used to solve for the near field with the generators of the right and left elements removed and replaced by the source impedance. The obtained dipole model was then used to build an equivalent dipole model for a 13-element array. The near field observation rectangle is taken to extend across the entire array at a distance λ from the ground plane. The global error for this array using the dipole model that does *not* take into account mutual coupling is 24.27%. The method outlined above reduced this global error to just 6.57%. This reduction in error is consistent in all field components with

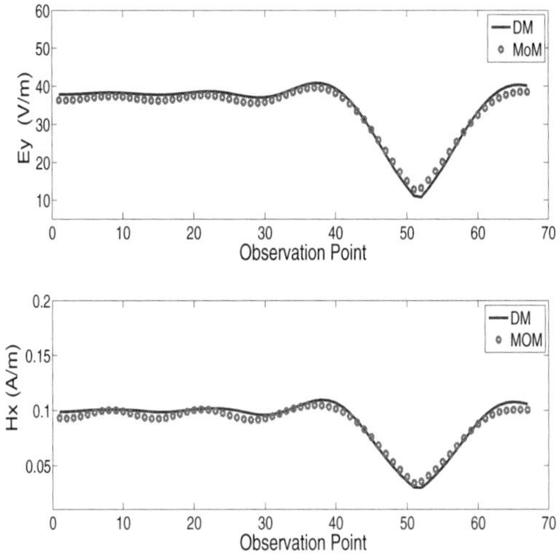

Figure 16.12 Comparison between the fields generated by the dipole model and the Mom for a 13-element linear array of patch antennas. All elements are excited by unit voltage.

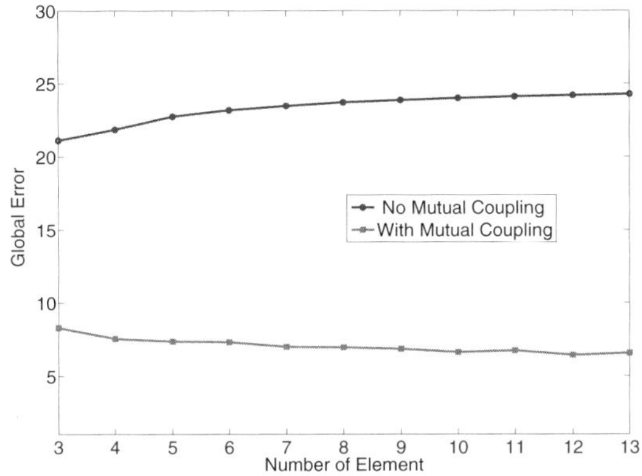

Figure 16.13 Global error comparison between the equivalent dipole model predictions of the near field for the situation when the single element dipole model is employed (no mutual coupling) and the modified dipole model (with mutual coupling) using the method of Section 16.4.8. The separation between the elements in all arrays is $d = 0.21\lambda$. The array configuration has E-plane coupling.

varying sizes of near field data and locations, indicating the validity of the simple picture presented in Section 16.4.8.2.

Figure 16.12 illustrates the comparison between MoM and the modified dipole model for the field components with maximum strength measured at a line midway across the length of the observation rectangle. This excellent agreement demonstrates the ability of the modified dipole model to reproduce the same quantitative variation of the field at various spatial location. The choice of the particular location of the observation is here restricted as described above only for the purpose of the demonstration. The authors have found that the modified dipole model, i.e., the one obtained originally with only two nearby elements effect taken into consideration, predicts correctly the extrapolated 13-element array field very accurately and consistently everywhere in the exterior domain.

Figure 16.13 illustrates the comparison between the dipole model without taking into effect mutual coupling and the modified model obtained using the method of this section in which various array sizes are assumed. In each array, a near field observation rectangle at a distance of one wavelength from the ground plane was used to calculate the error. The length of the rectangle was modified in each case to extend across the actual array size and enough number of samples was employed. It is clear from the results that the modified dipole model is consistently capable

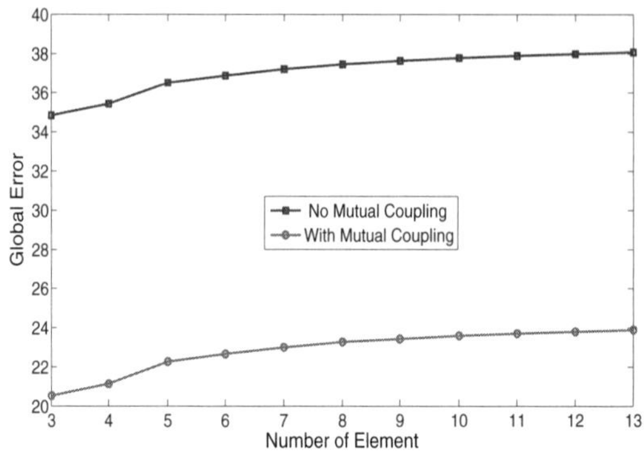

Figure 16.14 Global error comparison between the equivalent dipole model predictions of the near field for the situation when H-plane coupling is considered. The separation between the elements in all arrays is $d = 0.21\lambda$.

of capturing the effect of mutual coupling with increasing array size. Moreover, it is interesting to notice that the error level is nearly constant with the array size. Actually, we expect the method to work better when the array size gets larger because the effect of the asymmetry introduced by the edge elements becomes less important. It seems therefore that the error level around 6.5% in the modified dipole model is the *intrinsic limit* of the method within the approximations introduced by Hypothesis I and II.

In Figure 16.14 we study the mutual coupling in the same linear array but with H-plane coupling, instead of the E-plane coupling illustrated in Figure 16.5. To obtain these results, another (new) problem like the one in Figure 16.11 is introduced but with H-plane coupling. The QPSO algorithm is employed to find the modified equivalent dipole model for this configuration. We compare in Figure 16.14 the predictions of this model with the single element (no-mutual-coupling) model. It is clear that the method still works for this different coupling configuration. Indeed, the H-plane coupling represents a physically different mode of electromagnetic coupling and hence cannot be analyzed using the dipole model obtained with the configuration of E-plane coupling.

The comparison between the results of E- and H-plane coupling suggest that care must be taken in applying the method proposed in this part. Indeed, every new environment surrounding the antenna under consideration in the array will require its own modified equivalent dipole model. These models are functions not just of frequency and the separation between the elements, but also may crucially depend on the rotation of one element with respect to the other. By building a library of dipole models corresponding to the various modes of coupling needed in the application at hand, it is possible then to study the problem of arbitrary arrays using the corresponding dipole model obtained previously and for varying number of elements.

Finally, we end this section with few words about numerical efficiency. If the total overall cost of obtaining the equivalent dipole model, which is mainly due to the global optimization process, is taken into account in the comparison regarding the computational cost, then the full-wave numerical solution of arrays as large as the 13-element array we presented above is much less expensive. However, the value of obtaining the dipole model relies on two aspects, the theoretical advantage of the conceptual simplicity of infinitesimal dipoles as radiating structure (see Section 16.4.9 for elaboration on this point), and the economy in use within larger applications. The latter concerns the fact that the dipole model can now be used as part of a larger engine concerned with extensive study of an array under consideration, for example, the statistical analysis of a radar system or a direction-of-arrival algorithms in smart antennas. In this case, the availability of simple analytical models accurate enough for the entire exterior region of the array cuts dramatically the cost of the essentially repetitive calculations in this type of statistical studies. On the other hand, if the number of elements in the array grows to very large number, then the full-wave analysis will at some point exceeds even the optimization process cost, and in such large-and-complex arrays (not considered in this part), we expect that the dipole method will prove advantageous.

16.4.9 Spectral Analysis of the Antenna Array Near Field Using the Equivalent Dipole Model

16.4.9.1 Introduction

The detailed investigations of the foregoing parts of this part have established unequivocally the validity of the modified equivalent dipole model for an efficient computation of the near field at various locations close enough to the radiating elements. Indeed, (16.10) gives the array near field in a closed-form analytical expression written as a weighted sum of free-space Green's functions. We would like to conclude our work by sketching a general outline for a theory of the near field in the spectral domain that takes into account knowledge of an equivalent dipole model for some array under consideration.

The spectral analysis we have in mind is the decomposition of the radiated field into propagating and nonpropagating parts. As was suggested by the authors in Chapter 4, it is this decomposition that plays a fundamental role in establishing the correct principles of operation of general antenna systems. The physics of the near field can be effectively encoded in the manifold ways in which the "static" nonpropagating field is dynamically being converted into a propagating field as we move progressively away from the radiating elements. As was pointed out in Chapter 4, for an antenna embedded in free space, there is a freedom in choosing the orientation of the observation coordinate system, and the crux of the dynamic content of the decomposition into propagating and nonpropagating fields was traced back into this freedom of choice. However, in the array environment being studied in this part, there is a natural choice for the orientation of the observation frame, specifically that in which the z-axis is directed perpendicular to the array plane. In the remaining parts of this section, we analyze the fields in the spectral domain using this particular orientation. The study of the manner in which the near field spectral composition varies with other choices of the orientation of the observation frame will be taken up in other publications.

As will be shown in the coming sections, one of the major stages in the spectral analysis involves the computation of the spatial Fourier transform of the antenna current distribution. The analysis in Chapter 4 was conducted mainly for a single antenna system. In the next section, we extend the treatment to an array configuration and make use of the equivalent dipole model of the present part to eliminate completely the need to perform the spatial Fourier transform in the computation of the propagating and nonpropagating parts. Our goal is to demonstrate the power of the dipole model method in analyzing the fields of general antenna array systems both in the spatial and spectral domain.

16.4.9.2 Development of the Spectral Expansion of the Array Near Field in Terms of the Equivalent Dipole Model

The electric field radiated by this current distribution is given by the classic dyadic Green's function theorem

$$\mathbf{E}(\mathbf{r}) = i\omega\mu \int_V d^3r' \, \bar{\mathbf{G}}(\mathbf{r}, \mathbf{r}') \cdot \mathbf{J}(\mathbf{r}'), \tag{16.12}$$

Therefore, the electromagnetic fields radiated by the antenna can be totally determined by knowledge of the dyadic Green's function (16.9) and the current distribution on the antenna array $\mathbf{J(r)}$ with total support V. We can decompose the former into two parts, one pure propagating and the other evanescent. This task can be accomplished by using the Weyl expansion [35]

$$\frac{e^{ikr}}{r} = \frac{ik}{2\pi} \int_{-\infty}^{\infty} \int_{-\infty}^{\infty} dpdq \frac{1}{m} e^{ik(px+qy+m|z|)}, \tag{16.13}$$

where

$$m(p,q) = \begin{cases} \sqrt{1-p^2-q^2} & ,p^2+q^2 \leq 1 \\ i\sqrt{p^2+q^2-1} & ,p^2+q^2 > 1 \end{cases}. \tag{16.14}$$

Substituting the Weyl expansion (4.4) into the dyadic Green's function as given by (16.9) and making use of (16.12), we obtain after interchanging the order of integration

$$\mathbf{E}(\mathbf{r}) = \frac{-\omega k \mu}{8\pi^2} \int_{-\infty}^{\infty} dpdq \frac{\bar{\mathbf{I}}k^2 - \mathbf{KK}}{k^2 m} \cdot \tilde{\mathbf{J}}(\mathbf{k})\, e^{i\mathbf{K}\cdot\mathbf{r}}, \tag{16.15}$$

where $\tilde{\mathbf{J}}(\mathbf{K})$ is the spatial Fourier transform of the source distribution

$$\tilde{\mathbf{J}}(\mathbf{K}) = \int_V d^3r'\, \mathbf{J}(\mathbf{r}')\, e^{-i\mathbf{K}\cdot\mathbf{r}'}. \tag{16.16}$$

The spectral variable (wavevector) is given by

$$\mathbf{K} = \hat{x}kp + \hat{y}kq + \hat{z}\operatorname{sgn}(z)\, km, \tag{16.17}$$

where sgn stands for the signum function. From this it readily follows that the propagating and nonpropagating parts are given, respectively, by the expressions

$$\mathbf{E}_{\rm pr}(\mathbf{r}) = \frac{-\omega k \mu}{8\pi^2} \int_{p^2+q^2<1} dpdq \frac{\bar{\mathbf{I}}k^2 - \mathbf{KK}}{k^2 m} \cdot \tilde{\mathbf{J}}(\mathbf{k})\, e^{i\mathbf{K}\cdot\mathbf{r}}, \tag{16.18}$$

$$\mathbf{E}_{\rm ev}(\mathbf{r}) = \frac{-\omega k \mu}{8\pi^2} \int_{p^2+q^2>1} dpdq \frac{\bar{\mathbf{I}}k^2 - \mathbf{KK}}{k^2 m} \cdot \tilde{\mathbf{J}}(\mathbf{k})\, e^{i\mathbf{K}\cdot\mathbf{r}}. \tag{16.19}$$

In order to specify the region of validity of the expansion (16.15), we need to provide some specifications of the antenna system. We assume that the radiating system consists of a planner array of N elements. The current distribution on the nth array element is given by $\mathbf{J}_n(\mathbf{r})$ and its Fourier transform by $\tilde{\mathbf{J}}_n(\mathbf{K})$. That is, we can write

$$\mathbf{J}(\mathbf{r}) = \sum_{n=1}^{N} \mathbf{J}_n(\mathbf{r}), \quad \tilde{\mathbf{J}}(\mathbf{K}) = \sum_{n=1}^{N} \tilde{\mathbf{J}}_n(\mathbf{K}). \tag{16.20}$$

Assume further that the entire array is contained between the planes $z = L > 0$ and $z = L$, where the global observation coordinate system is chosen such that the z-axis is directed perpendicular to the array plane as mentioned in the introductory part of the section. Now, the expansion (16.15) is valid only in the region $z > L$ and $z < -L$, i.e., the region exterior to the infinite slab $-L \leq z \leq L$.[2] Notice also that the assumption $\mid z \mid> L$ was implicit in writing (16.17).

In its general form, the spectral integrations implied by the equations (16.15), (16.16), (16.18), (16.19), and (16.20) cannot be easily assessed in terms of computational complexity. Indeed, as given, the Fourier transform of the current distribution, $\tilde{\mathbf{J}}(\mathbf{K})$, is a generally complex function of $\mathbf{K}$ that cannot be known a priori but depends on the specific array topology and excitation. The only thing that can be stated with confidence at this level, we believe, is that the computation of the 3-dimensional spatial Fourier transform mentioned above is a computationally demanding process, which forms only one stage, while the next step is to use this obtained Fourier transform to compute the spectral integrals themselves as appearing in (16.15).

In order to mitigate these computational difficulties, we now make use of the equivalent dipole model method developed for strongly interacting antenna arrays in order to completely eliminate the need to perform the spatial Fourier transform.

Assume that for a given global error, a dipole model DM for the antenna array was found. Let the number of dipoles for the nth antenna element be M_n. The dipole moment and position for the lth dipole in the nth array element are given by $\mathbf{p}_{nl}$ and $\mathbf{r}_{nl}$, respectively. The equivalent dipole model source is then given by the form

$$\mathbf{J}^{\mathrm{DM}}(\mathbf{r}) = \sum_{n=1}^{N} \sum_{l=1}^{M_n} \mathbf{p}_{nl} \delta(\mathbf{r} - \mathbf{r}_{nl}). \tag{16.21}$$

We immediately remind the reader that this "equivalent" current does *not* replace the current given by (16.20). The latter is a *unique* current distribution obtained by solving a well-posed boundary value problem. The sense of 'equivalence' implied here is that the fields radiated by (16.20) and (16.21) are the same only in some exterior region. Let us define this region as the domain $z > |L'|$, where $L' \geq L$. The ideal situation in which L' is almost equal to L is seldom encountered in practice

2 Cf. Chapter 4.

where the number of dipoles is kept low to facilitate optimization. However, the value of the tolerable global error will ultimately dictate how much L' deviates from L, a topic that is a matter of further empirical research. In this section we will be concerned exclusively with the theoretical formulation of the problem.

By substituting (16.21) into (16.15), we arrive into the following expressions for the propagating and nonpropagating parts

$$\begin{aligned}\mathbf{E}_{\mathrm{pr}}^{\mathrm{DM}}(\mathbf{r}) = \tfrac{-\omega k\mu}{8\pi^2} \sum_{n=1}^{N}\sum_{l=1}^{M_n} \int_{p^2+q^2<1} dpdq \\ \times \tfrac{\bar{\mathbf{I}}k^2-\mathbf{KK}}{k^2 m} \cdot \mathbf{p}_{nl} e^{i\mathbf{K}\cdot(\mathbf{r}-\mathbf{r}_{nl})},\end{aligned} \quad (16.22)$$

$$\begin{aligned}\mathbf{E}_{\mathrm{ev}}^{\mathrm{DM}}(\mathbf{r}) = \tfrac{-\omega k\mu}{8\pi^2} \sum_{n=1}^{N}\sum_{l=1}^{M_n} \int_{p^2+q^2>1} dpdq \\ \times \tfrac{\bar{\mathbf{I}}k^2-\mathbf{KK}}{k^2 m} \cdot \mathbf{p}_{nl} e^{i\mathbf{K}\cdot(\mathbf{r}-\mathbf{r}_{nl})}.\end{aligned} \quad (16.23)$$

Here, we assume that all the dipoles are located in the region $|z| \leq L$, i.e., we assume that $|z_{nl}| \leq L$ for all n and l.

In the region of the validity of the dipole model, we finally have

$$\left.\begin{aligned}\mathbf{E}_{\mathrm{pr}}^{\mathrm{DM}}(\mathbf{r}) \simeq \mathbf{E}_{\mathrm{pr}}(\mathbf{r}) \\ \mathbf{E}_{\mathrm{ev}}^{\mathrm{DM}}(\mathbf{r}) \simeq \mathbf{E}_{\mathrm{ev}}(\mathbf{r})\end{aligned}\right\} \text{ for } |z| \geq L'. \quad (16.24)$$

By comparing (16.18) and (16.19) with (16.22) and (16.23), we find that (16.24) implies that the use of the equivalent dipole model lead to the elimination of the spatial Fourier transform appearing in the original expressions in the exterior region $|z| \geq L'$.

16.4.9.3 Discussion of the Results

The new expressions (16.22) with (16.23) now contain simple and universal forms for the spectral content of the fields. In other words, the complete functional form of the spectral integrands' dependence on the antenna configuration and excitation is now fully given by the manner of the simple appearance of the dipole moments $\mathbf{p}_{mn}$ and positions $\mathbf{r}_{mn}$ in (16.22) with (16.23). As such, a significant advance compared with the original scenario of (16.18) and (16.19) has been made by using the equivalent dipole model. This is not merely a computational gain, but also a theoretical one. The reason is that the spectral composition of the radiated fields as a mixture of propagating and nonpropagating modes is presently spelled out in terms of the same functional forms (strictly speaking, the spectral integrands are second-order polynomials in $\mathbf{K}$ multiplied by some complex exponential of $\mathbf{K}$ and divided by m) valid for *arbitrary* antenna arrays. What makes the variation of the spectral composition of one particular array to another is simply the numerical values of the model parameters $\mathbf{p}_{nl}$ and positions $\mathbf{r}_{nl}$.

We turn now to the assessment of the computational complexity of the expressions derived using the equivalent dipole model. From Appendix 16.6.3, we know that for a given arbitrary location of the observation point, we need in general six different numerical integrals in order to evaluate the full array of propagating and nonpropagating dyadic Green's function. A glance at (16.21) shows that for the full array, we have a total number of dipoles given by

$$N_d = \sum_{n=1}^{N} M_n. \tag{16.25}$$

Therefore, for each observation point, a total number of $6N_d$ numerical integrations is needed in order to obtain all of the six propagating and nonpropagating electric field components using the equivalent dipole mode. The same basic procedure can be extended to compute the magnetic field components once the spectral decomposition in terms of the electric field has been attained.[3] As can be seen from this rough discussion, a definite knowledge of the computational demands of the spectral analysis of the electromagnetic near fields for general arrays is made possible by the use of the equivalent dipole method for strongly interacting antenna arrays.

The previous analysis was conducted with the assumption of the most general dipole model for an arbitrary planner antenna arrays. If the method developed in the previous sections for computing the near fields of strongly interacting elements is adopted, then the same dipole models is used for all array elements. In this case, $M_n = M$, for all $n = 1, 2, ..., N$, and M is the fixed number of dipole obtained when an active element radiates in the presence of two inactive nearby elements. The positions of the dipoles at their respective locations within the array environment are obtained by simple translation operations. Unfortunately, it does not appear in this case that the number of numerical integrals needed in evaluating the field decomposition into propagating and nonpropagating parts can be reduced. The reason is that any change in the location of the observation point with respect to the source will require a new evaluation of the integral as can be seen from a glance

3 Cf. Chapter 4.

at (16.71) and (16.72). The total number of such integrals is in this case given by $6M$.

16.4.10 Summary

We presented an alternative approach to the analysis of electromagnetic interactions in antenna arrays in which there is no need to solve Maxwell's equations. Ideally, the method accepts only a set of near-field data and then uses the QPSO algorithm to search for a dipole model that can reproduce the same fields everywhere in the exterior region. The main contribution of the section is providing a simple theory accounting for electromagnetic interaction by interpreting coupling between nearby elements in the array as a multiple scattering effect. The application of this idea to the dipole model method led to dramatic reduction in the error caused by electromagnetic interaction on the performance of the dipole model originally obtained by neglecting the effect of such interactions. The proposed method appears to work very well with an arbitrary number of linear array elements for a given mode of coupling (for example, E- or H-plane coupling, etc.). We concluded by carrying out a spectral analysis of the near fields by describing its decomposition into propagating and nonpropagating parts along a direction normal to the antenna array under consideration. The analysis pointed out that one of the most computationally demanding stages in completing this analysis, namely the calculation of the spatial Fourier transform of the entire antenna array, can be totally eliminated by using the equivalent dipole method in its domain of validity. This led to the derivation of a simple universal analytical form for the spectral composition of arbitrary antenna arrays.

16.5 PERTURBATIVE APPROACH TO THE COMPUTATION OF MUTUAL COUPLING IN LARGE ANTENNA ARRAYS

16.5.1 Introduction

We now return to addressing mutual coupling through the general exact definition of the mutual coupling current Green's function (ACGF) introduced in Section 16.2. Following the intuitive but not exact approach though the infinitesimal dipole model, our objective now is to base the *quantitative* analysis of electromagnetic coupling on a solid basis by deriving a perturbation series for the mutual coupling

ACGF. This series can serve as the ultimate foundation for tackling mutual interactions in generic electromagnetic systems and represents the core of the theory developed in this chapter.

There are two major advantages of using perturbation theory in studying mutual coupling. It can be shown that the general perturbation series to be derived below provides

- an algorithmic method to compute mutual coupling for large systems without the need to invert the electromagnetic operator of the full problem, and
- it allows the construction of a physically intuitive map electromagnetic energy exchange in the entire array, in which it becomes possible to know from which part of the antenna array a desired effect of mutual coupling has originated and which path through the array this effect has traversed through its journey from the source to the final observation point.

16.5.2 The Basic Idea of Perturbation Theory

We consider that individual (isolated) antenna elements have already been designed and characterized. After that, the elements are placed in an array environment where they begin to mutually interact with each other. By knowing the original (isolated) performance of the device on one hand, and the forward interactions between the elements on the other hand, *can we predict accurately the effect of mutual coupling without actually solving the entire array problem as a single electromagnetic system*? The answer we find here is *yes*, and the method of obtaining the solution is perturbation theory. In a nutshell, we decompose the full array electromagnetic operator into a sum of isolated-element operator and interaction operator. The full operator is then inverted using the perturbation series and only few terms are retained. In this case, the effect of coupling can be approximated very accurately by few basic multiplication and addition operations involving small operators.

In order to develop the method for a general setting, the recent concept of antenna current Green's function (ACGF) Chapters 8 and 9 is employed for the fundamental characterization of the antenna as a system in space. The ACGF, which serves as the exact transfer function in space, will be developed for the mutual coupling context in Section 16.5.3 together with the derivation of the perturbation series. In Section 16.5.4, a method of moment procedure and a special perturbation code were written to verify the theory is used to supply the relevant proof of concept for two-element array. The basic idea can be extended with some efforts to handle arbitrary large arrays but the details will be presented in the future. Finally,

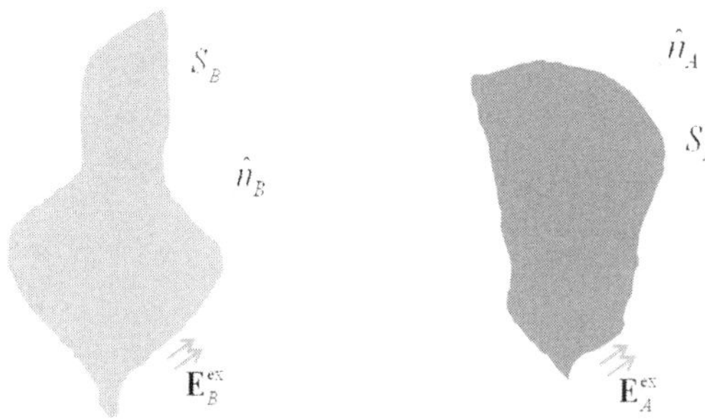

Figure 16.15 Two generic mutually interacting antennas are shown with arbitrary surfaces S_A and S_B. The unit outward normals to the surfaces are $\hat{n}_A$ and $\hat{n}_B$. The externally applied fields can interact with either a local portion of the surface (as in the Tx mode) or the entire surface area (as in the Rx mode).

the conclusion provides a summary of the new advantages supplied by the work developed here.

16.5.3 Derivation of the Perturbative Series

In Figure 16.15, we show a schematic illustration of two interacting antenna elements A and B. The antennas are assumed to have an arbitrary smooth surfaces S_A and S_B and are both possibly excited by fields $\mathbf{E}_A^{\text{ex}}$ and $\mathbf{E}_B^{\text{ex}}$, respectively, which may be defined either locally (transmitting mode) or globally (receiving mode). Our goal is to describe the interaction of these two antennas in terms of the ACGF formalism using the general conceptions outlined in Section 16.2.1.

Assume that both antennas support a perfect electric conductor boundary condition (PEC).[4] We can write the electric field integral operator for antennas A and B as $\mathcal{L}_A$ and $\mathcal{L}_B$. They are defined by the equation

$$\mathcal{L}\mathbf{J}(\mathbf{r}) = -i\omega\mu \int_S ds' \left[\bar{\mathbf{I}} + \frac{1}{k^2}\nabla\nabla\cdot\right] \frac{e^{ik|\mathbf{r}-\mathbf{r}'|}}{4\pi|\mathbf{r}-\mathbf{r}'|} \cdot \mathbf{J}(\mathbf{r}'). \tag{16.26}$$

4 The generalization to arbitrary boundary conditions is possible but length and will be reported elsewhere.

When the two-antenna system receives excitation fields $\mathbf{E}_A^{\text{ex}}$ and $\mathbf{E}_B^{\text{ex}}$ it responds by generating two self-consistent current distributions $\mathbf{J}_A(\mathbf{r})$ and $\mathbf{J}_B(\mathbf{r})$ supported by the surfaces S_A and S_B, respectively. By enforcing the PEC boundary conditions on the two antennas, the following operator equations must be obeyed

$$\begin{aligned}\hat{n}_A \times \mathbf{E}_A^{\text{ex}}(\mathbf{r}) + \hat{n}_A \times \mathcal{L}_B \mathbf{J}_B(\mathbf{r}) &= -\hat{n}_A \times \mathcal{L}_A \mathbf{J}_A(\mathbf{r}), \\ \hat{n}_B \times \mathbf{E}_B^{\text{ex}}(\mathbf{r}) + \hat{n}_B \times \mathcal{L}_A \mathbf{J}_A(\mathbf{r}) &= -\hat{n}_B \times \mathcal{L}_B \mathbf{J}_B(\mathbf{r}).\end{aligned} \tag{16.27}$$

Rearranging, we can write these two equations in the form

$$\begin{pmatrix} \hat{n}_A \times \mathcal{L}_A & \hat{n}_A \times \mathcal{L}_B \\ \hat{n}_B \times \mathcal{L}_A & \hat{n}_B \times \mathcal{L}_B \end{pmatrix} \begin{pmatrix} \mathbf{J}_A \\ \mathbf{J}_B \end{pmatrix} = -\begin{pmatrix} \hat{n}_A \times \mathbf{E}_A^{\text{ex}} \\ \hat{n}_B \times \mathbf{E}_B^{\text{ex}} \end{pmatrix}. \tag{16.28}$$

More compactly,

$$\tilde{\mathcal{L}}\,[\mathbf{J}] = [\mathbf{E}^{\text{ex}}], \tag{16.29}$$

where

$$\begin{aligned}\tilde{\mathcal{L}} &:= -\begin{pmatrix} \hat{n}_A \times \mathcal{L}_A & \hat{n}_A \times \mathcal{L}_B \\ \hat{n}_B \times \mathcal{L}_A & \hat{n}_B \times \mathcal{L}_B \end{pmatrix}, \\ [\mathbf{J}] &:= \begin{pmatrix} \mathbf{J}_A \\ \mathbf{J}_B \end{pmatrix}, \quad [\mathbf{E}^{\text{ex}}] := \begin{pmatrix} \hat{n}_A \times \mathbf{E}_A^{\text{ex}} \\ \hat{n}_B \times \mathbf{E}_B^{\text{ex}} \end{pmatrix}.\end{aligned} \tag{16.30}$$

Next, decompose the full operator into the direct sum

$$\tilde{\mathcal{L}} = \tilde{\mathcal{L}}_0 + \tilde{\mathcal{L}}_{\text{int}}, \tag{16.31}$$

where

$$\begin{aligned}\tilde{\mathcal{L}}_0 &:= \begin{pmatrix} -\hat{n}_A \times \mathcal{L}_A & 0 \\ 0 & -\hat{n}_B \times \mathcal{L}_B \end{pmatrix}, \\ \tilde{\mathcal{L}}_{\text{int}} &:= \begin{pmatrix} 0 & -\hat{n}_A \times \mathcal{L}_B \\ -\hat{n}_B \times \mathcal{L}_A & 0 \end{pmatrix}.\end{aligned} \tag{16.32}$$

Therefore, if the solution to the problem (16.28) exists, it is given by

$$[\mathbf{J}] = \tilde{\mathcal{L}}^{-1}[\mathbf{E}^{\text{ex}}] = \left(\tilde{\mathcal{L}}_0 + \tilde{\mathcal{L}}_{\text{int}}\right)^{-1}[\mathbf{E}^{\text{ex}}]. \tag{16.33}$$

The operator $\tilde{\mathcal{L}}_0$ can be thought of as the *self-interaction operator* of the antenna system. It is written in terms of the operators of the *isolated* antennas, i.e., when each exists in homogeneous infinite environment. The other operator $\tilde{\mathcal{L}}_{\text{int}}$ represents the interaction operator of the two-antenna system. It convey information about how the two elements couple to each other electromagnetically. Further insights into the structure of this operator will be developed below.

Suppose that the isolated antenna problems are solvable, i.e., the following inverse operator exist

$$\tilde{\mathcal{L}}_0^{-1} = \begin{pmatrix} \left(-\hat{n}_A \times \mathcal{L}_A\right)^{-1} & 0 \\ 0 & \left(-\hat{n}_B \times \mathcal{L}_B\right)^{-1} \end{pmatrix}. \tag{16.34}$$

Equation (16.33) can then written as

$$[\mathbf{J}] = \tilde{\mathcal{L}}_0^{-1} \left(\tilde{I} + \tilde{\mathcal{L}}_0^{-1} \tilde{\mathcal{L}}_{\text{int}}\right)^{-1} [\mathbf{E}^{\text{ex}}], \tag{16.35}$$

where $\tilde{I}$ is the identity operator corresponding to (16.30).

The ACGF of the two-antenna system is readily obtained as

$$\left[\bar{\mathbf{F}}\left(\mathbf{r}, \mathbf{r}'\right)\right] = \tilde{\mathcal{L}}_0^{-1} \left(\tilde{I} + \tilde{\mathcal{L}}_0^{-1} \tilde{\mathcal{L}}_{\text{int}}\right)^{-1} \left[\delta\left(\mathbf{r}, \mathbf{r}'\right)\right], \tag{16.36}$$

where

$$\left[\bar{\mathbf{F}}\left(\mathbf{r}, \mathbf{r}'\right)\right] := \begin{pmatrix} \bar{\mathbf{F}}_{AA}\left(\mathbf{r}, \mathbf{r}'\right) & \bar{\mathbf{F}}_{AB}\left(\mathbf{r}, \mathbf{r}'\right) \\ \bar{\mathbf{F}}_{BA}\left(\mathbf{r}, \mathbf{r}'\right) & \bar{\mathbf{F}}_{BB}\left(\mathbf{r}, \mathbf{r}'\right) \end{pmatrix}, \tag{16.37}$$

and

$$\begin{aligned} \begin{pmatrix} \mathbf{J}_A\left(\mathbf{r}\right) \\ \mathbf{J}_B\left(\mathbf{r}\right) \end{pmatrix} &= \int_{S_A+S_B} ds' \\ &\times \begin{pmatrix} \bar{\mathbf{F}}_{AA}\left(\mathbf{r}, \mathbf{r}'\right) & \bar{\mathbf{F}}_{AB}\left(\mathbf{r}, \mathbf{r}'\right) \\ \bar{\mathbf{F}}_{BA}\left(\mathbf{r}, \mathbf{r}'\right) & \bar{\mathbf{F}}_{BB}\left(\mathbf{r}, \mathbf{r}'\right) \end{pmatrix} \cdot \begin{pmatrix} \hat{n}_A \times \mathbf{E}_A^{\text{ex}}\left(\mathbf{r}'\right) \\ \hat{n}_B \times \mathbf{E}_B^{\text{ex}}\left(\mathbf{r}'\right) \end{pmatrix}. \end{aligned} \tag{16.38}$$

The excitation vector in (16.36) is defined by

$$\left[\delta\left(\mathbf{r}, \mathbf{r}'\right)\right] := \begin{pmatrix} \delta_A\left(\mathbf{r}, \mathbf{r}'\right) \\ \delta_B\left(\mathbf{r}, \mathbf{r}'\right) \end{pmatrix}, \tag{16.39}$$

where $\delta_A\left(\mathbf{r}, \mathbf{r}'\right)$ and $\delta_B\left(\mathbf{r}, \mathbf{r}'\right)$ stand for vector surface delta excitations applied at antennas A and B, respectively.

It is possible to give more specific form for the origin of the various ACGFs appearing in (16.36). Let us write the inverse operator in the detailed form

$$\tilde{\mathcal{L}}^{-1} = \begin{pmatrix} \mathcal{M}_{AA} & \mathcal{M}_{AB} \\ \mathcal{M}_{BA} & \mathcal{M}_{BB} \end{pmatrix} \tag{16.40}$$

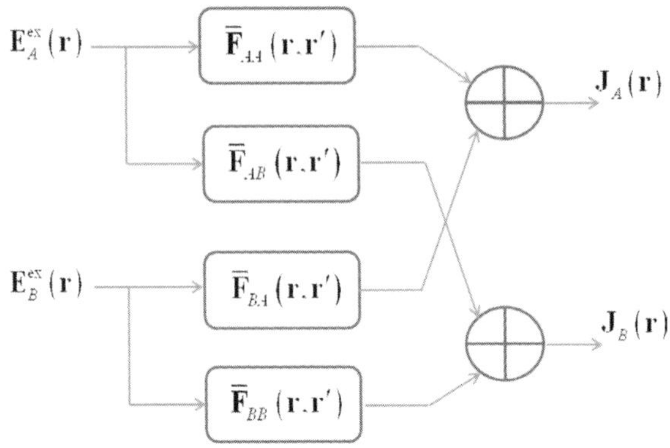

Figure 16.16 Flow chart representation of mutual coupling from the viewpoint of the antenna current Green's function formalism.

It then follows that

$$\begin{aligned}\bar{\mathbf{F}}_{AA}(\mathbf{r},\mathbf{r}') &= \mathcal{M}_{AA}\delta_A(\mathbf{r},\mathbf{r}'),\\ \bar{\mathbf{F}}_{BA}(\mathbf{r},\mathbf{r}') &= \mathcal{M}_{BA}\delta_A(\mathbf{r},\mathbf{r}'),\\ \bar{\mathbf{F}}_{AB}(\mathbf{r},\mathbf{r}') &= \mathcal{M}_{AB}\delta_B(\mathbf{r},\mathbf{r}'),\\ \bar{\mathbf{F}}_{BB}(\mathbf{r},\mathbf{r}') &= \mathcal{M}_{BB}\delta_B(\mathbf{r},\mathbf{r}').\end{aligned} \tag{16.41}$$

In other words, $\mathbf{F}_{AA}(\mathbf{r},\mathbf{r}')$ and $\mathbf{F}_{BB}(\mathbf{r},\mathbf{r}')$ are the ACGFs of the self-interactions of antennas A and B respectively. On the other hand, $\mathbf{F}_{AB}(\mathbf{r},\mathbf{r}')$ and $\mathbf{F}_{BA}(\mathbf{r},\mathbf{r}')$ completely characterize the mutual interactions between the two antennas.

Figure 16.16 illustrates the formulation above in a block diagram representation of the system viewpoint attained by our deployment of the antenna current Green's function formalism to describe mutual coupling in antenna arrays. Indeed, the isolated element design performance, i.e,, the individual device characteristics acquired by testing the antenna before placing it into the array environment, is fully given by the Green's functions (transfer functions) $\mathbf{F}_{AA}(\mathbf{r},\mathbf{r}')$ and $\mathbf{F}_{BB}(\mathbf{r},\mathbf{r}')$. The effect of coupling between the elements, the new factors arising from the coexistence of the elements in close proximity to each other, is rigorously and exactly determined by the Green's functions $\mathbf{F}_{AB}(\mathbf{r},\mathbf{r}')$ and $\mathbf{F}_{BA}(\mathbf{r},\mathbf{r}')$. However, the reader should not confuse these functions with the mutual coupling transfer functions defined in Section 16.2.1. Although the two sets comprised of the interaction and mutual coupling Green's functions are related to each other, yet they are not fully identical. The difference in meaning will be demonstrated shortly.

When there is no mutual coupling, i.e., when each antenna works in isolation, there is obviously no mutual interaction. The full Green's function of the antenna can be simply put as

$$\left[\bar{\mathbf{F}}\left(\mathbf{r},\mathbf{r}'\right)\right] := \begin{pmatrix} \bar{\mathbf{F}}_{AA}\left(\mathbf{r},\mathbf{r}'\right) & 0 \\ 0 & \bar{\mathbf{F}}_{BB}\left(\mathbf{r},\mathbf{r}'\right) \end{pmatrix}, \tag{16.42}$$

which provides no more information than the individual entries of the matrix. When mutual coupling is present, the problem is different and has to be solved in terms of the full operator (16.35). The new ACGF of the two-elements system will change into

$$\left[\bar{\mathbf{F}}'\left(\mathbf{r},\mathbf{r}'\right)\right] := \begin{pmatrix} \bar{\mathbf{F}}'_{AA}\left(\mathbf{r},\mathbf{r}'\right) & \bar{\mathbf{F}}_{AB}\left(\mathbf{r},\mathbf{r}'\right) \\ \bar{\mathbf{F}}_{BA}\left(\mathbf{r},\mathbf{r}'\right) & \bar{\mathbf{F}}'_{BB}\left(\mathbf{r},\mathbf{r}'\right) \end{pmatrix}, \tag{16.43}$$

where the off diagonal element signifies interactions. Following the definition of mutual coupling proposed in Section 16.2.1, the mutual coupling ACGF is given by

$$\begin{aligned} \left[\delta\bar{\mathbf{F}}\left(\mathbf{r},\mathbf{r}'\right)\right] &= \left[\bar{\mathbf{F}}'\left(\mathbf{r},\mathbf{r}'\right)\right] - \left[\bar{\mathbf{F}}\left(\mathbf{r},\mathbf{r}'\right)\right] \\ &= \begin{pmatrix} \bar{\mathbf{F}}'_{AA} - \bar{\mathbf{F}}_{AA} & \bar{\mathbf{F}}_{AB} \\ \bar{\mathbf{F}}_{BA} & \bar{\mathbf{F}}'_{BB} - \bar{\mathbf{F}}_{BB} \end{pmatrix} \\ &= \begin{pmatrix} \delta\bar{\mathbf{F}}_{AA}\left(\mathbf{r},\mathbf{r}'\right) & \bar{\mathbf{F}}_{AB}\left(\mathbf{r},\mathbf{r}'\right) \\ \bar{\mathbf{F}}_{BA}\left(\mathbf{r},\mathbf{r}'\right) & \delta\bar{\mathbf{F}}_{BB}\left(\mathbf{r},\mathbf{r}'\right) \end{pmatrix}. \end{aligned} \tag{16.44}$$

In other words, the mutual coupling ACGF is not merely the interaction (off diagonal) ACGF components, but includes also the perturbations on the main diagonal entries. As we will see below, these perturbations themselves are related to the interaction terms but in a rather complicated manner.

In order to proceed further, it is necessary to restrict the class of operators we are working with to bounded operators. The electric-field integral operator (16.26) is unbounded, however it can be approximated by a bounded operator, e.g., the method of moment as will be shown below. In general, if the electromagnetic operator of interest is unbounded, we replace it by a suitable bounded operator approximation. This will be supposed in what follows without change in notation for simplicity.

If the norm of a bounded operator L satisfies $\|L\| < 1$, then we have

$$(1 - L)^{-1} = \sum_{n=0}^{\infty} L^n, \tag{16.45}$$

which is the well known absolutely convergent geometric series. Let us assume that

$$\left\|\tilde{\mathcal{L}}_0^{-1}\tilde{\mathcal{L}}_{\text{int}}\right\| < 1, \tag{16.46}$$

which presents a sufficient condition for the perturbation series to hold true. In this case, we expand

$$\left(\tilde{I}+\tilde{\mathcal{L}}_0^{-1}\tilde{\mathcal{L}}_{\text{int}}\right)^{-1}=\sum_{n=0}^{\infty}\left(\tilde{\mathcal{L}}_0^{-1}\tilde{\mathcal{L}}_{\text{int}}\right)^{n}. \tag{16.47}$$

Therefore, the current distribution of the antenna system (16.35) can be written perturbatively as

$$\begin{aligned}[\mathbf{J}] &= \tilde{\mathcal{L}}_0^{-1}\left[\mathbf{E}^{\text{ex}}\right]+\left(\tilde{\mathcal{L}}_0^{-1}\right)^{2}\tilde{\mathcal{L}}_{\text{int}}\left[\mathbf{E}^{\text{ex}}\right] \\ &+\left(\tilde{\mathcal{L}}_0^{-1}\right)^{3}\left(\tilde{\mathcal{L}}_{\text{int}}\right)^{2}\left[\mathbf{E}^{\text{ex}}\right]+\left(\tilde{\mathcal{L}}_0^{-1}\right)^{4}\left(\tilde{\mathcal{L}}_{\text{int}}\right)^{3}\left[\mathbf{E}^{\text{ex}}\right]+\cdots.\end{aligned} \tag{16.48}$$

Note that the first term of this series $[\mathbf{J}_0] := \tilde{\mathcal{L}}_0^{-1}\left[\mathbf{E}^{\text{ex}}\right]$ is simply the current distribution on non-interacting (isolated) antennas.

Define the change in current due to interaction by $[\delta\mathbf{J}] := [\mathbf{J}] - [\mathbf{J}_0]$. We find then the mutual coupling effect on the current distribution to be given by the series

$$\begin{aligned}[\delta\mathbf{J}] &= \left(\tilde{\mathcal{L}}_0^{-1}\right)^{2}\tilde{\mathcal{L}}_{\text{int}}\left[\mathbf{E}^{\text{ex}}\right]+\left(\tilde{\mathcal{L}}_0^{-1}\right)^{3}\left(\tilde{\mathcal{L}}_{\text{int}}\right)^{2}\left[\mathbf{E}^{\text{ex}}\right] \\ &+\left(\tilde{\mathcal{L}}_0^{-1}\right)^{4}\left(\tilde{\mathcal{L}}_{\text{int}}\right)^{3}\left[\mathbf{E}^{\text{ex}}\right]+\cdots.\end{aligned} \tag{16.49}$$

Following the definition of the mutual coupling ACGF proposed in Section 16.2.1, we conclude that

$$\begin{aligned}&\left[\delta\bar{\mathbf{F}}\left(\mathbf{r},\mathbf{r}'\right)\right] \\ &=\left[\left(\tilde{\mathcal{L}}_0^{-1}\right)^{2}\tilde{\mathcal{L}}_{\text{int}}+\left(\tilde{\mathcal{L}}_0^{-1}\right)^{3}\left(\tilde{\mathcal{L}}_{\text{int}}\right)^{2}+\cdots\right]\cdot\left[\delta\left(\mathbf{r},\mathbf{r}'\right)\right].\end{aligned} \tag{16.50}$$

In a more suggestive form, (16.50) can be expressed as

$$\left[\delta\bar{\mathbf{F}}\left(\mathbf{r},\mathbf{r}'\right)\right]=\sum_{n=1}^{\infty}\delta\bar{\mathbf{F}}^{(n)}\left(\mathbf{r},\mathbf{r}'\right), \tag{16.51}$$

where

$$\delta\bar{\mathbf{F}}^{(n)}\left(\mathbf{r},\mathbf{r}'\right):=\left(\tilde{\mathcal{L}}_0^{-1}\right)^{n+1}\left(\tilde{\mathcal{L}}_{\text{int}}\right)^{n}\left[\delta\left(\mathbf{r},\mathbf{r}'\right)\right]. \tag{16.52}$$

This is our main result. It provides a complete characterization of mutual coupling in antenna arrays by tracing back the origin of the mutual coupling transfer function into various orders of corrections expressed in terms of the isolated antenna operator $\tilde{\mathcal{L}}_0$ and the interaction operator $\tilde{\mathcal{L}}_{\text{int}}$. Each correction of order n is given by $\delta\bar{\mathbf{F}}^{(n)}(\mathbf{r},\mathbf{r}')$ defined by (16.52).

The main advantages of the perturbation series (16.50) are the following

1. The perturbation technique accepts only the ACGF of the *isolated* antennas plus the interaction operator of the array. In other words, it is expressed in terms of the intuitive engineering understanding of complex systems as formed by independent parts plus interactions between them.
2. The perturbation technique involves only the *forward* interaction operator. There is no new inversion operation required to estimate mutual coupling effects. The only operations needed are multiplication and addition. The accuracy of the computations can be controlled by varying the number of perturbations needed.
3. The perturbation technique provide a full physical mechanism explaining the origin of each correction in the observable antenna array due to mutual coupling interactions.

We would like now to express the individual ACGFs in (16.44) in terms of a perturbation series similar to that above. Define the mutual coupling operator L_c as

$$L_c := \sum_{n=1}^{\infty} L_c^{(n)} = L_c^{(1)} + L_c^{(2)} + \cdots, \tag{16.53}$$

where

$$L_c^{(n)} = \left(\tilde{\mathcal{L}}_0^{-1}\right)^{n+1} \left(\tilde{\mathcal{L}}_{\text{int}}\right)^n. \tag{16.54}$$

From (**??**) we find

$$\tilde{\mathcal{L}}^{-1} = \tilde{\mathcal{L}}_0 + \tilde{L}_c. \tag{16.55}$$

The first-order perturbation term $L_c^{(1)} = \left(\tilde{\mathcal{L}}_0^{-1}\right)^2 \tilde{\mathcal{L}}_{\text{int}}$ can be computed directly for the two-antenna system. The result is

$$\begin{aligned} L_{c_{11}}^{(1)} &= L_{c_{22}}^{(1)} = 0, \\ L_{c_{12}}^{(1)} &= \left(\hat{n}_A \times \mathcal{L}_A\right)^{-1} \left(\hat{n}_A \times \mathcal{L}_A\right)^{-1} \left(-\hat{n}_A \times \mathcal{L}_B\right), \\ L_{c_{21}}^{(1)} &= \left(-\hat{n}_B \times \mathcal{L}_B\right)^{-1} \left(-\hat{n}_B \times \mathcal{L}_B\right)^{-1} \hat{n}_B \times \mathcal{L}_A. \end{aligned} \tag{16.56}$$

Similarly, the first-order perturbation term $L_c^{(2)} = \tilde{L}_0^{-3}\left(\tilde{L}_{\text{int}}\right)^2$ can be computed directly for the two-antenna system. The result is

$$\begin{aligned} L_{c11}^{(2)} &= (\hat{n}_A \times \mathcal{L}_A)^{-3}(-\hat{n}_A \times \mathcal{L}_B)\,\hat{n}_B \times \mathcal{L}_A, \\ L_{c22}^{(2)} &= (-\hat{n}_B \times \mathcal{L}_B)^{-3}\,\hat{n}_B \times \mathcal{L}_A\,(-\hat{n}_A \times \mathcal{L}_B), \\ L_{c21}^{(2)} &= L_{c21}^{(2)} = 0. \end{aligned} \tag{16.57}$$

Note that the diagonal elements in (16.56) are zero while the off diagonal element in (16.57) vanishes. In terms of the ACGFs, we use (16.56) and (16.57) in (16.41) to obtain

$$\begin{aligned} &\bar{\mathbf{F}}_{AA}^{(1)}(\mathbf{r},\mathbf{r}') \\ &= (\hat{n}_A \times \mathcal{L}_A)^{-3}(-\hat{n}_A \times \mathcal{L}_B)\,\hat{n}_B \times \mathcal{L}_A\,\delta_A(\mathbf{r},\mathbf{r}'), \\ &\bar{\mathbf{F}}_{AB}^{(1)}(\mathbf{r},\mathbf{r}') \\ &= (\hat{n}_A \times \mathcal{L}_A)^{-1}(\hat{n}_A \times \mathcal{L}_A)^{-1}(-\hat{n}_A \times \mathcal{L}_B)\;\delta_A(\mathbf{r},\mathbf{r}'), \\ &\bar{\mathbf{F}}_{BA}^{(1)}(\mathbf{r},\mathbf{r}') \\ &= (-\hat{n}_B \times L_B)^{-1}(-\hat{n}_B \times \mathcal{L}_B)^{-1}\,\hat{n}_B \times \mathcal{L}_A\,\delta_B(\mathbf{r},\mathbf{r}'), \\ &\bar{\mathbf{F}}_{BB}^{(1)}(\mathbf{r},\mathbf{r}') \\ &= (-\hat{n}_B \times \mathcal{L}_B)^{-3}\,\hat{n}_B \times \mathcal{L}_A\,(-\hat{n}_A \times \mathcal{L}_B)\;\delta_B(\mathbf{r},\mathbf{r}'). \end{aligned} \tag{16.58}$$

Note that the effect of mutual coupling on the self-interactions of antennas A and B, i.e., the ACGFs $\bar{\mathbf{F}}_{AA}^{(1)}(\mathbf{r},\mathbf{r}')$ and $\bar{\mathbf{F}}_{BB}^{(1)}(\mathbf{r},\mathbf{r}')$ emerges only starting from the *second*-order perturbation operator $L_c^{(2)}$. On the other hand, the interaction ACGFs $\bar{\mathbf{F}}_{AB}^{(1)}(\mathbf{r},\mathbf{r}')$ and $\bar{\mathbf{F}}_{BA}^{(1)}(\mathbf{r},\mathbf{r}')$ start gaining contributions from the *first*-order term $L_c^{(1)}$. We will not write down explicitly expressions for the higher order terms. The computations, although straightforward, become very tedious for arrays containing more than two elements.

16.5.4 Numerical Examples

To implement the perturbation method, we apply a Galerkin-type finite dimensional approximation of the forward interaction operators of the problem, where a triangular basis function is used to model the current distribution on thin wires. Gaussian quadrature is employed to compute all of the impedance matrix integrals except the self- and near- self terms, which are integrated analytically. The accuracy of our approximation is verified by comparison with the commercial software WIPL-D, which uses a higher-order basis function.

We consider a numerical example for a two-element antenna array comprised of two parallel thin-wires. The two antennas have identical radii of 0.001λ and lengthes 0.5λ. Antenna A is energized by a delta source to compute its current Green's function as in Chapter 9, while antenna B is kept passive. The input impedance (seen at the input port located in the middle of antenna A) is computed using the perturbation method and the results are compared with the inversion of the full MoM matrix in Figure 16.17. Excellent agreement is observed as long as the two antennas are not touching each other.

An important quantity helpful in characterizing the numerical aspect of the perturbation algorithm is the *spectral radius*. The spectral radius of a matrix A is defined as $\rho(A) := \max_i(|\lambda_i|)$, where λ_i are the eigenvalues of A. It can be shown (see [92]) that the validity of the perturbation series (16.47) is guaranteed only when this spectral radius is less than unity. This explains some slight discrepancy between the exact and perturbation results observed at very small inter-element separation. In Figure 16.18, we compute the spectral radius of the perturbation interaction matrix $\tilde{\mathcal{L}}_0^{-1}\tilde{\mathcal{L}}_{\text{int}}$ as a function of the separation between the antennas. As expected, $\rho(\tilde{\mathcal{L}}_0^{-1}\tilde{\mathcal{L}}_{\text{int}})$ is close to unity when the two antennas touch each other. At this extreme case the perturbation series is divergent. Note that while the perturbation series converges if and only if the spectral radius is less than unity, the hypothesis (16.46) is only a sufficient condition. Therefore, in the study of a more complex array problem using the perturbation theory developed in this part, the spectral radius should always be monitored.

The number of terms needed to ensure accurate results with the perturbation series strongly depends on both the spectral radius and the details of the antennas. For the results given in Figure 16.17, we used a constant number of 30 terms for the entire computation. However, a much smaller number of terms is needed while the separation between the elements increases. When the perturbation method is used to analyze a large array of fixed elements, one can determine the minimum number of terms needed according to the accuracy of the results required for the applications at hand. In Figure 16.19, we show the variation in the error in predicting Z_{in} with the order of perturbation. As expected, for smaller separation between the antennas, more terms in the perturbation are needed. The actual number of sufficient terms depends on the level of acceptable error, which in turn is determined by the application.

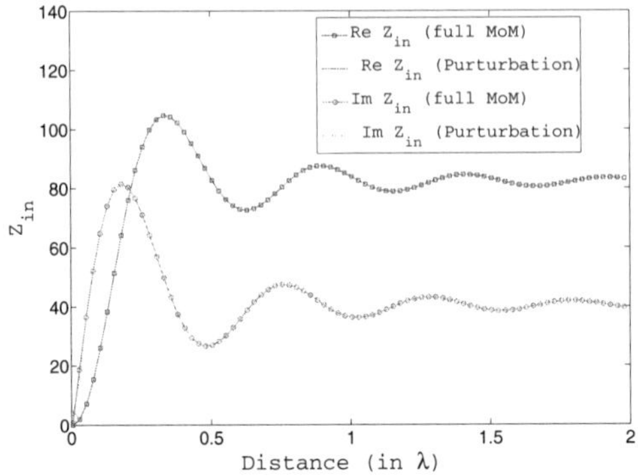

Figure 16.17 Comparison between the full MoM matrix inversion and perturbation input impedance results for two-element array having identical radii of 0.001λ and lengthes 0.5λ. The horizontal axis gives the variable distance between the elements while the impedance is measured in ohms.

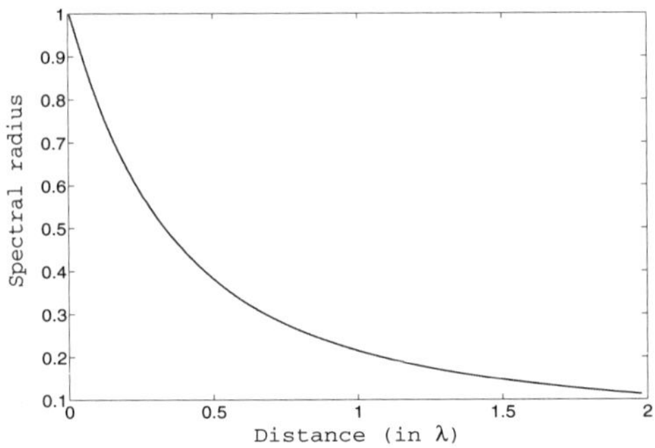

Figure 16.18 The spectral radius of the operator $\tilde{\mathcal{L}}_0^{-1}\tilde{\mathcal{L}}_{\text{int}}$.

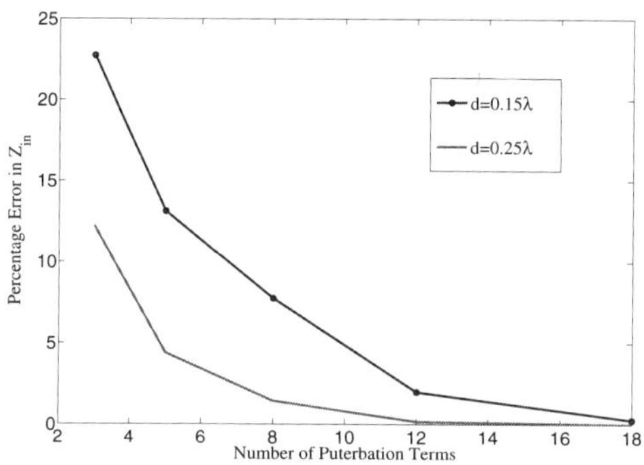

Figure 16.19 Percentage error in the input impedance predicted using the perturbation series as a function of the number of terms (order of perturbation). The separation between the two antennas is given by d. The spectral radius at $d = 0.15\lambda$ and $d = 0.25\lambda$ is 0.7168 and 0.5840, respectively.

16.5.5 Summary

The main advantages of the general method introduced in this part are the following. First, the perturbation technique accepts only the ACGF of the *isolated* antennas plus the interaction operator of the array. In other words, it is expressed in terms of the intuitive engineering understanding of complex systems as formed by independent parts plus interactions between them. Second, the perturbation technique involves only the *forward* interaction operator. There is no new inversion operation required to estimate mutual coupling effects, which makes full-wave analysis of very large arrays feasible. Third, the perturbation technique naturally provides full physical mechanisms explaining the origin of each correction in the observable antenna array due to mutual coupling interactions.

16.6 APPENDICES

16.6.1 The Optimization Process

Assume a set of infinitesimal electric dipoles $\{\chi_i\}_{i=1}^{N}$, where N is the number of dipoles and χ_i is a seven-element vector representing the parameters of the ith dipole given by

$$\chi_i = \left[\begin{array}{ccccccc} \mathrm{Re}\{p_i\} & \mathrm{Im}\{p_i\} & \alpha_i & \beta_i & x_i & y_i & z_i \end{array} \right]^T . \tag{16.59}$$

Here, the position of the *i*th dipole is given by x_i, y_i, and, z_i, which are constrained by the actual antenna size. p_i is the complex dipole moment, with orientation given by the direction cosines α_i and β_i defined with respect to the *x*- and *y*- axis, respectively. The third directional cosine can be obtained from

$$\gamma_i^2 = 1 - \alpha_i^2 - \beta_i^2 . \tag{16.60}$$

The components of the *i*th dipole moments are then given as

$$p_{ix} = p_i \alpha_i, \quad p_{iy} = p_i \beta_i, \quad p_{iz} = p_i \gamma_i . \tag{16.61}$$

Notice that by employing directional cosines in the formulation, equations (16.60) and (16.61) eliminate two degrees of freedom from the total number associated with each dipole as used with the representation of [133] and [134]. Moreover, by restricting the dipole type to be electric only, we attain further reduction by one degree of freedom for each dipole. This gives seven variables per dipole and a total of 7*N* for the entire optimization problem.

We define the cost measure of the optimization process as the function

$$\begin{aligned} F = w_E \sum_{n=1}^{N_{\mathrm{ops}}} \sum_u \left| \mathbf{E}_A^u (\mathbf{r}_n) - \mathbf{E}_{\mathrm{DM}}^u (\mathbf{r}_n) \right|^2 \\ + w_H \sum_{n=1}^{N_{\mathrm{ops}}} \sum_u \left| \mathbf{H}_A^u (\mathbf{r}_n) - \mathbf{H}_{\mathrm{DM}}^u (\mathbf{r}_n) \right|^2 \end{aligned} \tag{16.62}$$

where N_{ops} is the number of observation (samples) points of the near-field data set and the sum over u takes the three cartesian components x, y, and z. The position vector of the *n*th sampling point is $\mathbf{r}_n$. Here

$$w_E = \alpha \left(1 \Big/ \sum_{n=1}^{N_{\mathrm{ops}}} \sum_u \left| \mathbf{E}_A^u (\mathbf{r}_n) \right|^2 \right), \tag{16.63}$$

$$w_H = (1 - \alpha) \left(1 \Big/ \sum_{n=1}^{N_{\mathrm{ops}}} \sum_u \left| \mathbf{H}_A^u (\mathbf{r}_n) \right|^2 \right). \tag{16.64}$$

The factor $0 \leq \alpha \leq 1$ is a normalization coefficient and gives the contributions of electric and magnetic field to the objective function, respectively. In this part, we fix $\alpha = 0.5$.

In general, the cost measure is highly nonlinear, with a landscape full of local minima. This makes the optimization problem very difficult unless a powerful global search method is used. The quantum particle swarm optimization (QPSO) is used to perform the optimization process needed to find the best dipole model for the problem at hand. This algorithm, in the version developed in [22], contains only one control parameter g. In all the optimization used throughout this part, we fixed this parameter at $g = 3$.

16.6.2 Post-Processing Evaluation Measure

We introduce now the error criterion adopted in the quantitative evaluation of the performance of the obtained dipole model. The global error is defined as the arithmetic average of the global error in the electric and magnetic fields $(e_E + e_H)/2$. Here, the individual errors are defined in the following manner

$$e_{\Psi} = \sqrt{\frac{\sum\limits_{u=x,y,z} \sum\limits_{n=1}^{N_{\text{ops}}} \left|\Psi_u^{\text{Actual}}(n) - \Psi_u^{\text{DM}}(n)\right|}{\sum\limits_{u=x,y,z} \sum\limits_{n=1}^{N_{\text{ops}}} \left|\Psi_u^{\text{Actual}}(n)\right|}}, \tag{16.65}$$

where Ψ stands for either E or H, and N_{ops} is the size of the observation data set. The definition of a global error, that is, a single number that reflects the collective behavior of all the six field components at all observation points is certainly not a trivial matter. First of all, there is no a priori way to decide which error measure is "optimum" in any meaningful manner. Next, we notice that a good global error should satisfy some desirable characteristics. First, it must be reasonably small when the six field components appear visually in good agreement with the actual fields. Second, it must be consistent in the sense that the error monotonically increase when there are reliable reasons to believe that the dipole model fields are deviating from the actual fields. Third, the error must be roughly the same for any location (outside the forbidden region) and size of the observation near-field data set. It appears to the authors that the RMS definition in (16.65) satisfy these demands.

16.6.3 Computations of the Spectral Integrals

In this Appendix, we provide the minimum background necessary for a numerical computations of the spectral integrals (16.22) and (16.23). From the Weyl expansion (4.4) and (16.9), the dyadic Green's function in the spectral domain is written as

$$\bar{\mathbf{G}}(\mathbf{r}) = \frac{-\omega k\mu}{8\pi^2}\int_{-\infty}^{\infty} dpdq \frac{\bar{\mathbf{I}}k^2 - \mathbf{KK}}{k^2 m} e^{i\mathbf{K}\cdot\mathbf{r}}. \tag{16.66}$$

By direct computation, we find

$$\bar{\mathbf{G}}(\mathbf{r}) = \frac{-\omega k\mu}{8\pi^2}\int_{-\infty}^{\infty} dpdq \frac{1}{k^2 m}\bar{\mathbf{F}}(p,q)\, e^{i\mathbf{K}\cdot\mathbf{r}}, \tag{16.67}$$

where

$$\bar{\mathbf{F}}(p,q) = \begin{pmatrix} 1-p^2 & -pq & -\mathrm{sgn}(z)\,pm \\ -pq & 1-q^2 & -\mathrm{sgn}(z)\,qm \\ -\mathrm{sgn}(z)\,pm & -\mathrm{sgn}(z)\,qm & 1-m^2 \end{pmatrix}. \tag{16.68}$$

The Weyl expansion (4.4) can be reduced into a one-dimensional integral, and after separating the propagating and nonpropagating parts, we obtain

$$g_{\mathrm{pr}}(\mathbf{r},\mathbf{r}') = \frac{ik}{4\pi}\int_0^1 du J_0\left(k\rho_s\sqrt{1-u^2}\right) e^{ik|z-z'|u}, \tag{16.69}$$

$$g_{\mathrm{ev}}(\mathbf{r},\mathbf{r}') = \frac{k}{4\pi}\int_0^\infty du J_0\left(k\rho_s\sqrt{1+u^2}\right) e^{-k|z-z'|u}, \tag{16.70}$$

where $\rho_s = \sqrt{(x-x')^2 + (y-y')^2}$. Here x', y', and z' give the source coordinates.

Using the reduced forms (16.69) and (16.70), it can be demonstrated by direct calculations that the dyadic Green's function (16.66) can be expressed in terms of only six different basic integrals. These are

$$A_n = \int_0^1 du J_0\left(k\rho_s\sqrt{1-u^2}\right) u^n e^{iku|z-z'|}, \tag{16.71}$$

$$B_n = \int_0^\infty du J_0\left(k\rho_s\sqrt{1+u^2}\right) u^n e^{-ku|z-z'|}, \tag{16.72}$$

where $n = 0, 1, 2$. The calculations are straightforward but lengthy, and since no numerical results are presented here, we omit the details.

Integrals (16.71) and (16.72) can be evaluated analytically for special cases, e.g., when $z = z'$ or $x = x'$ and $y = y'$. For general observation point, it appears that no closed-form solution in terms of simple analytical function can be found and numerical integration in this case is therefore necessary.

16.7 CONCLUSION

This chapter presented a general theory of electromagnetic mutual coupling in antenna array systems with several applications. The theory itself works ultimately within the perspective of the antenna current Green's function (ACGF) formalism introduced in Part II, where the antenna as such is replaced by its exact transfer function in space and treated in the spirit of system theory as a transformation from input (excitation field) to outputs (radiating currents). We were able to isolate the mutual coupling transfer function and show that it constitutes a complete description of the system from the viewpoint of electromagnetic interactions. In order to bring this concept into an intuitive form and to facilitate computations, a perturbation approach was developed from scratch, proving that in general the mutual coupling ACGF can be approximated by a converging series that involves only the isolated elements' ACGFs plus small forward interaction operators. This puts the study of large antenna arrays and their mutual coupling in the light of the physically appealing pictures of atoms plus their interactions. In this case, each antenna behaves like an atom or unit cell with its isolated-element ideal behavior, while mutual coupling effects are incorporated perturbatively as successive levels of corrections imposed on the ideal behavior, where these corrective modifications emerge from forward interactions involving mainly neighboring elements. Since this method avoids the computationally daunting job of inverting the full electromagnetic operator of the entire (coupled) system, we expect that perturbation techniques will provide solid foundations for tackling full-wave analysis of large-and-complex antenna arrays.

The chapter also provided a simpler approach using the concept of infinitesimal dipole models to represent arbitrary antenna systems by analytical radiation formulas (the fields radiated by infinitesimal dipoles can be expressed in such manner). The advantages of the dipole model approach to mutual coupling is that 1) it does not require solving Maxwell's equations (since it works only with measured near field date), and 2) it motivates intuitively many of the ideas that will be developed rigorously using the ACGF formalism. For these reasons, the dipole model approach was introduced first in this chapter.

The dipole model approach also used the near field of the array to characterize mutual coupling and to distinguish between weak and strong mutual coupling scenarios. The near field contains more information than S-matrix parameters. In fact, we expect that future and current applications will require focus on the near zone in order to insure satisfactory performance of systems in complex and dense electromagnetic environments. As an application motivated by Part I on near-field theory, we presented here a method to analyze mutual coupling into its propagating and evanescent parts and provided new data concerning the physics of interactions in several examples, involving some information on the localization of electromagnetic energy in the near zone.

Chapter 17

Method for the Analysis of Localized Energy in Mutually-Coupled Antenna Systems

17.1 INTRODUCTION

This chapter addresses several fundamental issues within the topic of mutual coupling and energy transfer in generic electromagnetic systems, with focus given here on antennas as exemplary cases of such systems. The new approach proposed here takes its point of departure from the view that mutual coupling is essentially a near-field phenomenon, even though one may, in a rather generic manner, include interactions via far-field and guided propagating modes. In our opinion, the physical problem found in the setting of two antennas interacting with each other at a relatively short electric distance is basically one involving energy exchange through the near field in between the two elements.

The electromagnetic near field represents one of the most complex structures in the radiation problem and it seems that the topic has not received a comprehensive and sustained attention in the literature. Many mutual coupling studies have been published based on special proposals focusing mainly on the far fields and the circuit parameters (S-matrix). While the latter two performance measures are essential to many application, they don't exhaust the complexity of the antenna problem for several reasons. First, far fields and circuit parameters are themselves functions of the near field; they are determined by the physical processes occurring in the near-zone region; therefore, proper understanding of such popular performance characteristics may benefit from a closer examination of the near field structure. Second, there are many applications (current or future) that work directly with the near field; examples are close-range detection of unknown or buried objects, imaging experiments (especially sub-wavelength imaging), near-field communications, and wireless electromagnetic energy transfer, just to mention few now common research and industrial topics. Third, the near field is most likely to be the main site of energy storage in the antenna system. Since many fundamental issues in operating antennas depend crucially on how energy is stored in the region surrounding the source, e.g., bandwidth, quality factor, efficiency, etc, not to mention the utilization of free space for energy handling and manipulations (a topic not fully exploited yet), we anticipate that studies of the near field will gain some momentum in the near future in both fundamental and applied research.

Existing tools in the near field literature suffer from the shortcoming that they focus mainly on the *total* field. However, as was suggested previously by the authors, the interesting features in the near field arise when we look at how energy is divided into propagating and nonpropagating parts since this is actually what is involved in all existing and future devices. In other words, knowing the value of the total near field does not provide much new information. However, computing how the just-mentioned total field value splits into moving and non-moving energies usually illuminates the understanding of the concrete problem because it provides a proper visualization of the actual dynamic process of field propagation. The specific examples of mutual coupling in antenna arrays given below will illustrate this new insight.

For these reasons, the work presented here aims at developing new fundamental techniques suitable for analyzing the structure of the near field in antenna systems experiencing both strong and weak mutual coupling. For a general view of existing methods addressing the topic of mutual coupling from the perspective of far-field and circuit parameters, see for example [29], [121]. Antenna-Antenna interaction theory was developed some time ago in [55], [56], [122], [123]. Utilization of this theory in developing antenna field measurement methods based on careful understanding of electromagnetic coupling in the near and far zones can be found in [122], [35].

In order to develop new tools for studying mutual coupling from the perspective of the near field, two recent theoretical advances will be employed here, the near-field theory in the spectral domain Chapter (4) and the antenna current Green's function formalism Chapters (8 and 9). The new approach presented here is different from previous methods since we focus on the localization structure of the interaction energy right from the beginning. The geometrical shape of the interacting antennas will be included in the formulation by attempting to provide an analysis of energy coupling and transfer, where this analysis will be focused neither exclusively on coupling between two ports (the standard scattering matrix method [29]) nor between reference planes enclosing the interacting antennas (the generalized scattering matrix method [56]); instead, *each* pair of points on the *entire* surfaces of the two coupled antennas will be taken into account and the effect of shape, size, geometry, inter-element spacing will be incorporated fundamentally into the method. The analysis will provide information about how antenna-antenna interaction energy is divided into propagating (energy flow) and evanescent (localized energy) and how various parts of the antennas contribute to these energy processes. For example, a new coupling coefficient, the *energy localization coefficient*, will be introduced to measure how much of the total interaction energy between two antennas is realized in the form of stationary (localized) energy. This in turn provides direct knowledge of the structure of energy localization in real-life antenna systems. The topic then can be investigated in great details both conceptually and for many practical examples. One of the main features of the new method is that it was developed with attention devoted from the beginning to simple implementation utilizing existing the method of moment (MoM) architectures. Indeed, we show that the new analysis can be readily integrated into available codes.

Moreover, we avoid the technical complexity of the original formulations of Part I and II by working directly with a simplified model. Indeed, we will avoid relying exclusively on the antenna current Green's function in this investigation and work instead with the electromagnetic operators of the problem. As was shown in Chapter 8, the antenna Green's function can be constructed from the operators using distribution theory, and hence there is no loss of generality here. The gain is a more concise and focused presentation of the results given here. On the other hand, we will concentrate from the beginning on a method of moment (MoM) formulation of the proposed technique. The goal of this choice is to exploit certain technical aspects in the method of moment that don't seem easy to generalize when working on the exact level of operators or antenna Green's functions. In particular, the *full-vectorial* formulation of the near-field theory developed in Chapter 4 will not be heavily used here since it turns out that one can work with the *scalar* problem by exploiting the integration-by-parts trick in moving the differential operators from the free-space Green's functions to the current basis functions when evaluating the MoM impedance matrix. This allows the study of the near-field structure using the simpler spectral integrals of the *scalar* Green's functions, which turned out to be expressible in simple analytical closed form as is shown in this work.

The chapter is structured in the following manner. Section 17.2 provides an overall evaluation of the problem of antenna-antenna interactions and reviews existing methods. The new approach developed here will be briefly described and compared with the generalized scattering matrix approach. Section 17.3 will give a brief formulation of the setting of two interacting antennas in terms of the exact electromagnetic operators. Moreover, the definition of the mutual coupling antenna current Green's function proposed in Chapter 16 will be outlined although will not figure prominently in what follows since the results will be presented in conjunction with the interaction operators rather than the Green's function itself. Section 17.4 will outline the main ideas of the chapter. The required background from the near-field theory Chapter 4 will be simplified and adapted to the purpose of our purpose here. Detailed technical implementation of the analysis method will be given in Section 17.5 where the derivation of the MoM matrix decomposition into propagating and nonpropagating parts will be illustrated for linear wire antennas. The main results are the expressions (17.30) and (17.31), giving the MoM matrix elements decompositions. There, no numerical spectral integration for computing the total propagating and nonpropagating modes are needed since they turn out to be given in simple analytical form. The new derived MoM expressions can be readily included in existing codes. In Section 17.6, a set of carefully chosen numerical examples are provided to illustrate the physics of mutual coupling in typical antenna array configurations from the viewpoint of the near field. Section 17.7 show how the basic method can be applied to study the structure of the radiated field of an antenna by using another receiving antenna as a probe. (These results applies not only to the near field, but extends to the far zone.) We illustrate the approach by defining a factor measuring the degree of localization in the radiation field and shows several examples where these quantitative indicators can be used to infer information about the physical process in the antenna interactions with other objects. Finally, we end with a brief overview on possible applications and the conclusion.

17.2 DESCRIPTION OF THE PROBLEM OF ANTENNA-ANTENNA ENERGY TRANSFER PROBLEM AND COMPARISON WITH EXISTING METHODS

The basic setting of energy transfer system is shown in Figure 17.1, where two antennas A and B with separation d are shown. Here, the port signals are denoted at local reference planes by a_0 and b_0 for Antenna B and a_0' and b_0' for Antenna A. Since interest in energy coupling applications focuses on obtaining a characterization of the received signal in one antenna when the other is used as a source, the most natural choice for characterizing antenna-antenna interaction has been the coupling

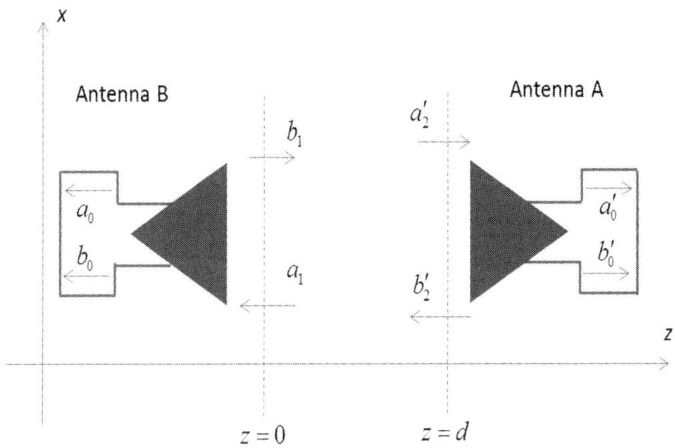

Figure 17.1 Schematic diagram of conventional antenna-antenna interaction problem analyzed using the generalized scattering matrix approach.

ratio b_0'/a_0 [56], which is reducible to the conventional scattering parameter S_{12} (after properly choosing the load impedance) [29].

To our best knowledge, the most comprehensive investigation of the problem of antenna-antenna coupling is due to Kerns, who developed the *generalized scattering matrix* approach to deal with the analysis of energy coupling in generic antenna systems. The basic theory can be found in [56], with detailed experimental investigation of the proposed formalism. The method was further developed and simplified by Yaghjian where coupling was expressed in terms of the far fields patterns for antennas not interacting in the deep near-zone region [123]. More recently, the formulation in [123] was used to study energy coupling for some currently popular applications such as RFID [136].

The antenna-antenna interaction approach of Kerns can be viewed as a massive generalization of the classic waveguide microwave circuit theory [29]. Indeed, as can be seen form Figure 17.1, the method defines fictitious (mathematical) surfaces at planes $z = 0$ and $z = d$ where "incident" and "outgoing" waves a_1, b_1 and a_1', b_1' are referenced, similar to what is done at the normal waveguide sections with port signals a_0, b_0 and a_0', b_0'. In order to mathematically extract a useful definition of "outgoing" and "incoming" waves, Kerns used the Weyl expansion to expand the fields into ([56], eq. (1.2-14))

$$\begin{aligned}\mathbf{E}(\mathbf{r}) &= \int_{\mathbb{R}^2} dk_x dk_y e^{i\mathbf{k}_t\cdot\mathbf{r}} \\ &\times \left[\mathbf{B}(k_x,k_y)\, e^{+ik_z|z|} + \mathbf{A}(k_x,k_y)\, e^{-ik_z|z|}\right],\end{aligned} \tag{17.1}$$

where **A** and **B** are "transverse field spectra" for the "outgoing" and "ingoing" wave components. Here, the integrations is on all real values of k_x and k_y while k_z can be either pure real (propagating modes) or pure imaginary (evanescent modes) and $\mathbf{k}_t = \hat{x}k_x + \hat{y}k_y$. In the above expansion, the plane $z = 0$ is chosen as a reference plane for describing the splitting of the fields into outgoing and incoming waves. By defining a scattering matrix for each spectral mode indexed by k_x and k_y, and then summing over all modes to obtain the total fields, Kerns was able to put antenna-antenna interactions in the framework of microwave circuit theory.

The alternative approach to the analysis of antenna-antenna interaction developed here uses the Weyl expansion but in a very different way compared with [56]. Figure 17.2 illustrates the basic configuration of interaction between two generic antennas. Here, we consider energy exchange between every point on the geometric surface of the antennas, rather than mathematical reference planes as in the generalized scattering method of Figure 17.1. The major difficulty encountered in implementing the approach of Figure 17.2 is that now we no longer can fix in advance the direction of the z-axis along which the fields are expanded into their spectral mixture of propagating and nonpropagating modes. Indeed, in order to reflect the general changing shape of the interacting antennas, the direction of z-axis must also change in general between every pair of interacting points (see the small local frame in Figure 17.2). The solution of this problem, which will be detailed in Section 17.4, exploits recent physical understanding of the antenna near zone in terms of dynamic splitting of the field into propagating and evanescent modes.[1] Indeed, it was shown by the authors that a 3D rotation matrix $\bar{\mathbf{R}}$ is needed in addition to the location $\mathbf{r}$ in the near zone in order to fully specify the spectral content of the field (the decomposition into homogeneous and inhomogeneous waves). The key idea in Figure 17.2 will be to orient the local frame described by this 3D matrix at each pair of points $\mathbf{r}_A$ and $\mathbf{r}_B$ in the interacting antennas such that the axis along which we decompose the coupled field into propagating and evanescent modes is aligned along the line joining the mentioned two points. In this sense, the analysis method proposed here can provide information about the localization of energy between variable regions on the interacting generic antennas (depending the geometrical shapes, distance, etc.) and how exchange energy is localized in the space between.

1 Cf. Chapter 4.

The major differences between our method and Kerns' are the following:

1. The Weyl expansion is used in (17.1) mainly as a *computational* method to compute *total* fields. In contrast, the proposed method is based on *separating* and explicitly differentiating the propagating field from the total field by emphasizing how the *composition* of interaction is split into propagating and evanescent modes in a way that is reflected by the geometry of the interacting antennas.
2. The method in [56] assumes fictitious mathematical reference planes enclosing the interacting antennas in order to complete the analysis into outgoing and incoming waves. In contrast, this method does not introduce new reference planes; instead, *we analyze interactions in terms of the geometrical surface of the antenna involved, with focus on how changing the antenna array geometrical details affect mutual energy coupling and exchange* for applications involving antenna design and synthesis.
3. The methods of Kerns [56] include evanescent modes in the definitions of *both* incoming and outgoing waves. In our approach, the basic emphasis is on localization of electromagnetic energy and hence the focus here shifts to how energy exchange is mediated by physically different types of fields, the propagating and nonpropagating modes. In other words, *instead of defining coupling in terms of ratio involving outgoing and incoming waves, we define new coupling coefficients in terms degree of localization or, equivalently, the relative strength of evanescent modes energy with respect to the total interaction field energy.*

Therefore, our method complements and expands the generalized scattering matrix approach by providing new information not addressed in the older techniques. Moreover, *the present approach essentially includes the traditional scattering matrix method since by simply recombining our propagating and nonpropagating sub-operators, the classic coupling ratio S_{12} is recovered* (see Section 17.5).

17.3 BASIC MUTUAL ENERGY COUPLING AND TRANSFER FORMULATION

Consider two interacting antenna elements A and B. The antennas are assumed to have arbitrary smooth surfaces S_A and S_B and are both possibly excited by fields $\mathbf{E}_A^{\mathrm{ex}}$ and $\mathbf{E}_B^{\mathrm{ex}}$, respectively, which may be defined either locally (transmitting

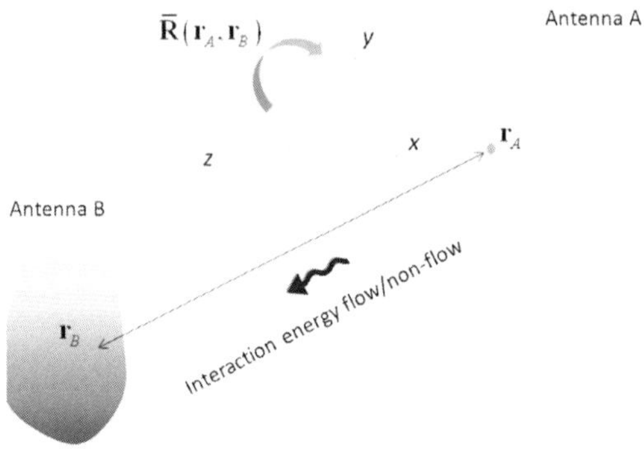

Figure 17.2 A spectral analysis of mutual coupling between two antennas A and B showing direct line-of-sight energy exchange between two points $\mathbf{r}_A$ and $\mathbf{r}_B$. In order to describe the spectral decomposition of the interaction, a local coordinate xyz system must be specified in order to compute the division of the net energy exchange into propagating and nonpropagating (evanescent) parts. The orientation of this local frame is described by the 3D rotation matrix $\bar{\mathbf{R}}$.

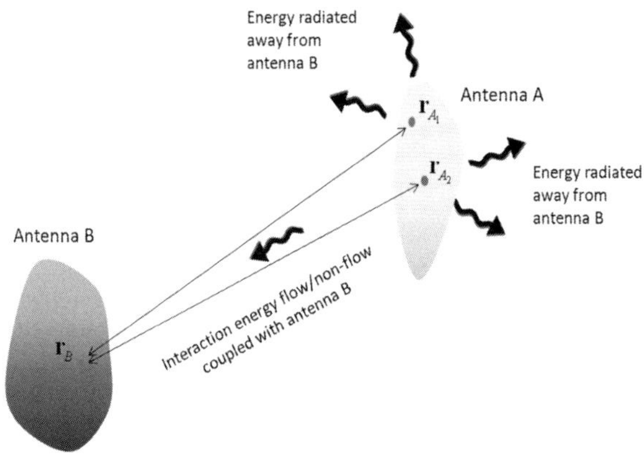

Figure 17.3 Analysis of interaction between two antennas A and B into line-of-sight (direct) energy exchange and "spilled" energy radiated away from the source. Two points $\mathbf{r}_{A_1}$ and $\mathbf{r}_{A_2}$ on antenna A may couple in a direct fashion with a point $\mathbf{r}_B$ on antenna B. However, part of the energy of antenna A will not be coupled with any point on antenna B and is lost to the surrounding space. (This is represented by the wavy arrows.) In this method, all numerical results are concerned mainly with the direct interaction energy between two antennas.

mode) or globally (receiving mode). Assume further that both antennas support a perfect electric conductor boundary condition (PEC).[2] We can write the electric field integral equation (EFIE) operator for antennas A and B as $\mathcal{L}_A$ and $\mathcal{L}_B$. They are defined by the equation [23], [38], [39], [34]

$$\mathcal{L}\mathbf{J}(\mathbf{r}) = -i\omega\mu \int_S ds' \left[\bar{\mathbf{I}} + \frac{1}{k^2}\nabla\nabla\cdot\right] \frac{e^{ik|\mathbf{r}-\mathbf{r}'|}}{4\pi|\mathbf{r}-\mathbf{r}'|} \cdot \mathbf{J}(\mathbf{r}'). \tag{17.2}$$

Here, $\bar{\mathbf{I}}$ stands for the unit dyad and k is the surrounding medium wavenumber.

When the two-antenna system receives excitation fields $\mathbf{E}_A^{\text{ex}}$ and $\mathbf{E}_B^{\text{ex}}$, it responds by generating two self-consistent current distributions $\mathbf{J}_A(\mathbf{r})$ and $\mathbf{J}_B(\mathbf{r})$ supported by the surfaces S_A and S_B, respectively. By enforcing the PEC boundary conditions on the two antennas, the following operator equations must be obeyed

$$\begin{aligned} \hat{n}_A \times \mathbf{E}_A^{\text{ex}}(\mathbf{r}) + \hat{n}_A \times \mathcal{L}_B\mathbf{J}_B(\mathbf{r}) &= -\hat{n}_A \times \mathcal{L}_A\mathbf{J}_A(\mathbf{r}), \\ \hat{n}_B \times \mathbf{E}_B^{\text{ex}}(\mathbf{r}) + \hat{n}_B \times \mathcal{L}_A\mathbf{J}_A(\mathbf{r}) &= -\hat{n}_B \times \mathcal{L}_B\mathbf{J}_B(\mathbf{r}), \end{aligned} \tag{17.3}$$

where $\hat{n}_{A,B}$ stands for outward unit normal vectors on antennas A and B. Rearranging, we can write these two equations in the form

$$\tilde{\mathcal{L}}\,[\mathbf{J}] = [\mathbf{E}^{\text{ex}}], \tag{17.4}$$

where

$$\begin{aligned} \tilde{\mathcal{L}} &:= -\begin{pmatrix} \hat{n}_A \times \mathcal{L}_A & \hat{n}_A \times \mathcal{L}_B \\ \hat{n}_B \times \mathcal{L}_A & \hat{n}_B \times \mathcal{L}_B \end{pmatrix}, \\ [\mathbf{J}] &:= \begin{pmatrix} \mathbf{J}_A \\ \mathbf{J}_B \end{pmatrix}, \quad [\mathbf{E}^{\text{ex}}] := \begin{pmatrix} \hat{n}_A \times \mathbf{E}_A^{\text{ex}} \\ \hat{n}_B \times \mathbf{E}_B^{\text{ex}} \end{pmatrix}. \end{aligned} \tag{17.5}$$

Next, decompose the full operator into the direct sum

$$\tilde{\mathcal{L}} = \tilde{\mathcal{L}}_0 + \tilde{\mathcal{L}}_{\text{int}}, \tag{17.6}$$

where

$$\begin{aligned} \tilde{\mathcal{L}}_0 &:= \begin{pmatrix} -\hat{n}_A \times \mathcal{L}_A & 0 \\ 0 & -\hat{n}_B \times \mathcal{L}_B \end{pmatrix}, \\ \tilde{\mathcal{L}}_{\text{int}} &:= \begin{pmatrix} 0 & -\hat{n}_A \times \mathcal{L}_B \\ -\hat{n}_B \times \mathcal{L}_A & 0 \end{pmatrix}. \end{aligned} \tag{17.7}$$

2 The generalization to arbitrary boundary conditions is possible but lengthy and therefore will be reported elsewhere.

Therefore, if the solution to the problem (17.4) exists, it is given by

$$[\mathbf{J}] = \tilde{\mathcal{L}}^{-1}[\mathbf{E}^{\text{ex}}] = \left(\tilde{\mathcal{L}}_0 + \tilde{\mathcal{L}}_{\text{int}}\right)^{-1}[\mathbf{E}^{\text{ex}}]. \tag{17.8}$$

The operator $\tilde{\mathcal{L}}_0$ can be though of as the *self-interaction operator* of the antenna system. It is written in terms of the operators of the *isolated* antennas, i.e., when each exists in homogeneous infinite environment. The other operator $\tilde{\mathcal{L}}_{\text{int}}$ represents the *interaction operator of the two-antenna system.*[3] It conveys information about how the two elements couple to each other electromagnetically. Further insights into the structure of this operator will be developed below.

Suppose that the isolated antenna problems are solvable, i.e., assume that the following inverse operator exists

$$\tilde{\mathcal{L}}_0^{-1} = \begin{pmatrix} (-\hat{n}_A \times \mathcal{L}_A)^{-1} & 0 \\ 0 & (-\hat{n}_B \times \mathcal{L}_B)^{-1} \end{pmatrix}. \tag{17.9}$$

Equation (17.8) can then written as

$$[\mathbf{J}] = \tilde{\mathcal{L}}_0^{-1}\left(\tilde{I} + \tilde{\mathcal{L}}_0^{-1}\tilde{\mathcal{L}}_{\text{int}}\right)^{-1}[\mathbf{E}^{\text{ex}}], \tag{17.10}$$

where $\tilde{I}$ is the identity operator corresponding to (17.5).

For completeness, we review briefly how the antenna current Green's function (ACGF), which provide complete description of the array as an electromagnetic system, can be obtained from operators defined above. Indeed, the ACGF of the two-antenna system is readily obtained as[4]

$$\left[\bar{\mathbf{F}}(\mathbf{r},\mathbf{r}')\right] = \tilde{\mathcal{L}}_0^{-1}\left(\tilde{I} + \tilde{\mathcal{L}}_0^{-1}\tilde{\mathcal{L}}_{\text{int}}\right)^{-1}[\delta(\mathbf{r},\mathbf{r}')], \tag{17.11}$$

where

$$\left[\bar{\mathbf{F}}(\mathbf{r},\mathbf{r}')\right] := \begin{pmatrix} \bar{\mathbf{F}}_{AA}(\mathbf{r},\mathbf{r}') & \bar{\mathbf{F}}_{AB}(\mathbf{r},\mathbf{r}') \\ \bar{\mathbf{F}}_{BA}(\mathbf{r},\mathbf{r}') & \bar{\mathbf{F}}_{BB}(\mathbf{r},\mathbf{r}') \end{pmatrix}, \tag{17.12}$$

and

$$\begin{aligned} \begin{pmatrix} \mathbf{J}_A(\mathbf{r}) \\ \mathbf{J}_B(\mathbf{r}) \end{pmatrix} &= \int_{S_A+S_B} ds' \\ &\times \begin{pmatrix} \bar{\mathbf{F}}_{AA}(\mathbf{r},\mathbf{r}') & \bar{\mathbf{F}}_{AB}(\mathbf{r},\mathbf{r}') \\ \bar{\mathbf{F}}_{BA}(\mathbf{r},\mathbf{r}') & \bar{\mathbf{F}}_{BB}(\mathbf{r},\mathbf{r}') \end{pmatrix} \cdot \begin{pmatrix} \hat{n}_A \times \mathbf{E}_A^{\text{ex}}(\mathbf{r}') \\ \hat{n}_B \times \mathbf{E}_B^{\text{ex}}(\mathbf{r}') \end{pmatrix}. \end{aligned} \tag{17.13}$$

3 Cf. Chapter 16.
4 Cf. Chapter 16.

The excitation vector in (17.11) is defined by

$$[\delta(\mathbf{r},\mathbf{r}')] := \begin{pmatrix} \delta_A(\mathbf{r},\mathbf{r}') \\ \delta_B(\mathbf{r},\mathbf{r}') \end{pmatrix}, \tag{17.14}$$

where $\delta_A(\mathbf{r},\mathbf{r}')$ and $\delta_B(\mathbf{r},\mathbf{r}')$ stand for vector surface delta excitations applied at antennas A and B, respectively.

I n order to simplify the presentation, throughout the rest of this chapter the ACGF (17.11) will not be used to express the final results. However, by employing the techniques developed in Chapter 9, transition the operator-based results given below to the corresponding ACGF-based form should not present a problem.

17.4 INITIAL FORMULATION OF THE PROBLEM OF MUTUAL ENERGY COUPLING AND TRANSFER FROM THE NEAR-FIELD PERSPECTIVE

Consider Figure 17.2. We are interested in studying the process of electromagnetic interaction between antenna A and antenna B. As suggested above, this process can be reduced to the study of the *interaction operator* $\mathcal{L}_{\text{int}}$ defined in the second equation of (17.7). It can be seen from this formula that each entry in the off-diagonal part of the interaction operator is itself written in terms of the EFIE operator (17.2). In order to understand the structure of the interaction operator, we carefully examine the *interaction fields* defined as

$$\begin{aligned}[\mathbf{E}_{\text{int}}] &= \begin{pmatrix} \mathbf{E}_{\text{int}}^A \\ \mathbf{E}_{\text{int}}^B \end{pmatrix} := \tilde{\mathcal{L}}_{\text{int}}[\mathbf{J}] \\ &= \begin{pmatrix} 0 & -\hat{n}_A \times \mathcal{L}_B \\ -\hat{n}_B \times \mathcal{L}_A & 0 \end{pmatrix} \begin{pmatrix} \mathbf{J}_A \\ \mathbf{J}_B \end{pmatrix} \\ &= \begin{pmatrix} -\hat{n}_A \times \mathcal{L}_B \mathbf{J}_B \\ -\hat{n}_B \times \mathcal{L}_A \mathbf{J}_A \end{pmatrix}. \end{aligned} \tag{17.15}$$

In other words, $\mathbf{E}_{\text{int}}^A = -\hat{n}_A \times \mathcal{L}_B \mathbf{J}_B$ is the tangential fields on antenna A radiated by the current on antenna B, and similarly for $\mathbf{E}_{\text{int}}^B = -\hat{n}_B \times \mathcal{L}_A \mathbf{J}_A$. In terms of the EFIE operator (17.2), we find

$$\begin{aligned}&\mathbf{E}_{\text{int}}^A(\mathbf{r}_A) \\ &= -i\omega\mu \int_{S_B} ds_B \left[\bar{\mathbf{I}} + \tfrac{1}{k^2}\nabla\nabla\cdot\right] \frac{e^{ik|\mathbf{r}_A-\mathbf{r}_B|}}{4\pi|\mathbf{r}_A-\mathbf{r}_B|} \cdot \mathbf{J}_B(\mathbf{r}_B).\end{aligned} \tag{17.16}$$

A similar equation can be written for $\mathbf{E}_{\text{int}}^{B} = -\hat{n}_B \times \mathcal{L}_A \mathbf{J}_A$. Consider the integrand of (17.16). The current on antenna B at position $\mathbf{r}_B$ acts as a source. Its effect is communicated to position $\mathbf{r}_A$ on antenna A via the Green's function of free space. This Green's function, however, is neither a pure propagating nor pure nonpropagating channel. It is in fact a mixture of *both* as can be seen by a Fourier analysis of the field (*viz* the plane-wave spectrum or the Weyl expansion, see for example [35]). *Therefore, physically speaking, it is the rather subtle manner in which the free-space Green's function divides into propagating and evanescent modes that determines how much of the interaction between the points* $\mathbf{r}_A$ *and* $\mathbf{r}_B$ *is mediated by either by pure flow of energy or a localized (non-moving) energy zone.* In Section 17.5, the details of our approach will be given with respect to the method of moment (MoM). Throughout the remaining parts of this section, we will illustrate the conceptual structure of the proposed analysis of interaction.

The free-space scalar Green's function is given by [35]

$$g(\mathbf{r}, \mathbf{r}') := \frac{e^{ik|\mathbf{r}-\mathbf{r}'|}}{4\pi |\mathbf{r}-\mathbf{r}'|}. \tag{17.17}$$

Using the Weyl expansion ((17.23) below), (17.17) can be expanded into propagating and nonpropagating parts as follows

$$g(\mathbf{r}, \mathbf{r}') = g_{\text{ev}}\left(\mathbf{r}, \mathbf{r}'; \bar{\mathbf{R}}\right) + g_{\text{pr}}\left(\mathbf{r}, \mathbf{r}'; \bar{\mathbf{R}}\right). \tag{17.18}$$

Here, we notice, as was reported in Chapter 4, that the decomposition of the free space Green's function into propagating and evanescent modes depends in general on the choice of the coordinate system. In order to capture the dynamic content of engineering problems in a systematic fashion (more on this below), it was proposed by the authors that a local coordinate system be introduced into the problem with relative orientation (with respect to the ordinal global coordinate system) described by a 3D rotational matrix $\bar{\mathbf{R}}$.[5] Note that although the *total* Green's function $g(\mathbf{r}, \mathbf{r}')$ on the LHS of (17.18) is obviously independent of the choice of the local frame, the *individual* propagating and nonpropagating parts *do* depend explicitly on $\bar{\mathbf{R}}$. *Physically, we say that the near field tends to split differently into propagating and evanescent modes along various directions in the space around the antenna.*

For problems of mutual interactions between several antennas, each pair of points on the interacting elements will have its own "line-of-sight" direction of energy exchange, i.e., that straight line connecting the two points. Moreover, as

5 Cf. Chapter 4.

suggested by Figure 17.2, we expect from the physics of the interaction problem that the proper locus of the spectral analysis of the near field should be that along the line carrying the vector $\mathbf{r}_A - \mathbf{r}_B$ connecting the source point $\mathbf{r}_B$ and observation point $\mathbf{r}_A$. Therefore, we need to rotate the local coordinate system such that the z-axis of the local frame coincides with the line along $\mathbf{r}_A - \mathbf{r}_B$. The rotation matrix that accomplishes this will be denoted by $\bar{\mathbf{R}}(\mathbf{r}_A, \mathbf{r}_B)$. By substituting the decomposition (17.18) into (17.16), we obtain

$$\mathcal{L}_{A,B} = \mathcal{L}_{\mathrm{ev}}^{A,B}\left(\bar{\mathbf{R}}\right) + \mathcal{L}_{\mathrm{pr}}^{A,B}\left(\bar{\mathbf{R}}\right). \tag{17.19}$$

In (17.19), the rotation matrix was reproduced explicitly on the RHS (however, one of the contributions of this chapter is the demonstration in Section 17.5 that the decomposition of the *total* interaction operator into propagating and nonpropagating parts can be effectuated independently of the this rotation). Next, the overall interaction operator for the two-antenna system is decomposed in the following manner

$$\mathcal{L}_{\mathrm{int}} = \underbrace{\begin{pmatrix} 0 & -\mathcal{L}_{\mathrm{ev}}^{B}\left(\bar{\mathbf{R}}\right) \\ -\mathcal{L}_{\mathrm{ev}}^{A}\left(\bar{\mathbf{R}}\right) & 0 \end{pmatrix}}_{\mathcal{L}_{\mathrm{ev}}^{\mathrm{int}}} + \underbrace{\begin{pmatrix} 0 & -\mathcal{L}_{\mathrm{pr}}^{B}\left(\bar{\mathbf{R}}\right) \\ -\mathcal{L}_{\mathrm{pr}}^{A}\left(\bar{\mathbf{R}}\right) & 0 \end{pmatrix}}_{\mathcal{L}_{\mathrm{pr}}^{\mathrm{int}}}. \tag{17.20}$$

An explication of the physical meaning and importance of these propagating and evanescent interaction sub-operators will be given in conjunction with the numerical examples of Section 17.6.

Finally, using a suitable norm to estimate the "size" of the operator [92] (the details will be given below), we may define the localization of the interaction energy as the percentage of the size of the evanescent operator $\mathcal{L}_{\mathrm{ev}}^{A,B}$ with respect to the size of the full operator, i.e., we define

$$\kappa := \frac{\left\|\mathcal{L}_{\mathrm{ev}}^{\mathrm{int}}\right\|}{\left\|\mathcal{L}^{\mathrm{int}}\right\|}. \tag{17.21}$$

This factor κ will be called the *energy coupling localization coefficient* of antenna-antenna interaction. It provides a quick quantitative view on how much of the total interaction between the two antennas is mediated by nonpropagating modes. In other words, it indicates how the near field between the two antennas tends to localize due to mutual coupling effects. *Stored* energy, for instance, is probably related to this phenomenon of localization.[6] The larger the value of κ, the stronger will be the tendency of the energy exchange to be located in the space in between the two antennas.

17.5 A NUMERICAL MODEL USING THE METHOD OF MOMENT

17.5.1 Basic MoM Formulation

We will employ the Method of Moment (MoM) to implement the theory developed in Section 17.4. Moreover, the antenna types to be considered in the following numerical examples of Section 17.6 are all thin-wire antennas. However, the basic technique applies to *generic* antenna systems and is *not* restricted to wire antennas. The choice of thin-wires made here is for simplicity.

Triangular pulse basis functions were deployed to approximate current distributions on a discretized thin-wire version of the EFIE (17.2) [38]. Testing is done using the Galerkin method to reduce the EFIE into a matrix equation with the complex amplitudes of the triangular basis functions as unknowns. An accurate Gauss quadrature method is used to compute the off-diagonal elements of the MoM matrix, while the singular and near-singular terms were derived analytically.

The MoM matrix element expression arising from Galerkin testing of the EFIE is given by [38]

$$\begin{aligned} z_{mn} = & \int\limits_{S_m}\int\limits_{S_n} ds' ds\, \mathbf{f}_m(\mathbf{r}) \cdot \tfrac{e^{ik|\mathbf{r}-\mathbf{r}'|}}{4\pi|\mathbf{r}-\mathbf{r}'|} \mathbf{f}_n(\mathbf{r}') \\ & -\tfrac{1}{k^2} \int\limits_{S_m}\int\limits_{S_n} ds' ds\, \nabla \cdot \mathbf{f}_m(\mathbf{r}) \cdot \tfrac{e^{ik|\mathbf{r}-\mathbf{r}'|}}{4\pi|\mathbf{r}-\mathbf{r}'|} \nabla \cdot \mathbf{f}_n(\mathbf{r}'). \end{aligned} \tag{17.22}$$

We first carefully examine the structure of the MoM matrix element expression (17.22). The two terms on the RHS are actually formally identical: they both involve the formula $\rho_{mn}\exp(ik|\mathbf{r}-\mathbf{r}'|)/4\pi|\mathbf{r}-\mathbf{r}'|$, where in the case of the first term we have $\rho_{mn} = \mathbf{f}_m(\mathbf{r}) \cdot \mathbf{f}_n(\mathbf{r}')$, while for the second term $\rho_{mn} = \nabla \cdot \mathbf{f}_m(\mathbf{r}) \nabla \cdot \mathbf{f}_n(\mathbf{r}') = -\omega^2 \rho_m \rho_n$. (In the latter relation, the equation of continuity $\nabla \cdot \mathbf{J}(\mathbf{r}) = i\omega\rho(\mathbf{r})$ was used.) In both cases, the quantity ρ_{mn} stands for the mutual interaction *charge* between the source $\mathbf{r}'$ and the observation $\mathbf{r}$. Formally, this interaction is mediated by the ***scalar*** Green's function $\exp(ik|\mathbf{r}-\mathbf{r}'|)/4\pi|\mathbf{r}-\mathbf{r}'|$ although the MoM formulation used here is for the *full*-wave *vectorial* case. This observation will

6 Cf. Chapter 4.

considerably simplify our quest for computing the spectral composition of the electromagnetic coupling between interacting antennas.

The Weyl expansion shows that the total scalar Green's function can be divided into the sum of two parts, one as pure propagating waves and the other as evanescent, hence nonpropagating part. Explicitly, we write [35]

$$\frac{e^{ikr}}{r} = \frac{ik}{2\pi} \int_{-\infty}^{\infty} \int_{-\infty}^{\infty} dpdq \frac{1}{m} e^{ik(px+qy+m|z|)}, \tag{17.23}$$

where

$$m(p,q) = \begin{cases} \sqrt{1-p^2-q^2} & ,p^2+q^2 \leq 1 \\ i\sqrt{p^2+q^2-1} & ,p^2+q^2 > 1 \end{cases} . \tag{17.24}$$

Here we used the notation $r = |\mathbf{r} - \mathbf{r}'|$. The propagating and nonpropagating (evanescent) parts are given, respectively, by the expressions

$$g_{\text{ev}}(\mathbf{r},\mathbf{r}') = \tfrac{ik}{8\pi^2} \int_{p^2+q^2>1} dpdq \tfrac{1}{m} e^{ik[p(x-x')+q(y-y')]} \times e^{im|z-z'|}, \tag{17.25}$$

$$g_{\text{pr}}(\mathbf{r},\mathbf{r}') = \tfrac{ik}{8\pi^2} \int_{p^2+q^2<1} dpdq \tfrac{1}{m} e^{ik[p(x-x')+q(y-y')]} \times e^{im|z-z'|}. \tag{17.26}$$

As was pointed out in Chapter 4, the *non*propagating part of the Green's function is related to *energy localization* in the space surrounding the antennas. We have explained in Section 17.4 how the changing direction along which we effect the decomposition into propagating and evanescent modes will determine the fine structure of the energy exchange between the two antennas. The observation just made will now be exploited by concentrating our spectral analysis on the *scalar* Green's function of the problem.

17.5.2 Analytical Evaluation of the Scalar Green's Function Dynamic Spectral Representations

Similar to (17.19), we separate the matrix element into propagating and nonpropagating parts

$$z_{mn} = z_{mn}^{\text{ev}} + z_{mn}^{\text{pr}}, \tag{17.27}$$

where

$$\begin{aligned} z_{mn}^{\mathrm{pr}} = & \frac{ik}{4\pi^2} \int_{S_m} ds' ds \mathbf{f}_m(\mathbf{r}) \cdot \int_{S_n} \mathbf{f}_n(\mathbf{r}') \\ & \times \int_{p^2+q^2<1} dpdq \, \tfrac{1}{m} \, e^{i\left(\bar{\mathbf{R}}^T \cdot \mathbf{K}\right) \cdot \mathbf{r}} \\ & - \frac{i}{4k\pi^2} \int_{S_m} ds' ds \nabla \cdot \mathbf{f}_m(\mathbf{r}) \cdot \int_{S_n} \nabla \cdot \mathbf{f}_n(\mathbf{r}') \\ & \times \int_{p^2+q^2<1} dpdq \, \tfrac{1}{m} \, e^{i\left(\bar{\mathbf{R}}^T \cdot \mathbf{K}\right) \cdot \mathbf{r}} \end{aligned} \tag{17.28}$$

$$\begin{aligned} z_{mn}^{\mathrm{ev}} = & \frac{ik}{4\pi^2} \int_{S_m} ds' ds \mathbf{f}_m(\mathbf{r}) \cdot \int_{S_n} \mathbf{f}_n(\mathbf{r}') \\ & \times \int_{p^2+q^2>1} dpdq \, \tfrac{1}{m} \, e^{i\left(\bar{\mathbf{R}}^T \cdot \mathbf{K}\right) \cdot \mathbf{r}} \\ & - \frac{i}{4k\pi^2} \int_{S_m} ds' ds \nabla \cdot \mathbf{f}_m(\mathbf{r}) \cdot \int_{S_n} \nabla \cdot \mathbf{f}_n(\mathbf{r}') \\ & \times \int_{p^2+q^2>1} dpdq \, \tfrac{1}{m} \, e^{i\left(\bar{\mathbf{R}}^T \cdot \mathbf{K}\right) \cdot \mathbf{r}} \end{aligned} \tag{17.29}$$

To proceed further, we will use results from Chapter 5. Using the analytical expression (5.20) and (5.21) in (17.28) and (17.29), respectively, we find

$$\begin{aligned} z_{mn}^{\mathrm{pr}} = & \frac{ik}{4\pi^2} \int_{S_m} ds' ds \, \mathbf{f}_m(\mathbf{r}) \cdot \int_{S_n} \mathbf{f}_n(\mathbf{r}') \, \frac{e^{ik|\mathbf{r}-\mathbf{r}'|}-1}{r} \\ & - \frac{i}{4k\pi^2} \int_{S_m} ds' ds \nabla \cdot \mathbf{f}_m(\mathbf{r}) \cdot \int_{S_n} \nabla \cdot \mathbf{f}_n(\mathbf{r}') \, \frac{e^{ik|\mathbf{r}-\mathbf{r}'|}-1}{r}, \end{aligned} \tag{17.30}$$

$$\begin{aligned} z_{mn}^{\mathrm{ev}} = & \frac{ik}{4\pi^2} \int_{S_m} ds' ds \mathbf{f}_m(\mathbf{r}) \cdot \int_{S_n} \mathbf{f}_n(\mathbf{r}') \, \tfrac{1}{r} \\ & - \frac{i}{4k\pi^2} \int_{S_m} ds' ds \nabla \cdot \mathbf{f}_m(\mathbf{r}) \cdot \int_{S_n} \nabla \cdot \mathbf{f}_n(\mathbf{r}') \, \tfrac{1}{r}. \end{aligned} \tag{17.31}$$

Therefore, the final expressions of the MoM matrix decomposition does not depend on the rotation matrix $\bar{\mathbf{R}}$. However, this conclusion applies only when we consider decomposition into the *total* propagating and *total* evanescent part. Since this is the only type of spectral decomposition considered here, the results (17.30) and (17.31) are enough for our practical interests, see Section 17.6 for numerical examples.

Finally, given the new matrices z_{ev} and z_{pr}, we can estimate the size of the sub-operators by using standard matrix operator norms. For example, we may define the energy localization factor κ as follows

$$\kappa := \frac{\|z^{\mathrm{ev}}\|}{\|z\|}. \tag{17.32}$$

Here, $\|\|$ stands for some matrix norm, see for example [92]. Concrete details about the definition (17.21) will be given in Section 17.6.

17.5.3 Verification of the Code

The relations (17.30) and (17.31) show that the same basic MoM formulation can be used with simple modification to compute the spectral structure of the interaction operator. Indeed, *no new numerical integrals are needed for the spectral analysis since these are evaluated analytically as given by* (5.20) and (5.21). The only additional computation is an increase in the number of the MoM sub-matrices integrals since now we have two integrations, one corresponding to the evanescent mode interaction, the other to the propagating mode case as can be seen after a glance at (17.30) and (17.31). *These additional integrals, however, have the same structure of the classic MoM integrations and present no considerable difficulty for adapting them to existing codes.*

A computer code was written to implement the theory and method developed above. Since the results presented here are new, verification proceeds in two carefully separated stages.

- Stage I: Conventional MoM scripts to discretize the current and evaluate the impedance matrix integrals using the classic scalar Green's function as mediator of interaction (Section 17.5.1). Verification of these code scripts was made by comparison with other standard full-wave EM solvers, for example WIPL-D which uses higher-order basis functions [39].
- Stage II: Separation of the scalar Green's function into propagating and nonpropagating dynamic parts depending on the location of the two coupled points on the two interacting antennas. This part was done analytically in 17.5.2. It is verified by numerical integration to confirm the analytical expression and also by substitution.

In writing the new code, the use of *independent classes* or subroutines in implementing the two stages of the code above allows direct combination of the two, each verified independently, to form the new code without any modification in the separate codes themselves. This is possible because, as can be seen from (17.22), the interaction propagator (the Green's function) appears in the MoM matrix element expression *independently* of the current basis function and the integration region. Therefore, the new expressions (17.30) and (17.31) can be readily programmed without any essential changes with respect to the already verified Stages I and II.

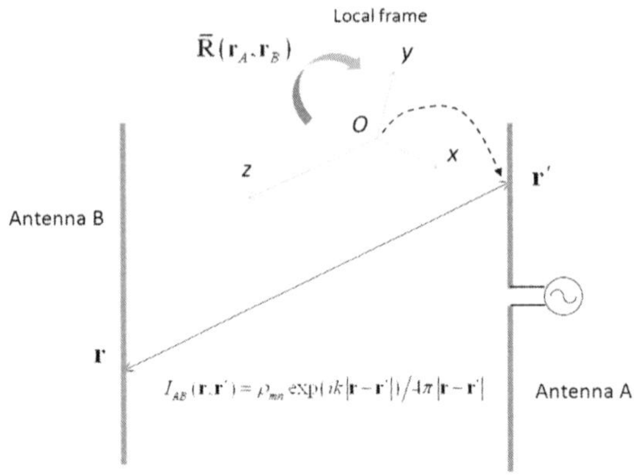

Figure 17.4 A spectral analysis of interaction between two linear wire antennas A and B. The two antennas have identical radius of 0.001λ and length 0.5λ. The spacing between the elements is given by d.

17.6 NUMERICAL EXAMPLES USING LINEAR WIRE ANTENNA ARRAYS

We first consider a basic two-element antenna array composed of two parallel thin wires A and B. Figure 17.4 illustrates the configuration of the interaction problem. Antenna A will be excited by a delta source gap placed at the center while antenna B (with no generator) will couple electromagnetically with antenna B. The overall coupling is represented by the operator (17.15), which involves the sum (integral) of all mutual interactions between points $\mathbf{r}'$ on antenna A and points $\mathbf{r}$ on antenna A. As we showed in Section 17.5, if the MoM is used to compute energy coupling, then the *mediator* of electromagnetic interaction (the "electromagnetic propagator"), can be formally captured by the *scalar* exchange $I_{AB}(\mathbf{r}, \mathbf{r}') = \rho_{mn} \exp(ik|\mathbf{r} - \mathbf{r}'|)/4\pi|\mathbf{r} - \mathbf{r}'|$. We rotate the local coordinate frame xyz such that its z-axis coincides with line of the vector $\mathbf{r}' - \mathbf{r}$ while the origin O is at the source point $\mathbf{r}'$. In this case, the formulas (17.30) and (17.31) become applicable.

The interaction operator is approximated by the MoM matrix

$$Z_{\text{int}} = \begin{pmatrix} 0 & Z_{AB} \\ Z_{BA} & 0 \end{pmatrix}. \tag{17.33}$$

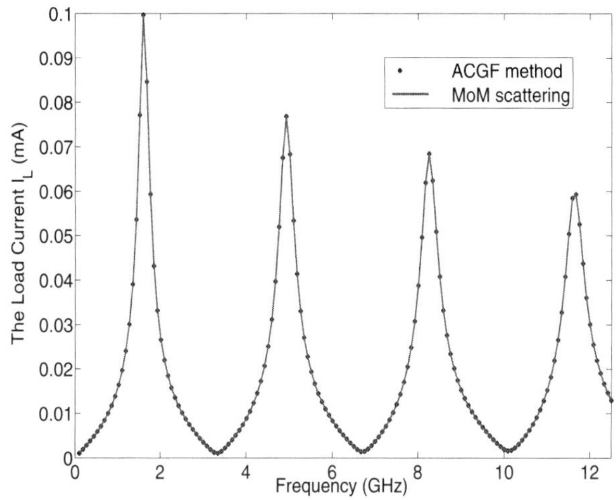

Figure 17.5 The distribution of the interaction self energy of the propagating part. The two antennas of Figure 17.4 have identical radius of 0.001λ and length 0.5λ. The spacing between the elements is given by 0.25λ.

The mnth matrix element is expanded as $Z_{AB,mn} = Z^{\text{pr}}_{AB,mn} + Z^{\text{ev}}_{AB,mn}$. Therefore, the corresponding mutual coupling energy is given by

$$|Z_{AB,mn}|^2 = \left|Z^{\text{pr}}_{AB,mn}\right|^2 + \left|Z^{\text{ev}}_{AB,mn}\right|^2 \\ +2\text{Re}Z^{\text{pr}}_{AB,mn}Z^{\text{ev}*}_{AB,mn}. \tag{17.34}$$

The terms $\left|Z^{\text{pr}}_{AB,mn}\right|^2$ and $\left|Z^{\text{ev}}_{AB,mn}\right|^2$ give the self energies of the propagating and evanescent parts, respectively. The third term $2\text{Re}Z^{\text{pr}}_{AB,mn}Z^{\text{ev}*}_{AB,mn}$, however, is a new energy process: it is the net exchange energy between the propagating and evanescent fields. It was shown by generalizing the Poynting theorem that this term does actually behave as energy and possess a suitable flux (Poynting) density flow.[7] Moreover, in contrast to the self energy, it can be either positive, zero, or negative.

Figures 17.5 and 17.6 illustrate the self energy of the pure propagating and evanescent parts of the interaction energy between antennas A and B when the spacing between the two antennas is 0.25λ. The two parts, however, exchange energy

7 Cf. Chapter 4.

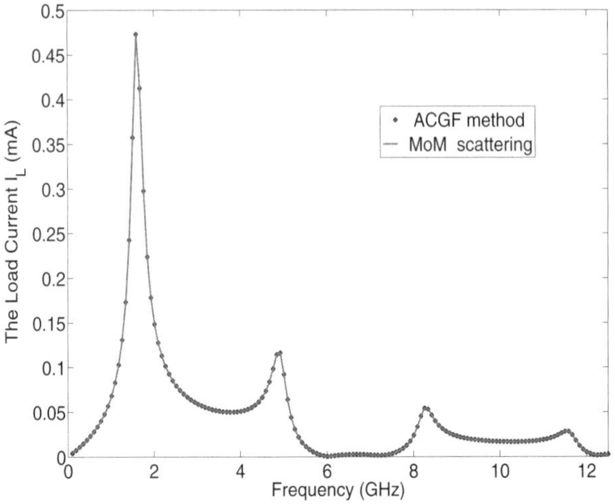

Figure 17.6 The distribution of the interaction self energy of the evanescent part. The array is the same as described in the caption of Figure 17.5.

and this net coupling between the propagating and nonpropagating components is quantified in Figure 17.7. In all these figures, the axis labeled 'Antenna A or B' list the MoM segments used in the discretization of the antenna system. There are several interesting features to draw from these data. First, the propagating part is maximum along the line in the horizontal plane given by $y = x$, i.e., for all interactions connecting each segment with the direct segment corresponding to it on the other antenna obtained by a line orthogonal to the parallel antennas. This is more obvious in Figure 17.6, where the self energy of the evanescent part exhibit a minimum at exactly the same line. On the other hand, the latter figure also shows something more: there exist two local maxima of the self evanescent energy at two different lines in the horizontal plane. In Figure 17.8 we show which parts of the two wires contribute to these maxima in the interaction energy. On the left diagram, the direct coupling of segments connected by the horizontal lines delimits the *maximal* transfer of pure propagating energy. In the right diagram of Figure 17.8, we see two complementary or symmetric couplings contributing to the two maxima in the self evanescent energy of Figure 17.6. *These parts of the two wires can then provide*

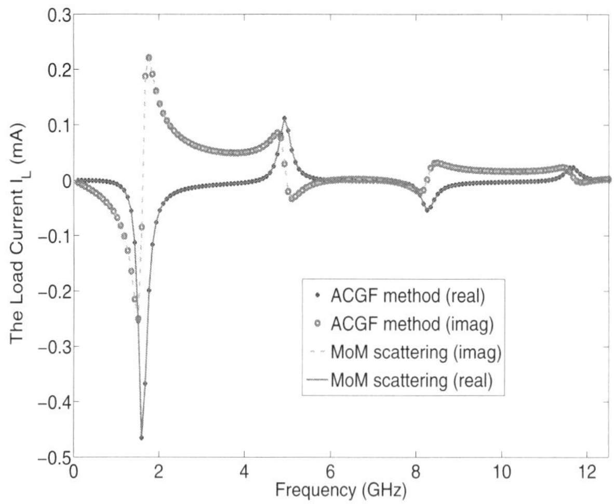

Figure 17.7 The distribution of the interaction exchange energy between the evanescent and propagating parts. The array is the same as described in the caption of Figure 17.5.

clues about the origin of the localization *of electromagnetic energy between the two wires due to mutual coupling.*

Finally, we notice from Figure 17.7 that the exchange energy between the propagating and evanescent parts is always negative. Moreover, the qualitative shape of this exchange energy distribution between the various segments of the two-antenna system follows a curious pattern that is intermediate between Figures 17.5 and 17.6. Indeed, the local maximum coincides with the local maximum of Figure 17.5 of the pure propagating parts, but the two nearby local minima are in opposition to the two local maxima of Figure 17.6. This suggests that the exchange energy is somehow ambiguous, in the sense of having the form of a *decaying propagating mode*, which may be interpreted as belonging to either propagating or nonpropagating energy depending on the problem and the context.

In order to compare with a case of weak mutual coupling, Figures 17.9, 17.10, and 17.11 provide the corresponding results when the spacing between the elements is as large as 0.5λ. As can be seen in this case, the main qualitative features of strong mutual coupling manifested in Figures 17.5, 17.6, and 17.7 are absent here, mainly the local maxima and minima in the self propagating and nonpropagating parts.

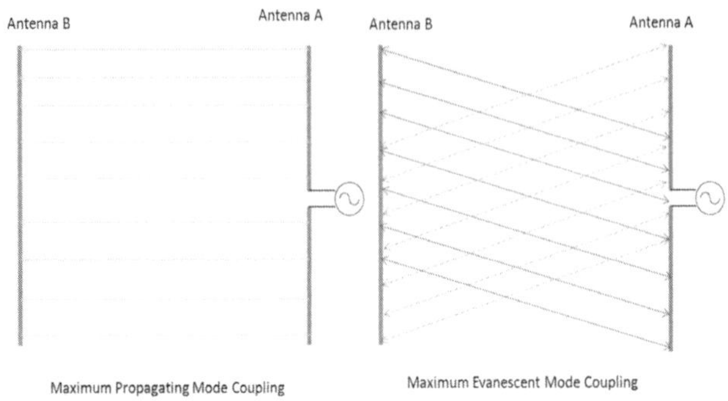

Figure 17.8 A schematic diagram illustrating the parts of the two-wire system with contributing to the maximum propagating and evanescent mode coupling. Here, the separation between the elements is 0.25λ In the left diagram, the direct lines highlight those pairs on the two wires contributing to maximum pure propagating energy transfer. In the right diagram, two complementary sets of pairs provide the distribution of parts contributing to maxima in the pure evanescent mode energy transfer.

This indicates that the spectral content of the near-field structure (and consequently energy localization) is strongly influenced by mutual coupling. It appears that a localization of the near field produced by the interaction between two antennas depends on a critical separation between the elements.

To further confirm these conclusions, we carried another numerical experiment in which the separation between the elements is reduced to just 0.1λ. Figures 17.12, 17.13, and 17.14 show the results in this case. Comparison between Figures 17.5 and 17.12 for the self energy of the propagating parts shows no qualitative difference. The same applies roughly to Figures 17.7 and 17.14 concerning the exchange energy between the propagating and evanescent parts, although the peak (local maximum) in the latter figure shows marked increase in concentration. There is, however, a qualitative change in Figures 17.6 and 17.13 concerning the self energy of the evanescent part of the interaction. In the 0.1λ array, stronger mutual coupling introduced a new peak or maximum at the position where only a minimum existed before, creating in the process two new local nearby minima. The new maximum in the evanescent energy shown in Figure 17.13 coincides with the maximum of the propagating energy of Figure 17.5 (represented geometrically in the left of Figure 17.8). This makes the interpretation of the results a bit more difficult since now both the nonpropagating and propagating parts possess maxima at the same location.

A change in the length of one antenna gives some clues about the nature of the coupling between the two elements. The length of antenna A is kept at 0.5λ

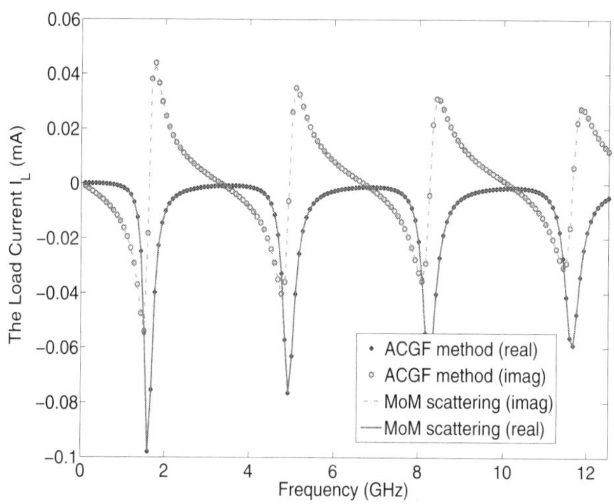

Figure 17.9 The distribution of the interaction self energy of the propagating part. The two antennas of Figure 17.4 have identical radius of 0.001λ and length 0.5λ. The spacing between the elements is given by 0.5λ.

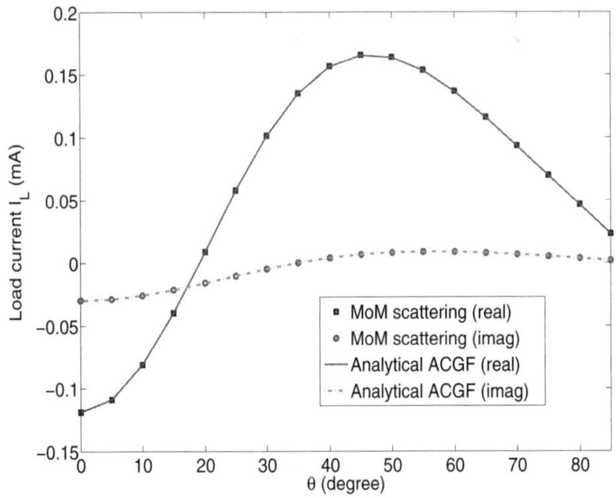

Figure 17.10 The distribution of the interaction self energy of the evanescent part. The array is the same as described in the caption of Figure 17.9.

but antenna B was shortened into 0.15λ while remaining parallel to antenna A. Figures 17.15, 17.16, and 17.17 give the results for this case where the separation between the two element is 0.2λ. The result look qualitatively similar to the case of Figures 17.5, 17.6, and 17.7 with the important difference that the maximum of the self energy of the propagating part of the interaction is no longer along the line $y = x$. Instead, only a *portion* of antenna A interacts with the entirety of antenna B in the mode of maximum propagating energy transfer. This geometrical relation is illustrated in Figure 17.18. There we see that the region of antenna A highlighted in the figure is the part that contribute the most to the propagating energy transferred to antenna B by the process of mutual coupling. As we can see, the direct lines mediating this transfer are no longer orthogonal to the wire as was the case in the left diagram of Figure 17.8. *Therefore, we don't expect that maximum energy transfer through the propagating modes alone should follow simple geometrical rules, but rather it depends on the separation distance and relative orientation of the elements.* The methods proposed help help automate the study of field localization due to mutual coupling by providing detailed pictures about the distribution of

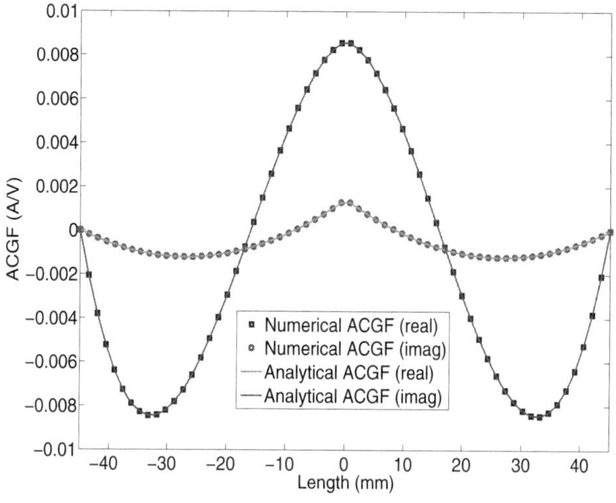

Figure 17.11 The distribution of the interaction exchange energy between the evanescent and propagating parts. The array is the same as described in the caption of Figure 17.9.

propagating and nonpropagating energies among the various parts of the antennas mutually interacting with each other.

17.7 MEASUREMENT OF THE LOCALIZATION OF THE RADIATED FIELD

Localization of the radiated field is *not* reducible to the value of the *total* field. Instead, it is about how *portion* of the energy contained in the total field gets to stay around certain regions in space.[8] While the total field is always time varying, it is not true that any time-varying field is moving energy in space. For example, a standing wave in transmission lines is time varying but does not carry energy away from the source. The localized field produced by an antenna can be compared with the standing waves of circuits. Although the situation in field theory is much more complex, see for example Chapter 4, the basic idea is to separate the *non*propagating part of the total field and ascribe localization of energy to this portion alone. In this

8 Cf. Chapter 4.

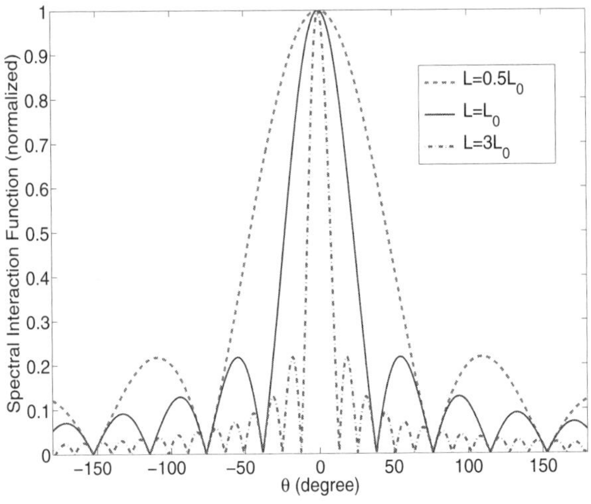

Figure 17.12 The distribution of the interaction self energy of the propagating part. The two antennas of Figure 17.4 have identical radius of 0.001λ and length 0.5λ. The spacing between the elements is given by 0.1λ.

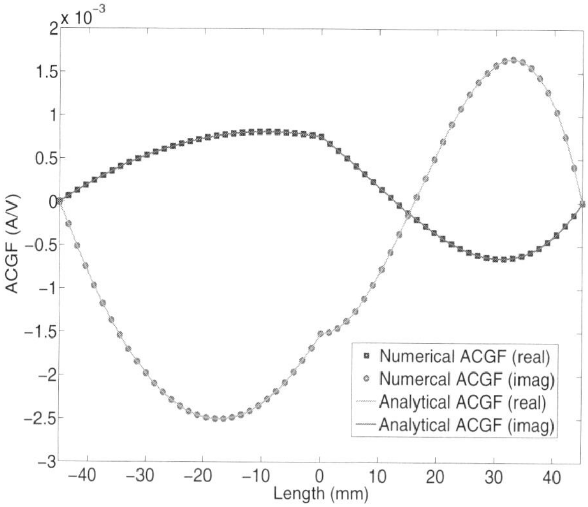

Figure 17.13 The distribution of the interaction self energy of the evanescent part. The array is the same as described in the caption of Figure 17.12.

section, we outline how the methods developed above can be used to implement an approach to perform a computation of this localization. We still work with the basic configuration of Figure 17.8 but now think of antenna B as a *probe* used for measurement while antenna A is the source of the field we would like to characterize in terms of localization. By varying the distance between the two antennas, the spectral decomposition of the interaction operator can be utilized to provide some rough idea about how much of the field received by the probe is localized in space. We don't focus on only the received signal at the port of antenna B, but examine the interaction of the *entire* receiving element with the illumination field by examining the MoM matrix of the interaction operator.

As was suggested by the numerical examples in Section 17.6, interaction energy certainly tends to get localized in a more complex manner when the separation between the elements gets smaller. However, in order to quantify this localization in a simpler manner, a *relative and global* measure has to be introduced. That is, we need to average the overall decompositions of the interaction MoM matrix into propagating and nonpropagating modes and then estimate the relative weight of each energy type with respect to the others. This we accomplish by computing

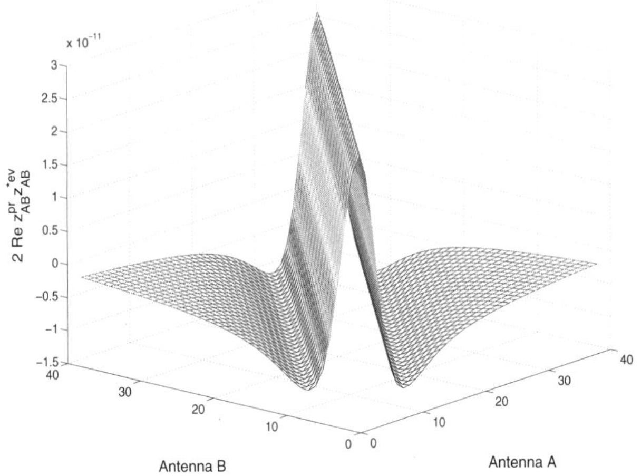

Figure 17.14 The distribution of the interaction exchange energy between the evanescent and propagating parts. The array is the same as described in the caption of Figure 17.12.

a suitable *norm* in order to roughly approximate the "size" of the operator, as is usually done in theoretical numerical analysis. The most convenient such norm in our case is the Frobenius norm defined as [92]

$$\|A\|_F := \sqrt{\sum_{m=1}^{M}\sum_{n=1}^{N} |a_{mn}|^2}, \tag{17.35}$$

which naturally corresponds to the energy definitions typical in physics and engineering.

We define the following three coefficients quantifying the near-field coupling

$$\begin{aligned}
\kappa_{\mathrm{pr}} &:= \frac{\|Z_{AB}^{\mathrm{pr}}\|_F^2}{\|Z_{AB}^{\mathrm{pr}}\|_F^2 + \|Z_{AB}^{\mathrm{ev}}\|_F^2 + \left\|2\mathrm{Re} Z_{AB}^{\mathrm{ev}}\left(Z_{AB}^{\mathrm{pr}}\right)^H\right\|_F^2}, \\
\kappa_{\mathrm{ev}} &:= \frac{\|Z_{AB}^{\mathrm{ev}}\|_F^2}{\|Z_{AB}^{\mathrm{pr}}\|_F^2 + \|Z_{AB}^{\mathrm{ev}}\|_F^2 + \left\|2\mathrm{Re} Z_{AB}^{\mathrm{ev}}\left(Z_{AB}^{\mathrm{pr}}\right)^H\right\|_F^2}, \\
\kappa_{\mathrm{pr/ev}} &:= \frac{\left\|2\mathrm{Re} Z_{AB}^{\mathrm{ev}}\left(Z_{AB}^{\mathrm{pr}}\right)^H\right\|_F}{\|Z_{AB}^{\mathrm{pr}}\|_F^2 + \|Z_{AB}^{\mathrm{ev}}\|_F^2 + \left\|2\mathrm{Re} Z_{AB}^{\mathrm{ev}}\left(Z_{AB}^{\mathrm{pr}}\right)^H\right\|_F^2}.
\end{aligned} \tag{17.36}$$

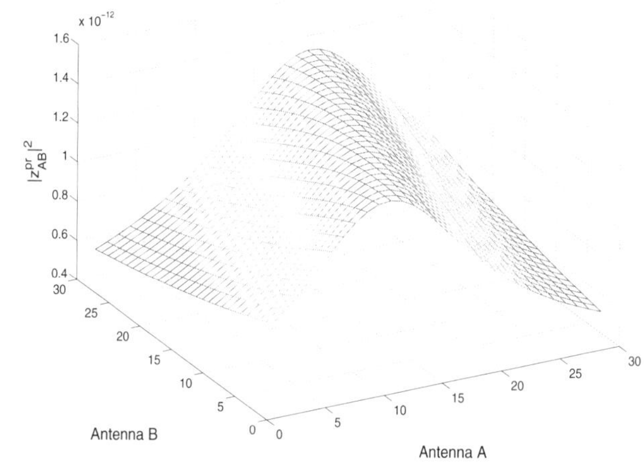

Figure 17.15 The distribution of the interaction self energy of the propagating part. The two antennas of Figure 17.4 have identical radius of 0.001λ but different lengthes. Antenna A is 0.5λ while antenna B is 0.15λ. The spacing between the elements is given by 0.2λ.

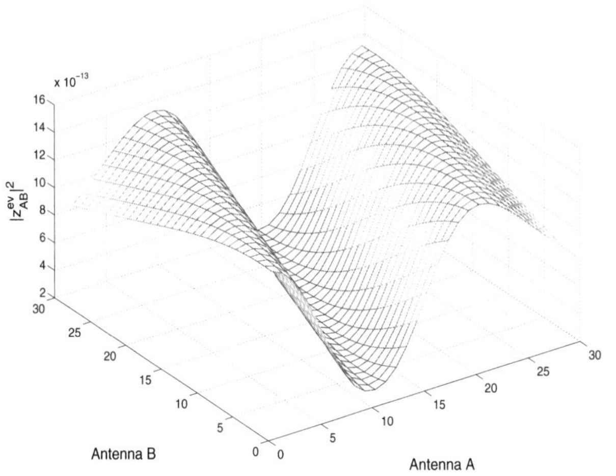

Figure 17.16 The distribution of the interaction self energy of the evanescent part. The array is the same as described in the caption of Figure 17.15.

Here, κ_{pr} measures the total propagating energy relative to the total interaction energy. Similarly, κ_{ev} measures the total nonpropagating or evanescent energy relative to the total interaction energy. The third coefficient $\kappa_{\text{pr/ev}}$ provides information about the exchange of energy between the propagating and nonpropagating parts.[9] As defined above, κ_{pr} and κ_{ev}, and $\kappa_{\text{pr/ev}}$ are always positive, even though the exchange energy between the evanescent and propagating parts (when taken term by term) can take negative values. In any case, the relation $\kappa_{\text{pr}} + \kappa_{\text{ev}} + \kappa_{\text{pr/ev}} = 1$ always holds.

Figure 17.19 shows the results for two-element antenna array with equal length of 0.5λ. The evanescent mode energy κ_{ev} reaches its maximum limit at 1.0 periodically at integer multiples of wavelength. At those locations, the propagating mode interaction energy drops to zero. Moreover, Figure 17.20 indicates that throughout all locations, the exchange of energy between the propagating and nonpropagating modes can be neglected.

If we view antenna B as a probe used to measure the structure of the field produced by antenna B, then it is clear from Figure 17.19 that around the locations

9 The superscript H appearing in (17.36) is the Hermitian of the matrix.

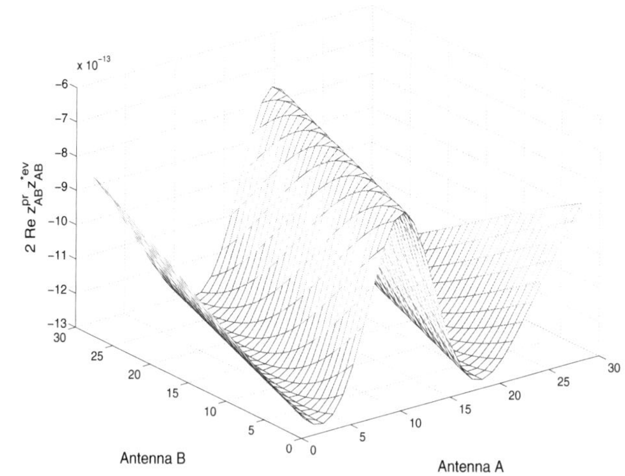

Figure 17.17 The distribution of the interaction exchange energy between the evanescent and propagating parts. The array is the same as described in the caption of Figure 17.15.

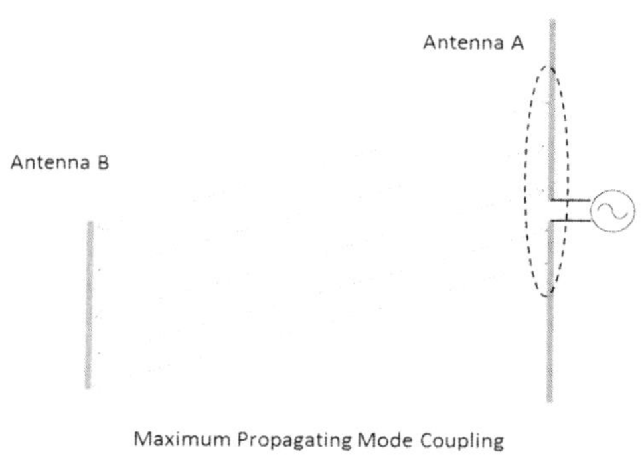

Figure 17.18 A schematic diagram illustrating the parts of the two-wire system with contributing to the maximum propagating and evanescent mode coupling. Here, the separation between the elements is 0.25λ In the left diagram, the direct lines highlight those pairs on the two wires contributing to maximum pure propagating energy transfer. In the right diagram, two complementary sets of pairs provide the distribution of parts contributing to maxima in the pure evanescent mode energy transfer.

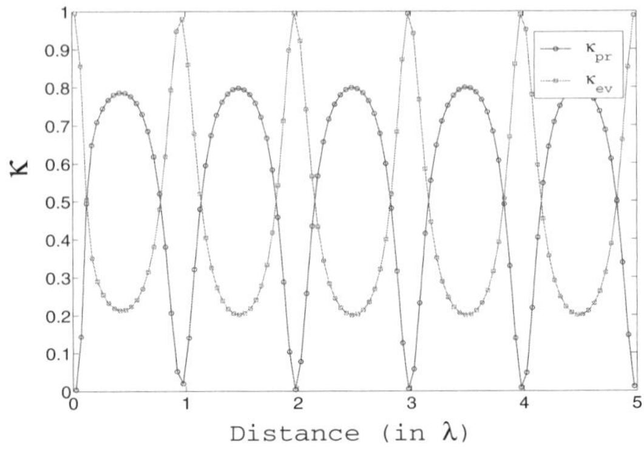

Figure 17.19 The localization coefficients of the two-element array in Figure 17.8. The elements specifications are as in the caption of Figure 18.8. The horizontal axis provides the separation distance between the elements in wavelength. The factors k_{pr} and k_{ev} are defined in (17.36).

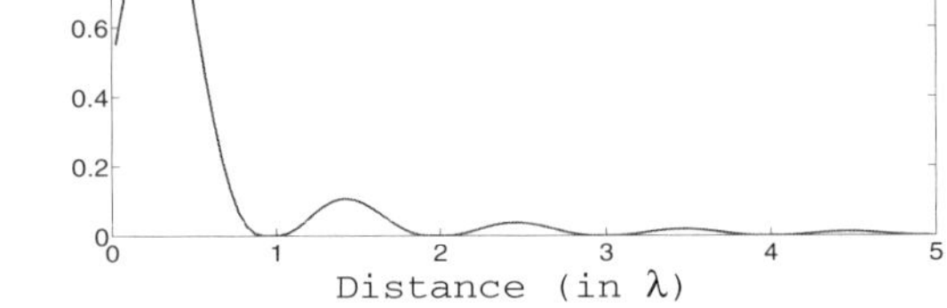

Figure 17.20 The exchange coefficients characterizing the interchange of energy between the propagating and nonpropagating modes for the system described in the caption of Figure 17.19.

$d = n\lambda, n = 0, 2, 3, ...$, the field radiated by the source A tends to get localized. The spatial extension of the localization width defined in terms of half-power level is about 0.392λ and is roughly the same at all locations $d = n\lambda, n = 0, 2, 3, ...$ We remind the reader that although the *total* field is not zero in this region, most of its spectral content is composed of *non*propagating modes. As we move outside the localization region, the propagating part grows from its zero total value and contributes to the moving energy flux before entering into another localization region, reaching a maximum of $\%80$ at the locations $d = (n + 1/2)\lambda, n = 0, 1, 2, ...$ In particular, this *cyclic* process continues indefinitely, regardless to how large the distance from the source can be. We are therefore witnessing a phenomenon that goes beyond mutual coupling between antenna elements close to each other. For example, even in the far zone, the pattern shown in Figure 17.19 still holds. It gives a characterization of energy flow away from the source in terms of series of full localization followed by partial localization of energy, followed by a collapse into complete localization, and so on.

Although the basic results shown here are for wire antennas, the method is general and can be used in conjunction with MoM to study the structure of the radiation field of arbitrary complex antennas.

17.8 POTENTIAL APPLICATIONS

The main findings presented above were formulated as a general methodology in order to reflect both its scope as basic research in antenna theory and also its wide applicability to a large range of practical problems encountered in the fields of radiation and coupling of electromagnetic energy. It is possible, however, to sketch briefly some of the potential uses of the current work in existing device applications without entering into details, which can be addressed by us or others in future publications.

Studies of Bandwidth. It is well known that one of the effective methods to achieve enhanced bandwidth in EM devices is the careful exploitation of mutual coupling. For example, if two narrow resonant peaks are forced to interpenetrate, then under certain circumstances one may obtain a combined resonance exhibiting considerable improvement in bandwidth compared with individual peaks that entered originally into its composition. This important engineering idea can now be analyzed in depth by studying the localization structure of the radiation field. Indeed, if it is known that the fields produced by two individual devices are localized in certain regions, then mutual coupling may considerably affect the nature of this localization by either destroying it or changing its spatial extension. It is not obvious *a priori* what will happen when devices interact given the complexity of the near-field problem. However, the intention in this chapter was to focus directly on the interaction operator of two antennas and to devise methods to understand its structure exactly in terms of localization in the space around the radiating elements. This makes the approach suitable for empirical studies aiming at documenting and characterizing the possible scenarios in which considerable improvement in bandwidth can result from manipulating localization through mutual coupling.

Detection problems. Many problems involve characterizing the electromagnetic signature of unknown objects, for examples early detection of tumors or geophysical detection of buried targets, or even basic radar and weather prediction. Antennas are used in this case for sending and receiving signals. The received signal is usually collected at a single location, the port. This signal is proportional to the total field and may not always reflect the fine variations in the spectral composition of the near field, as in detection of close targets. The methods proposed here can provide new pictures of the field forming additional output variables in the detection algorithm that can be integrated in order to enhance our knowledge of the nature of coupling and interaction with nearby objects. In particular, since we have characterized the interaction operator in terms of its dynamic decomposition into propagating and nonpropagating modes, it is probable that a fixed electromagnetic signature that was not seen when relying on, say, the scattering parameters S_{12}, can be found in the way the interaction operator splits into its spectral components. For example, the total field may not show a resonance indicating the presence of the target, but this resonance can be located in the propagating and nonpropagating parts. Figures 17.5–17.16 provided insight into how new propagating/nonpropagating field resonances and peaks emerges and varies considerably with distance between elements. We suggest that systematic use of the field dynamic decomposition developed above may provide the difficult problems of electromagnetic detection with some new physical degrees of freedom not yet fully recognized and exploited for research.

Energy resources. The study of energy localization is important for some new applications, such as information retrieval through the stored field and energy storage. A localized field can be used for either locating information in a complex energy processing system or to store energy outside the circuit component. Such potentials have not been realized at a commercial level but they represent some fascinating avenues for experimentation and innovation that require a more sophisticated approach going beyond traditional total-field characterizations.

17.9 CONCLUSION

The chapter introduced methods to characterize mutual energy coupling and transfer in interaction between antenna elements. Using a theory of electromagnetic near fields recently introduced by the authors, the basic idea was to express the mutual coupling problem in terms of an interaction operator and then studying the structure of this operator in order to gain some understanding of the physical processes involved. It was found that when formulated in terms of the method of moment (MoM), it is possible to reduce the analysis to the scalar Green's function problem and simplify considerably the computation of the spectral content of the interaction channel. The results obtained show how the coupling operator between two antennas can be judicially decomposed into propagating and nonpropagating parts, providing in the meantime valuable information about the physical distribution of moving/nonmoving energy linkages between the interacting elements. We have shown that detailed maps describing interactions can be given in which the various parts of the mutually coupled antennas exchange energy either via propagating or nonpropagating fields. A study of energy localization was also provided where one antenna was used as a probe coupled with the source. Some results demonstrated interesting localization phenomena in the radiated fields. Potential applications (energy engineering, NF communications, detection algorithms) of the methods proposed here were suggested and related to the various parts of the formulation. The formulation can be readily integrated into existing MoM codes.

Chapter 18

Applications to MIMO and Spatial Diversity Systems

This chapter provides some investigations of spatial-diversity based communications, in particular Multiple Input, Multiple Output (MIMO) systems, which have reemerged in recent years as one of the major backbone infrastructures employed by wireless communication technology. The choice of MIMO system as the topic of the concluding chapter of this book has its own symbolic significance. The system is indeed the quintessential application of spatial concepts to electromagnetism, and for this reason, together with mobile communications, MIMO is generally referred to as *spatial diversity*. Therefore, it is very natural to test how the various theoretical constructs evolved throughout the earlier chapters of this book may all sum up to make a definite contribution to our understanding of this field. Unfortunately, due to limitation of size, it is not possible to give complete treatment of the subject here. Instead, what follow are brief sketches with some concrete applications that are somehow fragmentary in nature though can be organized in a more focused form, which will be done by the authors somewhere else.

In order give some minimal presentation of the MIMO topic, we focus in this chapter on two themes. The first is how to perform a complete electromagnetic derivation of the MIMO channel matrix using the ACGF alone. This however, will only treat a special case that is simple enough yet sufficient for our present purposes. The main goal is to show that using the concept of Green's functions or transfer functions in space it is possible to complete eliminate electromagnetic fields and currents from the final expression of the MIMO channel matrix. All what appears in the end are the ACGFs of the Tx, Rx, and the response functions of any significant scatterer in the propagation environment.

The second theme is the introduction of a new but simpler type of Green's function, the far-field cross-correlation Green's function, which is conceptually in harmony with the ACGF but technically very different. The main intention here is first to give yet another demonstration of the enormous utility of the concept of Green's function in general as tool for clear thinking, and also to supply the reader with new types of applications and design methods that still circle around the major topic of this book, the spatial structure of electromagnetic fields.

18.1 OVERVIEW OF GENERIC MIMO ANTENNA ARRAYS

A generic MIMO system is shown in Figure 18.1 where we present only the electromagnetic hardcore, i.e., the input field excitations at the N_T ports of the transmitting array, the electromagnetic channel, and the N_R receiving antenna elements. The propagation channel consists of free space plus a set of N_s scatterers usually located close to the receiving array. The electromagnetic responsitivity of the scatterer is described by a polarizability tensor $\bar{\alpha}_n, n = 1, 2, ..., N_s$. The fields impinging on the antennas will produce a vector of received signals $y_n, n = 1, 2, ..., N_R$.

The general idea behind MIMO systems is how to exploit the spatial diversity of the environment in order to boost up the channel capacity to carry information without additional constraints on the bandwidth. This can be achieved, for example, by the ability of the receiver to resolve illumination fields coming from different directions. The Tx array may send up to N_T differently encoded data streams. The existence of multipath propagation in the channel, typically caused by the presence of scatterers, usually results in producing several illumination fields impinging on the Rx array through different directions. If the receiver can distinguish between these various illuminations, then differently encoded data can be recovered.

The most difficult step in the analysis and design of MIMO systems is to obtain accurate and realistic models for the coupling between the transmitted signals x_n and the received signals y_n. This information is summarized in the channel transfer matrix H, which is an $N_T \times N_R$ complex array, whose general nmth element H_{nm} gives the coupling between the input port of the nth Tx antenna and the mth Rx antenna's port. The entire performance of MIMO systems depends very crucially on the properties of this matrix H. For instance, the channel capacity, the most popular performance measure, is computed directly as a function of H.

The problem of how to obtain the channel transfer matrix is completely electromagnetic in nature. It involves the energization of a set of antennas in the

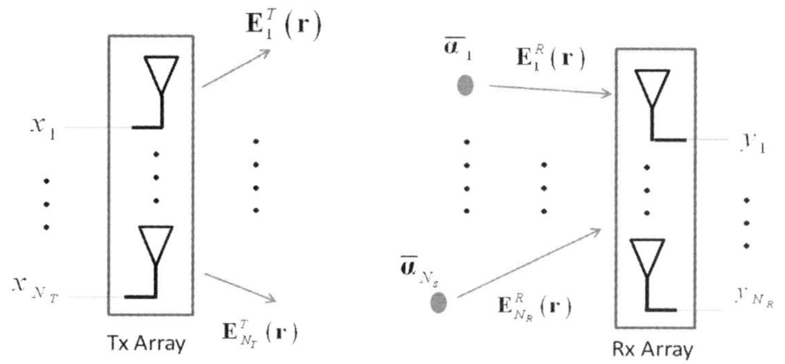

Figure 18.1 Generic scheme for MIMO systems.

transmitting terminal by signals $\mathbf{x}$, the subsequent radiation of these elements into free space, the propagation of the radiated fields through possible scattering objects, and the interaction of the illumination field with the receiving antennas giving rise to the production of the received signals $\mathbf{y}$. The relationship between the two sets of signals is given by

$$\mathbf{y}(\omega) = H(\omega)\,\mathbf{x}(\omega) + \mathbf{n}, \tag{18.1}$$

where $\mathbf{n}$ is a vector of additive noise produced by the antennas, circuits, and propagation environment. Although the channel matrix is obtained by solving a *deterministic* electromagnetic problem, in many cases the actual details of the objects in the propagation path of fields radiated by the Tx terminals are not known in advance, resulting in the need to consider a *random* channel matrix H. Some of the contributions of this work will be the elucidation of a connection between the deterministic electromagnetic problems and the statistical nature of MIMO systems.

Let us now consider more carefully how the MIMO system shown in Figure 18.1 can be modeled using the system approach of linear operator theory. Indeed, all the physical processes occurring in Figure 18.1 involves linear integro-differential operators derived from Maxwell's equations, explaining ultimately why a linear relation as (18.1) was written from the outset. However, it appears to the authors that it has not been sufficiently observed in the literature of MIMO systems that the individual building blocks of Figure 18.1 can be put in terms of relatively independent transfer functions connected in cascade to each other. The real importance of this observation is the fact that by writing the overall MIMO channel matrix H in

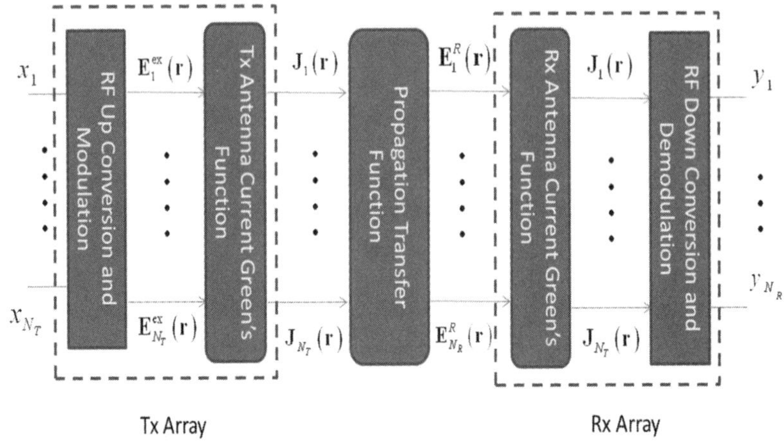

Figure 18.2 Analysis of general MIMO system into a series connection of linear operations.

terms of the appropriate electromagnetic transfer functions, it becomes possible to get a deeper insight into the operational details of the total process in ways that have been obscured hitherto by the focus on network parameters.

In Figure 18.2 we show the full details of the MIMO system from the electromagnetic viewpoint. Start with a set of N_T signals to be transmitted by the Tx array into the Rx array. A typical RF circuit will modulate the signals and perform RF up-conversion as a preparation for sending the information to the radiating antennas. In general, the signal x_n will be converted into some excitation field $\mathbf{E}_n^{\text{ex}}(\mathbf{r})$. This field is commonly assumed to be that of the propagating mode of the waveguide connected to the input port of the nth antenna. However, in this work the excitation field need not be restricted to a waveguide field, but can be any field variation whatsoever. The set of excitations fields $\mathbf{E}_n^{\text{ex}}(\mathbf{r}), n = 1, 2, ..., N_T$, will interact with the N_T Tx antennas and produce current distributions $\mathbf{J}_n^{N_T}(\mathbf{r}), n = 1, 2, ..., N_T$. As was shown in Chapter 8, this process can actually be described in an exact manner using the antenna current Green's function (ACGF) of the array, which plays a role similar to the impulse response in time domain but generalized in this case to deal with the *spatial* problem in three dimensions. The excited current distributions will radiate in the propagation environment, resulting in a set of fields $\mathbf{E}_n^T(\mathbf{r}), n = 1, 2, ..., N_T$ impinging on the N_s scatterers. The scattered and direct fields will be collected in a set of N_R fields $\mathbf{E}_n^R(\mathbf{r}), n = 1, 2, ..., N_R$, impinging on the N_R elements of the Rx system. Again, the process of interaction can be described by the ACGF of the receiving array, resulting in a set of N_R received current values $\mathbf{J}_n^R(\mathbf{r}), n = 1, 2, ..., N_R$ at the output ports. The currents will then be processed to produce the received signals $y_n, n = 1, 2, ..., N_R$.

Few general remarks about this complex process are in order. First, In this chapter, the first and last stages, i.e., RF up/down conversion and modulation/demodualtion are not treated as part of the basic electromagnetic process of MIMO systems. Indeed, we may start at the Tx terminal with the excitations fields $\mathbf{E}_n^{\text{ex}}(\mathbf{r}), n = 1, 2, ..., N_T$ and end at the Rx site with the current sources $\mathbf{J}_n^R(\mathbf{r}), n = 1, 2, ..., N_R$ connected with the proper equivalent Norton impedances. The signal processing and circuitry involved in those two discarded first and last blocks in Figure 18.2 are not directly involved with the spatial diversity aspects of MIMO systems. However, loading at the Rx terminal plays some role and will be taken into consideration. Moreover, we always assume that matching circuits are automatically implemented in the various stages of the system. Second, the transfer function of each block in Figure 18.2 is independent of the others, i.e., we assume as typical in system theory that mutual coupling between systems is ignored. That does *not* imply that mutual couplings effects don't arise inside each block. In fact, we will explicitly take relevant mutual couplings into consideration whenever necessary. *The basic idea will be to group all elements with significant mutual coupling into the same block and treat the resulting ensemble as a collective whole with one independent transfer function.*

18.2 COMPLETE ELECTROMAGNETIC THEORY OF GENERIC MIMO SYSTEMS

In this section, we take into account the general understanding of electromagnetic structures attained through Pat II of this book, where we viewed the electromagnetic phenomenon in terms of system theory. That is, the entire set of microwave and antenna parts is analyzed into cascade connections of various sub-modules, each modeled through its proper "transfer function" or Green's function. The purpose of this section is to provide a complete derivation of the MIMO channel matrix using only those Green's functions and without any explicit reference to fields and currents. The idea is that, within the ACGF framework, those fields and currents are merely "inputs" and "outputs" of some invariant electromagnetic systems (the Tx antenna, the propagation channel, the Rx antenna, scaterers, etc.). Therefore, if interest is focused only on the terminal relationship between transmitted signal and final received signal, which is the ultimate goal in every communication system, then the final result expressing this eventual relationship must be stated in terms of suitable fixed *system functions*. Up to now, the few works addressing the electromagnetic aspects of MIMO systems that can be found in literature appear to rely on models involving fields and currents. To our best knowledge, what follows is a complete derivation of the channel matrix relying completely on the invariant aspects of the electromagnetic structures is attempted. The derivation will provide also a full account of the impact of mutual coupling on the channel matrix.

18.2.1 Model for the Tx Array

In terms of the ACGF, we can write the current distribution produced by the tangential component of a field excitation $\mathbf{E}_t^{\mathrm{ex}}(\mathbf{r}')$ as

$$\mathbf{J}^T(\mathbf{r}) = \int_U ds' \bar{\mathbf{F}}^T(\mathbf{r},\mathbf{r}') \cdot \mathbf{E}_t^{\mathrm{ex}}(\mathbf{r}'), \tag{18.2}$$

where U is the area inside which the tangential component of the excitation field $\mathbf{E}_t^{\mathrm{ex}}(\mathbf{r})$ interacts with the surface of the antenna supporting the electromagnetic boundary condition of the problem.[1] The current distribution on the Tx antenna array is denoted by $\mathbf{J}^T(\mathbf{r})$ while the corresponding current Green's function is $\bar{\mathbf{F}}^T(\mathbf{r},\mathbf{r}')$.

In general, we need N_T equations like (18.2) in order to fully describe the N_T current distributions of the radiating array. However, due to mutual coupling between the elements, an increase in the complexity of the final expressions is inevitable due to cross interaction between the closely spaced antennas. We will follow here the proposal outlined in Chapter 16, which handles mutual coupling effects within the general framework of the ACGF by introducing a *new transfer function of mutual coupling* $\delta\bar{\mathbf{F}}_1^T(\mathbf{r},\mathbf{r}')$ defined as follows

$$\delta\bar{\mathbf{F}}_n^T(\mathbf{r},\mathbf{r}') := \bar{\mathbf{F}}_n^T(\mathbf{r},\mathbf{r}') - \bar{\mathbf{F}}_n^{T,0}(\mathbf{r},\mathbf{r}') = \sum_{m=1}^{N_T} \delta\bar{\mathbf{F}}_{nm}^T(\mathbf{r},\mathbf{r}'). \tag{18.3}$$

Here, $\bar{\mathbf{F}}_n^T(\mathbf{r},\mathbf{r}')$ is the ACGF of the Tx array responsible for generating the current $\mathbf{J}_n(\mathbf{r})$ on the nth element. On the other hand, $\bar{\mathbf{F}}_n^{T,0}(\mathbf{r},\mathbf{r}')$ is the *isolated* ACGF of the nth element, i.e., the nth element's ACGF when *no* other antenna is present. In the latter case (no mutual coupling), an antenna will respond only to the incident

1 The exact shape of this area is determined by the nature and type of the physical port.

field illuminating its own surface, i.e., no direct coupling with the other $N_T - 1$ excitations fields will appear in the isolated case. It is clear that $\delta\bar{\mathbf{F}}^T_{nm}(\mathbf{r},\mathbf{r}')$ is the part of the mutual coupling ACGF responsible of connecting the field excitation at the nth port with the current induced on the mth antenna, in other words, the *cross*-coupling between the elements. In effect, all such coupling or mutual interactions are moved into the mutual coupling transfer function $\delta\bar{\mathbf{F}}^T_n(\mathbf{r},\mathbf{r}')$ defined by (18.3). Therefore, in contrast to $\bar{\mathbf{F}}^{T,0}_n(\mathbf{r},\mathbf{r}')$, $\bar{\mathbf{F}}^T_n(\mathbf{r},\mathbf{r}')$ will accept as input excitations *all* input fields $\mathbf{E}^{\text{ex}}_n(\mathbf{r}')$, $n = 1, 2, ..., N_T$. In fact, we can write

$$\mathbf{J}^T_n(\mathbf{r}) = \int_{S^T_n} ds'\bar{\mathbf{F}}^{T,0}_n(\mathbf{r},\mathbf{r}')\cdot\mathbf{E}^{\text{ex}}_{n,t}(\mathbf{r}') + \sum_{m=1}^{N_T}\int_{S^T_m} ds'\delta\bar{\mathbf{F}}^T_{nm}(\mathbf{r},\mathbf{r}')\cdot\mathbf{E}^{\text{ex}}_{m,t}(\mathbf{r}'), \tag{18.4}$$

where the subscript 't' stands for the tangential component of the field while the superscript 'T' for the transmitting array. The first term in (18.4), on the other hand, involves $\bar{\mathbf{F}}^{T,0}_n(\mathbf{r},\mathbf{r}')$, which is physically explained as the *self*-coupling part of mutual coupling.

Figure 18.3 provides a graphical elucidation of the expressions (18.4) for the internal structure of the nth element of the Tx array as understood in terms of the language of the ACGF formalism. The first term in (18.4) can be interpreted as the *self*-interaction of the nth element with its own tangential component of the excitation field $\mathbf{E}^{\text{ex}}_{n,t}(\mathbf{r}')$. On the other hand, the remaining $N_T - 1$ terms represent the *cross*-interactions received by the nth element from fields exciting the *other* elements of the arrays, namely, $\mathbf{E}^{\text{ex}}_{m,t}(\mathbf{r}')$, $m = 1, 2, ..., N_T, m \neq n$. Each cross-interaction is mediated by the ACGF $\delta\bar{\mathbf{F}}^T_{nm}(\mathbf{r},\mathbf{r}')$, which provides complete information about how the field impinging on the mth element will influence the current induced at the nth element's port. Note that these cross-interaction transfer functions are valid for arbitrary excitation field. In expression (18.4), we notice further that each Tx antenna element can take any smooth geometric shape with enclosing surface S^T_m. Therefore, since no restriction is imposed on either the shape of the antennas or the nature of the excitation field, the expression (18.4) can be considered the most general model for the Tx array in MIMO systems from the electromagnetic viewpoint. All the other elements can be modeled in exactly the same way shown in Figure 18.3. The *total* MIMO Tx array current is then given by

$$\mathbf{J}^T(\mathbf{r}) = \sum_{n=1}^{N_T}\mathbf{J}^T_n(\mathbf{r}). \tag{18.5}$$

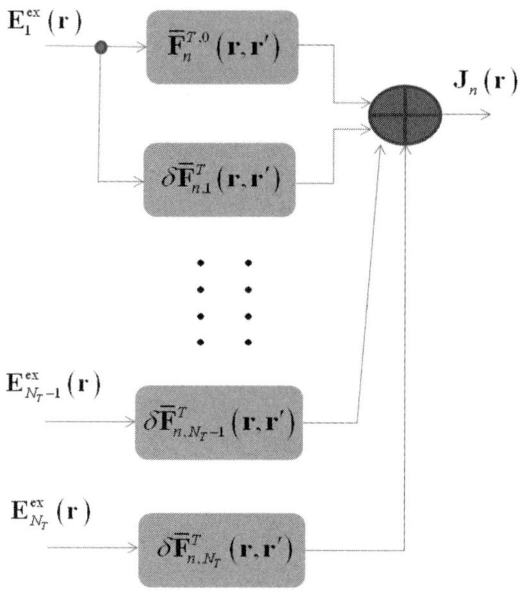

Figure 18.3 A system model for the nth element of the Tx array using the ACGF method.

It should be noted that $\mathbf{J}_n(\mathbf{r}) = 0, \ \forall \mathbf{r} \in S_m^T, m \neq n$. In other words, the nth current has a nonzero compact support only on its own surface S_n. From (18.4) and (18.5) we find

$$\mathbf{J}^T(\mathbf{r}) = \underbrace{\sum_{n=1}^{N_T} \int_{S_n^T} ds' \bar{\mathbf{F}}_n^{T,0}(\mathbf{r}, \mathbf{r}') \cdot \mathbf{E}_{n,t}^{\text{ex}}(\mathbf{r}')}_{\text{Self Interactions}} + \underbrace{\sum_{n=1}^{N_T} \sum_{m=1}^{N_T} \int_{S_m^T} ds' \delta \bar{\mathbf{F}}_{nm}^T(\mathbf{r}, \mathbf{r}') \cdot \mathbf{E}_{m,t}^{\text{ex}}(\mathbf{r}')}_{\text{Mutual Coupling Interactions}}. \tag{18.6}$$

which is the complete expression of the current induced on the MIMO Tx array expressed in terms of input excitation fields and the TX isolated and mutual coupling ACGFs.

18.2.2 Model for the Propagation Channel

The communication channel separating the Tx and Rx arrays was shown in Figure 18.1. It consists of free space plus a set of scatterers. In general the shape and nature of the materials entering into the composition of each scattering object can assume an almost unrestricted range of possibilities, rendering the formulation of the MIMO problem at the most general level impossible in principle. However, in practice we tend to restrict the structural variability of the environment in a manner permitting the construction of a realistic model that can describe the essential physical properties of the systems. Since in a theory of electromagnetic spatial diversity we are mostly concerned with the establishment of multiple paths of propagation, we will assume that the MIMO environment scattering obstacles are *point* scatterers (the exact definition will be given shortly). In this way, each object can interact with the Tx array field and produce a new field that will illumine the Rx array from its own angle. In case larger objects than point scatterers are present, those can be approximated by a set of point scatterers using the ACGF as will be shown later.

Consider the nth point scatterer. The current $\mathbf{J}_n^{\text{Sc}}(\mathbf{r})$ induced on the nth scatterer is obtained by the following relation

$$\mathbf{J}_n^{\text{Sc}}(\mathbf{r}) = \bar{\mathbf{P}}_n \cdot \mathbf{E}^T(\mathbf{r}_n) \delta(\mathbf{r} - \mathbf{r}_n), \tag{18.7}$$

where $\bar{\mathbf{P}}_n$ is the *polarizability tensor* of the nth point scatterer described by a 3×3 matrix for the most general case. The location of the scatterer is $\mathbf{r}_n$ and δ stands for the 3D Dirac delta function. In this expression, the scatterer is treated as a particle with a material tensor $\bar{\mathbf{P}}_n$ specifying how it will respond to an external electric field. Since the scattering object is point-like, only the value of the field at the location of the particle $\mathbf{r}_n$ is pertinent to the electromagnetic model. The fact that the induced current is supported by the point-like particle is indicated by the use of the Dirac delta function to mathematically describe the functional form of the scatterer's current.

The field $\mathbf{E}^T(\mathbf{r})$ radiated by the Tx array is given by the well-known expression

$$\mathbf{E}^T(\mathbf{r}) = \int_{S^T} ds\, \bar{\mathbf{G}}(\mathbf{r}, \mathbf{r}') \cdot \mathbf{J}^T(\mathbf{r}'), \tag{18.8}$$

where $\bar{\mathbf{G}}(\mathbf{r}, \mathbf{r}')$ is the dyadic Green's function of free space. Here, the surface S^T refers to the sum of all surfaces of the individual Tx antenna elements, i.e., $S^T = \bigcup_{n=1}^{N_T} S_n^T$.

The total field received by the Rx array will then consist of two parts: 1) direct field coming from the Tx array, and 2) the field scattered by the N_S objects in the environment. Since each scattering current given by (18.7) will radiate also in the free space environment, the same expression used in computing the direct field of the Tx array (18.8) will also be used to compute the scattered fields. The illumination field impinging on the Rx array can then be written most generally as

$$\mathbf{E}^R(\mathbf{r}) = \underbrace{\int_{S^T} ds\, \bar{\mathbf{G}}(\mathbf{r}, \mathbf{r}') \cdot \mathbf{J}^T(\mathbf{r}')}_{\text{Direct Field}} + \underbrace{\sum_{n=1}^{N_s} \bar{\mathbf{G}}(\mathbf{r}, \mathbf{r}_n) \cdot \left[\bar{\mathbf{P}}_n \cdot \mathbf{E}^T(\mathbf{r}_n)\right]}_{\text{Scattered Field}}. \tag{18.9}$$

18.2.3 Model for the Rx Array

The model of the receiving part of the MIMO system is the reverse of the model of the Tx array. Here, the illumination field found in (18.9) is the exciting field interacting with a set of N_R Rx antennas, leading to the generation of N_R signals at the output ports. Let us describe the ACGF of the nth Rx antenna element by $\bar{\mathbf{F}}_n^R(\mathbf{r}, \mathbf{r}')$ defined on the compact support S_n^R. The total surface of the receiving array is then given by $S^R = \bigcup_{n=1}^{N_R} S_n^R$. The individual (isolated) element ACGF is denoted by $\bar{\mathbf{F}}_n^{R,0}(\mathbf{r}, \mathbf{r}')$. Therefore, the Rx element mutual coupling ACGF is given by

$$\delta\bar{\mathbf{F}}_n^R(\mathbf{r},\mathbf{r}') := \bar{\mathbf{F}}_n^R(\mathbf{r},\mathbf{r}') - \bar{\mathbf{F}}_n^{R,0}(\mathbf{r},\mathbf{r}') = \sum_{m=1}^{N_R} \delta\bar{\mathbf{F}}_{nm}^R(\mathbf{r},\mathbf{r}'). \tag{18.10}$$

The current on nth Rx element at location $\mathbf{r}$ is found through the relation

$$\mathbf{J}_n^R(\mathbf{r}) = \int_{S_n^R} ds'\bar{\mathbf{F}}_n^{R,0}(\mathbf{r},\mathbf{r}')\cdot\mathbf{E}_{n,t}^R(\mathbf{r}') + \sum_{m=1}^{N_T}\int_{S_m^R} ds'\delta\bar{\mathbf{F}}_{nm}^R(\mathbf{r},\mathbf{r}')\cdot\mathbf{E}_{m,t}^R(\mathbf{r}'), \tag{18.11}$$

which is the analog of (18.4). The total current on the Rx array is written as

$$\mathbf{J}^R(\mathbf{r}) = \sum_{n=1}^{N_R}\mathbf{J}_n^R(\mathbf{r}), \tag{18.12}$$

with the proviso $\mathbf{J}_n^R(\mathbf{r}) = 0,\ \forall\,\mathbf{r}\in S_m^R,\ m\neq n$. Therefore, from (18.11) and (18.12) we conclude

$$\begin{aligned}\mathbf{J}^R(\mathbf{r}) = &\underbrace{\sum_{n=1}^{N_R}\int_{S_n^R} ds'\bar{\mathbf{F}}_n^{R,0}(\mathbf{r},\mathbf{r}')\cdot\mathbf{E}_{n,t}^R(\mathbf{r}')}_{\text{Self Interactions}}\\ &+\underbrace{\sum_{n=1}^{N_R}\sum_{m=1}^{N_R}\int_{S_m^R} ds'\delta\bar{\mathbf{F}}_{nm}^R(\mathbf{r},\mathbf{r}')\cdot\mathbf{E}_{m,t}^R(\mathbf{r}')}_{\text{Mutual Coupling Interactions}}.\end{aligned} \tag{18.13}$$

which is the most general relation at the Rx array side.

18.2.4 Derivation of the Channel Transfer Function of General MIMO System

We are in a position now to put together all the results obtained so far regarding the modeling of the individual parts of general MIMO systems in order to derive a complete expression for the channel matrix H. We assume that the propagation channel consists of free space plus point scatters with definite polarizability tensors as in Section 18.2.2. Moreover, in order to simplify the presentation, it is assumed that the input and output ports at the Tx and Rx arrays, respectively, are point-like. The generalization to arbitrary complex ports is straightforward.

An input $N_T \times 1$ vector of information signal $\mathbf{x}$ is fed into the MIMO system by the RF up-conversion stage and converted into a set of excitation fields $\mathbf{E}_n^{ex}, n = 1, 2, .., N_T$ as follows

$$\mathbf{x}(\omega) = \begin{pmatrix} x_1(\omega) \\ x_2(\omega) \\ \cdot \\ \cdot \\ x_{N_T}(\omega) \end{pmatrix} \xrightarrow{\text{RF Up - Conversion}} [\mathbf{E}^{\text{ex}}(\omega)] = \begin{pmatrix} \mathbf{E}_1^{ex}(\omega) \\ \mathbf{E}_2^{ex}(\omega) \\ \cdot \\ \cdot \\ \mathbf{E}_{N_T}^{ex}(\omega) \end{pmatrix} \tag{18.14}$$

The excitation fields in the MIMO system can be expressed as

$$[\mathbf{E}^{\text{ex}}(\omega)] = \begin{pmatrix} x_1(\omega)\,\mathbf{a}_1^{\text{ex}}\delta(\mathbf{r} - \mathbf{r}_1^{\text{ex}}) \\ x_2(\omega)\,\mathbf{a}_2^{\text{ex}}\delta(\mathbf{r} - \mathbf{r}_1^{\text{ex}}) \\ \cdot \\ \cdot \\ x_{N_T}(\omega)\,\mathbf{a}_{N_T}^{\text{ex}}\delta\left(\mathbf{r} - \mathbf{r}_{N_T}^{\text{ex}}\right) \end{pmatrix}, \tag{18.15}$$

where the vectors $\mathbf{a}_n^{ex}(\omega)$ give the components of the external excitations tangential to the Tx antenna surfaces S_n^T. Here, $\mathbf{r}_n^{ex}$ are the positions of the delta-like excitation fields $\mathbf{E}_n^{ex}, n = 1, 2, .., N_T$. These excitation vector fields are in general frequency dependent and arise from the excited mode field associated with the waveguide used to feed the corresponding Tx antenna.

These excitations generate a set of N_T current distributions on the Tx array antennas given by

$$\left[\mathbf{J}^T(\mathbf{r})\right] = \begin{pmatrix} \mathbf{J}_1^T(\mathbf{r}) \\ \mathbf{J}_2^T(\mathbf{r}) \\ \cdot \\ \cdot \\ \mathbf{J}_{N_T}^T(\mathbf{r}) \end{pmatrix} = \begin{pmatrix} x_1(\omega)\,\bar{\mathbf{F}}_1^T(\mathbf{r}, \mathbf{r}_1^{ex}) \cdot \mathbf{a}_1^{ex}(\omega) \\ x_2(\omega)\,\bar{\mathbf{F}}_2^T(\mathbf{r}, \mathbf{r}_2^{ex}) \cdot \mathbf{a}_2^{ex}(\omega) \\ \cdot \\ \cdot \\ x_{N_T}(\omega)\,\bar{\mathbf{F}}_{N_T}^T\left(\mathbf{r}, \mathbf{r}_{N_T}^{ex}\right) \cdot \mathbf{a}_{N_T}^{ex}(\omega) \end{pmatrix}, \tag{18.16}$$

where use was made of (18.2) with the data (18.14).

18.2.5 The Case of No Scattering Objects in the Propagation Environment

For simplicity, and in order to further grasp the electromagnetic structure of the MIMO channel matrix, we first develop the derivation when only homogeneous and isotropic free space exists between the Tx and Rx antennas.

We will try to write down the expression of the mnth term of the channel matrix, i.e., H_{mn}, based on the electromagnetic model outlined above. In order to provide maximal clarity, each contribution to the net transfer between the nth Tx port and the mth Rx port will be spelled out separately in detail. By the end, all contributions will be summed up in order to provide the net outcome H_{mn}.

18.2.5.1 Direct Coupling Path

Here, ANT_n^{Tx} is directly coupled with ANT_m^{Tx}. The no-mutual-coupling ACGF $\bar{\mathbf{F}}_n^{T,0}(\mathbf{r},\mathbf{r}')$ connects the excited signal at the mth port with the received signal in the nth port. The current induced on the nth antenna due to a source $x_n(t)$ on its port with mode field a_n^{ex} is given by

$$\mathbf{J}_n(\mathbf{r}) = \bar{\mathbf{F}}_n^{T,0}\left(\mathbf{r},\mathbf{r}_{\text{n}}^{\text{ex}}\right)\cdot\mathbf{a}_n^{\text{ex}}x_n(\omega)\,. \tag{18.17}$$

This current will give rise to radiated field

$$\mathbf{E}_n^T(\mathbf{r}) = \int_{S_n^T} ds'\,\bar{\mathbf{G}}(\mathbf{r},\mathbf{r}')\cdot\left[\bar{\mathbf{F}}_n^{T,0}\left(\mathbf{r}',\mathbf{r}_{\text{n}}^{\text{ex}}\right)\cdot\mathbf{a}_n^{\text{ex}}x_n(\omega)\right]. \tag{18.18}$$

The signal received by the mth antenna is

$$\mathbf{J}_m^R(\mathbf{r}_m) = \int_{S_m^R} ds'\bar{\mathbf{F}}_m^{R,0}(\mathbf{r}_m,\mathbf{r}')\cdot\int_{S_n^T} ds''\,\bar{\mathbf{G}}(\mathbf{r}',\mathbf{r}'')\cdot\left[\bar{\mathbf{F}}_n^{T,0}\left(\mathbf{r}'',\mathbf{r}_{\text{n}}^{\text{ex}}\right)\cdot\mathbf{a}_n^{\text{ex}}\right]x_n(\omega)\,. \tag{18.19}$$

The receiver circuit will sample this current based on the relation

$$y_m(\omega) = \mathbf{b}_m^{\text{Rx}}\cdot\mathbf{J}_m^R(\mathbf{r}_m)\,, \tag{18.20}$$

where $\mathbf{b}_m^{\text{Rx}}(\omega)$ is the transfer function of the receiving circuit. The final direct contribution to the channel matrix is given by

$$H_{mn}^{0,0} = \int_{S_m^R} ds'\int_{S_n^T} ds''\,\mathbf{b}_m^{\text{Rx}}\cdot\bar{\mathbf{F}}_m^{R,0}(\mathbf{r}_m,\mathbf{r}')\cdot\bar{\mathbf{G}}(\mathbf{r}',\mathbf{r}'')\cdot\bar{\mathbf{F}}_n^{T,0}\left(\mathbf{r}'',\mathbf{r}_{\text{n}}^{\text{ex}}\right)\cdot\mathbf{a}_n^{\text{ex}}. \tag{18.21}$$

Formally, the expression (18.21) explicates the following physical mechanism

$$\mathbf{r}_{\text{n}}^{\text{ex}}\xrightarrow{\text{Excite the nth Tx ANT}}\mathbf{r}''\xrightarrow{\text{Radiation by nth Tx ANT}}\mathbf{r}'\xrightarrow{\text{Reception by mth Rx ANT}}\mathbf{r}_m \tag{18.22}$$

Formula (18.21) provides the MIMO channel matrix when mutual coupling is ignored in both the Tx and Rx arrays.

18.3 CROSS CORRELATION IN ANTENNA SYSTEM: GENERALIZED APPROACH THROUGH GREEN'S FUNCTIONS

Starting from this section and until the end of the chapter, we begin to formulate a new concept for spatial diversity systems, the far-field cross-correlation Green's function, which is independent of the ACGF. Both types of Green's functions share in common the fact that they capture fundamental aspects in the spatial structure of electromagnetic phenomena in general, and the relation between fields and currents in particular.

18.3.1 Introduction

By relying on spatial considerations, an enhancement in the performance of MIMO systems is usually attributed to the so-called *diversity gain* of the system. It is now commonly believed by researchers that among the most important factors determining this diversity gain of MIMO systems there stands the cross correlation between the far fields radiated by the transmitting terminal [151]. The basic idea here is to use the reciprocity theorem that relates the far field of a transmitting antenna to its receiving characteristics. If identical elements are used for both transmit and receive arrays, and the angles of incidence of the illuminating fields are assumed (statistically) to be uniformly distributed, then the degree of correlation in the receiver signals can be directly related to the cross correlation of the far fields radiated by the elements of the transmit array [152]. For this reason, there has been a great interest in finding methods that allow the computation of this important performance parameter. However, since cross correlation involves integration of far field patterns over the entire radiation solid angle, which may require full 3-dimensional measurement of the radiation field for both cross and copolar components, there has been an attempt to reduce the computation of the correlation into measurement of scattering parameters at the terminals of the transmit array. Some expressions relating the far-field cross correlation to the scattering parameters have been derived and used extensively for the design of MIMO system. However, we will show in this chapter that such reduction of correlation to scattering parameters is not possible in principle. In particular, specific basic examples will be given to show how the scattering parameter expressions fail to predict the cross correlation in the case of no-mutual-coupling arrays (e.g. electrically-small antennas, see Section 18.3.5.2), in addition to being unable to deal with scattering objects in the receiver environment.

We present here new consideration of the electromagnetic aspects of spatial diversity and MIMO systems. In particular, we will first devise a strategy to reduce the computation of the far-field cross correlation into expressions involving only the current distribution on the antenna elements. The goal is to provide a method helping in understanding how small current segments and their polarization at the radiating antenna surface contribute directly to the far-field cross correlation. It will be shown here that the cross-correlation expression can be put in a form resembling the method of moment impedance matrix but with the proviso that the free-space Green's function is being replaced here by a new transfer function acting like a "Green's function for the cross correlation." The physical interpretation of this new Green's function turns out to be simply the cross correlation of the far fields radiated by two infinitesimal dipoles, and hence the motivation for the new term 'correlation Green's function' becomes quite natural in this context. It is shown that if this correlation Green's function is computed, then the total cross correlation is obtained merely by straightforward superposition as in the method of moment (MoM) or the radiation integrals of electromagnetics. The general expression, incidently, proves that cross correlation depends on the *entire* current on the antenna surfaces, and consequently *it is not possible in general to reduce the calculation of cross correlation into measurement of scattering parameters since the latter are relevant only to the location of antenna current at the physical ports.*

The following investigations are organized in the following manner. Section 18.3.2 provides the derivation of the fundamental relations in this work. In order to bring the far-field cross correlation to the design level, we express the total cross correlation coefficient in terms of the source current distributions on the two antennas. It is found that one can completely characterize correlation in the far zone in terms of current segments at the antennas themselves only if a new dyadic function is introduced, namely the *cross-correlation Green's function*. In particular, all angular integrations are moved into this cross correlation function while the contribution of each current segment depends just on its position and direction. This permits us to focus on the *design* aspects of MIMO systems, for example, how the shape and location of each part of the antennas contribute to the total diversity gain of the system. In this way, it is possible to understand how modifications in the antenna shape, position, and orientation should be introduced in order to manipulate the MIMO system performance.

Section 18.3.3 supplies an in-depth analysis of the results obtained in Section 18.3.2. In Section 18.3.3.1, we outline briefly the salient features of the new cross-correlation Green's function by comparing it with the familiar free-space Green's function. Both similarities and differences between the two fundamental quantities are emphasized in order to gain some understanding of the physical processes involved in spatial diversity systems. Since the cross-correlation Green's function is the value of cross correlation (in the far zone) between two *infinitesimal* dipoles, and given the fact that arbitrary current distributions can be expanded as sum of infinitesimal dipoles, the new correlation Green's function is expected to play a fundamental role in the electromagnetic analysis and design of MIMO systems. Section 18.3.3.2 builds on the observations of Section 18.3.3.1 in order to develop a comparative scheme between the classic method of moment (MoM) in applied electromagnetics and the problem of cross correlation in spatial diversity applications. It will be shown that the structure of the MoM impedance matrix element is identical to the cross-correlation expression derived in Section 18.3.2 provided we replace the free-space Green's function by the new cross-correlation Green's function. This observation can then help building algorithms for processing cross correlation in arbitrary antenna arrays using the already existing numerical infrastructure of standard MoM routines. It will be shown that if the new cross-correlation Green's function is computed in a separate routine, then the rest of the evaluation of the total cross correlation for arbitrary antenna elements can be completed in a way formally identical to MoM. Moreover, certain peculiarities in the cross correlation algorithm make the latter more efficient that the MoM procedure, and such advantages are mentioned and discussed here.

A critical review of the now common approach to cross correlation through S-parameters is given in Section 18.3.4. It will be shown that three major difficulties known to the authors render the use of correlation coefficients based solely on S-parameters of questionable value. The divergence between the exact value of the cross correlation and its estimation using port measurements is mentioned and discussed, though no in-depth investigation of the errors in the derivation of the S-parameters formula found in literature will be given here. The expression of cross correlation based on the currents derived in Section 18.3.2 will be shown later to agree exactly with the far-field-based computation and to be free of all the difficulties associated with the procedure based on measurements at the array ports. It seems to the authors that reliance on port measurements is not the best strategy in the design process of MIMO antennas since it provides only information about the currents at the ports, while our analysis has shown that cross correlation depends significantly on the position and polarization of *all* current segments in the antenna. Moreover, even in lossless antennas, the S-parameters formula actually produces wrong numerical results in many important cases, which strongly suggests that this formula is not exact but merely an approximation of cross correlation in general.

Section 18.3.5 supplies a series of numerical studies and design examples in order to verify the proposed theory and outline some of its potential applications in spatial diversity and MIMO systems. We start in Section 18.3.5.1 by verifying the form of the cross-correlation Green's function by direct comparison with the exact expression in terms of the far field. This will show that the formula of cross correlation in terms of the current distributions is valid for arbitrary antennas provided that the size of these antennas is finite. The case of infinite antenna source, i.e, antenna systems where the radiating source extends to infinity, requires special investigation that is beyond the scope of this discussion. However, all typical MIMO systems involve a finite number of antennas each with finite size, rendering the expressions derived below together with their validation sufficient for most practical purposes. In Section 18.3.5.2, we give numerical studies of the cross correlation between infinitesimal dipoles with variable distances and polarizations. Some general observations that help understand and guide the design process for more complicated arrays and/or antennas will be clearly spelled out throughout the discussion. In order to assess the status of our knowledge following the basic theory outlined above, we provide in Section 18.3.6 a design methodology to demonstrate how the derived expressions for cross correlation in terms of the antenna currents can be used in concrete cases to find the optimum locations of the radiating elements capable of producing a set of radiation patterns with minimum total cross correlation (maximum diversity gain). The methodology will be illustrated by applying it to linear and circular arrays with a few small dipoles but is applicable to arbitrary antenna arrays. In particular, we demonstrate the tradeoff existing between the density of the array and its diversity gain, which motivates the need to do a proper design for antenna arrays if the size of the system is to be reduced. Due to the limitations of space, only a simple random search method will be developed here in order to illustrate the physical and engineering aspects of antenna spatial diversity design problems. Finally, Section 18.3.7 will show that the computational and design techniques applied in Sections Section 18.3.5.2 and 18.3.6 for infinitesimal dipoles can be extended to arbitrary current distributions. The idea is to exploit the observation made earlier amounting to the fact that the new cross-correlation Green's function is nonsingular. This allows us to apply classic quadrature rules to convert the integrations in the cross-correlation formula derived in Section 18.3.2 into a sum of small number of correlations involving a set of equivalent infinitesimal dipoles. The procedure will be demonstrated by replacing the current on a dipole antenna by such set of infinitesimal dipoles and the results are numerically verified for arbitrary currents. This concludes the study with one of our main observations, namely that the computation and design of cross correlation in arbitrary array systems can be reduced to the problem of correlation between infinitesimal dipoles, and hence follows the fundamental importance of the cross-correlation Green's function introduced here.

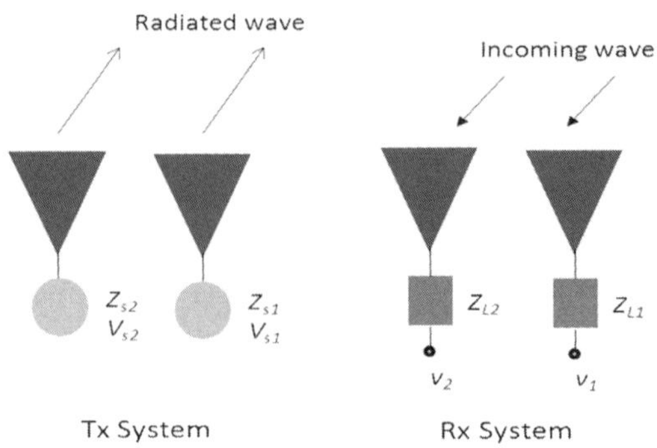

Figure 18.4 Tx array (left) and Rx array (right) in a generic 2×2 MIMO system.

18.3.2 Derivation of the Mutual Correlation Expression in Terms of the Antenna Currents

For a basic MIMO configuration consider a 2-element Tx and Rx array as shown in Figure 18.4. The generalization to arbitrary number of antenna elements is straightforward but will not be presented here for simplicity. When the same antenna type is used for both terminals, it is possible to apply reciprocity theorems to relate the receiving characteristics of the system to the far field radiation pattern of the transmitting terminal. Let the radiation patterns of antennas 1 and 2 be $\mathbf{E}_1(\theta,\varphi)$ and $\mathbf{E}_2(\theta,\varphi)$, respectively. The envelope cross correlation between the two antennas is defined as

$$\rho = \frac{\left|\int_{4\pi} d\Omega\, \mathbf{E}_1(\theta,\varphi)\cdot\mathbf{E}_2^*(\theta,\varphi)\right|}{\sqrt{\int_{4\pi} d\Omega\, \left|\mathbf{E}_1(\theta,\varphi)\right|^2}\sqrt{\int_{4\pi} d\Omega\, \left|\mathbf{E}_2(\theta,\varphi)\right|^2}}, \tag{18.23}$$

where the integration is with respect to the full solid angle Ω, $'*'$ denotes the complex conjugate operator, and $\left|\mathbf{E}_{1,2}(\theta,\varphi)\right|^2 := \mathbf{E}_{1,2}(\theta,\varphi)\cdot\mathbf{E}_{1,2}^*(\theta,\varphi)$.

When the two Tx array antennas are connected to their respective sources, current distributions $\mathbf{J}_1(\mathbf{r})$ and $\mathbf{J}_2(\mathbf{r})$ will be generated on the two antennas 1 and 2, respectively. Our goal in this section is to express the correlation coefficients (18.23) in terms of the currents $\mathbf{J}_1(\mathbf{r})$ and $\mathbf{J}_2(\mathbf{r})$ only. This will bring the spatial diversity problem into the design level available to the engineer at the Tx terminal side. Here, we assume that the volume (or surface) on which the two currents $\mathbf{J}_1(\mathbf{r})$ and $\mathbf{J}_2(\mathbf{r})$ are nonzero are bounded subregions of the entire space $\mathbb{R}^3$. A time-harmonic excitation $\exp(-i\omega t)$ is assumed and suppressed everywhere. The antennas are also assumed to be radiating in infinite isotropic and homogeneous space with magnetic permeability μ and dielectric permittivity ϵ.

We first compute the magnetic vector potential due to each antenna alone using the well-known radiation formula [37]

$$\mathbf{A}_{1,2}(\mathbf{r}) = \frac{\mu}{4\pi}\int_{V_{1,2}} d^3r\mathbf{J}_{1,2}(\mathbf{r}')\frac{e^{ik|\mathbf{r}-\mathbf{r}'|}}{|\mathbf{r}-\mathbf{r}'|}, \tag{18.24}$$

where $\mathbf{A}_{1,2}(\mathbf{r})$ are the vector potentials due to sources on antennas 1,2, with current supports on the volumes $V_{1,2}$. The wavenumber k is given by $k = \omega\sqrt{\mu\epsilon}$. In the far zone, the integrals (18.24) are replaced by the Fourier transform expressions [27]

$$\mathbf{A}_{1,2}(r,\hat{\mathbf{r}}) = \frac{\mu}{4\pi}\frac{e^{ikr}}{r}\int_{V_{1,2}} d^3r\mathbf{J}_{1,2}(\mathbf{r}')e^{ik\mathbf{r}'\cdot\hat{\mathbf{r}}}, \tag{18.25}$$

where $\hat{\mathbf{r}}(\theta,\varphi) := \hat{x}\sin\theta\cos\varphi + \hat{y}\sin\theta\sin\varphi + \hat{z}\cos\theta$ is the unit vector $\mathbf{r}/\|\mathbf{r}\|$. Since the spherical wave $e^{ik|\mathbf{r}-\mathbf{r}'|}\big/r$ does not depend on the observation angle, we follow the custom by referring to the *radiation pattern functions* defined as

$$\mathbf{A}_{1,2}(\hat{\mathbf{r}}) := (i\omega)^{-1}\int_{V_{1,2}} d^3r\mathbf{J}_{1,2}(\mathbf{r}')e^{ik\mathbf{r}'\cdot\hat{\mathbf{r}}}, \tag{18.26}$$

which is now a function of only the angles θ and φ. In the far zone, it is possible to express the electric field in terms of the vector potentials using the following relations [27]

$$\mathbf{E}_{1,2}(\hat{\mathbf{r}}) = i\omega\left[\mathbf{A}_{1,2}(\hat{\mathbf{r}}) - (\mathbf{A}_{1,2}(\hat{\mathbf{r}})\cdot\hat{r})\hat{r}\right]. \tag{18.27}$$

From (18.26) and (18.27) we find

$$\mathbf{E}_{1,2}(\hat{\mathbf{r}}) = \int_{V_{1,2}} d^3r\left[\mathbf{J}_{1,2}(\mathbf{r}') - (\mathbf{J}_{1,2}(\mathbf{r}')\cdot\hat{r})\hat{r}\right]e^{ik\mathbf{r}'\cdot\hat{\mathbf{r}}}. \tag{18.28}$$

It will be important for the remaining part of the derivation to extract the current distribution $\mathbf{J}_{1,2}$ from the bracket in the integrand of (18.27). To achieve this, we use the dyadic calculus and write

$$\mathbf{E}_{1,2}(\hat{\mathbf{r}}) = \int_{V_{1,2}} d^3r'\mathbf{J}_{1,2}(\mathbf{r}')\cdot\left[\bar{\mathbf{I}} - \hat{r}\hat{r}\right]e^{ik\mathbf{r}'\cdot\hat{\mathbf{r}}}, \tag{18.29}$$

where $\bar{\mathbf{I}}$ is the unit dyad. Indeed, one can verify by direct calculation that $\mathbf{J}_{1,2}(\mathbf{r}') \cdot [\bar{\mathbf{I}} - \hat{r}\hat{r}] = [\bar{\mathbf{I}} - \hat{r}\hat{r}] \cdot \mathbf{J}_{1,2}(\mathbf{r}') = \mathbf{J}_{1,2}(\mathbf{r}') - (\mathbf{J}_{1,2}(\mathbf{r}') \cdot \hat{r})\,\hat{r}$.

Using (18.29), we write the numerator of the cross correlation definition (18.23) in the following way

$$\begin{aligned} &\int_{4\pi} d\Omega\, \mathbf{E}_1(\hat{\mathbf{r}}) \cdot \mathbf{E}_2^*(\hat{\mathbf{r}}) \\ &\quad= \int_{4\pi} d\Omega \left[\int_{V_1} d^3r'\, \mathbf{J}_1(\mathbf{r}') \cdot [\bar{\mathbf{I}} - \hat{r}\hat{r}]\, e^{ik\mathbf{r}'\cdot\hat{\mathbf{r}}} \right. \\ &\quad\times \cdot \left. \int_{V_2} d^3r'' \,[\bar{\mathbf{I}} - \hat{r}\hat{r}] \cdot \mathbf{J}_2^*(\mathbf{r}'')\, e^{-ik\mathbf{r}''\cdot\hat{\mathbf{r}}} \right]. \end{aligned} \tag{18.30}$$

The next step relies on the observation that all the integrals in (18.30) are *finite* integrals, which applies only to antennas with *bounded* source regions (basically, all antennas of practical interest satisfy this condition). Therefore, we can interchange the order of integrations (e.g., see [90] for background in analysis) in (18.30) and arrive at the following form

$$\begin{aligned} &\int_{4\pi} d\Omega\, \mathbf{E}_1(\hat{\mathbf{r}}) \cdot \mathbf{E}_2^*(\hat{\mathbf{r}}) \\ &= \int_{V_1} d^3r' \int_{V_2} d^3r'' \\ &\times \mathbf{J}_1(\mathbf{r}') \cdot \left\{ \int_{4\pi} d\Omega\, [\bar{\mathbf{I}} - \hat{r}\hat{r}]\, e^{ik(\mathbf{r}'-\mathbf{r}'')\cdot\hat{\mathbf{r}}} \right\} \cdot \mathbf{J}_2^*(\mathbf{r}''). \end{aligned} \tag{18.31}$$

In deriving (18.31), we have used the identity $[\bar{\mathbf{I}} - \hat{r}\hat{r}] \cdot [\bar{\mathbf{I}} - \hat{r}\hat{r}] = [\bar{\mathbf{I}} - \hat{r}\hat{r}]$, which can be verified by expanding the multiplication.

Consequently, we have been able to carry the reduction of the complicated expression (18.23) for far-field cross correlation into the following form

$$\begin{aligned} &\int_{4\pi} d\Omega\, \mathbf{E}_1(\hat{\mathbf{r}}) \cdot \mathbf{E}_2^*(\hat{\mathbf{r}}) \\ &= \int_{V_1} d^3r' \int_{V_2} d^3r''\, \mathbf{J}_1(\mathbf{r}') \cdot \bar{\mathbf{C}}(\mathbf{r}', \mathbf{r}'') \cdot \mathbf{J}_2^*(\mathbf{r}''), \end{aligned} \tag{18.32}$$

where the function $\bar{\mathbf{C}}(\mathbf{r}', \mathbf{r}'')$ is defined as

$$\bar{\mathbf{C}}(\mathbf{r}', \mathbf{r}'') := \int_{4\pi} d\Omega\, [\bar{\mathbf{I}} - \hat{r}\hat{r}]\, e^{ik(\mathbf{r}'-\mathbf{r}'')\cdot\hat{\mathbf{r}}}. \tag{18.33}$$

The quantity $\bar{\mathbf{C}}(\mathbf{r}', \mathbf{r}'')$ will be called here *cross-correlation Green's function*. The reason for this terminology will be shown later (Section 18.3.5.1) when it is found that this function is the cross correlation of the far fields radiated by two infinitesimal dipoles placed at locations $\mathbf{r}'$ and $\mathbf{r}''$.

We can use the general formula (18.32) to compute the quantities appearing in the denominator of the envelope cross correlation (18.23) as follows

$$\begin{aligned}&\int_{4\pi} d\Omega \ \left|\mathbf{E}_1\left(\hat{\mathbf{r}}\right)\right|^2 \\ &\quad = \int_{V_1} d^3r' \int_{V_2} d^3r'' \ \mathbf{J}_1\left(\mathbf{r}'\right) \cdot \bar{\mathbf{C}}\left(\mathbf{r}', \mathbf{r}''\right) \ \cdot \mathbf{J}_1^*\left(\mathbf{r}''\right),\end{aligned} \tag{18.34}$$

$$\begin{aligned}&\int_{4\pi} d\Omega \ \left|\mathbf{E}_2\left(\hat{\mathbf{r}}\right)\right|^2 \\ &\quad = \int_{V_1} d^3r' \int_{V_2} d^3r'' \ \mathbf{J}_2\left(\mathbf{r}'\right) \cdot \bar{\mathbf{C}}\left(\mathbf{r}', \mathbf{r}''\right) \ \cdot \mathbf{J}_2^*\left(\mathbf{r}''\right).\end{aligned} \tag{18.35}$$

Therefore, the expressions (18.32), (18.33), (18.34), and (18.35) provide an exact evaluation of the envelop cross correlation (18.23) of an arbitrary pair of antennas expressed in terms of the current distributions on the antennas.

18.3.3 General Remarks on the Results

18.3.3.1 Comparison Between the Free-Space Green's Function and the cross-correlation Green's Function

As will become increasingly apparent throughout the rest of this chapter, the new Green's function (18.33) is the most fundamental electromagnetic quantity connecting the performance of the array in spatial diversity and MIMO systems with the underlying physical distribution of sources in the array elements' region. Therefore, it is worth spending some time on comparing the cross-correlation Green's function with the classic free-space Green's function in electromagnetic theory. This will turn out to be specially important when we show in Section 18.3.3.2 that cross correlation can be computed using the the basic scheme of the method of moment (MoM).

The classical dyadic Green's function in electromagnetic theory is given by

$$\bar{\mathbf{G}}\left(\mathbf{r}, \mathbf{r}'\right) = \left[\bar{\mathbf{I}} + \frac{1}{k^2}\nabla\nabla\cdot\right] \frac{e^{ik\left|\mathbf{r}-\mathbf{r}'\right|}}{4\pi\left|\mathbf{r}-\mathbf{r}'\right|}. \tag{18.36}$$

By comparing (18.36) with the cross-correlation Green's function (18.33), the following observations are immediate:

- Both the correlation and free-space Green's functions are *dyadic* functions, i.e., they are constituted by 3×3 (spatial) matrices. However, the dyadic structure of the correlation matrix has the simple form $\bar{\mathbf{I}} - \hat{r}\hat{r}$ while the free-space Green's function has the form $\bar{\mathbf{I}} - \nabla\nabla/k^2$, the latter being more complex since it involves differentiation with respect to all spatial coordinates.

- The free-space Green's function is highly singular. In contrast, the correlation Green's function is regular. In fact, it can be shown to be smooth.
- The free-space Green's function is given in analytic form. The correlation Green's function is up to now known to involve a numerical integration. But the integration is finite and the integrand is smooth, making its computation using standard numerical methods a straightforward task. The detailed expressions of these integrals are given in the Appendix. However, it is possible to evaluate the cross-correlation Green's function in analytical form. Since the derivation is lengthy, it is not provided here but will be addressed elsewhere.
- The free-space Green's function and the correlation Green's function are both *shift invariant*, i.e., they both depend only on the distance between the two source points. In other words, $\bar{\mathbf{C}}\left(\mathbf{r}', \mathbf{r}''\right) = \bar{\mathbf{C}}\left(\mathbf{r}' - \mathbf{r}''\right)$.

Fundamentally speaking, the existence of a cross-correlation Green's function, as in the case of any Green's function, finds its origin in the superposition principle of electromagnetics. Indeed, the rigorous relation (18.32) tells us that total far-field cross correlation between two current distributions is the sum of the mutual interactions between all *infinitesimal* sources entering into the constitution of the collective currents on the two antennas under consideration. This intuitive picture will be further elaborated in Section 18.3.5 when we verify the derivation of (18.32).

18.3.3.2 Comparison With the Method of Moment

The expression of cross correlation (18.32) has a striking similarity to the method of moment impedance matrix. Indeed, in replacing the cross-correlation Green's function $\bar{\mathbf{C}}\left(\mathbf{r}', \mathbf{r}''\right)$ by the free-space Green's function $\bar{\mathbf{G}}\left(\mathbf{r}', \mathbf{r}''\right)$ we formally obtain the general MoM impedance matrix expression for 3-dimensional electromagnetic problems solved by surface integral equations [23]. In order to see this, it suffices to replace the volumes V_1 and V_2 in (18.32) by the antenna surfaces S_1 and S_2 and compare the form of the integrals in both cases. Note that in MoM, the position $\mathbf{r}'$ is called the 'source' while $\mathbf{r}''$ is 'observation'. In the cross-correlation problem both points are due to sources, but in essence the computation is the same.

The new method proposed here to compute the MIMO system envelope cross correlation (18.23) can be directly utilized in existing MoM codes through the following proposed procedure

1. Write a subroutine to compute the cross-correlation Green's function $\bar{\mathbf{C}}(\mathbf{r}', \mathbf{r}'')$ for any two positions $\mathbf{r}'$ and $\mathbf{r}''$. This will require simple numerical integration using the expression (18.33).

2. Expand the currents on the two antennas using a suitable basis functions. That is, write

$$\mathbf{J}_{1,2}(\mathbf{r}') = \sum_{n=1}^{N_{1,2}} I_n^{1,2}\, \mathbf{f}_n^{1,2}(\mathbf{r}'), \tag{18.37}$$

where $\mathbf{f}_n$ stands for the current basis functions (real), e.g., RWG basis functions. The numbers of the MoM basis function on antennas 1 and 2 and the current excitation (amplitudes) are given by $N_{1,2}$ and $I_n^{1,2}$ (complex), respectively.

3. Use the expression (18.32) to write the far-field cross correlation in the form

$$\int_{4\pi} d\Omega\, \mathbf{E}_1(\hat{\mathbf{r}}) \cdot \mathbf{E}_2^*(\hat{\mathbf{r}}) = \sum_{m=1}^{N_1} \sum_{n=1}^{N_2} I_n^1\, I_n^{2*} \rho_{mn}, \tag{18.38}$$

where

$$\rho_{mn} := \int_{S_m} ds' \int_{S_n} ds''\, \mathbf{f}_m^1(\mathbf{r}') \cdot \bar{\mathbf{C}}(\mathbf{r}, \mathbf{r}') \cdot \mathbf{f}_n^2(\mathbf{r}''). \tag{18.39}$$

The array of real numbers ρ_{mn}, $(m = 1, 2, .., N_1; n = 1, 2, .., N_2)$, form an $N_1 \times N_2$ matrix, which we call the *source correlation matrix*. The physical interpretation of its elements is that each number ρ_{mn} provides the net cross correlation between the far field radiated by mth current element $\mathbf{f}_n^1$ on antenna 1 and the far field radiated by the nth current element $\mathbf{f}_n^2$ on antenna 2.

4. The mnth element in the MoM impedance matrix, however, can be put in the following form (missing conventional scaling factor) [23]

$$z_{mn} = \int_{S_m} ds' \int_{S_n} ds''\, \mathbf{f}_m^1(\mathbf{r}') \cdot \bar{\mathbf{G}}(\mathbf{r}, \mathbf{r}') \cdot \mathbf{f}_n^2(\mathbf{r}''). \tag{18.40}$$

Comparing (18.40) with (18.38), we find that the cross correlation of the far fields is equal to the sum of matrix elements ρ_{mn} having exactly the same formal structure of the MoM impedance matrix elements z_{mn} except that the free-space

Green's function (18.36) in the MoM expression must be replaced by the cross-correlation Green's function (18.33).

Regarding the numerical efficiency of this procedure, we notice the following

- The computation of the total cross correlation using the MoM scheme proposed above does *not* involve inverting the correlation matrix ρ_{mn}. Indeed, as the relation (18.38) shows, the total correlation is obtained by summing the elements of the matrix, a very simple numerical operation compared with typical MoM matrix inversion.
- In contrast to the classical MoM, computations of the correlation matrix elements via (18.39) don't involve any singularity treatment when the distance $|\mathbf{r}' - \mathbf{r}''|$ is small. This follows at once from the observation we made in Section 18.3.3.1 to the effect that the correlation Green's function (18.33) is nonsingular for all values of its arguments. Therefore, special subroutines to deal with singular and near singular integrations found in the literature of MoM are not needed here.
- The non-existence of differential operators in the expression of correlation Green's function (18.33) reduces down the cost of developing the computational scheme proper for the integrals (18.39). This is in contrast to typical MoM procedures based on the impedance matrix (18.40), which involve the free-space dyadic Green's function (18.36) containing double differentiation.

Finally, we mention that the method developed here for computing the total cross correlation using current elements defined by basis function as in (18.37) introduces a strategy for designing MIMO arrays by experimenting with the locations and strengths of various small current elements and observing how cross correlation varies according to the shapes of the antennas, levels and types of relative excitations, and mutual coupling. This not only provide a deeper insight into the electromagnetic infrastructure of MIMO systems, but also can help in actually automating the design process itself by using optimization methods. We will supply some basic numerical studies and design examples illustrating this proposal in Section 18.3.5.

18.3.4 Comparison with the S-Parameter Methods

In this section, we provide a few remarks about how the formula of cross correlation in terms of the cross-correlation Green's function compares with some recent methods that rely on using port measurement (S-parameters) to estimate far-field envelope correlation. The next few paragraphs don't attempt a comprehensive survey of the topic or a complete criticism, but aim only at pointing out some occasionally encountered difficulties with the methods based on port measurements together with some general comments.

It is clear from the expression (18.32) that the envelope cross correlation between the far fields of two arbitrary antennas depends on the full current of both antennas, i.e., it involves the values of the currents $\mathbf{J}_1(\mathbf{r})$ and $\mathbf{J}_2(\mathbf{r})$ for all points $\mathbf{r}_1$ and $\mathbf{r}_2$ on the antennas' volumes (or surfaces) V_1 (or S_1) and V_2 (or S_2), respectively. The S parameters of an antenna array, however, depend only on the currents observed on the antenna at the locations of the physical ports [26], [29], [37]. This observation suggests that *the cross correlation between two antennas in general cannot be reduced to measurement or calculation of S parameters*. In fact, it has been known since 1962 that there exists no one-to-one correspondence between the correlation matrix of the far fields produced by an array and the scattering matrix of this array. This analysis of the problem can be found in [151] and is based on conservation of energy and linear algebra.

Recently, it was proposed that an exact relation exists for lossless antennas in which the envelope cross correlation between two antennas can be determined uniquely through the S-parameters through the relation [128]

$$\rho \stackrel{?}{=} \frac{|S_{11}^* S_{12} + S_{21}^* S_{22}|}{\sqrt{1 - |S_{11}|^2 - |S_{21}|^2}\sqrt{1 - |S_{22}|^2 - |S_{12}|^2}}. \tag{18.41}$$

(The square roots in the denominator of (18.41) are introduced to match the definition (18.23).) Relation (18.41) has become popular because it replaces the computationally difficult evaluation of the envelope cross correlation using the original definition (18.23) by the standard measurement procedure of input parameters at the array terminals. However, the numerical inadequacy of expression (18.41) for general array configurations has been frequently observed by comparing its calculations with (18.23). In particular, it was pointed out in [129] that the relation (18.41) is not valid when the antenna array is lossy. However, it was not sufficiently noticed that the derivation of (18.41) is in fact incomplete. As we will see here, it is not only that the expression (18.41) is not valid for lossy antennas, but in fact the equation is not correct even for lossless antennas. Counter examples to (18.41) will be given in Section 18.3.5.

To the best of our knowledge, there are three main problems with the S-parameters approach to the estimation of cross correlation:

1. For antennas with no port coupling, i.e., antennas with $S_{12} = 0$, the formula (18.41) produces completely wrong results. Details will be given in Section 18.3.5.

2. When there is mutual coupling between two excited antennas but an electromagnetic coupling exists with a *third* object in the surrounding environment, e.g., a scatterer or another antenna, the S-parameters formula (18.41) is ambiguous and in general gives wrong results.

3. When port mutual coupling is not zero but weak, i.e., for small values of S_{12}, the formula (18.41) is found to be sometimes inaccurate.

Regarding point 1, the formula (18.41) predicts that cross correlation is exactly zero when $S_{12} = 0$. In Section 18.3.5 we provide numerical and theoretical evidence showing that non-mutually coupled antennas are in general correlated in the far zone. In addition to these examples, it is possible to realize an approximation of infinitesimal dipoles as very short dipoles and construct an implementation in typical EM solvers. Such constructed examples can be considered as belonging to the genera of electrically small antennas, which are very popular and widely used. By computing S_{12} in these systems, the value obtained is not exactly zero, but very small and can be safely neglected. The prediction of the envelope cross correlation in this case agrees with our formula and the original definition, while the S-parameter formula gives the wrong result.

Point 3 is somehow related to point 1 though the connection is not very clear. It seems from various numerical experiments that in the absence of scatterers, strongly coupled antennas show cross correlation pattern that is well approximated by the S-parameters formula (18.41). However, this agreement seems rather accidental because when the level of S_{12} decreases, examples of lossless antennas can be found where there is divergence between the original definition (18.23) and (18.41). Since (18.41) is believed to be *exact* for lossless antennas, this divergence in the results for some cases strongly suggests that the S-parameters formula is not in fact valid for arbitrary lossless arrays.

Point 2 seems to be still more interesting. Consider Figure 18.5 where we show a generic problem consisting of two antennas and a third object (scatterer or another antenna). Two scenarios are depicted at the top and bottom parts. In the top diagram, we are interested in measuring the cross correlation between antennas 1 and 2 using S-parameters. Assume that the three objects are mutually coupled, i.e., the measured S-parameters take into account a) mutual coupling between antennas 1 and 2, and b) the mutual coupling between antenna 1 and the object, and between antenna 2 and the same object. Next, consider the configuration in the

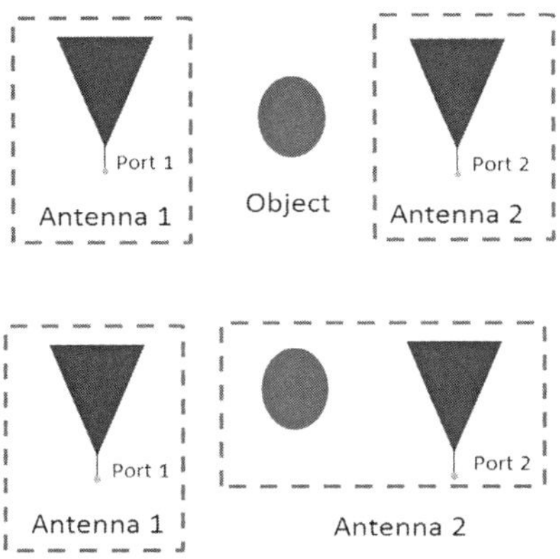

Figure 18.5 *Top*. Generic two-antenna system with a third object in the radiation environment. *Bottom*. The same problem in the top diagram but with the scattering object included as part of antenna 2. In both cases, strong mutual coupling between the antennas and objects is allowed and is reflected in the S-parameters measurement. Excitations and observations are permitted only at the locations of Ports 1 and 2.

bottom diagram. Here, we have exactly the same physical configuration as in the top diagram, but instead of asking for cross correlation between antennas 1 and 2 as defined there, we look for cross correlation between antenna 1 and a *new* antenna 2 comprised of the old antenna 2 plus the object. Since the radiated far field of the new antenna 2 is different from the old, the far-field cross correlation using (18.23) will be different in the two cases. *However, the S-parameter expression is always the same regardless to how we regroup interacting antennas*. Numerical experiments with various scattering objects show indeed that the S-parameter formula produces completely wrong results. What is more interesting here is that there is an essential theoretical ambiguity in the status of scattering parameters approach to cross correlation when scattering objects exist.

The abundance of counter examples to (18.41) even in lossless antennas indicate that the derivation of this formula cannot be considered complete. It seems that S-parameter estimation of cross correlation may occasionally work, but it is not correct in general and in many practical cases (existence of several antennas in the array or weak or zero mutual coupling) it fails. All these difficulties totally disappear when the general formula (18.32) in terms of the currents on the antenna is used.

The basic idea of using conservation of energy to derive a relation between the correlation matrix and the scattering matrix is to compute the total power flow radiating away from the arrays using the far fields themselves. Another computation using the scattering parameters is possible. Since power computed using the far field and power found from the energy absorbed by the ports have to agree (Poynting theorem), then by equating the two quantities a relation between the correlation matrix and the scattering matrix os obtained. However, it does not follow from this relation that an equation like (18.41) holds.

The technical argument is briefly the following. Start with a set of far fields F_n radiated by N antenna elements, taking into account the effect of mutual coupling. We also assume that standard waveguides supporting a fundamental mode are connected to each antenna terminal, for details see [29]. Let the vector excitation be $\mathbf{a}$. By evaluating the net power field flow, we can by expanding and rearranging find that the total power radiated by the array is given by an expression of the form $\mathbf{a}^H C\,\mathbf{a}$, where H stands for the Hermitian transpose and R is the correlation matrix of the array. The mnth element of this matrix, i.e., R_{mn}, is proportional to the cross correlation $\int_{4\pi} d\Omega\, \mathbf{F}_m\left(\hat{r}\right) \cdot \mathbf{F}_n\left(\hat{r}\right)$ between the far field $\mathbf{F}_m\left(\hat{r}\right)$ and $\mathbf{F}_n\left(\hat{r}\right)$ radiated by antennas m and n. For the complete expression including the scaling factors and their physical meaning, see [151]. At the same time, a basic argument from circuit theory shows that the power absorbed by the array at its input terminals equal $\mathbf{a}^H\left(I - S^H S\right)\mathbf{a}$ [29]. In the lossless case, we have by conservation of energy

$$\mathbf{a}^H\left(I - S^H S\right)\mathbf{a} = \mathbf{a}^H C\,\mathbf{a}, \tag{18.42}$$

which is an exact relation valid for any $\mathbf{a}$. However, it does not mathematically follow from (18.42) that $I - S^H S = C$. In fact, it was already shown in [151] that in general

$$I - S^H S \neq C. \tag{18.43}$$

Relation (18.41) is based on the incorrect equality $I - S^H S = R$, and therefore we conclude that the expression reducing the envelope correlation to the S parameters is not in general valid. Numerical counter examples to (18.41) will be given in Section 18.3.5. The goal of this discussion is showing that cross correlation in spatial diversity systems like MIMO cannot be reduced to expressions involving only S parameters. In fact, the derivation of the general formula (18.32) above shows that the complete current distribution on the entire interacting antennas should be known in order to evaluate envelope cross correlation.

18.3.5 Verification, Numerical Studies, and Design Examples of Basic Spatial Diversity Antenna Array Configurations

18.3.5.1 Verification of the cross-correlation Green's Function Formula

Although the derivation of the expression (18.32) is exact and rigorous, we will present here a brief verification of the basic result. Incidently, we will also prove at the same time that the quantity introduced in (18.33) is indeed a Green's function. The basic idea is to compute cross correlation for two infinitesimal dipoles and then compare the outcome with the direct definition (18.23).

Consider two infinitesimal dipoles located at points $\mathbf{r}'$ and $\mathbf{r}''$ with currents $\mathbf{J}_1(\mathbf{r}) = \hat{\alpha}_1 a_1 \delta(\mathbf{r}-\mathbf{r}_1)$ and $\mathbf{J}_1(\mathbf{r}) = \hat{\alpha}_1 a_1 \delta(\mathbf{r}-\mathbf{r}_2)$, respectively. Here, $\hat{\alpha}_{1,2}$ and are two unit vectors and $a_{1,2}$ are the complex dipole moments for sources $\mathbf{J}_{1,2}$, respectively. As usual we denote the far fields radiated by the two currents by $\mathbf{E}_{1,2}(\hat{\mathbf{r}})$. Using (18.32), the far-field cross correlation is immediately given by

$$\int_{4\pi} d\Omega\, \mathbf{E}_1(\hat{\mathbf{r}}) \cdot \mathbf{E}_2^*(\hat{\mathbf{r}}) = a_1 a_2^* \, \hat{\alpha}_1 \cdot \bar{\mathbf{C}}(\mathbf{r}_1, \mathbf{r}_2) \cdot \hat{\alpha}_2 \,, \tag{18.44}$$

where the sifting property of the Dirac delta function was used. The expression (18.44) clearly shows that the tensor $\bar{\mathbf{C}}(\mathbf{r}', \mathbf{r}'')$ does represent the Green's function for the far-field cross-correlation problem. Remember that the Green's function of a linear problem is defined simply as the response to an infinitesimal source. As can be seen from Figure 18.6, our problem consists of two 'inputs', the sources at the two general positions $\mathbf{r}'$ and $\mathbf{r}''$, and one 'output', the far-field cross correlation. When the inputs are reduced to point sources acting at specific positions $\mathbf{r}_1$ and $\mathbf{r}_2$, we obtained merely the value of the general function $\bar{\mathbf{C}}(\mathbf{r}', \mathbf{r}'')$ evaluated at these two special positions, i.e., we obtained a quantity proportional to $\bar{\mathbf{C}}(\mathbf{r}_1, \mathbf{r}_2)$. This proves that $\bar{\mathbf{C}}(\mathbf{r}', \mathbf{r}'')$ is indeed the Green's function of the cross-correlation problem.

To verify (18.44), consider for simplicity two vertically-oriented infinitesimal dipoles as in Figure 18.7(top). From the well-known expressions of the far-field radiation of infinitesimal dipoles together with linear array theory, the fields due to the two antennas are given by [27]

$$\mathbf{E}_1 = \hat{\theta} a_1 \sin\theta, \;\; \mathbf{E}_2 = \hat{\theta} a_2 \sin\theta e^{ikd \sin\theta \sin\varphi}. \tag{18.45}$$

The cross correlation is then

$$\int_{4\pi} d\Omega\, \mathbf{E}_1(\hat{\mathbf{r}}) \cdot \mathbf{E}_2^*(\hat{\mathbf{r}}) = a_1 a_2^* \int_{4\pi} d\Omega \;\sin^2\theta \, e^{-ikd \sin\theta \sin\varphi}. \tag{18.46}$$

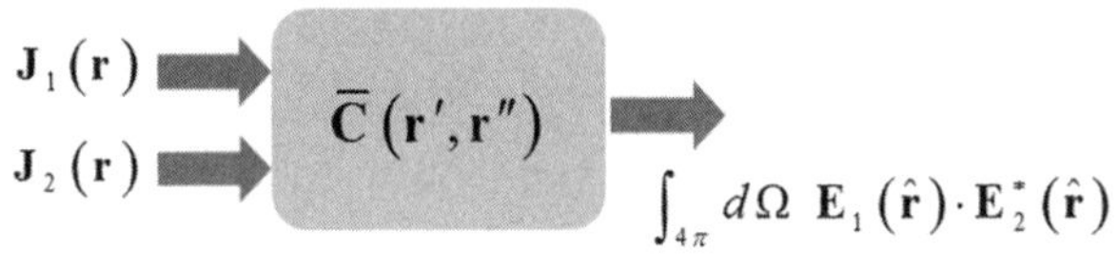

Figure 18.6 Representation of the cross-correlation problem in spatial diversity arrays as a system with the cross-correlation Green's function (18.33) acting as its "transfer function."

However, according to (18.44), we have $\int_{4\pi} d\Omega\, \mathbf{E}_1(\hat{\mathbf{r}}) \cdot \mathbf{E}_2^*(\hat{\mathbf{r}}) = a_1 a_2^* \hat{z} \cdot \bar{\mathbf{C}}(\mathbf{r}_1, \mathbf{r}_2) \cdot \hat{z} = a_1 a_2^* c_{zz}(d)$, where d is the separation between the two dipoles (see Figure 18.7). Using the expression of c_{zz} in Appendix 18.3.8, we find

$$\begin{aligned} &\int_{4\pi} d\Omega\, \mathbf{E}_1(\hat{\mathbf{r}}) \cdot \mathbf{E}_2^*(\hat{\mathbf{r}}) \\ &\quad = a_1 a_2^* \int_{4\pi} d\Omega\, \left(1 - \cos^2\theta\right) e^{ik(\mathbf{r}_1 - \mathbf{r}_2)\cdot\hat{r}}. \end{aligned} \tag{18.47}$$

Employing the obvious relations $\mathbf{r}_1 - \mathbf{r}_2 = -\hat{y}d$, $1 - \cos^2\theta = \sin^2\theta$, and $\hat{r} = \hat{x}\sin\theta\cos\varphi + \hat{y}\sin\theta\sin\varphi + \hat{z}\cos\theta$, it follows that (18.46) and (18.47) are identical. Therefore, the expression (18.44) yields the same result of the original definition (18.23). The verification for all other different orientations of the infinitesimal dipoles is essentially the same but lengthy and so the details are omitted. This proves that for any two infinitesimal dipoles, the computation of the far-field cross correlation produces the correct result. Finally, for arbitrary source distributions, we just note that the far-field correlation is simply the sum of the contributions of all mutual cross correlations between the infinitesimal currents forming the total source. Consequently, the cross-correlation formula (18.32) is now generally verified.

18.3.5.2 Numerical Study of Cross Correlation between Two Infinitesimal Dipoles

Consider two vertically-oriented electrically small dipoles as in Figure 18.7. Figure 18.8 shows the result for the cross correlation coefficient (18.23) using the formula (18.44) but with the sign retained. For the two small antennas, $S_{12} = 0$ and therefore according to the S-parameters formula (18.41) we have $\rho = 0$. Therefore, *the prediction of the far-field cross correlation based on the S-parameters of the array is completely wrong*. The calculation of ρ in Figure 18.8 using the cross-correlation Green's function (18.33) fully agrees with the original definition (18.23) computed directly via the far fields radiated by the dipoles.

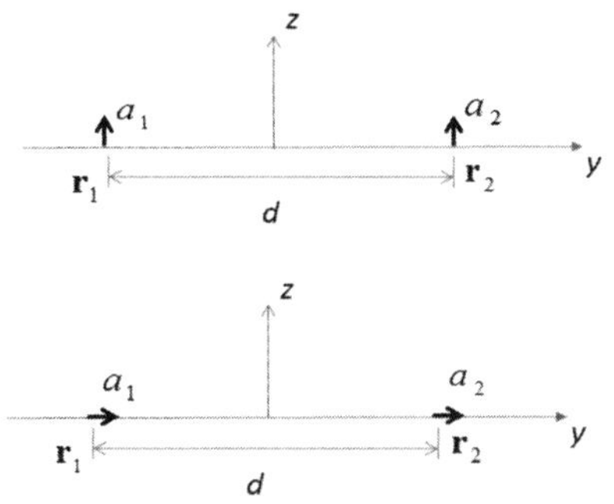

Figure 18.7 Two infinitesimal dipoles $\mathbf{J}_1(\mathbf{r}) = \hat{z}a_1\delta(\mathbf{r} - \mathbf{r}_1)$ and $\mathbf{J}_1(\mathbf{r}) = \hat{z}a_2\delta(\mathbf{r} - \mathbf{r}_2)$. *Top.* Vertically-oriented array. *Bottom.* Horizontally-oriented array.

We show in the same figure the calculation for two horizontally oriented dipoles. It is possible to observe from Figure 18.8 that the horizontal case far-field envelope correlation is higher in magnitude than the vertically-oriented array but up till separation of 0.45λ, where correlation in the later case reached zero. After that, the horizontal polarization array cross correlation becomes less than the vertical case. We observe also from Figure 18.8 that at periodic separations (roughly speaking, at $0.438\lambda, 0.97\lambda, 1.48\lambda, 1.98\lambda, 2.49\lambda, 2.98\lambda, 3.49\lambda$, and so on), the cross correlation between the two dipoles becomes exactly zero (in fact, it switches sign in a cyclic fashion). These critical locations are interesting and give first hints about the possibility of designing antenna arrays with minimum cross correlation.

We next consider the scenario when antenna 1 is inclined with an angle a with respect to the positive y-direction in Figure 18.7(top). The computation using the expression (18.44) are shown in Figure 18.9. It is observed that as the first dipole is further rotated toward $a = 0$, the state of complete orthogonality with respect to the other antenna, cross correlation decreases and eventually drops to very small value at the exact orientation $a = 0$. Similar numerical results can be shown for other relative inclinations, suggesting that the far-field cross correlation of two orthogonal dipoles is very low. This information can also help in designing antenna elements to reduce far-field cross correlation.

However, note that the far fields of two perpendicular small dipoles need not be orthogonal everywhere on the radiation sphere. The results only suggest that the net cross correlation of orthogonal dipoles is very small though not always exactly zero. For example, consider the array configuration shown in Figure 18.10. The

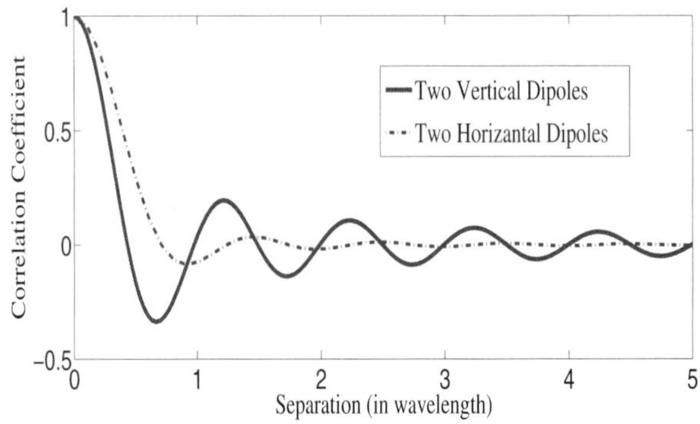

Figure 18.8 Envelope correlation coefficient (with the sign retained) for the two vertically/horizantally-oriented infinitesimal dipoles.

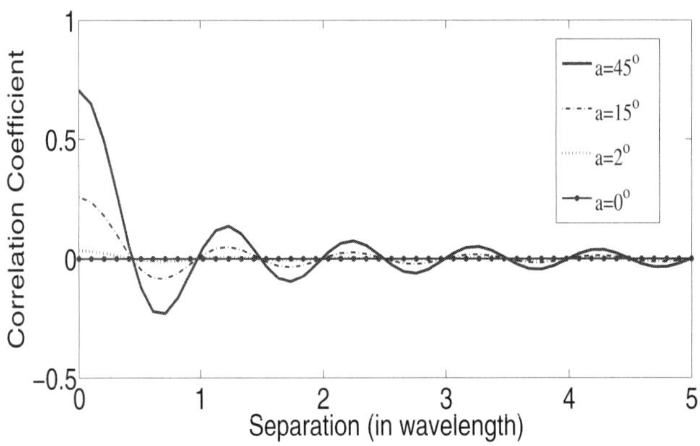

Figure 18.9 Envelope correlation coefficient (with the sign retained) for two infinitesimal dipoles with variable relative orientation. Antenna 1 in Figure 18.7(top) is inclined with an angle a measured with respect to the positive y-direction while antenna 2 is kept vertically oriented.

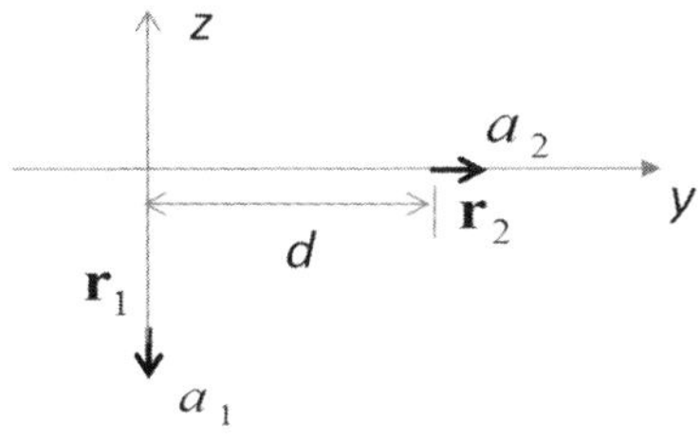

Figure 18.10 Two perpendicular infinitesimal dipoles $\mathbf{J}_1(\mathbf{r}) = \hat{z}a_1\delta(\mathbf{r}-\mathbf{r}_1)$ and $\mathbf{J}_1(\mathbf{r}) = \hat{y}a_2\delta(\mathbf{r}-\mathbf{r}_2)$ whose far-field cross correlation in Figure 18.17 is not exactly zero.

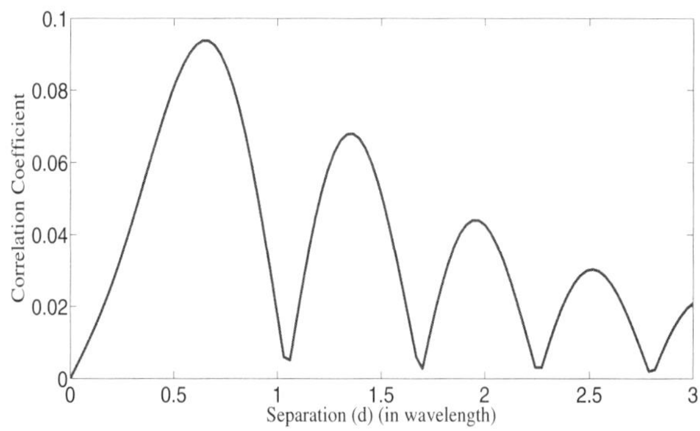

Figure 18.11 Cross correlation between the two dipoles in Figure 18.10.

computation of the envelope cross correlation is shown in Figure 18.11. It is clear that although the correlation is not high, it is certainly not negligible. Therefore, although it seems that typically for perpendicular dipoles their far fields tend to be uncorrelated or very weakly correlated, caution is needed in order not to generalize this observation to all cases. It appears that cross correlation is exactly zero when the field radiated by one antenna is orthogonal to the current of the other. It is possible to see that this is *not* the case in the example of Figure 18.10, which may explain why the corresponding cross correlation in Figure 18.11 is not exactly zero. General mathematical proofs of this observation will not be given here.

18.3.6 Design Methodology for Best Diversity Gain

It is possible using the technique presented above to examine the engineering problem of designing optimum antenna arrays capable of achieving certain diversity performance as determined by the cross correlation between the far fields. In order to do so, we need to define a measure for the diversity gain attained through a proper design of the radiating element.

First, we define the cross correlation matrix of an antenna array composed of N elements as the $N \times N$ matrix R whose mnth element R_{mn} is given by

$$R_{mn} := \int_{4\pi} d\Omega\, \mathbf{E}_m(\hat{\mathbf{r}}) \cdot \mathbf{E}_n^*(\hat{\mathbf{r}}). \tag{18.48}$$

Consider then an antenna array described by the correlation matrix R. We define the *diversity gain* as

$$G = \frac{tr\left(R^2\right)}{\|R\|_{Fr}}, \tag{18.49}$$

where tr is the matrix trace (the sum of the diagonal elements) and $\|R\|_{Fr}$ is the Frobenius norm defined as $\|R\|_{Fr} := \sum_{m,n} |\rho_{mn}|^2$. This definition is very natural since it measures the degree to which the correlation matrix R tends to have zero off-diagonal element. For the best performance, i.e., when all the array elements are mutually uncorrelated, the diversity gain will be unity, which corresponds to the best performance from the viewpoint of MIMO and spatial diversity. In general, $0 < G \leq 1$. The definition (18.49) is similar to the correlation metrics in [153].

We will first motivate the design approach by studying the impact of the *array density* on the diversity gain. The results of Section 18.3.5 suggest the following general rules:

1. Cross correlation between vertical and horizontal current elements tend to decrease with increasing distance.
2. Cross correlation between horizontal current elements tend to dominate for short distances compared with vertical current segments. For larger separation, the converse is observed.
3. Cross correlation between perpendicular current elements tend to be either vanishing or small.

These are very general observations that may help in designing arrays for MIMO applications. However, it is possible go further in quantifying the MIMO system

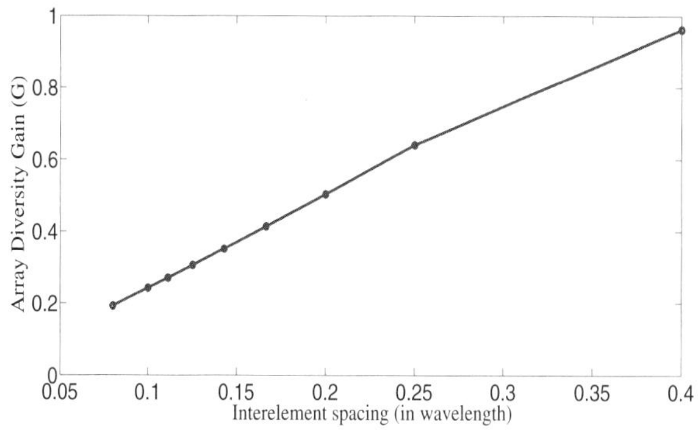

Figure 18.12 Diversity gain for array of linear vertically-polarized infinitesimal dipoles as in Figure 18.7(top) with uniform spacing between the radiating elements.

performance by showing how the *density* of the array elements per unit wavelength affects the diversity gain. We consider a linear array with uniform spacing of infinitesimal dipoles all oriented vertically. Figure 18.12 shows the variation of the array diversity with the density of the array expressed in terms of the inter-element spacing. It is clear that performance deteriorates when more elements are utilized within the same length span. Therefore, there is a fundamental tradeoff in MIMO systems between size and diversity gain: *the smaller the total antenna size, the higher is the density of the array, the worse the diversity gain becomes.*

A view on the design approach will be now given using a random search method. We suggest that positions and orientations of each dipole to be changed randomly until a good diversity gain is achieved. In general, it would be preferable to formulate the problem using a more complex search algorithm, for example, global optimization, convex optimization, etc. However, in this chapter we will develop the most direct approach using random search and leave more elaborate design tools for future work.

Consider an antenna array comprised of a set of N infinitesimal dipoles at positions $\mathbf{r}_n$ and orientations $\hat{\alpha}_n$. The random search algorithm works by changing the data $\mathbf{r}_n$ and $\hat{\alpha}_n$ such that the trial arrays all fit within a given size limitation and a geometrical form. The diversity performance of the search trials is monitored by observing the correlation gain G defined in (18.49). When the best performance is attained, the algorithm is terminated.

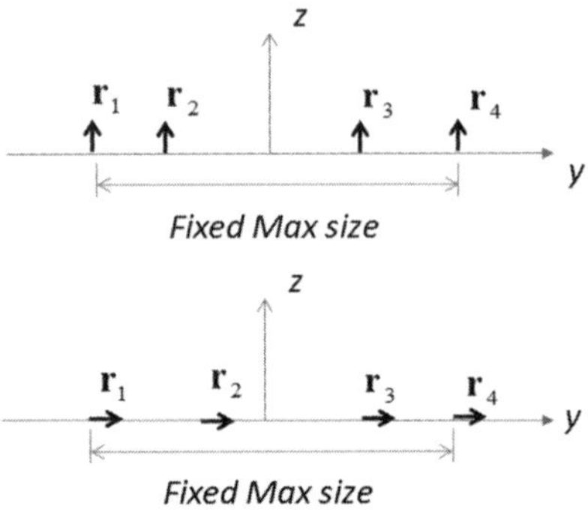

Figure 18.13 The design topology for linear arrays of vertically-polarized sources (top) and horizontally-polarized sources (bottom). In each case, the dipole positions $\mathbf{r}_n$ is varied using the optimization algorithm in order to maximize the diversity gain (18.49).

We provide basic examples involving linear and circular antenna arrays but the method is applicable to arbitrary topologies. Figure 18.13 shows the schematics of two possible arrays, one for vertical polarization (top) while the other is for horizontal polarization (bottom). The maximum size of the array is fixed at one wavelength and a set of four dipoles is considered for the design. The average density in this case is 0.25λ between two antennas. The positions of the dipoles are given by the vectors $\mathbf{r}_n$, $n = 1, 2, 3, 4$. The y-component of each $\mathbf{r}_n$ is changed according to the formula

$$r_n^y = y_{\min} + w\left(y_{\max} - y_{\min}\right), \tag{18.50}$$

where $[y_{\min}, y_{\max}]$ is the allowable search range and w is a random variable uniformly distributed between 0 and 1.

Figure 18.14 shows the results of the above random search algorithm for the vertically-polarized array. The best performance obtained in 100 trials is $G = 0.705$. For the horizontally-polarized case, Figure 18.15 provides the design results, where the best diversity gain is lower at the level $G = 0.5768$. For both cases, the design parameters, i.e., the positions of antennas, are given in the captions of the two figures. These results can be further improved by using a local search algorithm with initial points at the positions just obtained using the random search method.

As another example, we consider this time the direction of the dipoles as a design parameters. For simplicity, we force all the dipoles to be parallel to each other but instead of being restricted to either vertical or horizontal directions, the

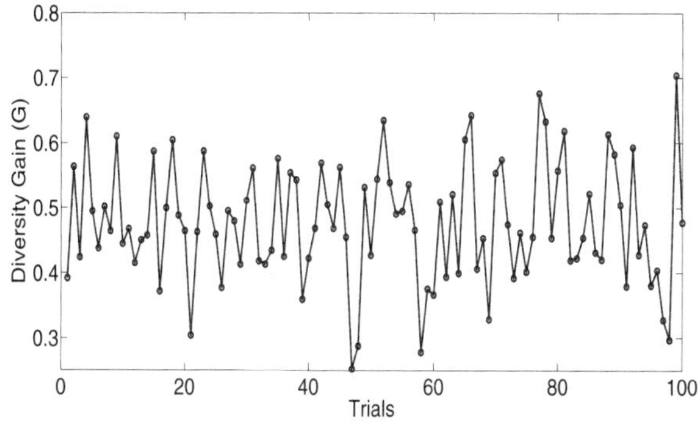

Figure 18.14 The diversity gain for a linear array of vertically-polarized infinitesimal dipoles using direct random search algorithm. The number of dipoles is 4 per maximum size of unit wavelength, i.e., with an average inter-element spacing of 0.25λ. The best solution obtained in 100 trials is with positions (not to scale in the diagram) given by: $\mathbf{r}_1 = 0.0826\lambda, \mathbf{r}_2 = 0.5118\lambda, \mathbf{r}_3 = 0.7196\lambda, \mathbf{r}_4 = 0.9962\lambda$, with a corresponding diversity gain of about 0.705.

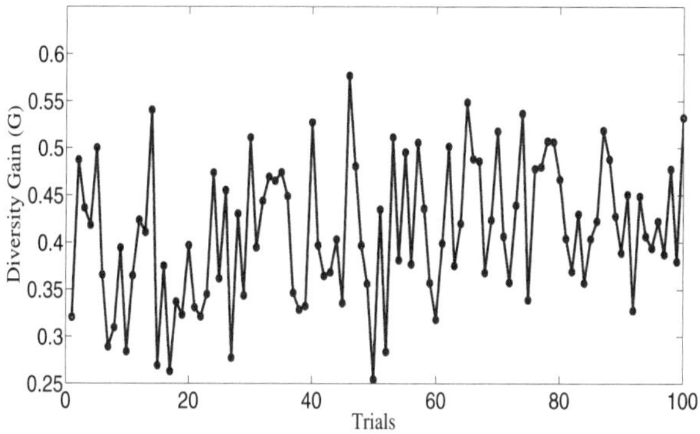

Figure 18.15 The diversity gain for a linear array of vertically-polarized infinitesimal dipoles using direct random search algorithm. The number of dipoles is 4 per maximum size of unit wavelength, i.e., with an average inter-element spacing of 0.25λ. The best solution obtained in 100 trials is with positions (not to scale in the diagram) given by: $\mathbf{r}_1 = 0.0170\lambda, \mathbf{r}_2 = 0.1209\lambda, \mathbf{r}_3 = 0.5801\lambda, \mathbf{r}_4 = 0.9283\lambda$, with a corresponding diversity gain of about 0.5768.

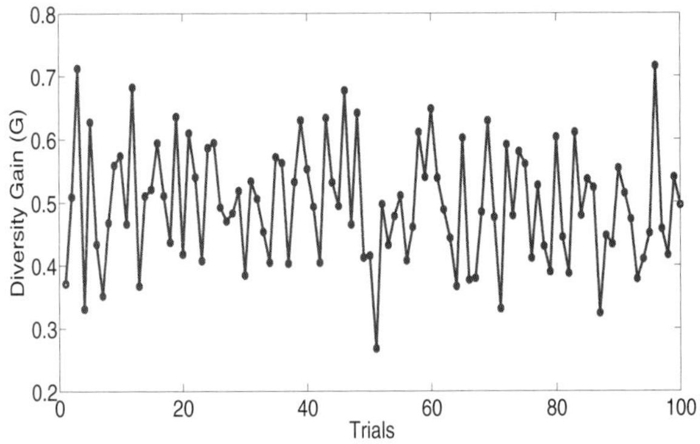

Figure 18.16 The diversity gain for a linear array of parallel infinitesimal dipoles with varying current polarization. All the dipole currents are restricted to be parallel to each other but this common direction is varied from one trial to another together with the positions. The trials are obtained using direct random search algorithm. The number of dipoles is 4 per maximum size of unit wavelength, i.e., with an average inter-element spacing of 0.25λ. The best solution obtained in 100 trials is with positions (not to scale in the diagram) given by: $\mathbf{r}_1 = 0.0298\lambda, \mathbf{r}_2 = 0.4972\lambda, \mathbf{r}_3 = 0.8667\lambda, \mathbf{r}_4 = 0.8944\lambda$,. The common orientation of the four dipoles currents is the unit vector $\hat{x}0.2995 + \hat{y}0.3351 + \hat{z}0.8933$. The best gain is $G = 0.7158$.

entire set of radiating elements is re-oriented in each search trial along the direction of the vector $\hat{x}\alpha_x + \hat{y}\alpha_y + \hat{z}\alpha_z$, where $\alpha_x, \alpha_y, \alpha_z$ are statistically-independent random variables uniformly distributed between 0 and 1. Figure 18.16 provides us with the results and the design data for this case. The best performance in 100 random trials was obtained when the currents on the dipoles were all oriented in the direction of the unit vector $\hat{x}0.6782 + \hat{y}0.2160 + \hat{z}0.7024$. In this case, the best attained diversity gain is $G = 0.7158$, which is better than the those achieved with vertical and horizontal polarization in Figures 18.14 and 18.15.

In general, the overall results in Figure 18.16 when compared with Figures 18.14 and 18.15 suggest that varying the orientation of the radiating elements in a linear array improves the diversity gain results. Motivated by this observation, we provide a final example in which the direction of the current of *each* dipole is varied *independently* of the others. To achieve this, we change in each trial the direction of the current using a statistically-independent uniformly distributed random variable such that each component of the direction vector of each dipole current is changed independently of the others. The results are shown in Figure 18.17, with the spectacular diversity gain $G = 0.9087$. The positions and orientations of each antenna element belonging to this best configuration are given in the caption. We conclude that in linear antenna arrays, if there are no restrictions on the orientation of the individual radiating elements, then it is best to allow the individual directions of the current in the array to vary independently through the design process. However, in practice it is sometimes not easy to change the orientation individually, for example because of the inflexibility of the feed circuit, and so the results of uniform orientation presented in Figures 18.14-18.16 remain important even with the superior results of Figure 18.17.

Before ending this section, we consider an example of circular array as shown in Figure 18.18. All the antenna elements are oriented in the z-direction pointing outward to the plane containing the array. The radius of the circle a will determine the maximum size of the array, i.e., the circumference $2\pi a$. The position of each element in the xy-plane is given by $a \cos \phi_n, a \sin \phi_n, n = 1, 2, 3, 4$. Here, we vary the angles ϕ_n as statistically-independent random variable uniformly distributed between 0 and π. Figure 18.19 shows the design results for this array topology. We choose the radius as 0.25λ, resulting in an average array density of about 0.4λ. The best gain in 100 trials is $G = 0.6073$. The angles of the array elements are given in the caption.

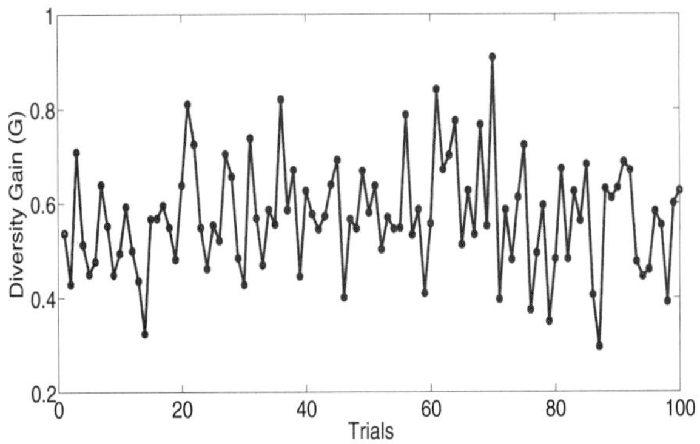

Figure 18.17 The diversity gain for a linear array of infinitesimal dipoles with varying current polarization. The currents positions and directions of the dipoles are allowed to change independently of each other in each trial. The trials are obtained using direct random search algorithm. The number of dipoles is 4 per maximum size of unit wavelength, i.e., with an average inter-element spacing of 0.25λ. The best solution obtained in 100 trials is with positions (not to scale in the diagram) given by: $\mathbf{r}_1 = 0.0298\lambda, \mathbf{r}_2 = 0.4972\lambda, \mathbf{r}_3 = 0.8667\lambda, \mathbf{r}_4 = 0.8944\lambda,$. The orientations of the four dipole currents are $\hat{x}0.4607 + \hat{y}0.0112 + \hat{z}0.1425$, $\hat{x}0.1706 + \hat{y}0.2345 + \hat{z}0.9397$, $\hat{x}0.7944 + \hat{y}0.7070 + \hat{z}0.0003$, $\hat{x}0.8885 + \hat{y}0.6942 + \hat{z}0.4388$. The best gain is $G = 0.9087$.

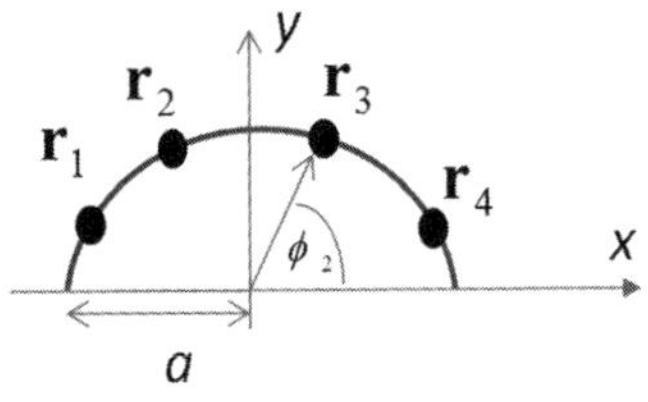

Figure 18.18 The design topology for circular array of vertically-polarized sources. In each case, the dipole positions $\mathbf{r}_n$ is varied by changing the angle ϕ_n using the optimization algorithm in order to maximize the diversity gain (18.49).

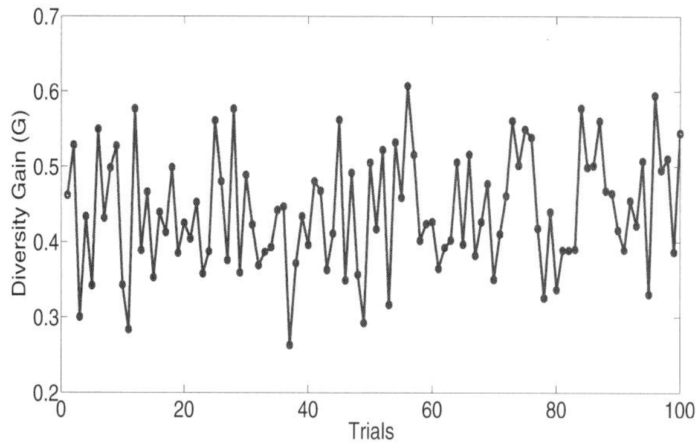

Figure 18.19 The diversity gain for a the circular array of Figure 18.18. All the dipoles are oriented along the z-direction outward to the xy plane. The trials are obtained using direct random search algorithm. The average inter-element spacing of 0.4λ. The best solution obtained in 100 trials is with positions (not to scale in the diagram) given by: $\phi_1 = 3.4474^o, \phi_2 = 63.2383^o, \phi_1 = 158.0790^o, \phi_1 = 177.8924^o$. The best gain is $G = 0.6073$.

18.3.7 Generalization to Arbitrary Antennas

In this section, we show how the techniques and methods of the previous parts can be applied to more complex antenna types. We first outline the method through which knowledge of cross correlation among infinitesimal dipoles can be used to estimate cross correlation in arbitrary complex radiating elements. Next, we apply the method to an example consisting of long thin wire. The same method can be applied to any antenna type without modification but only results for wire cases are shown here for brevity. Further examples and design methods will be treated in future work.

The technique developed above managed to first express the far-field cross correlation between a set of infinitesimal dipoles in the simple form of the cross-correlation Green's function, and second to reduce the problem of several point sources to the superposition of all pairwise interactions between the dipoles. In light of the general expressions given by (18.32) and (18.39), it is even possible to reduce the computation for the case of *arbitrary* current distribution into the infinitesimal dipole scenario. The key idea here is based on the observation in Section 18.3.3.1 that the cross-correlation Green's function (18.33) is nonsingular. Therefore, it is always possible to approximate the integral in (18.32) by a finite sum evaluated at discrete points using the quadrature methods familiar from the context of the method of moment. Indeed, we may approximate (18.39) by

$$\rho_{mn} = \sum_{l_1=1}^{N_1} \sum_{l_2=1}^{N_2} \gamma_{l_1} \gamma_{l_2} \mathbf{f}_m^1 \left(\mathbf{r'}_{l_1}\right) \cdot \bar{\mathbf{C}} \left(\mathbf{r'}_{l_1}, \mathbf{r''}_{l_2}\right) \cdot \mathbf{f}_n^2 \left(\mathbf{r''}_{l_2}\right), \tag{18.51}$$

where N_1 and N_2 are the number of quadrature points associated with the current basis functions $\mathbf{f}_m^1$ and $\mathbf{f}_n^2$, respectively. The data comprised of the positions $\mathbf{r}_l$ and weights γ_l can be found from standard quadrature routines. These rules are available for both one-dimensional, two-dimensional, and three-dimensional regions, and hence are enough to handle any correlation integral in electromagnetics.

After evaluating the individual elements of the correlation matrix ρ_{mn}, it is possible to compute the cross correlation between any group of antennas by simple matrix partitioning. We will illustrate the general idea by a practical example. Consider the antenna configuration in Figure 18.20(top). Antenna A is an infinitesimal dipole with current $\mathbf{f}^A(\mathbf{r}) = \hat{z}\delta(\mathbf{r} - \mathbf{r}_A)$ located at position $\mathbf{r}_A$, while antenna B is a short (not infinitesimal) thin-wire dipole. We further assume that the current distribution on this antenna B can be approximated by the familiar sinusoidal form [26], [27]

$$\mathbf{f}^B(z) = \hat{z} \sin\left[k\left(L - |z|\right)\right], \tag{18.52}$$

where $k = 2\pi/\lambda$ and $2L$ is the length of the antenna. The correlation integral then has the form

$$\rho_{AB} = \int_{S_A} ds' \int_{S_B} ds'' \, \mathbf{f}^A(\mathbf{r'}) \cdot \bar{\mathbf{C}}(\mathbf{r'}, \mathbf{r''}) \cdot \mathbf{f}^B(\mathbf{r''}), \tag{18.53}$$

where S_A and S_B are the surfaces of antennas A and B with local length parameters s' and s'', respectively. By the sifting property of the Dirac delta function, we have

$$\rho_{AB} = \int_{S_B} ds'' \, \hat{z} \cdot \bar{\mathbf{C}}(\mathbf{r}_A, \mathbf{r''}) \cdot \mathbf{f}^B(\mathbf{r''}). \tag{18.54}$$

Using the quadrature method, a set of points $\mathbf{r}_n$ residing in S_B are chosen to approximate the integral (18.53). By this method, the total cross correlation is simply given by the superposition rule

$$\rho_{AB} = \sum_{l=1}^{N} \gamma_l \hat{z} \cdot \bar{\mathbf{C}}(\mathbf{r}_A, \mathbf{r}_l) \cdot \mathbf{f}^B(\mathbf{r}_l), \tag{18.55}$$

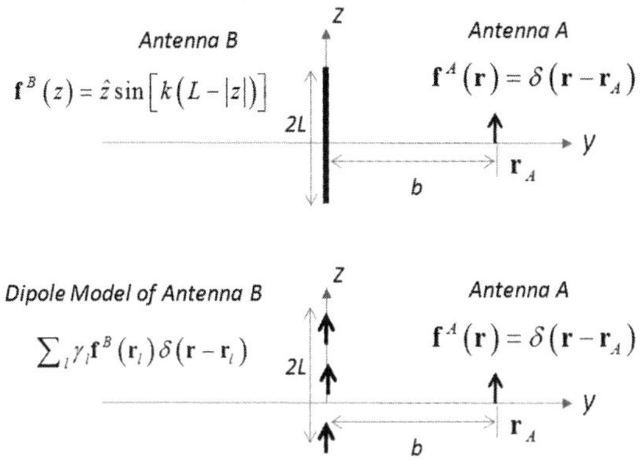

Figure 18.20 *Top.* Two-antenna system composed of linear thin-wire antenna B and an infinitesimal dipole A (antenna). The linear wire has length $2L$ and is separated from the small dipole located at $\mathbf{r}_A$ by distance *d*. *Bottom.* The same problem in the top diagram but with antenna B replaced by a finite number of infinitesimal dipoles. The dipoles are located at locations r_l, with the same orientation as the original antenna. The excitation of each equivalent dipole is $\gamma_l \mathbf{f}^B(\mathbf{r}_l)$. The computation of cross correlation between the two antennas using the equivalent dipole model in the bottom diagram produces the same result corresponding to the original problem in the top. We shown only three equivalent dipoles but the number of dipoles in each model can be varied according to the accuracy needed.

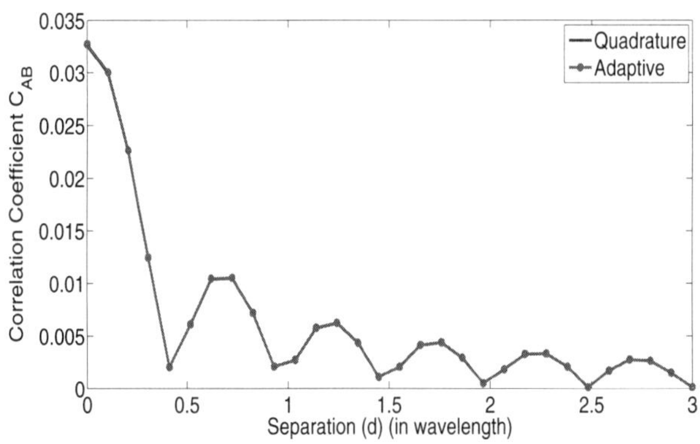

Figure 18.21 Computation of the cross correlation between antennas A and B in Figure 18.20(top) using the equivalent dipole model in Figure 18.20(bottom). The length of the wire antenna B is 0.05λ. The cross correlation (not normalized) is based on 10 infinitesimal dipoles.

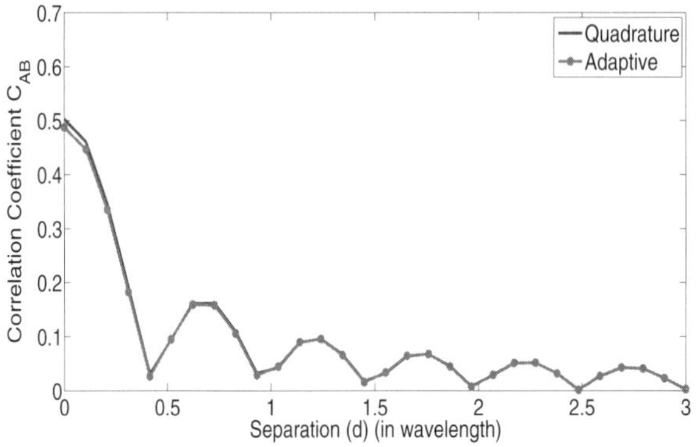

Figure 18.22 Computation of the cross correlation for the same problem of Figure 18.21 but with the length of the wire antenna B changed to 0.2λ. The cross correlation (not normalized) is based on 10 infinitesimal dipoles.

where N is the number of quadrature points $\mathbf{r}_l$ with weights γ_n.

The technique developed above is applied to the simple problem illustrated in Figure 18.20(top) using the equivalent dipole model in the bottom part of the same figure. The computation of the cross correlation (not normalized) is effected using the formula (18.55) and the results compared with the exact expression (18.53) computed using accurate adaptive integration routine. With 10-dipole model obtained through a simple Gauss quadrature method, excellent agreement was observed as shown in Figure 18.21. Note in general that the number of the dipoles needed to represent a given antenna for the purpose of computing its cross correlation strongly depends on the physical size of the antenna in question. In particular, the linear wire in the previous example has length 0.05λ. When the length is increased to 0.2λ, the results using again 10-dipole model are shown in Figure 18.22. Some discrepancy can be seen, especially at short separation. However, by using more dipoles, the agreement can always be improved.

In practical applications the sinusoidal current distribution (18.52) is in fact used as a basis function and the entire wire span is a just one segment in the overall dsicretization scheme, e.g., method of moment. It is always true in this case that the length of the segment is small compared with the wavelength in order to insure convergence of the full-wave numerical solver to correct results. Therefore, the same level of accuracy obtained in Figure 18.21 can be attained in typical calculations of cross correlation for practical spatial diversity applications.

18.3.8 Appendix: Detailed Components of the Far-Field Cross Correlation Dyadic Green's Function

In the following, set $\hat{\mathbf{r}}(\theta,\varphi) := \hat{x}\sin\theta\cos\varphi + \hat{y}\sin\theta\sin\varphi + \hat{z}\cos\theta$. The components of the cross-correlation Green's function (18.33) are given by

$$
\begin{aligned}
c_{xx}(\mathbf{r}',\mathbf{r}'') &= \int_0^{\pi}\int_0^{2\pi} d\theta d\varphi\, \sin\theta\left(1-\sin^2\theta\cos^2\varphi\right) e^{ik\left(\mathbf{r}'-\mathbf{r}''\right)\cdot\hat{\mathbf{r}}},\\
c_{xy}(\mathbf{r}',\mathbf{r}'') &= -\int_0^{\pi}\int_0^{2\pi} d\theta d\varphi \sin^3\theta\cos\varphi\sin\varphi\, e^{ik\left(\mathbf{r}'-\mathbf{r}''\right)\cdot\hat{\mathbf{r}}},\\
c_{xz}(\mathbf{r}',\mathbf{r}'') &= \int_0^{\pi}\int_0^{2\pi} d\theta d\varphi \sin^2\theta\cos\theta\cos\varphi\, e^{ik\left(\mathbf{r}'-\mathbf{r}''\right)\cdot\hat{\mathbf{r}}},\\
c_{yx}(\mathbf{r}',\mathbf{r}'') &= -\int_0^{\pi}\int_0^{2\pi} d\theta d\varphi\, \sin^3\theta\sin\varphi\cos\varphi\, e^{ik\left(\mathbf{r}'-\mathbf{r}''\right)\cdot\hat{\mathbf{r}}},\\
c_{yy}(\mathbf{r}',\mathbf{r}'') &= \int_0^{\pi}\int_0^{2\pi} d\theta d\varphi\, \sin\theta\left(1-\sin^2\theta\sin^2\varphi\right) e^{ik\left(\mathbf{r}'-\mathbf{r}''\right)\cdot\hat{\mathbf{r}}},\\
c_{yz}(\mathbf{r}',\mathbf{r}'') &= -\int_0^{\pi}\int_0^{2\pi} d\theta d\varphi\, \sin^2\theta\cos\theta\sin\varphi\, e^{ik\left(\mathbf{r}'-\mathbf{r}''\right)\cdot\hat{\mathbf{r}}},\\
c_{zx}(\mathbf{r}',\mathbf{r}'') &= -\int_0^{\pi}\int_0^{2\pi} d\theta d\varphi\, \sin^2\theta\cos\theta\cos\varphi\, e^{ik\left(\mathbf{r}'-\mathbf{r}''\right)\cdot\hat{\mathbf{r}}},\\
c_{zy}(\mathbf{r}',\mathbf{r}'') &= -\int_0^{\pi}\int_0^{2\pi} d\theta d\varphi\, \sin^2\theta\cos\theta\, \sin\varphi e^{ik\left(\mathbf{r}'-\mathbf{r}''\right)\cdot\hat{\mathbf{r}}},\\
c_{zz}(\mathbf{r}',\mathbf{r}'') &= \int_0^{\pi}\int_0^{2\pi} d\theta d\varphi\, \sin\theta\left(1-\cos^2\theta\right) e^{ik\left(\mathbf{r}'-\mathbf{r}''\right)\cdot\hat{\mathbf{r}}}.
\end{aligned}
\tag{18.56}
$$

18.4 CONCLUSION

The chapter provided a general and integrated study of MIMO systems. We first outlined how the ACGF can be used to effectively provide a complete and exact derivation of the MIMO channel matrix, effectively eliminating fields and currents in the final expression. The chapter also proposed a general approach to cross correlation in arbitrary antenna systems based on reducing the computation of far-field correlations to the radiating currents on antenna elements. It is found that the reduction requires the introduction of a fundamental function, the cross-correlation Green's function, which measures the degree of correlation between the fields produced by infinitesimal dipoles. For arbitrary source distribution, the total cross correlation is then shown to be merely the superposition of all the point-source mutual correlations obtained by means of this cross-correlation Green's function. The new approach is further elaborated by careful comparison with the classical Green's function and method of moment in applied electromagnetics. The estimation of cross correlation using S-parameters (measurement at the ports) was criticized and shown to be inadequate for general problems and for design purposes. Cross correlation expressions based on the currents were verified by direct comparison with the exact formula in terms of the far field and they avoid all the difficulties associated with the S-parameter expressions. Several numerical studies of basic array configurations were given. Some of the findings include demonstrating the inverse relation between the array density and the diversity gain, suggesting the need for a proper design of antenna arrays if the total size of the MIMO system is to be brought down. Several design studies and examples involving linear and circular arrays were given. The overall goal of the chapter was to establish the underlying principles of methods suitable for the design of antenna elements in MIMO systems such that desired diversity gain data can be attained. In particular, we showed how the position and orientation of each current segment contribute to the total diversity gain of the system and used this knowledge to devise a general design strategy for spatial diversity systems.

References

[1] Karl Poper, *Objective Knoweldge*, Oxford University Press, 1972.

[2] Naom Chomsky, *The Logical Strucutre of Lingusitic Theory*, Springer, 1975.

[3] Mario Bunge, *Ontology I: The Furniture of the World*, D. Reidel Publishing, 1977.

[4] Said M. Mikki and Yahia M. M. Antar, "Generalized analysis of the relationship between polarization, matching Q factor, and size of arbitrary antennas," *Proceedings of IEEE APS-URSI International Symposium*, Toronto, July 11–17, 2010.

[5] Said M. Mikki and Yahia M. M. Antar, "Critique of antenna fundamental limitations," *Proceedings of URSI-EMTS International Conference*, Berlin, August 16-19, 2010.

[6] Said M. Mikki and Yahia M. M. Antar, "On the spatial strucutre of the antenna electromagnetic near fields," *Proceedings of URSI General Assembly and Scientific Symposium*, Istanbul, August 13-20, 2011.

[7] Said M. Mikki and Yahia M. M. Antar, "Analysis of electromagnetic interactions in antenna arrays using the antenna current Green's function method," *Proceedings of IEEE APS-URSI International Symposium*, Spoken, July 3–8, 2011.

[8] Said M. Mikki and Yahia M. M. Antar, "Fundamental research directives in applied electromagnetic theory," *28th National Radio Science Conference*, (NRSC), April 26-28, 2011, National Telecommunication Institute, Egypt, 2011.

[9] Said M. Mikki and Yahia M. M. Antar, "A theory of antenna electromagnetic near field—Part I," to appear in *IEEE Trans. Antennas Propagat.*

[10] Said M. Mikki and Yahia M. M. Antar, "A theory of antenna electromagnetic near field—Part II," to appear in *IEEE Trans. Antennas Propagat.*

[11] Said M. Mikki and Yahia M. M. Antar, "Near field analysis of electromagnetic interactions in antenna arrays through equilant dipole models," to appear in *IEEE Trans. Antennas Propagat.*

[12] Said M. Mikki and Yahia M. M. Antar, "The antenna current Green's function formalismm—A paradigm for thr analysis of general electromagnetic systems—Part II," submitted to *IEEE Trans. Antennas Propagat.*

[13] Said M. Mikki and Yahia M. M. Antar, "On the fundamental relationship between the transmitting and receiving modes of general antenna systems: a new approach" accepted for publication in *IEEE Antennas and Wireless Communication Letters*.

[14] S. Henault, Y. M. M. Antar, S. Rajan, R. Inkol, and S. Wang, "The multiple antenna induced EMF method for the precise calculation of the coupling matrix in a receiving antenna array," *Progress In Electromagnetics Research* M, Vol. 8, 103-118, 2009.

[15] S. Henault, S. K. Podilchak, S. Mikki, Y. M. M. Antar, "A general methodology for mutual coupling estimation and compensation," submitted to *IEEE Trans. Antennas Propagat.*

[16] Said M. Mikki and Yahia M. M. Antar, "A New Technique for the Analysis of Energy Coupling and Exchange in General Antenna Systems," *IEEE Trans. Antennas Propagat.*, no. 12, December 2015.

[17] Said M. Mikki and Yahia M. M. Antar, "Aspects of Generalized Electromagnetic Energy Exchange in Antenna Systems: A New Approach to Mutual Coupling," EuCap 2015, Lisbon, April 12-17, 2015.

[18] Said M. Mikki and Ahmed A. Ksihk, "Electromagnetic wave propagation in dispersive negative group velocity media," *IEEE MTT-S International Symposium*, June 15–20, 2008.

[19] Said M. Mikki and Ahmed Kishk, "Electromagnetic wave propagation in nonlocal media: negative group velocity and beyond," *Progress In Electromagnetics Research B*, Vol. 14, 149-174, 2009.

[20] Said M. Mikki and Ahmed. A. Kishk, "Nonlocal Electromagnetic Media: A Paradigm for Material Engineering," in *Handbook of Microwave and Millimeter Wave Technologies*, INTECH, 2010.

[21] Said M. Mikki and Ahmed A. Ksihk, "Theory and applications of infinitesmal dipole models for computational electromagnetics," *IEEE Trans. Antennas Progat.*, vol. 54, pp. 2764–2775, May 2007.

[22] Said M. Mikki and Ahmed A. Kishk, "Quantum particle swarm optimization for electromagnetics," *IEEE Trans. Antennas Progat.*, vol. 54, pp. 2764–2775, Oct. 2006.

[23] R. F. Harrington, *Field Computation by Moment Methods*, A Classic Reissue, New Jersey: IEEE Press, 1992.

[24] Moris Kline, *Electromagnetic Theory and Geometrical Optics*, Princton, NJ: Van Nostrand, 1943.

[25] Sergi A. Schelkunoff, *Electromagnetic Waves*, Princton, NJ: Van Nostrand, 1943.

[26] Sergi A. Schelkunoff and Harolod T. Friis, *Antennas: Theory and Practice*, New York: Wiley, 1952.

[27] Carlo G. Someda, *Electromagentic Waves*, Chapman & Hall, 1998.

[28] Jean Van Bladel, *Electromagnetic Fields*, Wiley-Interscience (*IEEE* Re-Print), 2007.

[29] Robert Collin, *Field Theory of Guided Waves*, Wiley-Interscience (*IEEE* Re-Print), 2007.

[30] Robert S. Elliott, *Electromagnetics: History, Theory, and Applications*, Wiley-IEEE Press, 1 edition, 1999.

[31] Robert S. Elliott, *Antenna Theory & Design*, Wiley-IEEE Press; Revised Edition edition, (*IEEE* Re-Print), 2002.

[32] John D. Kraus and Ronald J. Marhefka, *Antennas For All Applications*, McGraw-Hill, 2001.

[33] David John Jackson, *Classical Electrodynamics*, John Wiley & Sons, third edition, 1999.

[34] Weng Cho Chew, *Waves and Fields in Inhomogenous Media*, New York: Van Nostrand Reinhold, 1990.

[35] Thorkild B. Hansen and Arthur D. Yaghjin, *Plane-Wave Theory of Time-Domain Fields*, The IEEE Press Series, 1999.

[36] Jean G. Van Bladel, *Singular Electromagnetic Fields and Sources* , first edition, The IEEE Press Series, 1996.

[37] Constantine A. Balanis, *Antenna Theory: Analysis and Design*, third edition, Wiley-Interscience, 2005.

[38] Walton C. Gibson, *The Method of Moement in Electromagnetics*, Chapman & Hall/CRC, 2008.

[39] Branko Kolundzija and Antonije Djordjevic, *Electromagnetic Modeling of Composite Mettallic and Dielectric Structures*, Artech House, 2002.

[40] Weng Cho Chew, Mei Song Tong, and Bin Hu, *Integral Equation Methods for Electromagnetic and Elastic Waves*, Morgan & Claypool Publishers, 2009.

[41] Leopold B. Felsen and Nathan Marcuvitz, *Radiation and Scattering of Waves, IEEE* Press (re-print edition), 1994.

[42] J. E. Hansen, *Spherical Near-Field Antenna Measurements*, London, U.K.: Peregrinus, 1988.

[43] Lukas Novotny and Bert Hecht, *Principles of Nano-Optics*, Cmbridge University Press, 2006.

[44] Jaechun Lee and Sangwook Nam, "Fundamental Aspects of Near-Field Coupling Small Antennas for Wireless Power Transfer," *IEEE Trans. Antennas Propagat.*, VOL. 58, NO. 11, pp. 3442-3449, November, 2010.

[45] Hubert Kalf, "On the expansion of a function in terms of spherical harmonics in arbitrary dimensions," *Bull. Belg. Math. Soc. Simon Stevin*, vol. 2, no. 4, pp. 361-380, 1995.

[46] C. H. Wilcox, "Generalization of theorems of Rellich and Atkinson," *Proc. Amer. Math. Soc.*, Vol 7, No 2, pp. 271-276, 1956.

[47] C. H. Wilcox, "An expansion theorem for the electromagnetic fields," *Communications on Pure and Appl. Math.*, vol. 9, pp. 115–134, 1956.

[48] C. H. Wilcox, "Debye Potentials", *Journal on Mathematics and Mechanics*, Vol 6, No 2, pp. 167–201, 1957.

[49] Ronald L. Fante, "Quality factor of general ideal antennas," *IEEE Trans. Antennas Propagat.*, vol. AP-17, no. 2, pp. 151-155, March 1969.

[50] L. J. Chu, "Physical limitations of omni-directional antennas," *J. Appl. Phys.*, vol. 19, pp. 1163-1175, December 1948.

[51] R. E. Collin and S. Rothschild, "Evaluation of antenna Q," *IEEE Trans. Antennas Propagat.*, vol. AP-12, pp. 23-21, January 1964.

[52] D. R. Rhodes, "A reactance thoerem," *Proc. R. Soc. Lond. A.*, vol. 353, pp.1-10, Feb. 1977.

[53] Arthur D. Yaghjian and Steve. R. Best, "Impedance, bandwidth, and Q of antennas," *IEEE Trans. Antennas Propagat.*, vol. 53, no. 4, pp. 1298-1324, April 2005.

[54] A. T. De Hoop, "The N-Port Receiving Antenna and its Equivalent Electrical Network," Philips Research Reports, vol. 30, pp. 302-315, 1975.

[55] D. M. Kerns, "Analytical techniques for the correction of nearfield antenna measurements made with an arbitrary but known measuring antenna," *URSI-IRE Meeting*, Washington, DC, Apr. 29-May 2, 1963.

[56] David M. Kerns, "Plane-wave scattering-matrix theory of antennas and antenna-antenna interactions: formulation and applications,“ *Journal of Research of the National Bureau of Standards—B. Mathematica Scineces*, vol. 80B, no. 1, pp. 5-51, January-March, 1976.

[57] B. L. Moiseiwitsch, *Integral Equations*, London and New York: Longman, 1977.

[58] Sadri Hassani, *Mathematical Physics: An Introduction to Its Foundations*, Springer, 2000.

[59] Bernard Schutz, *Geometrical Methods of Mathematical Physics*, Cambridge University Press, 1980.

[60] William M. Boothby, *An Introduction to Differentiable Manifolds and Riemannian Geometry*, Academic Press, 1975.

[61] S. P. Novikov and I. A. Taimanov, *Modern Geometric Structures and Fields*, American Mathematical Society (AMS), 2006.

[62] David V. Widder, *Advanced Calculus*, second edition, Dover, 1989.

[63] Richard Courant and Fritz John, *Introduction to Calculus and Analysis*, reprint of the first edition (1999), Springer, 1989.

[64] Bryan Rynne and M.A. Youngson, *Linear Functional Analysis*, Springer, 2007.

[65] Ram P. Kanwal, *Generalized Functions: Theory and Applications*, BirkhÃ¤user Boston, third edition, 2004.

[66] D. S. Jones, *The Theory of Generalised Functions*, Cambridge University Press, 2009.

[67] Walter Appel, *Mathematics for Physics and Physicists*, Princeton University Press, 2007.

[68] D. C. Champeney, *A Handbook of Fourier Theorems*, Cambridge University Press, 1987.

[69] O. D. Kellogg, *Foundations of Potential Theory*, Springer, 1929.

[70] David Colton and Rainer Kress, *Integral Equation Methods in Scattering Theory*, John Wiley & Sons, 1983.

[71] Ivar Stakgold, *Green's Functions and Boundary Value Problems*, John Wiley & sons, 1979.

[72] Brian S. Thomson, Judith B. Bruckner, Andrew M. Bruckner, *Elementary Real Analysis*, CreateS-pace, 2 edition, 2008.

[73] John M. Lee, *Introduction to Smooth Manifolds*, 1 edition, Springer, 2002.

[74] Vladimir I. Arnol'd, *Ordinary Differential Equations*, translated from the Russian by Roger Cooke, Springer, 1992.

[75] Richard S. Palais and Robert A. Palais, *Differential Equations, Mechanics, and Computation* American Mathematical Society (AMS), 2009.

[76] Vladimir I. Arnol'd, *Mathematical Methods of Classical Mechanics*, translated by K. Vogtmann and A. Weinstein, Springer, 1989.

[77] Lawrence Perko, *Differential Equations and Dynamical systems*, Springer, 1991.

[78] Ralph Abraham and Christopher D. Shaw, *Dynamics: The Geometry of Behavior*, HarperCollins Canada / Perseus Books, 1998.

[79] Roger Knobel, *An Introduction to the Mathematical Theory of Waves*, American Mathematical Socirty (AMS), 2000.

[80] P. M. Cohn, *Lie Groups*, Cambridge University Press, 1957.

[81] Andrew Baker, *Matrix Groups: An Introduction to Lie Group Theory*, Springer, 2006.

[82] Brian C. Hall, *Lie Groups, Lie algebra, and Representations*, Springer, 2003.

[83] Gregory S. Chirikjian and Alexander B. Kyatkin, *Engineering Applications of Noncommutative Harmonic Analysis: With Emphasis on Rotation and Motion Groups*, CRC Press, 2000.

[84] Andrew H. Wallace, *Differential Topology*, Dover, 2006.

[85] David B. Gauld, *Differential Topology*, Dover, 2006.

[86] Antal Majthay, *Foundation of Catastrophe Theory*, Longman Higher Education , 1985.

[87] Yukio Matsumoto, *An Introduction to Morse Theory*, American Mathematical Society (AMS), 2002.

[88] M. Abramowitz and I. A. Stegunn, *Handbook of Mathematical Functions*, Dover Publications, 1965.

[89] Philip Morse and Herman Fesbach, *Methods of Theortical Physics* (Volume I and II), McGraw-Hill, 1953.

[90] David Bressoud, *A Radical Approach to Real Analysis*, The Mathematical American Society of America (AMS), 1994.

[91] Roger Knobel, *An Introduction to the Mathematical Theory of Waves*, American Mathematical Socirty (AMS), 2000.

[92] Eugene Isaacson and Herbert Bishop Keller, *Analysis of Numerical Methods*, Dover, 1994.

[93] Ralph P. Boas, *Invitation to Complex Analysis* , Mathematical Association of America, 2010.

[94] Hermann Weyl, *Space, Time, Matter*, Dover, 1918.

[95] R. Feynman, *Lectures on Physics*, Addison-Wesley, 1964.

[96] Marc Lange, *An Introduction to the Philosophy of Physics: Locality, Fields, Energy, and Mass*, Wiley-Blackwell, 2002.

[97] V. M. Agranovich and V. L. Ginzburg, *Spatial Dispersion in Crystal Optics and the Theory of Excitons*, Interscience Pulishers (John Wiley & Sons), 1966.

[98] V. L. Ginzburg, *The Propagation of Electromagnetic Waves in Plasmas*, Pergamon Press, 1970.

[99] P. Halevi (ed.), *Spatial Dispersion in Solids and Plasmas*, Springer, 1992.

[100] Leon Brillouin, *Wave Propagation in Periodic Structures*, Dover, 1946.

[101] L. Brillouin, *Wave Propgation and Group Velocity* (New York: Academic Press), 1960.

[102] L. D. Landau and E. M. Lifshitz, *Electrodynamics of Continuous Media*, Pergamon Press, 1960.

[103] Yu.A. Il'inskii and L.V. Keldysh, *Electromagnetic Response of Material Media*, Springer, 1994.

[104] Martin Dressel and George Grüner, *Electrodynamics of Solids*, Cambridge University Press, 2002.

[105] D. B. Melrose and R. C. McPhedran, *Electromagnetic Processes in Dispersive Media*, Cambridge University Press, 2005.

[106] K. Cho, *Optical Response of Nanostructures*, Springer, 2003.

[107] K. Cho, *Reconstruction of Macroscopic Maxwell Equations: A Single Susceptibility Theory*, Springer, 2010.

[108] Ole Keller, *Quantum Theory of Near-Field Electrodynamics*, Springer, 2011.

[109] Motoichi Ohtsu and Kiyoshi Kobayashi, *Optical Near Fields: Introduction to Classical and Quantum Theories of Electromagnetic Phenomena at the Nanoscale*, Springer, 2004.

[110] Eberhard Zeidler, *Quantum Field Theory II: Quantum Electrodynamics*, Springer, 1 edition, 2008.

[111] D. P. Craig and T. Thirunamachandran, *Molecualr Quantum Electrodynamics: An Introduction to Radiation Molecule Interactions*, Dover, 1984.

[112] Chandra Roychoudhuri, A.F. Kracklauer, Kathy Creath (editors), *The Nature of Light: What is a Photon?* , CRC Press, 2008.

[113] P. W. Milonni, *Fast Light, Slow Light and Left-Handed Light*, Institute of Physics Publishing, 2005.

[114] Frederick Wooten, *Optical Properties of Solids*, Academic Press, 1972.

[115] Yutaka Toyozawa, *Optical Processes in Solids*, Cambridge University Press, 2003.

[116] Max Born and Kun Haung, *Dynamical Theory of Crystal Lattices*, Oxford University Press, London, 1st edition, 1954.

[117] G. Grosso and G. Parravicini, *Solid-State Physics*, Academic Press, 2000.

[118] A. Schuster, *An Introduction to the Theory of Optics*, London: E. Arnold, 1909.

[119] L. Mandel'shlam, *Complete Works*, vol. 5, Moscow: Izd. An SSSR, 1950.

[120] Antonios Gonis and William H. Butler, *Multiple Scattering in Solids*, Springer, 2000.

bibitemLee and Nam Jaechun Lee and Sangwook Nam, "Fundamental Aspects of Near-Field Coupling Small Antennas for Wireless Power Transfer," *IEEE Trans. Antennas Propagat.*, VOL. 58, NO. 11, pp. 3442-3449, November, 2010.

[121] C. Craeye and D. GonzÃ¡lez-Ovejero, "A review on array mutual coupling analysis," *Radio Sceince*, vol. 46, RS2012, doi:10.1029/2010RS004518, 2011.

[122] A. D. Yaghjian, "Upper-bound errors in far-field antenna parameters determined from planar near-field measurements, Part 1: Analysis," *Nat. Bur. Stand. Tech. Note* 667, Oct. 1975.

[123] Arthur D. Yaghjian, "Efficient Computation of Antenna Coupling and Fields Within the Near-Field Region," *IEEE Trans. Antennas Propagat.*, VOL. AP-30, NO. 1, January 1982.

[124] Yen-Sheng Chen, Shih-Yuan Chen, and Hsueh-Jyh Li, Senior Member, "Analysis of Antenna Coupling in Near-Field Communication Systems," *IEEE Trans. Antennas Propagat.*, VOL. 58, NO. 10, October 2010.

[125] T. Su and H. Ling, "On modeling mutual coupling in antenna arrays using the coupling matrix," *Microwave and Optical Technology Letters*, Dec. 2000.

[126] M. A. Jensen and J. W. Wallace, "A review of antennas and propagation for MIMO wirelss communications," *IEEE Trans. Antennas Propagat.*, vol. 52, no. 11, pp. 2810–2824, March 2012.

[127] W. Geyi, "Multi-antenna information theory," *Progress in Electromagnetic Research (PIER)*, 75, 11-50, 2007.

[128] S. Blanch, J. Romeu, and I. Corbella, "Exact representation of antenna system diversity perfomance from input parameter describtion," *Electronics Letters*,vol. 39, no. 9, pp. 705–707, May 2003.

[129] P. Hallbjorner, "The significance of efficiencies when using S-paramters to calculate received signal correlation from two antennas," *IEEE Antennas & Wireless Propagat. Letters*, vol.4, pp. 97–99, 2005.

[130] M. Wehr and G. Monich, "Detection of radiation leaks by spherically scanned field data," in *Proc. 10th Int. Zurich Symp. And Techn. Exhb. on EMC*, pp. 337-342, 1993.

[131] M. Wehr, A. Podubrin, and G. Monich, "Automated search for models by evolution strategy to characterize radiators," *in Proc. 11th Int. Zurich Symp. And Techn., Exhb. on EMC*, pp. 27-34, 1995.

[132] D. McNay, E. Michielssen, R. L. Rogers, S. A. Taylor, M. Akhtari, and W. W. Sutherling, "Multiple source localization using genetic algorithms," *J. Neurosci. Methods*, vol. 64, pp. 163-172, Feb. 1996.

[133] Joan-Rammon, Miquel Ribo, Josep-Maria Garrell, and Antonio Martin, "A genetic algorithm method for source identification and far-field radiated emissions predicted from near-field measurements for PCB characterization," *IEEE Trans. on Electromagn. Comp.*, vol. 43, no. 4, pp. 520-530, November 2001.

[134] Taninder S. Sijher and Ahmed A. Kishk, "Antenna modeling by infinitesimal dipoles using Genetic Algorithms," *Progress In Electromagnetic Research, PIER 52*, pp. 225–254, 2005.

[135] Dimitra I. Kaklamani and Hristos T. Anastassiu, "Aspects of the method of auxiliary sources (MAS) in computational electromagnetics," *IEEE Antenna's and Propagation Magazine*, Vol. 44, No. 3, p.48-84, June 2002.

[136] Yen-Sheng Chen, Shih-Yuan Chen, and Hsueh-Jyh Li, Senior Member, "Analysis of Antenna Coupling in Near-Field Communication Systems," *IEEE Trans. Antennas Propagat.*, VOL. 58, NO. 10, October 2010.

[137] S. I. Pekar, "Theory of electromagnetic waves in a crystal in which excitons are produced," *Sov. Phys. JETP*, vol. 6, p. 785, 1957.

[138] J. J. Hopfield, "Theory of the contribution of excitons to the complex dielectric constant of crystastals," *Physical Review B*, vol. 112, pp. 1555–1567, December 1958.

[139] J. J. Hopfield and D. G. Thomas, "Theoritcial and experimental effects of spatial dispersion on the optical properteis of crystals ," *Physical Review B*, vol. 132, pp. 563–572, October 1963.

[140] G. S. Agrawal, D. N. Pattanyak, and E. Wolf, "Electromangetic fields in spatial dispersive media," *Physical Review B*, vol. 10, pp. 1447–1475, August 1974.

[141] K. Cho,,"Nonlocal theory of radiation-matter interaction: Boundary-condition-less treatment of Maxwell equations,"*Prog. Theor. Phys. Suppl.*, No. 106, 225âŁ"233, 1991.

[142] Kikuo Cho, "General form of Maxwell equations for metamaterials study," *phys. stat. sol.* (b) 245, No. 12, pp. 2692âŁ" 2695,2008.

[143] V. M. Agranovich, Y. R. Shen, R. H. Baughman, and A. A. Zakhidov, "Linear and nonlinear wave propagation in negative refraction metamaterials," *Physical Review B*, vol. 69, pp. 165112–165112, April 2004.

[144] V. M. Agranovich, Yu. N. Gartstein, and A. A. Zakhidov, "Negative refraction in gyrotropic media," *Physical Review B*, vol. 73, pp. 045114–045126, January 2006.

[145] V. G. Veselago, "The electrodynamics of substances with simltanously negative values of ϵ and μ," *Sov. Phys. Uspekhi*, vol. 10, no. 4, pp. 509-514, 1968.

[146] F. R. Faxvog, C. N. Y. Chow, T. Bieber, and J. A. Carruthers , "Meassure pulse velocity greater than c in a Newon absorbing cell," *Appl. Phys. Lett.*, vol .17, no. 5, pp. 192–193, Sepetember 1970.

[147] S. Chu and S. Wong, "Linear pulse propagation in Absorbing Media,âŁž *Phys. Rev. Lett.*, vol. 48, pp. 738–741, 15 March 1982.

[148] L. J. Wang, A. Kuzmich, and A. Dogariu, "Gain-assisted superluminal light propagation," *Nature*, vol. 406, pp.277–279, 2000.

[149] Omar F. Siddiqui, Mo Mojahedi, and George Eleftheriades, "Periodically loaded transmission line with effective negative refractive index and negative group velocity," *IEEE Trans. Ant. Propag.*, vol. 51, no. 10, pp. 2619–2625, October 2003.

[150] J. F. Woodly and M. Mojahedi, "Negative group velocity and group delay in left-handed media," *Phys. Rev. E*, vol. 70, pp. 046603–046609, October 2004.

[151] Seymour Stein, "On cross coupling in multiple-beam antennas," *IRE Transactions on Antennas and Propagation*, pp. 548–557, September 1962.

[152] W. C. Jakes *et al*, *Microwave Mobile Communications*, John Wiley & Sons, 1974.

[153] Nelson Costa and Simon Haykin, *Multiple-Input Multiple Output Channel Models*, John Wiley & Sons, 2010.

[154] R. G. Vaughan and J. B. Andersen, "Antenna diversity in mobile communications," *IEEE Trans. Veh. Technology.*, 36, pp. 149–172, 1987.

[155] D'Arcy Wentworth Thompson, *On Growth and Form*, first published in 1917; Cambridge University Press, Abridged edition, 1961.

[156] Hans Driesch, *The Science and Philosophy of the Organism*, second edition, Black, 1929.

[157] Hans Driesch, *The History and Theory of Vitalism*, translated from German by C. K. Odgen, Macmilian and Co., 1914.

[158] C. H. Waddington, *The Strategy of Genes*, London : George Allen & Unwin, 1957.

[159] C. H. Waddington, *Principles of Development and Differentiation*, New York: Macmillan Company, 1966.

[160] C. H. Waddington, *The Evolution of an Evolutionist* , Cornel University Press, 1975.

[161] Rupert Sheldrake, *Morphic Resonance*, fourth edition, Inner Tradition, 2009.

[162] Rupert Sheldrake, *The Presence of the Past*, Collins, 1988.

[163] Rene Thom, *Structural Stability and Morphogenesis*, W. A. Benjamin, 1975.

[164] Rene Thom, *Mathematical Models of Morphogenesis*, 1983.

[165] Henry Bergson, *Time and Free Will: An Essay on the Immediate Data of Consciousness* , translators from French by F. L. Bogson (1910), Dover Publications , 1889.

[166] Henry Bergson, *Matter and Memory*, translators from French by N.M. Paul and W.S. Palmer, Dover Publications , 1896.

[167] Henry Bergson, *The Creative Evolution*, translated from French by Arthur Mitchell, Barnes & Noble, 1907.

[168] Gilles Deleuze, *Difference and Repetition*, translated from French by Paul Patton (1994), Columbia University Press, 1968.

[169] Don Ihde, *Heidegger's Technologies: Postphenomenological Perspectives*, Fordham University Press, 2010.

Index

Recent Titles in the Artech House Antennas and Electromagnetics Analysis Library

Christos Christodoulou, Series Editor

Adaptive Array Measurements in Communications, M. A. Halim

Advanced Computational Electromagnetic Methods and Applications, Wenhua Yu, Wenxing Li, Atef Elsherbeni, Yahya Rahmat-Samii, Editors

Advances in Computational Electrodynamics: The Finite-Difference Time-Domain Method, Allen Taflove, editor

Advances in FDTD Computational Electrodynamics: Photonics and Nanotechnology, Allen Taflove, editor; Ardavan Oskooi and Steven G. Johnson, coeditors

Analysis Methods for Electromagnetic Wave Problems, Volume 2, Eikichi Yamashita, editor

Antenna Design with Fiber Optics, A. Kumar

Antenna Engineering Using Physical Optics: Practical CAD Techniques and Software, Leo Diaz and Thomas Milligan

Antennas and Propagation for Body-Centric Wireless Communications, Second Edition, Peter S. Hall and Yang Hao, editors

Antennas and Site Engineering for Mobile Radio Networks, Bruno Delorme

Analysis of Radome-Enclosed Antennas, Second Edition, Dennis J. Kozakoff

Applications of Neural Networks in Electromagnetics, Christos Christodoulou and Michael Georgiopoulos

The Art and Science of Ultrawideband Antennas, Second Edition, Hans G. Schantz

AWAS for Windows Version 2.0: Analysis of Wire Antennas and Scatterers, Antonije R. Djordjević, et al.

Broadband Microstrip Antennas, Girsh Kumar and K. P. Ray

Broadband Patch Antennas, Jean-François Zürcher and Fred E. Gardiol

CAD of Microstrip Antennas for Wireless Applications, Robert A. Sainati

The CG-FFT Method: Application of Signal Processing Techniques to Electromagnetics, Manuel F. Cátedra, et al.

Computational Electrodynamics: The Finite-Difference Time-Domain Method, Third Edition, Allen Taflove and Susan C. Hagness

Electromagnetic Modeling of Composite Metallic and Dielectric Structures, Branko M. Kolundzija and Antonije R. Djordjević

Electromagnetic Waves in Chiral and Bi-Isotropic Media, I. V. Lindell, et al.

Electromagnetics, Microwave Circuit and Antenna Design for Communications Engineering, Peter Russer

Engineering Applications of the Modulated Scatterer Technique, Jean-Charles Bolomey and Fred E. Gardiol

Fast and Efficient Algorithms in Computational Electromagnetics, Weng Cho Chew, et al., editors

Frequency-Agile Antennas for Wireless Communications, Aldo Petosa

Fresnel Zones in Wireless Links, Zone Plate Lenses and Antennas, Hristo D. Hristov

Handbook of Antennas for EMC, Thereza MacNamara

Handbook of Reflector Antennas and Feed Systems, Volume I: Theory and Design of Reflectors, Satish Sharma, Sudhakar Rao, and Lotfollah Shafai, editors

Handbook of Reflector Antennas and Feed Systems, Volume II: Feed Systems, Lotfollah Shafai, Satish Sharma, and Sudhakar Rao, editors

Handbook of Reflector Antennas and Feed Systems, Volume III: Applications of Reflectors, Sudhakar Rao, Lotfollah Shafai, and Satish Sharma, editors

Introduction to Antenna Analysis Using EM Simulators, Hiroaki Kogure, Yoshie Kogure, and James C. Rautio

Iterative and Self-Adaptive Finite-Elements in Electromagnetic Modeling, Magdalena Salazar-Palma, et al.

LONRS: Low-Noise Receiving Systems Performance and Analysis Toolkit, Charles T. Stelzried, Macgregor S. Reid, and Arthur J. Freiley

Measurement of Mobile Antenna Systems, Second Edition, Hiroyuki Arai

Microstrip Antenna Design Handbook, Ramesh Garg, et al.

Microwave and Millimeter-Wave Remote Sensing for Security Applications, Jeffrey A. Nanzer

Mobile Antenna Systems Handbook, Third Edition, Kyohei Fujimoto, editor

Multiband Integrated Antennas for 4G Terminals, David A. Sánchez-Hernández, editor

New Foundations for Applied Electromagnetics: The Spatial Structure of Fields, Said Mikki and Yahia Antar

Noise Temperature Theory and Applications for Deep Space Communications Antenna Systems, Tom Y. Otoshi

Phased Array Antenna Handbook, Second Edition, Robert J. Mailloux

Phased Array Antennas with Optimized Element Patterns, Sergei P. Skobelev

Plasma Antennas, Theodore Anderson

Printed MIMO Antenna Engineering, Mohammad S. Sharawi

Quick Finite Elements for Electromagnetic Waves, Giuseppe Pelosi, Roberto Coccioli, and Stefano Selleri

Radiowave Propagation and Antennas for Personal Communications, Third Edition, Kazimierz Siwiak

Reflectarray Antennas: Analysis, Design, Fabrication and Measurement, Jafar Shaker, Mohammad Reza Chaharmir, and Jonathan Ethier